BIOCHEMISTRY OF
BACTERIAL GROWTH

Biochemistry of Bacterial Growth

EDITED BY

JOEL MANDELSTAM

Microbiology Unit
Department of Biochemistry
University of Oxford

KENNETH McQUILLEN

Sub-Department of Chemical Microbiology
Department of Biochemistry
University of Cambridge

SECOND EDITION

BLACKWELL SCIENTIFIC PUBLICATIONS

OXFORD LONDON EDINBURGH MELBOURNE

ISBN 0 632 08610 6   Cloth
ISBN 0 632 09840 6   Limp

First published 1968
Second edition 1973

Printed and bound in Great Britain by
William Clowes & Sons, Limited
London, Beccles and Colchester

Contributors

S. BAUMBERG, *Department of Genetics, University of Leeds*

E. A. DAWES, *Department of Biochemistry, University of Hull*

J. R. S. FINCHAM, *Department of Genetics, University of Leeds*

HARLYN O. HALVORSON, *Rosenstiel Medical Science Research Center, Brandeis University, Waltham, Massachusetts, U.S.A.*

D. KERRIDGE, *Sub-Department of Chemical Microbiology, Department of Biochemistry, University of Cambridge*

P. J. LARGE, *Department of Biochemistry, University of Hull*

KENNETH MCQUILLEN, *Sub-Department of Chemical Microbiology, Department of Biochemistry, University of Cambridge*

J. MANDELSTAM, *Microbiology Unit, Department of Biochemistry, University of Oxford*

P. E. REYNOLDS, *Sub-Department of Chemical Microbiology, Department of Biochemistry, University of Cambridge*

M. SCHAECHTER, *Department of Molecular Biology and Microbiology, Tufts University School of Medicine, Boston, Massachusetts, U.S.A.*

J. SZULMAJSTER, *Laboratoire d'Enzymologie, C.N.R.S., Gif-sur-Yvette, S. & O., France*

MICHAEL J. WARING, *Department of Pharmacology, University of Cambridge*

Contents

Contributors   v

Preface to Second Edition   viii

Preface to First Edition   ix

Introduction: abstract of the book   3

PART I

Section

   1   The Bacterial Cell: major structures   11

   2   Growth: cells and populations   20

   3   Class I Reactions: supply of carbon skeletons   26

   4   Class II Reactions: biosynthesis of small molecules 29

5 & 6   Class III Reactions: synthesis of macromolecules   34

   7   Genetics   44

8 & 9   Co-ordination and Differentiation   54

PART II

Chapter

   1   The Bacterial Cell: major structures   63
       P. E. REYNOLDS

   2   Growth: cells and populations   137
       M. SCHAECHTER

3 Class I Reactions: supply of carbon skeletons 160
E. A. DAWES AND P. J. LARGE

4 Class II Reactions: synthesis of small molecules 202
E. A. DAWES AND P. J. LARGE

5 Class III Reactions: the structure and synthesis of nucleic
acids 251
MICHAEL J. WARING

6 Class III Reactions: synthesis of proteins 316
KENNETH MCQUILLEN

7 Genetics 367
J. R. S. FINCHAM

8 Co-ordination of metabolism 423
S. BAUMBERG

9 Differentiation: sporogenesis and germination 494
HARLYN HALVORSON AND JEKISIEL SZULMAJSTER

APPENDICES

A Bacterial Classification 517
D. KERRIDGE

B Enzyme Mechanisms: functions of vitamins and co-enzymes 536
KENNETH MCQUILLEN

C Glossary 548

Index 555

Preface to Second Edition

In preparing this edition we have tried to take account of some of the many helpful suggestions made by reviewers or by colleagues. More specifically, we have transposed much of the material dealing with the regulation of specific metabolic pathways from Chapter 8 to Chapters 3 and 4. Apart from helping the reader, this will conform with the presentation of the subject that is now often adopted in practice. This has necessitated an entirely new treatment of Regulation, the chapter on this subject in the first edition having in any case become out of date during the time the book was in press. We have also eliminated the somewhat speculative concluding chapter, which neither students nor teachers seem to have found particularly useful. All the remaining chapters have been brought up to date and contain new material, but Part I has been retained largely in its original form.

Some readers, including one or two reviewers, appear to have misunderstood the purpose of the book and to have been unduly upset because some aspects of the subject were omitted or insufficiently stressed. Among topics listed were enzyme kinetics, the application of quantum mechanics and the diversity of fermentation reactions. We thought for a while of providing in this preface an exhaustive list of omissions for the guidance of such readers. We abandoned the idea with regret because we have no real belief that they will read this preface any more attentively than they read the previous one in which the opening sentence stated that this was not intended to be a comprehensive text book. It was, however, intended to be an experiment in the presentation of the subject. This is how most readers viewed it and we hope that it is from this standpoint that the usefulness of the book will continue to be judged.

JOEL MANDELSTAM
KENNETH McQUILLEN

Preface to First Edition

This book is not intended to be a comprehensive textbook on the biochemistry of bacteria. It has been written in the belief that the advances in biochemistry in the last ten years provide a basis for a fairly comprehensive description of bacterial life in biochemical terms, and that such a view of the bacterial cell can with advantage be presented to beginners. We also believe that the most recent concepts are as readily intelligible as the older and more basic ideas. For this reason we have, for example, thought it just as easy, and much more interesting, for the student to learn the modern view of replication of the bacterial chromosome before he learns the structural formulae of the nucleotides. Similarly we have presented the 'coding problem' in protein synthesis before introducing the chemical structures of the twenty common amino acids. As far as possible this method of approach has been followed throughout.

We have also attempted to build up from the start a coherent picture. Too often in the teaching of biochemistry the student is taken through one detailed aspect of the subject after another. Only at the end does he have all the information which will allow him to construct some sort of integrated picture. By this time his mind may be so clogged with details that the process of fitting them together is needlessly difficult. Our method of presentation will, we hope, avoid this danger and it has resulted in a book written in three parts. The introduction is a summary of the book in a few pages and it is based upon a very general account of what a bacterial cell does during growth. This is followed in Part I by a somewhat more detailed description of the same material; it presupposes very little knowledge apart from some basic chemistry. If we have been successful in our exposition, the student should, at this stage, have a clear picture—still in very general terms—of a bacterial cell as an integrated biochemical system. The detailed biochemistry will be found in Part II, the third and largest part of the book. We realize that the subject may seem to be so oversimplified by this treatment that the impression is given that everything in bacterial life is now explicable in biochemical terms apart from a few minor gaps. We have tried to avoid this by stressing, particularly in the conclusion, those phenomena which are as yet not reducible to biochemistry, and which are likely to be the growing points of the subject.

Finally, we have avoided the historical approach which, while it may be the most scholarly, is for the reader the dullest. It can, furthermore, reasonably be argued that it usually fails in its object. The significance of early discoveries and controversies is best appreciated by those who already know the subject fairly well, and not by beginners: it is for the latter that the book has been written.

In our attempt to co-ordinate the chapters of this book we have inflicted our views and prejudices upon the contributors to a considerable extent. This was particularly true during the preparation of Part I which has now been written and re-written so many times that it is impossible to attribute individual authorship to the sections. We are grateful to the authors for their tolerance and patience. We are also most appreciative of the willing help and co-operation of everyone at Blackwell's who was concerned with this book, in particular Mr Per Saugman, Mr John Robson and Miss Yvonne Prince.

JOEL MANDELSTAM
KENNETH McQUILLEN

Introduction

Introduction

Abstract of the Book

This chapter is a highly condensed introduction to the biochemical events that underlie bacterial growth. It is, at the same time, intended to be a summary description of the contents of the rest of the book.

For a model system it will be convenient to choose an unspecified bacterium that can grow in a medium containing glucose as the carbon source. Its nitrogen requirements are satisfied by ammonium ions and its sulphur requirements by sulphate. Magnesium and phosphate are essential and it needs trace amounts of other metals (e.g. iron). It can be considered as a 'generalized bacterial cell' and we shall attribute to it a mixture of the properties found in several different kinds of bacteria. It should be regarded as an abstraction in much the same way as the 'average man'. Real bacteria will be considered in the main section of the book and some of the ways in which they differ from the model will become apparent.

The organism is represented schematically in Figure 1. It is rod-shaped and has a rigid outer wall that maintains and supports the membrane that it encloses. The wall is made of a polymer substance, the peptidoglycan, and the membrane contains proteins and lipids. These coats surround the cytoplasm, which consists mainly of polymers: deoxyribonucleic acid (DNA); ribonucleic acid (RNA); proteins and polysaccharides. In terms of dry weight the polymers account for about 90% of the cell (see Figure 2). The remaining 10% of the cell is made up of a large variety of small molecules: amino acids, nucleotides, vitamins, fats. Although these constitute so small a fraction of the cell mass, they are metabolically of great importance.

Not only do the macromolecules make up the bulk of the bacterial cell, they also give it the characteristics that distinguish it from all other types of bacteria. The small molecules, on the other hand, are common to all types of bacteria and, indeed, to other forms of life.

When the organism is in a suitable environment or growth medium, more of all these materials is produced and in due course the cell divides into two daughter cells indistinguishable from one another and from their parent. The subject of this book is a description of the way in which the simple organic and inorganic constituents of the medium are transformed into new cell material with its enormous diversity of molecular species.

3

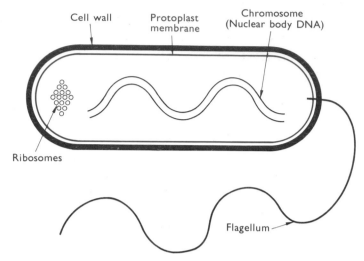

Figure I. Diagrammatic representation of a bacterial cell.

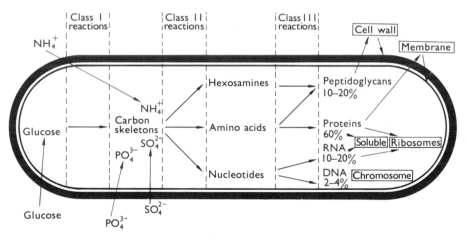

Figure 2. Flow diagram of synthesis of bacterial constituents.

Classification of biochemical reactions

The number of chemical reactions involved in growth is unknown but is probably of the order of a thousand. Of these a few may occur spontaneously but the vast majority have to be catalysed by specific proteins, the *enzymes*. Each of these catalyses a specific reaction such as the addition or removal of water, or hydrogen, or 1-C residues, or amino groups, etc. (see Appendix B, p. 536). For our purposes enzyme reactions can be grouped into three classes.

Class I (degradative reactions)

There is first a complex of enzymes which degrades glucose to smaller aliphatic carbon compounds. This class will be called degradative enzymes. The net process is exergonic, i.e. produces energy. It also results in the supply of carbon skeletons for synthetic reactions.

Class II (biosynthetic reactions)

From these carbon skeletons a further series of enzymes catalyses the formation of the small molecules which are the basic components of the macromolecules. Many of these intermediates (amino acids, nucleotides, hexosamines) contain nitrogen which is derived from the NH_4^+. Some contain sulphur which comes from SO_4^{2-}. At the same time some small molecules (vitamins, cofactors) are synthesized but are not incorporated into macromolecules; rather, they are needed for the proper functioning of the enzymes. The enzymes producing all these substances will be called biosynthetic, Class II. As a group they largely require energy and are therefore endergonic. The energy is produced by the Class I reactions.

Class III (biosynthetic reactions)

A further series of enzymes then converts the basic small molecules into macromolecules. When enough of these have been synthesized the cell divides. Since the distinctive character of the cell is determined by its macromolecules, much of this book will be concerned with the mechanisms by which these complicated structures are reproduced so exactly.

The genetic information for copying the cell is carried in its DNA which is the 'blueprint' for the whole cell, that is, *all* the information determining what the biochemical machinery shall be and how it will be put together is encoded in the DNA. When the cell divides each of the daughter cells must, apart from anything else, receive a complete copy of the 'blueprint'. It is essential that the DNA molecules should be copied correctly at every division because, as in any highly organized system, a random error will almost certainly be damaging. It is only *very* rarely that such an error will be advantageous. We have thus two separate problems to consider. Firstly, how an exact copy of the DNA is made and then, when this has happened, how the information in it is translated into the other types of molecules.

The DNA molecule is a very long polymer made up of four kinds of nucleotide joined through their phosphates. All four contain the sugar *deoxy*ribose and are represented by dA, dG, dC and dT because of the four different bases (see Figure 3, p. 17 for their chemical structures). The properties of these nucleotides are such that dA and dT have an affinity for each other and so have dG and dC. Thus if we have a chain as follows:

—dA—dA—dG—dT—dC—dG—

then free nucleotides will tend to line up in accordance with their pairing properties, thus:

—dA—dA—dG—dT—dC—dG—   Original chain

dT   dT   dC   dA   dG   dC   Complementary sequence

Cells possess an enzyme (DNA polymerase) that links these nucleotides co-valently to give a polymer that will be exactly complementary to the original template. This complementary chain can itself be copied, again in accordance with the pairing properties of the nucleotides, thus giving a strand identical with the original chain:

—dT—dT—dC—dA—dG—dC—   Complementary chain

—dA—dA—dG—dT—dC—dG—   Identical copy of original chain

The complementary chain can be regarded as the biochemical equivalent of a photographic negative. This explains in an over-simplified way the principle of DNA replication. In fact, the DNA exists as a double-stranded structure that is unravelled either before or during the copying process (see p. 38 and Chapter 5).

So far, then, we have accounted for the formation of a DNA molecule containing all the necessary information for the hundreds of enzymes which will catalyse the three classes of reactions we have described. Now these reactions are responsible for all the materials that the cell contains and for all the biochemical reactions it can carry out. Our problem is thus reduced essentially to that of understanding how the information in the DNA is translated into that of the enzyme proteins. The information for any particular species of protein is carried in a stretch of DNA which may contain more than 1000 nucleotides and which is known as the *structural gene* for that protein. The enzymes and other proteins consist of chains containing twenty kinds of amino acid. The number of amino acid residues and their order are different for each kind of protein. The problem is to find the way in which the four types of nucleotide in a stretch of DNA specify the 20 types of amino acid in a protein. This is formally analogous to finding out how the two-letter system (dots and dashes) of Morse code is translated into the ordinary alphabet of 26 letters and it is generally referred to as the coding problem.

The translation of DNA into protein occurs in a number of steps. First, the informational content of the structural gene is transferred to a strip of RNA known as the *messenger-RNA*. RNA is also a polymer consisting of four kinds of nucleotide, but in this nucleic acid they all contain ribose instead of deoxyribose and they have U instead of T. They are represented by A, G, C and U. However, pairing can occur between these nucleotides and those of DNA. In the presence of the appropriate enzyme (RNA polymerase) and a DNA template, the four ribonucleotides are polymerized into a complementary copy of the DNA strand:

dA—dA—dG—dT—dC—dG— DNA template

U—U—C—A—G—C— Complementary RNA

Thus DNA fulfills *two* separate template functions. The first, mediated by DNA polymerase, is to serve as a template for its own replication: the second is to act as a template for the production of the complementary messenger-RNA.

The messenger-RNA acts as a template on which amino acid residues are assembled in correct sequence before being linked together. The solution of the coding problem is that the code is triplet, i.e. that a sequence of three nucleotides codes for each amino acid:

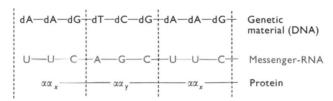

Here the DNA triplet, dA—dA—dG is transcribed into the complementary messenger-RNA triplet, U—U—C which is translated as the amino acid x ($\alpha\alpha_x$), etc. Each succeeding triplet causes the insertion of the next appropriate amino acid until the protein is complete. The whole chain might easily contain 300 amino acid residues. The assembly of proteins is, however, more complex than this description suggests and involves the participation of other types of molecules.

The remaining macromolecules to be considered are the polysaccharides and peptidoglycans. The biosynthesis of these is generally simpler than that of the proteins. Some, like glycogen, are polymers containing only one type of sugar residue. Theoretically a single enzyme could string together a chain of such residues to form a polymer. In fact most polysaccharides are more complex, but even so only a few enzymes are required for the synthesis of any one of them. The peptidoglycans are somewhat similar, containing two kinds of amino sugar occurring in regular alternation. They also have short side chains of amino acids but their assembly is achieved by a fairly small number of enzymes.

The number and types of small molecules that a cell can make and degrade are determined more or less completely by its content of enzymes.

Genetics

So far all the processes we have outlined may be considered as taking place in a single bacterium. They involve the conveyance of information in the DNA to the rest of the cell material. However, information can also be conveyed *inter-cellularly*. In bacteria this can be effected in one of three ways: (a) by a mating process; (b) by transformation—the direct uptake by one cell of free DNA liberated from another cell, by lysis or otherwise; (c) by transduction. Here an infective virus particle during its formation picks up some of the DNA of the host cell and then transfers it to the next bacterial cell it infects.

Growth and the regulation of biosynthesis

We can now summarize the events that take place when some viable cells are placed into growth medium. Some of their enzymes degrade glucose, some synthesize basic molecules and yet others assemble macromolecules including more of all the enzymes. With more of all these catalysts thus available, the same processes will continue but at an accelerated rate, giving yet more enzyme, and more cells. The rate of synthesis is thus proportional to the amount of cell material present and this leads to an exponential rate of growth that continues until something in the environment becomes limiting. When this happens some types of bacteria simply cease to grow but others form spores which are heat-resistant and can lie in a dormant state for many years. Subsequently, if the environment becomes favourable, these can germinate and begin to grow again. Sporulation and germination are among the most primitive forms of cell differentiation and are consequently of considerable interest.

Returning to the actively growing cell, let us consider its internal economy. It has to produce 20 types of amino acids for its proteins, four types of nucleotides containing deoxyribose for DNA, four more containing ribose for RNA and also a variety of co-factors and lipids. The synthesis of any one of these substances may easily involve ten or more specifically catalysed steps carried out by enzymes of Class II. In addition there is a considerable number of intermediates produced from glucose by the Class I enzymes.

For efficient growth all the basic materials and all the macromolecules derived from them have to be produced in the correct proportions. Under natural conditions, bacteria are probably often in competition for a limited amount of nutrient. The consequence is that a very efficient regulation mechanism has evolved. Since virtually all metabolic steps are enzymically catalysed, in considering metabolic regulation, we have really to consider regulation of enzymic function. There are two ways in which this can occur. One is by alteration of the *amount* of any particular enzyme, the other is by alteration of the *rate* at which it functions. Both types of regulation are found in the bacterial cell and their combined effect ensures that the cell is geared to get the maximum yield of protoplasm from its environment and to do so in the minimum time.

Part I

Section 1
The Bacterial Cell: Major Structures

Bacterial cells occur in all sorts of different shapes and sizes depending on the kind of organism and on the way in which it has been grown, but for many purposes it is possible to disregard these variations and to consider the common properties of the 'generalized bacterial cell'. Thus although some bacteria are spherical or curved or spiral, the majority are rod-shaped and are about 1 μm wide and 2 μm in length (1 μm = 0·001 mm). A single bacterial cell may thus have a volume of 10^{-12} ml and contain $2·5 \times 10^{-13}$ g of dry matter (equivalent to a molecular weight of $1·5 \times 10^{11}$). But this bacterium is not just an undifferentiated blob of 'protoplasm'. It is a highly organized structure with organelles corresponding in function to many of those found in higher organisms. The hereditary material (DNA) is embedded in the *cytoplasm* which, surrounded by the *cell membrane*, is called the protoplast. Outside this lies the *cell wall*.

Cell walls and membranes

The wall is fairly rigid and gives shape and protection to the cell. It amounts to about 10% of the weight of the entire cell. Always there is present in it peptidoglycan, and this seems to be what makes the wall rigid. The peptidoglycans are made of chains of amino sugars, N-acetylglucosamine alternating with N-acetylmuramic acid. Short peptides are linked to the muramic acid residues and separate chains may be joined by these peptides to form the two-dimensional structure needed for a wall (Figure 1).

The amino sugar, muramic acid, has not been found in any biological polymers other than the peptidoglycans of the cell walls of bacteria and the closely related blue-green algae. These peptides are also interesting in that they contain unusual amino acids. Besides L-alanine they contain D-alanine and D-glutamic acid, the so-called 'unnatural' isomers which are not present in proteins. Most

11

species also contain diaminopimelic acid which, like muramic acid, is restricted in nature to these peptidoglycans.

Other polymers which may occur in cell walls include teichoic acids, lipopolysaccharides and lipoproteins (see Chapter 1).

In species in which the wall is simple peptidoglycan it is sometimes possible to digest it away with an enzyme called *lysozyme* which occurs in secretions such as tears and sweat and also in white of egg. Enzymic digestion of the wall releases the protoplast, but this is likely to burst unless given some osmotic protection. This is because the concentration of intra-cellular solutes has an effect

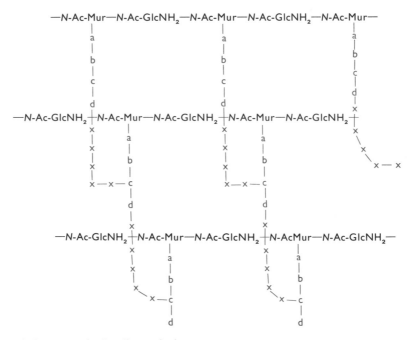

Figure 1. Structure of cell wall peptidoglycan.
Alternating amino sugars (N-acetyl-glucosamine and N-acetyl-muramic acid) form a backbone with a peptide (a-b-c-d) attached to a muramic acid residue. These peptides may be linked through further amino acids (x-x-x-).

equivalent to 5–25 atmospheres pressure. A solution of sucrose (10–20% w/v) will usually prevent this lysis of protoplasts. No matter what the shape of the cell, the naked protoplast on release from the cell wall assumes a spherical shape. It is bounded by a very delicate membrane called variously the *plasma membrane*, the *protoplast membrane*, or the *cytoplasmic membrane*. This structure consists predominantly of protein and lipid, as do all biological membranes, and it has a thickness of about 8 nm—this is dictated by the dimensions of the molecules of which it is composed. The membrane is the main permeability barrier of the cell since the wall is freely penetrated by most molecules except very large ones. Some substances pass into and out of cells by passive diffusion but many are transported by highly specific systems which require energy and are located in the cell membrane. The name *permease* or sometimes *translocase* is given to such a system. As far as passive diffusion is concerned, smaller molecules and substances of high lipid solubility penetrate through membranes more easily than do larger molecules and polar substances. For instance, 4-C sugars and some 5-C sugars may pass freely but other 5-C and all 6-C sugars (including glucose) may fail to penetrate by passive diffusion except very slowly. Bacterial cells are also generally impermeable to some small cations and to inorganic phosphate ions. These non-penetrating substances have to be actively transported into the organisms.

Proteins and nucleic acids

Three classes of polymers are found in all bacteria. These are the proteins, and the two kinds of nucleic acids. Viruses, on the other hand, contain protein and *either* DNA *or* RNA.

Structure of proteins

The proteins perform various functions, some catalytic and some structural, and it is this class of substance which determines the identity of an organism as of a particular kind. About half of the dry weight of a bacterium consists of protein but there may be more than 1000 different species of this material in any one cell. The size of these polymers ranges in molecular weight from a few thousands to millions but each has a definite, precise composition.

The constituents of proteins are the amino acids, and these have the general formula:

$$NH_2CH.COOH \quad \text{see Figure 2}$$
$$|$$
$$R$$

They occur with various frequencies and in various orders in the proteins. A simple protein may consist of a single polypeptide chain formed by the condensation of amino acids:

$$NH_2CH.COOH + NH_2CH.COOH + NH_2CH.COOH + \text{etc.}$$
$$| \qquad\qquad | \qquad\qquad |$$
$$R_a \qquad\qquad R_b \qquad\qquad R_c$$

$$\longrightarrow NH_2CH.CONH.CH.CONH.CH.CO- \quad \text{etc.}$$
$$| \qquad\qquad | \qquad\qquad |$$
$$R_a \qquad\qquad R_b \qquad\qquad R_c$$

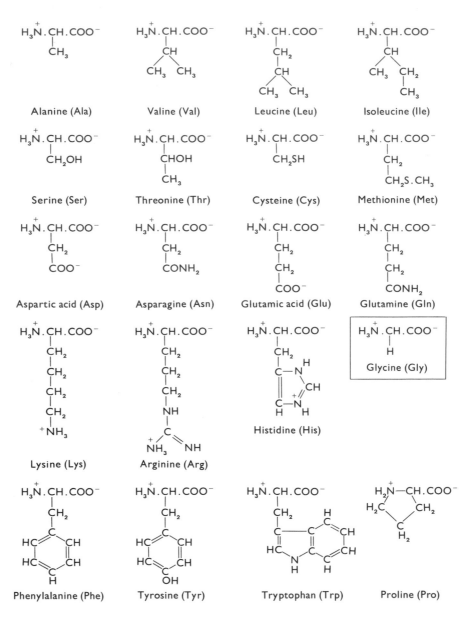

Figure 2. The 20 amino acids commonly occurring in proteins together with their abbreviations. All the amino acids have the L-configuration when they occur in proteins except glycine which is not optically active.

The sequence of amino acid residues in a protein or polypeptide is called its *primary structure*. Physical studies have shown that the chains do not exist in a straight, extended form but often coil into a spiral with $3\frac{2}{3}$ amino acid residues per turn, that is, eleven for each three turns of the helix. This α-helical structure us called the *secondary structure* and it is stabilized principally by so-called hydrogen bonds (H-bonds) between $\diagup\!\!\diagdown\!\!$NH and O$=$C$\diagup\!\!\diagdown$ residues in the chain.

Many proteins consist not of one but of several polypeptide chains joined together, and this is frequently achieved through —S—S— bridges. The amino acid cysteine possesses an —SH group in its side-chain and residues of cysteine in two chains may couple together thus:

$$
\begin{array}{l}
\text{NH}_2\text{CH.CONH.CH.CONH.CH.CONH—}\\
\quad\ \ |\qquad\qquad |\qquad\qquad |\\
\quad\ \ \text{R}_a\qquad\quad\ \text{R}_b\qquad\quad\ \text{CH}_2\\
\qquad\qquad\qquad\qquad\qquad\qquad |\\
\qquad\qquad\qquad\qquad\qquad\qquad \text{SH}\\
\qquad\qquad +\qquad\qquad\qquad\qquad\qquad\qquad\longrightarrow\\
\qquad\qquad\qquad\qquad\qquad\qquad \text{SH}\\
\qquad\qquad\qquad\qquad\qquad\qquad |\\
\qquad\qquad\qquad\qquad\qquad\qquad \text{CH}_2\\
\qquad\qquad\qquad\qquad\qquad\qquad |\\
\text{NH}_2\text{CH.CONH.CH.CONH.CH.CONH—}\\
\quad\ \ |\qquad\qquad |\\
\quad\ \ \text{R}_p\qquad\quad\ \text{R}_q
\end{array}
$$

$$
\begin{array}{l}
\text{NH}_2\text{CH.CONH.CH.CONH.CH.CONH—}\\
\quad\ \ |\qquad\qquad |\qquad\qquad |\\
\quad\ \ \text{R}_a\qquad\quad\ \text{R}_b\qquad\quad\ \text{CH}_2\\
\qquad\qquad\qquad\qquad\qquad\qquad |\\
\qquad\qquad\qquad\qquad\qquad\qquad \text{S}\\
\qquad\qquad\qquad\qquad\qquad\qquad |\\
\qquad\qquad\qquad\qquad\qquad\qquad \text{S}\\
\qquad\qquad\qquad\qquad\qquad\qquad |\\
\qquad\qquad\qquad\qquad\qquad\qquad \text{CH}_2\\
\qquad\qquad\qquad\qquad\qquad\qquad |\\
\text{NH}_2\text{CH.CONH.CH.CONH.CH.CONH.—}\\
\quad\ \ |\qquad\qquad |\\
\quad\ \ \text{R}_p\qquad\quad\ \text{R}_q
\end{array}
$$

Two cysteine residues in the same chain can likewise be oxidized to form a cystine residue and thus form an *intra*-chain bridge:

$$
\begin{array}{l}
\text{—NH.CH.CONH—}\ldots\ldots\ldots\ldots\text{—NH.CH.CONH—}\\
\qquad |\qquad\qquad\qquad\qquad\qquad\qquad\qquad |\\
\qquad \text{CH}_2\qquad\qquad\qquad\qquad\qquad\qquad \text{CH}_2\\
\qquad |\qquad\qquad\qquad\qquad\qquad\qquad\qquad |\\
\qquad \text{S}\text{———————————————}\text{S}
\end{array}
$$

Other bonds can be formed between the R-groups of amino acid residues, e.g. electrostatic bonds between the —COO$^-$ of an acidic amino acid and the —$^+$NH$_3$ group of a basic one. These side-group interactions cause the polypeptide chain to contort into a three-dimensional *tertiary structure* characteristic of the particular sequence of amino acid residues. It will be apparent that secondary and tertiary structures are interrelated.

Sometimes a biologically functional protein consists of an aggregate of units held together by non-covalent bonds. This arrangement, which is somewhat analogous to that found in a crystal, is called the *quaternary structure* of the protein.

Thus all proteins which have been investigated consist of specific sequences of amino acids joined together as polypeptide chains which may in turn be linked to each other. The whole molecule has a precise three-dimensional conformation which is necessary for its biological activity. Non-protein components may be associated with the polypeptide chains, as, for instance, the haem group in cytochromes. Many proteins have catalytic functions and, as has been said, about 1000 of such enzymes may occur in a single bacterial cell. The *catalytic site* or *active centre* of an enzyme is the region where the substrate molecule combines, and it is usually very highly specific (e.g. being able to distinguish between glucose and galactose or between aspartic acid and glutamic acid). This is due to the conformation of specific amino acid residues forming the active centre of a specific enzyme.

Other proteins have structural functions in cell walls and membranes, and the machinery for propelling motile bacteria consists wholly of protein. The motion is brought about by *flagella*, and each flagellum is composed of molecules of the protein flagellin which have a molecular weight of about 40,000. A solution of this protein can be caused to aggregate artificially into structures apparently identical to natural flagella. These are 2–5 μm in length, 12–30 nm in diameter, and consist of the molecules of flagellin (*c.* 5 nm diameter spheres) arranged like strings of beads with up to a dozen strands spiralling round each other.

The formation of these organs of motility seems likely to be spontaneous if the appropriate protein units are available at the appropriate site in the cell.

Deoxyribonucleic acid (DNA)

Nucleic acids are polymers of nucleotides, and each nucleotide consists of a nitrogenous base, a sugar and phosphate. They are linked through the phosphates. In DNA the sugar is 2-deoxyribose (dRib) and the bases are the purines, adenine (Ade) and guanine (Gua), and the pyrimidines, cytosine (Cyt) and thymine (Thy). Thus the DNA chain can be represented:

$$
\begin{array}{ccccccccccc}
-\text{dRib} & -\!\textcircled{P}\!- & \text{dRib} & -\!\textcircled{P}\!- & \text{dRib} & -\!\textcircled{P}\!- & \text{dRib} & -\!\textcircled{P}\!- & \text{dRib} & -\!\textcircled{P}\!- \\
| & & | & & | & & | & & | \\
\text{Cyt} & & \text{Ade} & & \text{Thy} & & \text{Ade} & & \text{Gua}
\end{array}
$$

The structures of the bases and this polydeoxyribonucleotide are shown in Figure 3.

Usually the DNA is found to occur as a double strand with the nucleotide sequence in one strand related to that in the other. The molecule forms a double

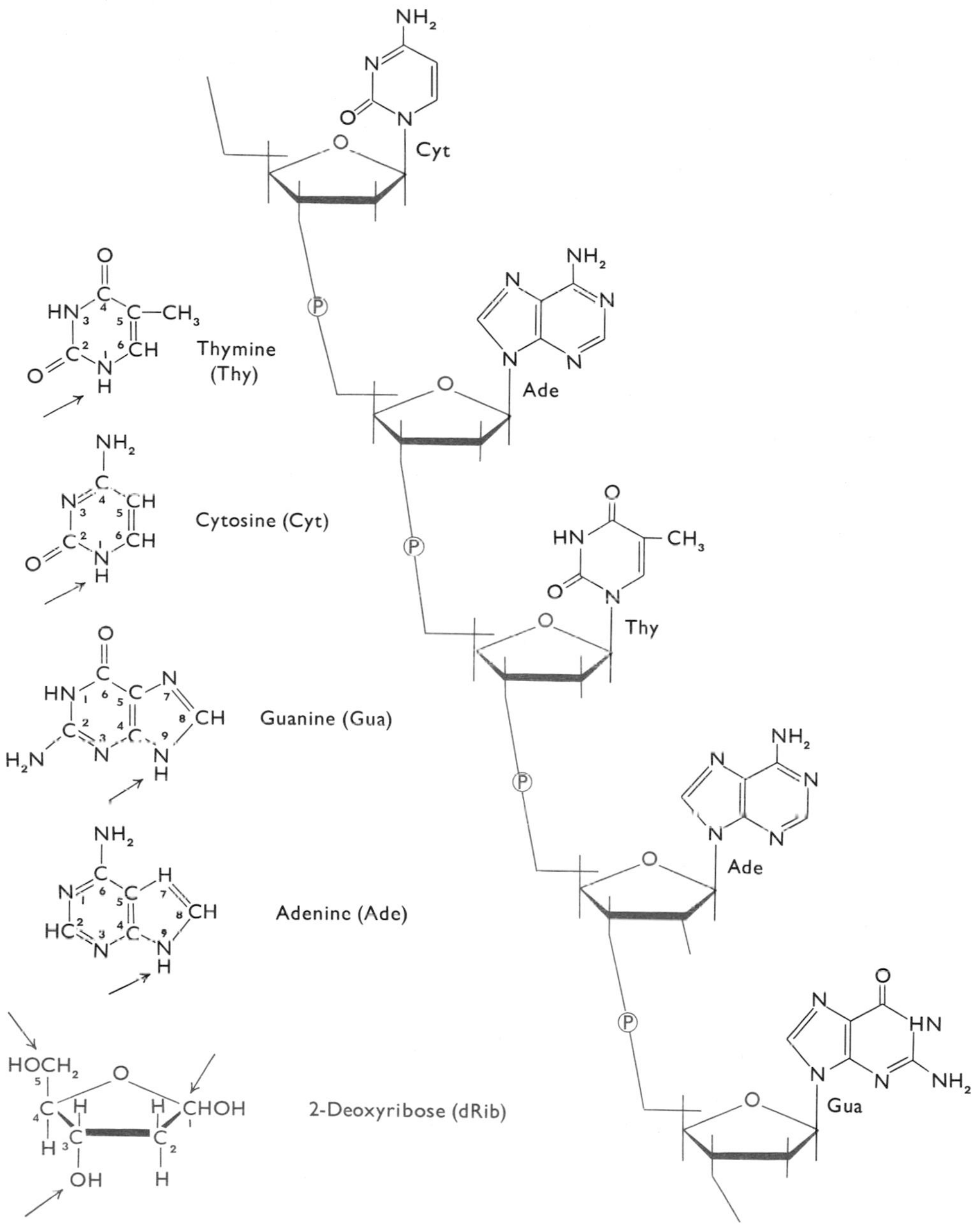

Figure 3. Structures of some purines and pyrimidines and of part of a DNA chain.

The pyrimidines are thymine and cytosine and these are linked via N-1 to the C-1 of deoxyribose. The purines are guanine and adenine and these are linked via N-9 to C-1 of deoxyribose. The sugars themselves are joined by phosphodiester bridges between C-3 and C-5.

helix in which an adenine nucleotide in one chain is side-by-side with a thymine nucleotide in the other, and similarly for guanine and cytosine:

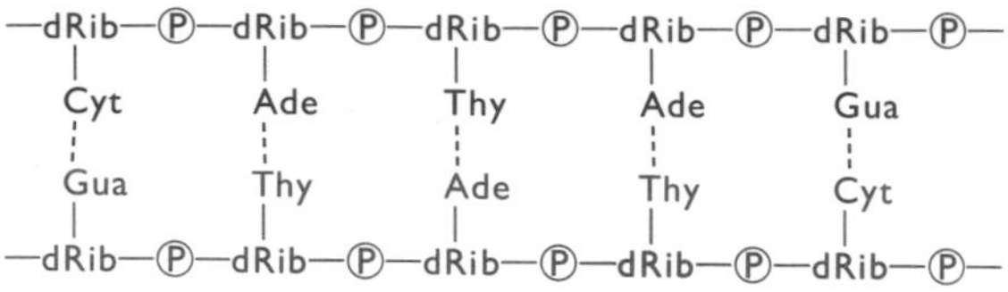

```
—dRib—Ⓟ—dRib—Ⓟ—dRib—Ⓟ—dRib—Ⓟ—dRib—Ⓟ—
     |        |        |        |        |
    Cyt      Ade      Thy      Ade      Gua
     ⋮        ⋮        ⋮        ⋮        ⋮
    Gua      Thy      Ade      Thy      Cyt
     |        |        |        |        |
—dRib—Ⓟ—dRib—Ⓟ—dRib—Ⓟ—dRib—Ⓟ—dRib—Ⓟ—
```

Such a structure is said to have complementary base-pairing, and this is possible because H-bonds (indicated by the dashes in the diagram above) can form specifically between adenine and thymine and between guanine and cytosine, thus stabilizing the double helix. It follows that an analysis of DNA usually shows that the amount of adenine is equal to that of thymine, and similarly with guanine and cytosine.

Electron microscopy and autoradiography suggest that the total DNA from a bacterium is in the form of a single giant 'molecule'. It is composed of a double helix about 1000 μm in length—this from a cell only about 2 μm in length. It has a molecular weight of perhaps 2.5×10^9 and is built up from some 8 million nucleotides. This hereditary material, representing a linear array of some thousands of genes, is stuffed into the cytoplasm but is never surrounded by a nuclear membrane as occurs in the nuclei of higher organisms.

Ribonucleic acid (RNA)
Ribonucleic acid resembles DNA in some but not in other respects. The sugar is D-ribose (Rib) and the bases are adenine, guanine and cytosine as in DNA,

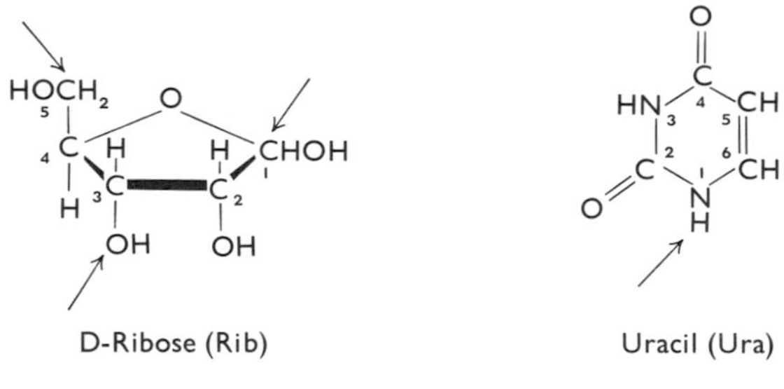

D-Ribose (Rib)   Uracil (Ura)

Figure 4. Structures of D-ribose and uracil.
The other purines and pyrimidine which occur in RNA are shown in Figure 3. The links between residues are the same in RNA as in DNA.

but the second pyrimidine is uracil (Ura) instead of thymine (5-methyl-uracil) which occurs in DNA (*cf.* Figures 3 and 4). Thus part of an RNA chain may be represented:

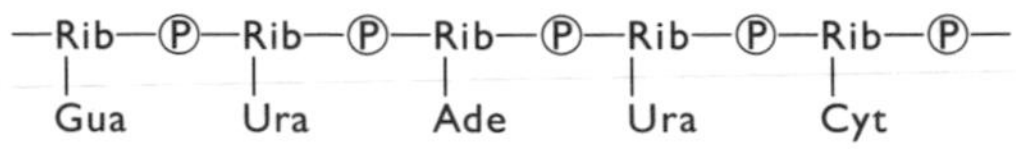

```
—Rib—Ⓟ—Rib—Ⓟ—Rib—Ⓟ—Rib—Ⓟ—Rib—Ⓟ—
    |       |       |       |       |
   Gua     Ura     Ade     Ura     Cyt
```

Three kinds of RNA (messenger, transfer and ribosomal; see p. 41) occur in all cells but none of these is double-stranded. The transfer-RNA's (tRNA) have molecular weights of about 25,000, but the other kinds may range up to more than a million. In a cell the ribosomal-RNA is combined with protein to form ribonucleoprotein particles known as *ribosomes*. These are found in the cytoplasm and many of them may be strung like beads on a strand of messenger-RNA (mRNA) to form a polyribosome (Figure 5).

Figure 5. Diagrammatic representation of a polyribosome.

Polyribosomes consist of several ribosomes attached to the same strand of mRNA. The ribosomes are ribonucleoprotein particles consisting of a smaller and a larger part. The polyribosomes are known also as polysomes and ergosomes.

Section 2
Growth: Cells and Populations

An organism, whether it be a bacterium or not, has an identity specified by its genetic make-up and largely functioning by virtue of the enzymes it contains. For the individual and the species to survive and be perpetuated, the specification and the functional systems must be maintained and reproduced. Growth can be considered at the level of the individual cell or at that of the population, although the latter depends, of course, on the former. When one organism becomes two, everything has been duplicated—the amount of cell wall, of membrane, of ribosomes, the DNA, the RNA, the proteins, the other cytoplasmic constituents such as ions, amino acids, intermediates of metabolism, and so on. Essentially a bacterial cell increases to double its size and then divides into two. This process is repeated again and again.

Exponential growth

The time taken for the number of organisms in a culture to double can be less than one hour for bacteria. In this time one cell increases to two, two increase to four—and 1000 enzyme molecules increase to 2000. Thus in a suitable environment a population will tend to increase exponentially for a time—1, 2, 4, 8, 16, 32, ... It is convenient to regard exponential growth as the 'normal' or 'ideal' state and then to consider changes from this state. The 'generalized bacterial cell' if put into its simple glucose/ammonium/salts medium and aerated at 37° will grow exponentially—or rather the culture will. The ions and glucose will pass from the medium through the membrane into the cytoplasm, some by passive diffusion and some carried by specific transport systems or permeases. Within the cell will be an array of enzymes, some free in solution, some organized on particulate structures, but each coded for by a specific structural gene. Absence or alteration of the gene will result in absence or alteration of the corresponding enzyme.

20

The Class I enzymes (see Section 3) will degrade the carbon substrate, glucose, to other substances including carbon dioxide and compounds whose chemical energy can be used by the organism. Class II and III enzymes will then catalyse the synthesis of the smaller compounds and the macro-molecular polymers which are characteristic of the organism. Some of these substances will be organized in the form of organelles or other structures, and when everything has been duplicated, the cell will divide. If the environment remains essentially the same, the process will be repeated. A small population in a large volume of growth medium will be in this state for some time. The passage of nutrients from the environment into the cell and the passage of end-products out of it will not materially alter the composition of the medium and under these conditions the composition of the bacteria will be as given in Figure 2, p. 4.

If exponential growth continued and if the mean generation time were 60 min, the population would increase in 48 hr from 1 to 281,474,976,710,656 and in 96 hr would reach about 10^{29}. This is equivalent to a mass of $10^{29} \times 2.5 \times 10^{-13}$ g or 2.5×10^{13} kg. Manifestly this could only occur if the bacteria were in an impossibly large volume of medium.

Growth curves

The *growth curve* of a culture of bacteria can be plotted as growth versus time. The abscissa may be numbers of organisms or mass or N-content or some other index of growth. It is often possible to use a turbidimeter or spectrophotometer to determine the optical density of a suspension of bacteria and by means of a calibration curve to convert this reading to a measure of growth. Figure 1 shows the curves obtained by plotting growth against time in various ways for a culture which, after a lag of 4 hr, begins to grow exponentially with a mean generation time of 1 hr in a medium containing 4 g of glucose per litre. It will be seen that while growth is exponential it is impractical to use a linear plot but that a straight line is obtained in a semi-log plot.

Continuous cultivation

It is possible to keep a culture growing exponentially more or less indefinitely by diluting it periodically with fresh medium. In a sense this is what occurs in a *chemostat* or *bactogen* (see p. 142, Chapter 2), where the addition of medium is continuous. But it can also be done with a batch culture in a flask. The optical density of the culture is measured and part of the suspension is removed and replaced by medium. Further readings are taken and when the culture density is restored the process is repeated. In this way it is possible to have a continuous supply of exponentially growing organisms (Figure 2).

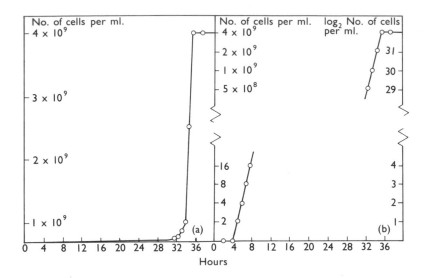

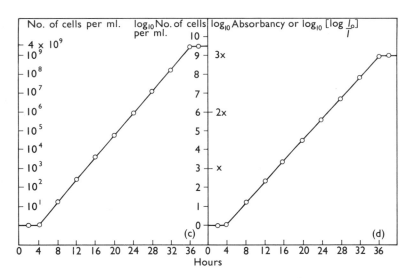

Figure 1. Growth curves.

A culture of bacteria (one organism per ml) after a lag of about 4 hours grows exponentially with a doubling time of one hour in a medium containing 4 mg of glucose per ml. This supports the development of 1 mg dry weight of cells—about 4×10^9 since each weighs 0.25×10^{-10} mg. (a) Linear plot of numbers versus time. (b) and (c) Semi-log plots (with exponential ordinate). These give a straight line but the scale has to be chosen appropriately or interrupted as in (b). Note that log can be to base 10 or to base 2. (d) Turbidity (absorbancy at some suitable wavelength, 450 nm, 600 nm, or 650 nm) can be used in place of numbers of organisms. So can dry weight, N-content, etc.

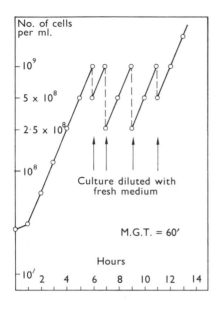

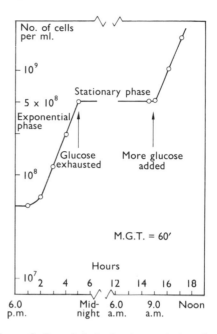

Figure 2. Maintenance of a batch culture in the exponential phase of growth.

Once a culture is growing exponentially, portions can be removed and replaced with fresh medium to preserve approximately steady-state conditions.

Figure 3. Growth limitation by restriction of the carbon substrate.

A culture grown with limiting glucose reaches its maximum density in a few hours and then remains in the stationary phase overnight. Addition of more glucose in the morning leads rapidly to an exponentially growing culture.

Limitation of growth

In a batch culture which is not treated in this way the growth eventually stops when a constituent of the medium is all consumed. For instance, 1000 ml of the synthetic medium with 0·4% glucose will yield about 1 g dry weight of bacteria (about 4×10^{12} organisms or 4×10^9 per ml). The bacteria will have grown exponentially with a mean generation time of one hour until the glucose was exhausted. Had a lower concentration of glucose been used, the yield of organisms would have been proportionately less. If more sugar were present a higher yield might have been obtained but at some concentration of glucose another component in the medium would be limiting, e.g. nitrogen or magnesium or sulphur. Moreover, it is possible that the oxygen supply might restrict growth or that the pH value of the solution might, as a result of the formation of acid products of metabolism, fall to a value at which growth no longer occurred. When growth is limited by exhaustion of the carbon source, it is frequently possible to cause it to resume more or less immediately by further addition of the substrate—even after the culture has been in the *stationary phase* for many hours. This can be exploited practically since it is often useful to grow an over-night culture of bacteria under such conditions and to add more substrate the following morning. This soon results in a culture which is growing exponentially (Figure 3).

The constitution of a bacterial cell is determined to a considerable extent by the medium in which it is growing. A 'rich medium' is one which provides many substances such as amino acids, purines and pyrimidines, and bacterial vitamins, which would otherwise have to be synthesized. This kind of environment tends to cause cells to grow more rapidly and to be larger than does a poor medium. Merely altering the carbon source from one substance to another may provoke changes in mean generation time and size of cells. Sometimes, one substrate (e.g. glucose) is preferred to another (e.g. galactose) and organisms will grow exponentially in the presence of both, using only one until it is ex-hausted and then, after a short lag, will use the other but with a different generation time and yielding cells of different size and composition (Figure 4).

Effect of temperature on growth

Temperature also affects the rate at which cultures grow, but probably has less effect on composition than does the nature of the growth medium. Many species grow best between 30° and 37° but, although some tolerate much higher or lower temperatures, only a few are truly *thermophilic* or *cryophilic* in the sense of growing better at high or low temperatures. Most of the reactions

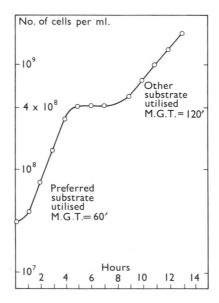

Figure 4. Biphasic growth curve.

In the presence of two carbon substrates one may be used preferentially and then, after a lag, the other may be used. The first usually gives a faster growth rate than the second.

occurring in living organisms are chemical reactions and even when catalysed by enzymes have a Q_{10} of about 2, i.e. the rate doubles for each $10°$ rise in temperature. However, the catalysts are themselves heat-labile so that the optimal temperature for growth results from a balance between the increased rates of reactions and the increased rate of thermal inactivation of the enzymes.

Growth of individual organisms

So far we have been considering the growth of populations of bacteria and it is, of course, much easier to do this than to study the growth of individual cells. Indeed, it is only comparatively recently that techniques other than straight-forward microscopy have been evolved which can be applied to single organisms or used to produce collections of cells all in exactly the same state. Normally, a culture of bacteria will contain organisms in every possible state ranging from those which have just divided to those which are just about to do so. Biochemical investigations using such material will, therefore, give the average results for a very heterogeneous population. In only a very few instances is it possible to assay the amount of a component such as an enzyme in a single cell, but micro-electrophoresis can give information about the electrical properties of the surface layers of individuals and autoradiography can be used to demonstrate the amount and location of radioactive atoms in individual bacteria. Because of the smallness of bacteria, the scope of these methods is limited.

Complementary to these approaches are those in which a whole population of organisms is brought into synchronous growth so that all divide simultaneously and pass through each stage of growth at the same time. It is then possible to use methods applicable only to large numbers. It has, for instance, been established that the synthesis of DNA occurs during practically the whole of the cell division cycle and that RNA and protein are also made continuously. The sequence of changes on moving from one growth medium to another, and hence from one generation time to another, can also be investigated. Since enzymes are proteins, since about half of the mass of a bacterial cell is protein, and since the synthesis of protein involves the ribonucleoprotein particles called ribosomes, it is characteristic of rapidly growing organisms that they contain large numbers of ribosomes. The immediate consequence of a *shift-down* to a poorer growth medium is a reduction in the rate of ribosome synthesis so that the content per cell falls. This results in a slower rate of protein synthesis and hence of growth. A *shift-up* to a richer medium leads to rapid ribosome formation, faster protein synthesis and an increased rate of growth. These effects are dealt with at greater length in Chapter 2.

Section 3
Class I Reactions: Supply of Carbon Skeletons

As we have seen, a bacterial cell, in order to grow, must be supplied with sources of carbon and of energy in addition to nitrogen and essential inorganic ions. Often the source of carbon serves also as the energy source, and this holds true for our 'generalized bacterial cell' which can utilize glucose for this dual purpose. The present chapter outlines the ways in which glucose is metabolized to furnish the basic components essential for synthesis of the macromolecules necessary for cellular growth. These reactions, which are referred to collectively as *intermediary metabolism*, comprise a series of enzymic degradations and syntheses carried out in a stepwise manner. In essence, the individual reactions are relatively simple, involving the addition or removal of hydrogen, water, ammonia, carbon dioxide or phosphate.

The macromolecules of the cell include proteins, nucleic acids (RNA and DNA) and the peptidoglycan of the cell wall; the essential units for their biosynthesis may be summarized as follows:

Proteins:	20 amino acids
RNA:	4 ribonucleotides
DNA:	4 deoxyribonucleotides
Peptidoglycan:	hexosamines + amino acids.

These component units must be synthesized before polymerization can take place. The glucose molecule, containing six carbon atoms, is subjected to stepwise degradation by the action of various enzymes resulting in the formation of smaller molecules of five, four, three or two carbon atoms. These carbon skeletons are then utilized, either directly or after further modification, as building blocks for the synthesis of macromolecules. We may thus distinguish the *catabolic* or degradative enzymes (Class I) which break down glucose and the *anabolic* or biosynthetic (Class II) enzymes which effect the conversion of either glucose itself or its degradation products to compounds such as amino sugars (hexosamines), amino acids and nucleotides for the subsequent syntheses of peptidoglycan, proteins and nucleic acids. The overall reactions catalysed by Class I enzymes additionally yield energy to the cell and are therefore said to be *exergonic*, while Class II enzymes catalyse processes which require the expenditure of energy and are referred to as *endergonic*. Class I enzymes thus furnish not only the carbon skeletons necessary for the synthesis of the basic components of macromolecules, but also the energy required for the transformations to occur under the influence of Class II enzymes.

Breakdown of glucose

The oxidation of glucose to carbon dioxide and water involves a considerable number of steps. At certain of these energy is produced and stored in a chemical

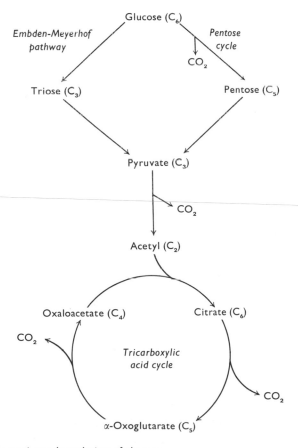

Figure I. Class I reactions: degradation of glucose.

     The breakdown of glucose to CO_2 via intermediates which are also precursors of other small molecules (see Figures 1–4, pp. 30–33). The Class I reactions also produce utilizable energy in the form of ATP.

form as the compound adenosine triphosphate (ATP). ATP is essential for cellular work of all kinds including endergonic biosynthetic reactions, mechanical work as in flagellar movement, and the osmotic work required for the maintenance of intracellular concentrations of metabolites.

Figure 1 illustrates two ways in which glucose may be degraded in bacteria; one of these results in the conversion of the C_6 sugar into two C_3 units, while the other permits the release of carbon dioxide with the formation of a pentose (C_5) sugar which, after further transformations, can also yield a C_3 compound. The first of these metabolic sequences is known as the *Embden-Meyerhof pathway*. It involves conversion of glucose to an isomeric C_6 sugar, fructose, followed by splitting to two C_3 units or trioses. The trioses are then oxidized and by further reactions yield pyruvate, so that two pyruvate molecules are derived from each molecule of glucose.

The other metabolic pathway differs in that removal of two hydrogen atoms from the glucose molecule first occurs and the resulting C_6 compound is converted to a pentose with the elimination of CO_2. The pentose can then yield pyruvate (C_3) as a result of further reactions. Thus the key C_3 compound in both these degradative sequences is pyruvic acid, a keto acid which serves as the starting point for entry to a cyclic process known as the *Krebs*, *citric* or *tricarboxylic acid cycle* which is the major aerobic mechanism for cellular oxidation and energy release. In this process pyruvic acid is first oxidized to an acetyl (C_2) unit (as will become apparent later, it is actually in combination with an organic co-factor molecule) with release of carbon dioxide, and the acetyl unit then combines with a C_4 organic acid, oxaloacetate, to form a C_6 compound, citrate. By a sequence of some nine reactions involving oxidations, the removal and addition of H_2O, and the loss of two molecules of CO_2, citrate is converted to a C_5 compound, α-oxo-glutarate, and subsequently to oxaloacetate, the C_4 acid which initiated the cycle. The net effect of one turn of the tricarboxylic acid cycle is thus the complete oxidation of a C_2 unit to CO_2, and H_2O; simultaneously, the oxidation processes make energy available to the cell.

Section 4
Class II Reactions: Biosynthesis
of Small Molecules

The various degradative (Class I) enzymes we have so far discussed convert glucose to smaller C_5, C_4, C_3 and C_2 aliphatic units, as summarized in Figure 1, p. 27. The complete oxidation to CO_2 and water of some C_2 units in the tricarboxylic acid cycle yields energy to the bacterial cell, some of which will be required for the conversion of C_6, C_5, C_4, C_3 and other C_2 units by Class II enzymes to the basic components necessary for the biosynthesis of macromolecules. It will be noticed that the tricarboxylic acid cycle, in addition to furnishing energy, also yields C_5 and C_4 units for biosynthesis. The relationship between these various units and the formation of amino acids, the building blocks of proteins, is illustrated on p. 30.

Amino acids

One of the interesting features of amino acid biosynthesis is the way in which various amino acids fall into well defined 'families' sharing common pathways of synthesis. This, perhaps, is not so surprising when we consider that such similarities of chemical structure as the possession of an aromatic nucleus, or a branched carbon skeleton exist among particular amino acids. Thus the biosynthesis of the three amino acids which possess an aromatic nucleus, namely tryptophan, phenylalanine and tyrosine, involves a triose and a number of reactions common to all three, which are discussed in detail in a later chapter. Similarly, the amino acids leucine and valine, which possess a branched carbon skeleton, have pyruvate as a precursor and share a common sequence of reactions for part of their individual biosynthetic pathways.

It will be apparent that an essential feature of amino acid biosynthesis from the degradation products of glucose must be the introduction of nitrogen in the form of amino groups. This is usually achieved by the addition of ammonia to a keto acid in the process known as *amination*. The simplest examples of this are to be seen in the conversion of pyruvate to alanine and α-oxoglutarate to glutamate:

$$\text{Pyruvate} + NH_3 \longrightarrow \text{Alanine}$$
$$\alpha\text{-Oxoglutarate} + NH_3 \longrightarrow \text{Glutamate}$$

29

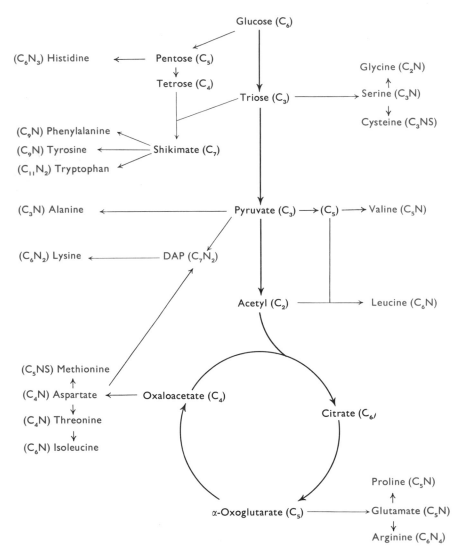

Figure 1. Class II reactions: synthesis of amino acids.
The carbon is supplied by glucose, the nitrogen by NH_4^+, and the sulphur by SO_4^{2-} (diamino-pimelate is represented by DAP).

Aspartate, a very important amino acid in several biosynthetic pathways, may be formed in some bacteria by the addition of ammonia to the unsaturated dicarboxylic acid fumarate:

$$\text{Fumarate} + NH_3 \longrightarrow \text{Aspartate}$$

An amino group may also be added to a keto acid by transfer from an amino acid, a process termed *transamination*: this would, of course, produce the keto acid corresponding to the original amino acid:

$$NH_2CH.COOH + O{=}C.COOH \rightleftharpoons NH_2CH.COOH + O{=}C.COOH$$
$$\quad\;\; |\qquad\qquad\quad |\qquad\qquad\qquad\;\; |\qquad\qquad\quad |$$
$$\quad R_a \qquad\qquad\; R_b \qquad\qquad\qquad R_b \qquad\qquad\; R_a$$

Aspartate is of considerable importance in the biosynthesis of five other amino acids and Figure 1 shows that three distinct pathways are involved leading to (a) threonine and isoleucine, (b) the sulphur containing amino acid methionine and (c) diaminopimelate and lysine.

By such biosynthetic pathways some twenty different amino acids are synthesized from degradation products of glucose and are then available for assembly into proteins by Class III enzymes (Chapter 6). Certain amino acids,

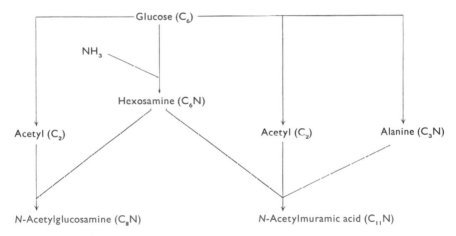

Figure 2. Class II reactions: synthesis of hexosamines.

The acetylated hexosamines are constituents of the peptidoglycans which are essential components of bacterial cell walls.

such as glutamate, alanine, diaminopimelate, and glycine, are required also for the synthesis of cell wall peptidoglycan where they occur in combination with amino sugars.

Hexosamines

We have already noted that the peptidoglycan of the bacterial cell wall contains various amino acids and amino sugars. The amino sugars are C_6 units, i.e.

hexosamines, and their synthesis requires the introduction of an amino group into a hexose. The reaction involved is a transamination between the amide of glutamate (glutamine) and a ketohexose, e.g.

$$\text{Ketohexose} + \text{Glutamine} \longrightarrow \text{Hexosamine} + \text{Glutamate}.$$

The hexosamines found in peptidoglycans include glucosamine and muramic acid and both have their amino groups acetylated as shown in Figure 2 which outlines their biosynthesis.

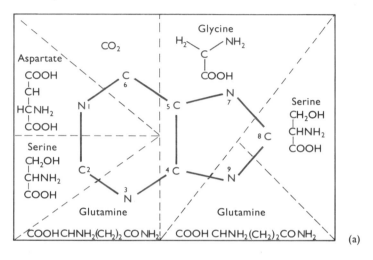

(a)

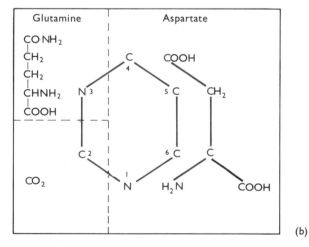

(b)

Figure 3. Origins of the atoms of purine rings (a) and pyrimidine rings (b).

Nucleotides

It remains to be seen how the basic components of the nucleic acids are produced. These are the purine and pyrimidine ribonucleotides and deoxyribonucleotides (see Figures 3 and 4, pp. 17 and 18 for their chemical structures). The deoxy-derivatives are formed from the ribonucleotides by reduction of the ribose to deoxyribose. The ribose itself is formed from glucose by loss of a carbon atom as CO_2. The purine ring can then be built up by stepwise addition to ribose phosphate. The carbon atoms come variously from CO_2 and the amino acids,

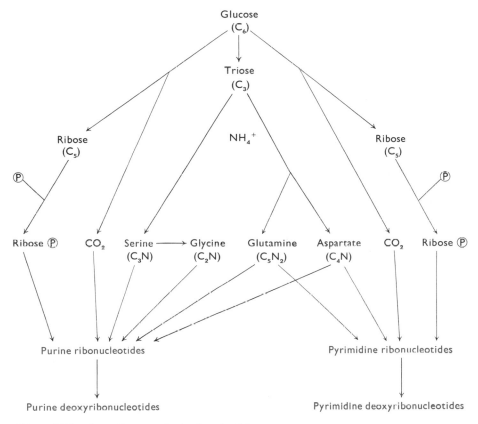

Figure 4. Class II reactions: synthesis of nucleotides.

glycine and serine; the nitrogen atoms are supplied by glycine, aspartate and glutamine. Pyrimidine carbon atoms come from CO_2 and aspartate while the nitrogen atoms are derived from aspartate and glutamine or ammonia. The pathways for the biosynthesis of these nucleotides are indicated in Figures 3 and 4.

Sections 5 and 6
Class III Reactions:
Synthesis of Macromolecules

Much of a bacterial cell consists of substances of high molecular weight and universally they are formed by condensation of large numbers of small units. These may all be the same so that a *homopolymer* results, or they may be of several kinds as in *heteropolymers*. Examples of the latter are proteins which are made from 20 different kinds of amino acids, and nucleic acids in which four different nucleotides occur. Polysaccharides, on the other hand, may be of either class of polymer. However, an even more important distinction is between those polymers in which regular repetition of units or groups of units occurs, and those in which there is no such repetition (Figure 1).

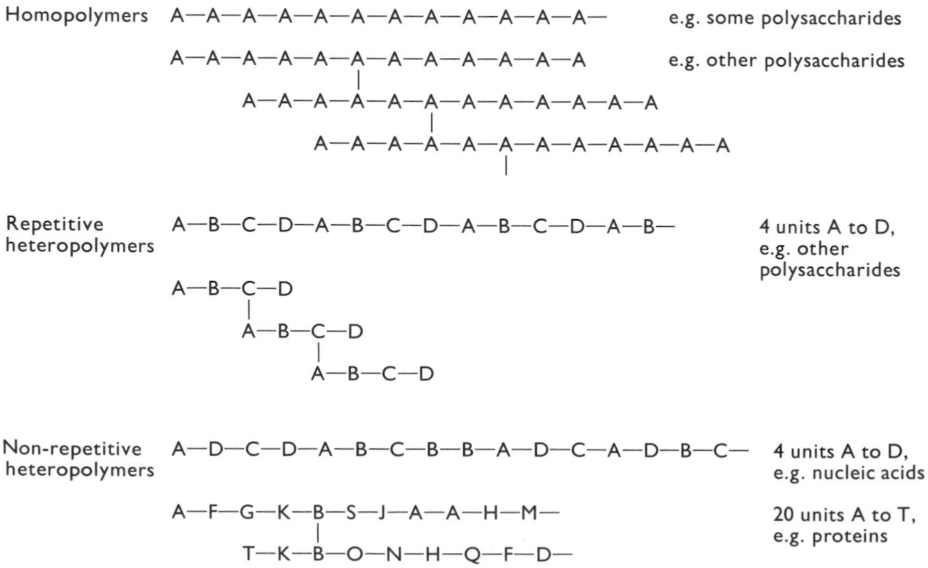

Homopolymers A—A—A—A—A—A—A—A—A—A—A—A—   e.g. some polysaccharides

A—A—A—A—A—A—A—A—A—A—A—A   e.g. other polysaccharides
          |
A—A—A—A—A—A—A—A—A—A—A—A
          |
A—A—A—A—A—A—A—A—A—A—A—A
          |

Repetitive   A—B—C—D—A—B—C—D—A—B—C—D—A—B—   4 units A to D,
heteropolymers   e.g. other
                                                                polysaccharides
A—B—C—D
    |
A—B—C—D
    |
A—B—C—D

Non-repetitive   A—D—C—D—A—B—C—B—B—A—D—C—A—D—B—C—   4 units A to D,
heteropolymers   e.g. nucleic acids

A—F—G—K—B—S—J—A—A—H—M—   20 units A to T,
         | e.g. proteins
T—K—B—O—N—H—Q—F—D—

Figure I. Types of polymer found in bacteria.

Polysaccharides and peptidoglycans are repetitive polymers whereas proteins and nucleic acids are not. This has profound consequences on the nature of the substances and on the mechanisms by which they are made. Firstly, all the molecules of any protein are identical to each other in composition, in sequence of units, and in molecular weight: the same is true of nucleic acids. Polysaccharides, however, may be polydisperse, i.e. a range of molecules may exist all having essentially the same repeating units but differing in size so that the term 'molecular weight' can only refer to a mean value. Secondly, the repetitive polymers can be made by successive action of a small number of enzymes whereas proteins and nucleic acids cannot—it would require an astronomically large number. For example, a polysaccharide composed of four different sugars could be put together by four different enzymes as follows:

```
Enz.
1:  A + B ——→ A—B
2:  A—B + C ——→ A—B—C
3:  A—B—C + D ——→ A—B—C—D
4:  A—B—C—D + A—B—C—D + etc. ——→ A—B—C—D
                                          |
                                       A—B—C—D
                                          |
                                             etc.
```

But in order to make a polynucleotide (nucleic acid) in which the four nucleotide units occur in a definite, specified, but non-repetitive, sequence, a large number of enzymes would be required. For instance, the sequence shown in Figure 1 would be made as follows:

```
Enz.
1:  A + D ——→ A—D
2:  A—D + C ——→ A—D—C
3:  A—D—C + D ——→ A—D—C—D
4:  A—D—C—D + A ——→ A—D—C—D—A
5:  A—D—C—D—A + B ——→ A  D  C  D  A  B        etc.
```

Moreover, hundreds or thousands of different nucleic acids exist in a bacterium, and there are about as many species of proteins. In these, any of the 20 amino acids may be neighbour to itself or to any of the other 19, so 400 (20^2) pairs are possible—and hence 400 enzymes would be necessary to make the specific unions. But even this would not be enough since a sequence ... K—B might be followed in some places by S and in others by O as is represented in Figure 1:

$$A—F—G—K—B + S ——→ A—F—G—K—B—S$$

but

$$T—K—B + O ——→ \qquad T—K—B—O$$

Clearly the existence of the necessary enzymes for the reactions:

$$B + S ——→ B—S \quad and \quad B + O ——→ B—O$$

would not ensure that the right amino acid was added in each instance—there
would be uncertainty unless different enzymes catalysed each reaction:

$$...G—K—B \ + \ S \longrightarrow ...G—K—B—S$$
and
$$T—K—B \ + \ O \longrightarrow \ T—K—B—O$$

It will be appreciated that enormously more than 20^2 enzymes would be necessary to cope with all the possibilities without ambiguity. And because each
enzyme is itself a specific protein, the absurdity of this as a mechanism is
apparent.

Instead, it was postulated and is now proven, that some kind of template
must specify the sequence of the units in the non-repetitive heteropolymers. If
the units are lined up in the correct order, there is little difficulty in envisaging a
mechanism for joining them together involving only a small number of enzymes.
Whether the polymeric product is to be DNA, or RNA or protein, the template
always turns out to be a nucleic acid. How this occurs will be described after
examples of repetitive polymer formation are mentioned.

REPETITIVE POLYMERS

Polysaccharides

The monomeric units of polysaccharides are sugars, amino sugars and sugar
acids, and the polymers have various functions and occur in various parts of
bacterial cells. For instance, glycogen, a homopolymer of glucose, is often
found as a storage material in the cytoplasm and some organisms such as pneumococci, surround their cell walls with a capsule made of a specific hetero-
polymer. In general, the mechanism of synthesis of these polysaccharides seems
to be similar. The carbohydrate units are added stepwise to an existing chain or
primer and a specific enzyme catalyses this reaction which frequently involves
the transfer of the unit from a 'carrier' nucleoside diphosphate such as UDP
(uridine diphosphate). Thus glycogen may be formed as follows:

$$Glc\text{-}Glc\text{-}Glc\text{-}Glc \ + \ UDP\text{-}Glc \longrightarrow Glc\text{-}Glc\text{-}Glc\text{-}Glc\text{-}Glc \ + \ UDP \text{ etc.}$$

The same enzyme then catalyses successive additions. Other nucleoside diphosphates may also act as carriers.

Peptidoglycans

Bacterial cell walls contain peptidoglycan (see Figure 1, p. 12) which has a
repetitive backbone of *N*-acetyl amino sugars and short peptides attached to it:

—N-Ac-GlcNH$_2$—N-Ac-Mur—N-Ac-GlcNH$_2$—N-Ac-Mur—
           | |
         Peptide   Peptide

It is made stepwise by a series of enzymic reactions and again UDP derivatives of the carbohydrate units participate (Figure 2).

$$N\text{-Ac-GlcNH}_2 \; + \; \text{UTP} \; \longrightarrow \; \text{UDP-}N\text{-Ac-GlcNH}_2 \; + \; \circledP$$

$$\text{UDP-}N\text{-Ac-Mur} \; + \; \text{L-ala} \; \xrightarrow{\text{ATP}} \; \text{UDP-}N\text{-Ac-Mur-L-ala}$$

$$\text{UDP-}N\text{-Ac-Mur-L-ala} \; + \; \text{D-glu} \; \xrightarrow{\text{ATP}} \; \text{UDP-}N\text{-Ac-Mur-L-ala-D-glu}$$

$$\text{UDP-}N\text{-Ac-Mur-L-ala-D-glu} \; + \; \text{DAP} \; \xrightarrow{\text{ATP}} \; \text{UDP-}N\text{-Ac-Mur-L-ala-D-glu-DAP}$$

$$\text{UDP-}N\text{-Ac-Mur-L-ala-D-glu-DAP} \; + \; \text{D-ala-D-ala} \; \xrightarrow{\text{ATP}}$$
$$\text{UDP-}N\text{-Ac-Mur-L-ala-D-glu-DAP-D-ala-D-ala}$$

i.e. UDP-N-Ac-Mur + amino acids $\xrightarrow[\text{ATP as energy source}]{\text{Stepwise addition}}$ UDP-N-Ac-Mur
                                                                  |
                                                                 Peptide

UDP-N-Ac-GlcNH$_2$ | UDP-N-Ac-Mur + existing peptidoglycan $\longrightarrow$
                       |
                      Peptide   More peptidoglycan + UDP

Figure 2. Biosynthesis of bacterial cell wall peptidoglycan.
    The subunits are (a) N-acetyl glucosamine and (b) N-acetyl muramic acid to which amino acids have been added stepwise to form a short peptide. These subunits are then polymerized to give peptidoglycan which has a backbone of alternating amino sugars (see Figure 1, p. 12).

Thus even this relatively complicated peptidoglycan structure is made by a handful of enzymes catalysing successive reactions and producing an unambiguous product of precise composition.

NON-REPETITIVE POLYMERS

DNA

Perhaps the reason why DNA occurs as a double helix with one strand complementary to the other (Figure 3) is because it is difficult or impossible to produce an exact copy of a single strand. No chemical method for this can readily be envisaged. However, the ability of adenine to pair with thymine, and of guanine

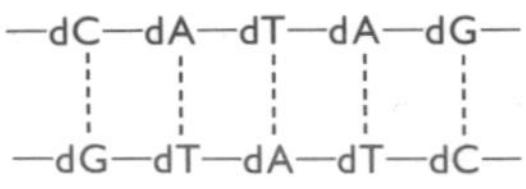

```
—dC—dA—dT—dA—dG—
  |    |    |    |    |
  |    |    |    |    |
—dG—dT—dA—dT—dC—
```

Figure 3. Complementary strands of DNA.
    dA, dG, dC and dT are the deoxyribonucleotides of adenine, guanine, cytosine and thymine.
In DNA which is double-stranded dA is always paired with dT and dG is paired with dC.

and cytosine to interact similarly, forms the basis of the way in which a *double* strand can be replicated (Figure 4). Essentially, a complementary copy is made of each of the original strands and division of the resulting four strands into two pairs of one old and one new will be seen to have resulted in replication of

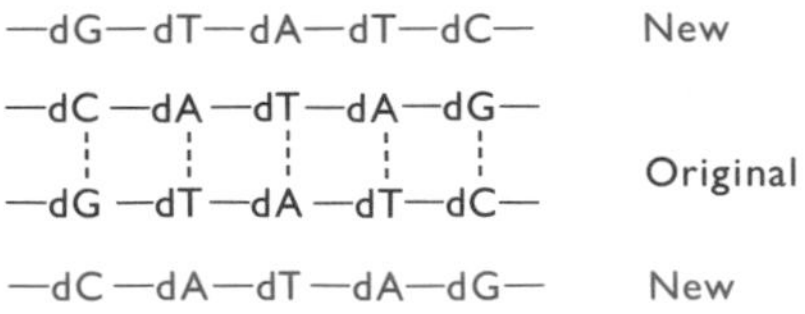

```
—dG—dT—dA—dT—dC—        New

—dC —dA —dT—dA—dG—
  |    |    |    |    |
  |    |    |    |    |                Original
—dG —dT —dA —dT—dC—

—dC —dA—dT —dA—dG—       New
```

Figure 4. Replication of double-stranded DNA.
    The original double-stranded DNA is base-paired (dA with dT and dG with dC). Each original strand acts as template for the formation of its complement.

the original pair. The synthesis begins at one end by separation of the two strands and proceeds sequentially to the other end of the original DNA chains (Figure 5).

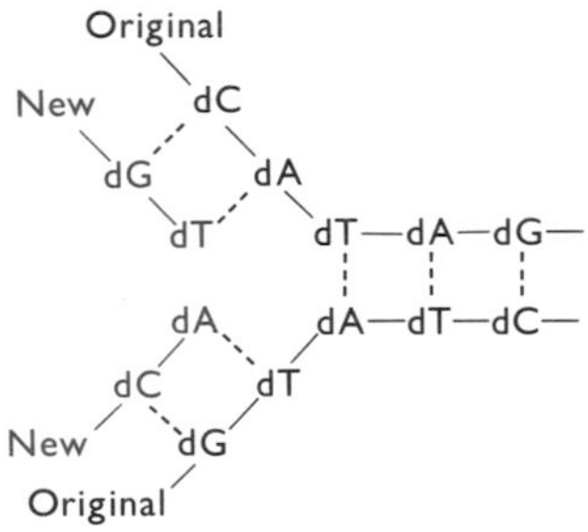

```
                  Original
                     \
       New            dC
          \        /
           dG      dA
             \      \
             dT    dT—dA—dG—
                      |    |    |
             dA    dA—dT—dC—
            /      /
          dC     dT
       New  \   /dG
             \ /
          Original
```

Figure 5. Replication of DNA.
    Synthesis begins at one end by separation of the two strands of the original double helix and proceeds sequentially to the other end. Each double-stranded product contains one old and one new chain.

The precursors which form the units to be polymerized are not the nucleotides themselves (e.g. Ade-dRib-$\textcircled{P}$ or dAMP) but the pyrophosphate derivatives of these, the deoxyribonucleoside triphosphates (Figure 6). Condensation occurs with the elimination of inorganic pyrophosphate and the production of polynucleotides (DNA) (Figure 7). An enzyme has been prepared and purified from bacteria which will catalyse a reaction of this kind *in vitro*. All four

Ade-dRib—Ⓟ—Ⓟ—Ⓟ   Cyt-dRib—Ⓟ—Ⓟ—Ⓟ
dATP   dCTP

Gua-dRib—Ⓟ—Ⓟ—Ⓟ   Thy-dRib—Ⓟ—Ⓟ—Ⓟ
dGTP   dTTP

Figure 6. The deoxyribonucleoside triphosphates.
The combination of base and sugar is called a nucleoside; the monophosphate of this is a nucleotide—here a deoxyribonucleotide.

deoxyribonucleoside triphosphates must be present, as must some DNA to act as a template for copying. The reaction can be formulated:

$$dATP + dGTP + dCTP + dTTP \xrightarrow[\text{DNA polymerase}]{\text{DNA template}} \text{More DNA of same kind} + Ⓟ—Ⓟ$$

In this way precise copies of the genes can be made.

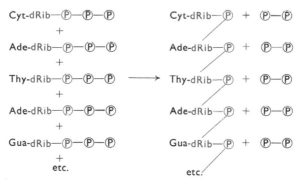

Figure 7. Formation of DNA.
Deoxyribonucleoside triphosphates condense with elimination of inorganic pyrophosphate to form polynucleotide (DNA).

RNA

The genetic material, DNA, has not only to be replicated exactly so that the species can be perpetuated but it must also be transcribed into functional RNA molecules—the ribosomal, amino acid transfer, and messenger RNA's mentioned earlier. In this process only one strand of the DNA is used as a template to specify the nucleotide sequence of the product but the mechanism is similar to that of DNA synthesis—the ribonucleoside triphosphates ATP, GTP, CTP and UTP (Figure 8) are needed, as is template DNA and a specific enzyme. It should be noted that it is ribonucleotides rather than deoxyribonucleotides

Ade-Rib—℗—℗—℗   Cyt-Rib—℗—℗—℗
ATP   CTP

Gua-Rib—℗—℗—℗   Ura-Rib—℗—℗—℗
GTP   UTP

Figure 8. The ribonucleoside triphosphates. Compare with Figure 6.

which are being polymerized and that one of the pyrimidines is uracil in place of thymine. However, uracil like thymine can form H-bonds to adenine so that a similar pairing mechanism operates (Figure 9).

—dC—dA—dT—dA—dG—
  | | | | | DNA template
—dG—dT—dA—dT—dC—

——C——A——U——A——G—   RNA product

Figure 9. Transcription of DNA into RNA.
Only one strand of DNA is transcribed into the complementary RNA strand, the operation involving base-pairing between dA and U, dG and C, dC and G, and between dT and A.

As in DNA synthesis, the triphosphates are condensed to polynucleotides (RNA) with elimination of inorganic pyrophosphate so that the reaction can be formulated:

$$\text{ATP} + \text{GTP} + \text{CTP} + \text{UTP} \xrightarrow[\text{RNA polymerase}]{\text{DNA template}} \text{RNA complementary to one strand of DNA} + ℗—℗$$

Proteins

In the making of proteins two main problems have to be solved—how the amino acids are joined together and how they are selected in the right sequence. The formation of a peptide bond between two amino acids involves condensation and elimination of water and is a reaction which is catalysed by a peptidase:

$$\underset{R_a}{H_2N.CH.COOH} + \underset{R_b}{H_2N.CH.COOH} \rightleftharpoons \underset{R_a}{H_2N.CH.CONH.}\underset{R_b}{CH.COOH} + H_2O$$

However, the equilibria of such reactions are strongly in favour of the right-to-left hydrolysis. In order to facilitate the forward reaction the amino acid has first to be 'activated' and this occurs by its reaction with ATP in a manner somewhat similar to that in which glucose is activated. In this instance, however, inorganic pyrophosphate is eliminated and an AMP-derivative of the amino acid is formed:

$$H_2N.CH.COOH + \text{P-Rib} \longrightarrow H_2N.CH.CO\sim\text{P-Rib}$$

(with R groups and Ade, P residues as shown)

($\sim$ P indicates a high-energy bond to a phosphate residue.)

If the amino acid or the amino acyl residue is represented by $\alpha\alpha$, the reaction may be expressed as follows:

$$\alpha\alpha + ATP \longrightarrow \alpha\alpha\sim AMP + \text{P—P}$$

For each of the 20 amino acids listed in Figure 2 (p. 14) there is a specific amino acid-activating enzyme which catalyses the formation of the amino acyl-AMP derivative. These could condense with the elimination of AMP to form peptides but this would not provide a mechanism for specifying the sequence in which the amino acid residues were joined together and, as has been seen, this sequence is all-important in making a specific protein.

It is abundantly clear that the ultimate control over specification in an organism resides in its genetic make-up, i.e. its DNA. It is believed that every protein which a cell is capable of making is represented by a gene, i.e. a piece of DNA. This structural gene specifies the amino acid sequence of the polypeptide(s) that makes up the protein.

The next problem, therefore, is how one linear polymer can specify another when the former (DNA) is composed of four species of nucleotide unit and the latter (protein) contains 20 species of amino acids. A one-for-one correspondence is clearly impossible and it is apparent that since there are only 16 (4×4) ways of combining any two of the four nucleotides (AA, AG, AC, AT, GA, GG, GC, etc.) the code cannot be a doublet one but must be a *triplet code* (or something more complex). There are 64 ($4 \times 4 \times 4$) possible triplet sequences which can be made (AAA, AAG, AAC, AAU, GAA, GAG, etc.) but this still does not explain *how* a trinucleotide sequence in a DNA can account for the positioning of an amino acid residue in a polypeptide.

In fact, the DNA does not participate directly in the process of protein synthesis but an active structural gene is transcribed into a messenger-RNA which does take part. There is a specific mRNA for each kind of protein which is being made, and the nucleotide sequence of the mRNA is directly related to that of the DNA template on which it was made so that it carries as much information in its sequence as does the original gene. However, there are no grounds for believing that there can be any kind of chemical 'recognition' between an amino acid and a trinucleotide sequence. The only recognitions so far established are the highly specific enzyme-substrate interactions (i.e. a specific protein recognizing a small molecule) and the polynucleotide/polynucleotide interaction based on H-bond formation between guanine and cytosine and between adenine and either thymine or uracil, i.e. base-pairing.

Together these provide a way in which a triplet code can operate. Each activated amino acid attached to AMP (e.g. $\alpha\alpha_x \sim AMP$) is transferred to a specific RNA molecule, a transfer-RNA (tRNA). This reaction is brought about by its activating enzyme which is able to recognize both the appropriate amino acid ($\alpha\alpha_x$) and the corresponding $t_x RNA$. The latter, by virtue of a specific trinucleotide sequence in its make-up, is able to recognize a complementary sequence in an mRNA and thus deliver the amino acid to its appropriate place so that a defined polypeptide sequence can result:

$$\alpha\alpha_x \sim AMP \; + \; t_x RNA \longrightarrow \alpha\alpha_x\text{-}t_x RNA \; + \; AMP$$

$$\alpha\alpha_x\text{-}t_x RNA \; + \; \alpha\alpha_y t_y RNA \; + \; etc. \xrightarrow{\text{Directed by mRNA}} \text{Specific polypeptide} \; + \; tRNA\text{'s}$$

However, the mRNA's do not occur floating free in the cytoplasm of cells; they have ribosomes associated with them and before messenger-RNA was discovered or even postulated it was known that proteins were synthesized on ribosomes. These ribonucleoprotein particles are about 20 nm × 17 nm, have a sedimentation coefficient of 70 S in the ultracentrifuge, and consist of a smaller (30 S) and a larger (50 S) component. Each of these subunits is approximately 60% RNA and 40% protein and it is believed that they orient the mRNA and the tRNA into appropriate juxtaposition, the 30 S ribosomes having an affinity for mRNA and both for tRNA (Figure 10).

A complex is formed with the composite 70 S ribosome attached at the beginning of the message (mRNA) and $t_1 RNA$ bearing the first amino acid

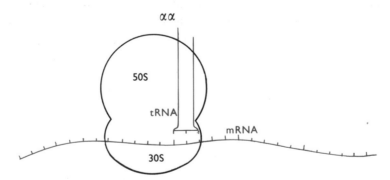

Figure 10. Interactions of ribosomes, mRNA and tRNA.

The 70 S ribonucleoprotein particles (ribosomes) are made up of a smaller unit (30 S) which has affinity for mRNA and a larger unit (50 S) which has affinity for tRNA. The latter has an amino acid ($\alpha\alpha$) attached to it and is thought to base-pair with a complementary triplet in the mRNA as indicated. Each amino acid transfer-RNA would have a specific complementary trinucleotide sequence.

bound both to the ribosome and to the mRNA—a triplet of nucleotides in t_1RNA being complementary to a triplet in mRNA. This first amino acid will be *N*-terminal in the polypeptide specified by this particular mRNA, the next triplet of which will select its appropriate $\alpha\alpha_2$-t_2RNA and line it up on the ribosome. The two amino acids condense with elimination of the tRNA. The process is then repeated with $\alpha\alpha_3$-t_3RNA. And so on.

The mRNA and ribosome may move relative to one another so that the ribosome traverses the strand from one end to the other, 'reading' the message as it goes and with the nascent polypeptide chain increasing in length until completed. Before one ribosome has moved far along an mRNA another may attach at the beginning and so on, so that *polyribosomes* (*polysomes, ergosomes*) are formed (Figure 11).

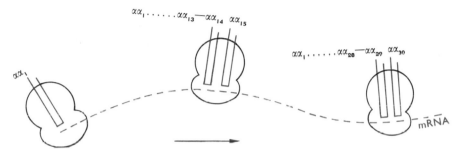

Figure 11. Polyribosomes during the course of protein synthesis.
A ribosome is thought to become attached to a strand of mRNA at the beginning of its message. The appropriate tRNA associates with it (Figure 10) and delivers the first amino acid ($\alpha\alpha_1$). The ribosome progesses along the mRNA with the nascent polypeptide increasing in length. Further ribosomes can become attached before the first has traversed the whole length of the mRNA. Polyribosomes are these associations of ribosomes attached to the same mRNA.

The amino acids have to be linked enzymically but probably only a single enzyme is required to form the peptide bond between successive amino acids, and possibly another is needed to release the completed product from the last tRNA and from the ribosome. The precise sequence of the amino acids has been dictated by the nucleotide sequence of the mRNA and only a relatively small number of enzymes have been required.

Subsequently several polypeptide chains may be linked together to form a native protein molecule or this may consist of a single chain.

Section 7
Genetics

The primary structures of all the proteins of the cell are encoded in the genetic material, which is ordinarily DNA. The DNA controls protein structure via messenger RNA which is transcribed directly from it. The proteins, in their turn, have structural and enzymic functions which determine the form and metabolism of the cell. The genetic material thus ultimately determines the nature of the organism. The whole of a cell's genetic information, carried in the DNA, is called the *genotype*, as distinct from the *phenotype* which is the complete assembly of characteristics exhibited at a given time. The distinction is necessary for two reasons. Firstly, many potentialities provided for in the genotype may remain latent. Thus the genotype may include the latent capacity to make an enzyme to hydrolyse lactose, but whether the enzyme is actually made will be determined by environmental factors which will therefore affect the phenotype. Secondly, in the case of diploid cells, the phenomenon of *dominance* may mask the expression of part of the genotype (see p. 53).

The entire genotype or *genome* of a bacterium is commonly carried in a single giant piece of nucleic acid which, in bacteria, usually forms a closed loop and is called a *chromosome*. The bacterial chromosome, which is an extremely long and fine thread, is folded to form a compact skein which can be seen (after appropriate staining) with the electron microscope as a *nuclear body*, or *chromatinic body*. Unlike cell nuclei of higher organisms it is not enclosed by a nuclear membrane.

The chromosome consists of a number of functionally distinct and non-overlapping segments, the *genes*. Each gene has a single function which can be varied or lost by mutation independently of the functions of other genes. The function in question is usually the specification of the amino acid sequence of a polypeptide chain, with one gene for each kind of polypeptide chain. Some genes, probably a small minority, have the function of specifying the structures of ribosomal and transfer RNA molecules.

Proof of the central position of DNA in heredity has come mainly from microbiological studies and it has been shown to be possible to transfer characters from one organism to another by three techniques of DNA transfer—

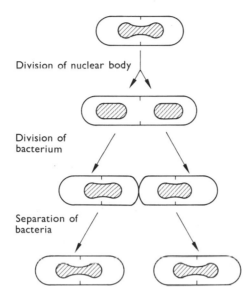

Figure 1. DNA replication and cell division.

First the nuclear body replicates and the chromosomes separate; then a cross wall forms between them; the cell divides and finally separates.

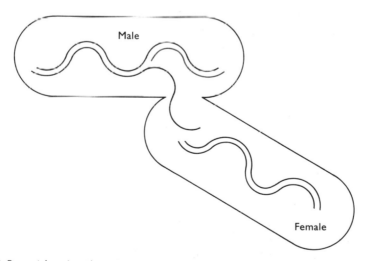

Figure 2. Bacterial conjugation.

The male and female cells adhere. Replication of the chromosome occurs in the male and one old strand is injected into the female while the other is retained. There is no reciprocal transfer from the female.

transformation, conjugation and *transduction*. The first of these involves direct treatment of bacteria with DNA purified from another related strain with different characters, e.g. DNA from a smooth strain of pneumococcus added to cells of an appropriate rough strain may 'transform' some of them—and the ability to make the enzyme which makes the capsular polysaccharide will be inherited by the progeny of the original treated cells.

Conjugation in bacteria has been detected in only a few species including *Escherichia coli* but has nevertheless been of enormous importance in the development of bacterial genetics. What are equivalent to male and female forms adhere to each other and there is a slow transfer of DNA from male to female, the outcome of which is that new genetic characters can be acquired by the acceptor cell from the donor (Figure 2).

Transduction involves the use of bacterial viruses or *bacteriophages*. A virus, unlike all living organisms from bacteria to mammals, contains one rather than both kinds of nucleic acid. Usually this is DNA but some bacterial, plant and animal viruses contain RNA instead. The nucleic acid is the genetic material of the virus and the rest of the structure may be just a protein container and a mechanism for getting the nucleic acid into a suitable host cell. Thus a DNA-containing bacteriophage is like a miniature syringe (Figure 3). The tip of the

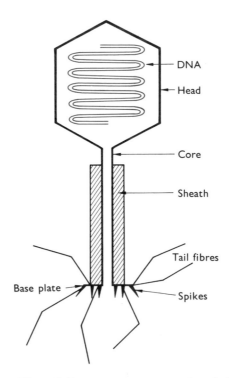

Figure 3. Diagrammatic representation of a bacteriophage.

phage tail sticks on a receptor in the wall of a sensitive bacterium and the DNA is injected into the cytoplasm. Here it acts like a group of genes and controls the formation of enzymes which in turn produce more phage DNA and *also more phage protein*. The phage DNA carries the specification for phage protein and the complete virus particles are assembled within the bacterium—perhaps

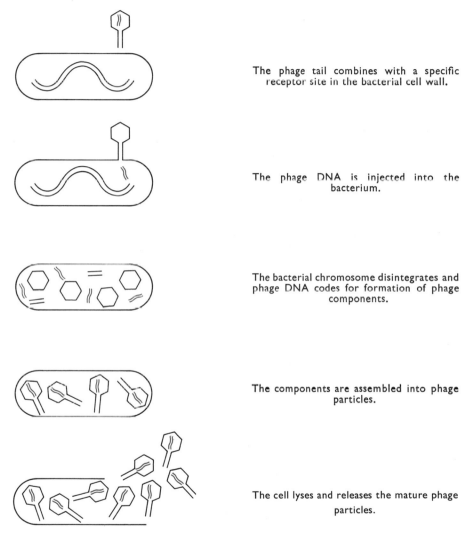

The phage tail combines with a specific receptor site in the bacterial cell wall.

The phage DNA is injected into the bacterium.

The bacterial chromosome disintegrates and phage DNA codes for formation of phage components.

The components are assembled into phage particles.

The cell lyses and releases the mature phage particles.

Figure 4. Infection of a bacterium by a lytic bacteriophage.

hundreds for each phage genome injected—and the bacterial cell lyses, some 30 minutes after infection, releasing the now mature infective bacteriophage particles (Figure 4).

Sometimes infection of a bacterium by a phage does not result in lysis. This

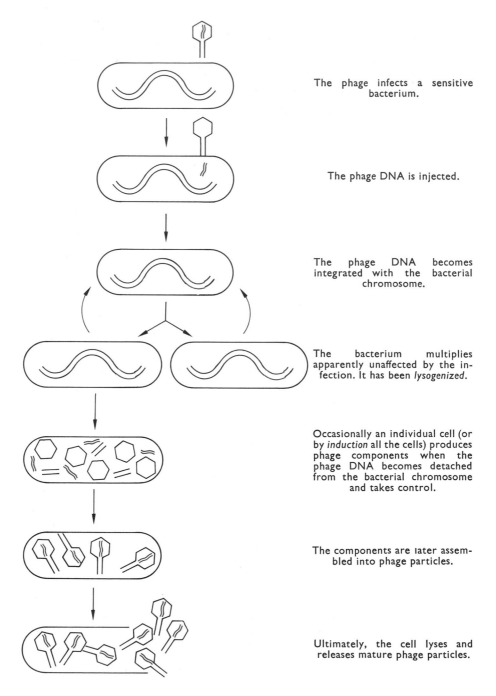

The phage infects a sensitive bacterium.

The phage DNA is injected.

The phage DNA becomes integrated with the bacterial chromosome.

The bacterium multiplies apparently unaffected by the infection. It has been *lysogenized*.

Occasionally an individual cell (or by *induction* all the cells) produces phage components when the phage DNA becomes detached from the bacterial chromosome and takes control.

The components are later assembled into phage particles.

Ultimately, the cell lyses and releases mature phage particles.

Figure 5. Infection of a bacterium by a temperate bacteriophage: lysogenization and induction. The probable integration mechanism is illustrated on p. 400.

can happen in a process called *lysogenization* in which the genome of the phage becomes associated with the bacterial chromosome and is replicated with it as if it were part of the host genome. This can continue for many generations and the bacteria are said to be in a lysogenic state. From time to time, in such a culture, the *temperate* phage will somehow begin to reproduce itself in a virulent fashion and produce lysis. This may happen spontaneously or it may be *induced* by ultra-violet light or by the addition of certain chemicals (Figure 5).

Now, many of the phage particles may have 'picked up' small portions of the genome of the host cell, which they carry as if these were part of their own DNA. They may then bring to the next host cell that they infect a character that it did not possess. This is called *transduction*—the transfer of a character from one bacterium to another by means of bacteriophage.

We have already seen that DNA possesses two of the properties necessary for the genetic material—the ability to specify the amino acid sequence in proteins and the ability to be replicated precisely so that the specification can be passed on (Sections 5 and 6, pp. 37–43). It must also be able to undergo *mutation* and the change must be heritable. Mutations are abrupt changes—loss, addition, alteration—of genetic characters. They occur 'spontaneously' at very low frequency—perhaps about 1 in 10^6 times per character per cell-generation or lower. This rate can be increased enormously by certain physical and chemical treatments—ultra-violet and X-irradiation, nitrous acid, nitrogen mustards, etc. These treatments are all known to cause modification to DNA structure and it is probable that 'spontaneous' mutation has a similar cause. Frequently the result of a mutation is to change one base for another in the DNA. The error is then repeated at every replication and it may manifest itself by the fact that the progeny of the mutated cell are now, because of a change in some enzyme, unable to synthesize an amino acid or a vitamin, or they can no longer grow on some particular sugar as a carbon source. The loss of a character is conventionally denoted by a negative sign. Common examples are trp^-, thr^-, ade^-, arg^-, which represent inability to synthesize trytophan, threonine, adenine or arginine; similarly gal^- and lac^- represent inability to grow on galactose or lactose. The corresponding positive characters are written trp^+, etc. The alternative forms of a gene, e.g. trp^+ and trp^-, are called *alleles* and the *allele* normally found is referred to as the *wild-type* allele.

In higher organisms there is an elaborate sequence of events concerned with replication and the partitioning of the genetic material between daughter cells. The DNA in the nucleus becomes prominent in the form of visible chromosomes which thicken and then split longitudinally. The membrane surrounding the nucleus disappears and a bundle of fibres, called the *spindle*, spans the cytoplasm. The paired chromosomes line up equatorially across the spindle whose fibres are now attached to them. One of each pair of chromosomes is drawn along the spindle fibres towards each pole, the cell begins to divide, the spindle disintegrates, and the two clusters of chromosomes are surrounded by new nuclear membranes. They become diffuse, and we are back to where we started. This whole process is known as *mitosis*, and it appears to be considerably more complex than is the replication in bacteria which do not possess a nuclear membrane, do not form a spindle, and have only a single circular chromosome (see Figure 8, p. 52). However, bacteria do have a great deal in common with higher organisms and it is possible to prepare genetic maps by establishing the relative positions of mutations in the genome. Our ability to do this depends on a little-understood process involving DNA. When two similar lengths of DNA come together (*in vivo*, but not *in vitro*) there is a tendency for *crossing-over* to occur and for the corresponding segments to be exchanged. The stretches of DNA must be generally, but need not be exactly, alike. The exchange process is known as *recombination*.

Immediately after conjugation, transformation or transduction the bacterium which is normally *haploid* (having only one copy of the genome) will be *diploid* or *partially diploid* because it will have an additional chromosome or portion of a chromosome. This is usually only transient and at cell division each daughter tends to receive one or other piece of DNA but not both. The pieces of *homologous* DNA become *segregated* from one another. However, recombination may have preceded this.

Let us consider a bacterium whose genome originally carried the closely linked characters thr^+ and ade^- (the thr^+ meaning that it has an active functional gene for making the amino acid threonine and the ade^- indicating an inactive mutated gene concerned with adenine synthesis). Hence the organism has to be supplied with adenine but can synthesize threonine. A piece of DNA from another cell with the allelic characters thr^- and ade^+ is then introduced. The recipient may receive a piece of DNA carrying thr^- or ade^+ or both. In each case a transient partial diploid is formed and after cell division (see Figure 6) four kinds of progeny are possible: thr^+ade^- (original); thr^-ade^+; thr^-ade^-; and thr^+ade^+. The majority of cells will be the original thr^+ade^- variety but some may have received and exchanged the incoming piece carrying thr^-ade^+ and these now will be able to make adenine but will have lost the ability to make threonine. Then there will be a small proportion of the recombinants thr^-ade^- (a double mutant requiring both threonine and adenine) and thr^+ade^+ (wild-type, able to grow without either threonine or adenine) (see Figure 6). Experimentally the last type of recombinant, having wild-type characters, is the easiest to count because only these recombinants will be able to grow when the cells are plated on a minimal medium, i.e. one containing neither threonine nor adenine. If the number of recombinant cells is compared with the number of original cells we obtain a measure of *recombination frequency*. This is an important measure because it indicates how far apart the genetic loci *thr* and *ade* are from one another. If we assume that recombination is equally likely to occur

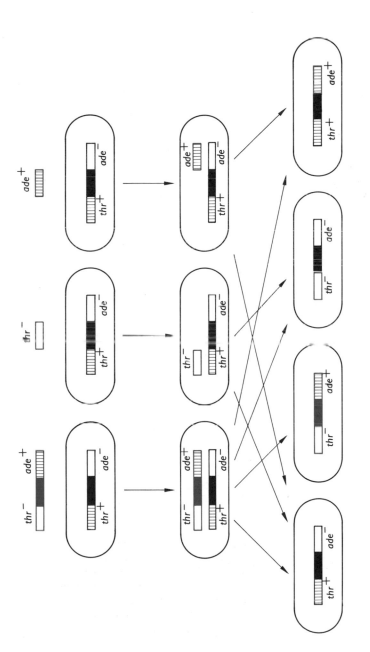

Figure 6. Formation of genetic recombinants.

anywhere along the chromosome, it follows that the further two loci are from one another, the more likely it is that crossing-over will occur at some point between them. The recombination frequency is thus a measure of the distance between loci. If a third mutation (say one concerned with ability to utilize galactose as carbon source) is studied it can be compared in relation to the other two markers. Should the recombination frequency between *thr* and *gal* be found to be roughly equal to the sum of those between *thr* and *ade* and between *ade* and *gal*, then the sequence of the loci on the chromosome must be *thr*, *ade*, *gal* (Figure 7).

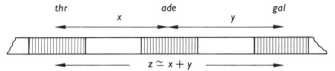

Figure 7. Linear mapping of genes by measurement of recombination frequencies.

The frequency value (x) for two of the genes *ade* and *thr* is first obtained (see text and Figure 6). Frequencies of recombination between *ade* and *gal* (y) and between *thr* and *gal* (z) are then measured in the same way. If z is roughly equal to the sum of $x+y$ then the sequence of the genes is as shown.

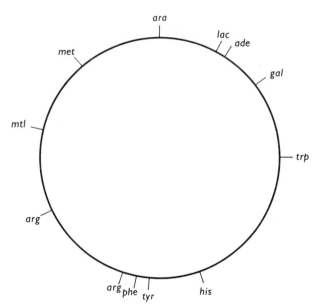

Figure 8. Genetic map of *Escherichia coli*.

The chromosome is circular and the symbols for those markers shown are as follows:

ara	arabinose	*tyr*	tyrosine
lac	lactose	*phe*	phenylalanine
ade	adenine	*arg*	arginine (2 loci shown)
gal	galactose	*mtl*	mannitol
trp	tryptophan	*met*	methionine
	his	histidine	

By these and other methods a large number of characters have been put into sequence and comprehensive genetics maps of *Escherichia coli* and *Salmonella typhimurium* have been prepared. Other observations, partly genetic and partly electron microscopical, show that the DNA from a bacterium is a single closed loop with no free ends (Figure 8).

In special circumstances the partial diploid state which we have said is normally transient may become stabilized and a cell may contain two alleles of the same gene. For instance, a *gal*$^-$ strain may by phage infection receive *gal*$^+$. The presence of the *gal*$^-$ gene does not affect the functioning of the *gal*$^+$, and so the cell is phenotypically *gal*$^+$ and can grow on galactose. The positive character is said to be *dominant* and the negative one *recessive*. The dominance of the positive character is a general rule in genetics and it is a key point in the explanation of regulation of enzyme synthesis.

Sections 8 and 9
Co-ordination and Differentiation

Regulation

A growing bacterial cell makes use—at a guess—of 1000–5000 enzymically catalysed reactions. It is essential that the rates of these should be properly co-ordinated, for a faulty adjustment in even one enzymic step is likely to have harmful consequences for the cell in which it occurs. Thus, suppose that we have a culture of exponentially growing cells in all of which there is a 'correct' (i.e. optimum) co-ordination of all enzymic steps. Now if in one of the cells a change occurs for some reason, which affects one Class II enzyme in such a way that an amino acid required for protein synthesis is underproduced to the extent of 20%, the supply of this amino acid will become rate-limiting for protein synthesis and hence for growth, the rate of which will consequently fall by 20%. Furthermore the enzymes and ribosome content of a cell are 'geared' to give the higher growth rate. In the new situation 20% of the cell's synthetic capacity is unused and the material and energy that went into making it initially are wasted.

If the cell had no regulatory mechanisms or if it failed to adjust the rates of its enzyme reactions, it would go on degrading glucose at the old rate and building up more building blocks than it could assimilate. These might leak from the cell and be used by other cells in the same culture. A single derangement could thus have three consequences: reduction in the *increase* of proto-plasm per unit time; reduction by a similar amount in the *yield* of protoplasm per unit of nutrient; utilization of part of the cell's synthetic capacity for producing materials that might be used for growth by its neighbours. It is clear that, with the passage of time, the progeny of the defective cell would form a smaller and smaller fraction of the total population. It is not surprising that in bacteria an elaborate system of control mechanisms has evolved to keep the enzymic processes in step with one another.

The rates of these reactions are controlled in two ways: in the first, the amount of an enzyme is unaltered but its activity is reduced; this is by specific *inhibition* (see below). In the second, the *amount* of enzyme is reduced. Frequently the same enzyme is subject to both types of control, i.e. it is reduced in amount and the enzyme that is formed functions at a reduced rate. The types of regulation can usefully be considered in relation to the classes of enzymes previously enumerated.

Regulation of Class I enzymes

Our generalized cell has the enzymic capacity for degrading glucose whatever the medium in which it is grown. Many enzymes are like this and are formed in quantity in any growth medium. The cell has, however, a group of latent enzymic capacities. These are formed *only when the substrate is present* and include the enzymes that degrade other carbohydrates, e.g. maltose, galactose, lactose. The process by which the presence of a substrate evokes the synthesis of the enzyme required to degrade it is called *induction*. Also, cells have the ability, when no sugars are available, to utilize certain amino acids as sources of carbon and energy. Again the enzymes are inducible. As a rule, induced enzymes convert substrates to compounds that can be utilized in the general metabolic pathways of the organisms. Thus lactose is split to glucose and galactose; maltose is split to glucose; serine is split to give pyruvate and this in turn is 'fed' into the tricarboxylic acid cycle; tryptophan can also be degraded to give pyruvate as a product.

The mechanism of induction in the case of the proteins induced by lactose can be satisfactorily explained on the basis of the scheme proposed by Jacob and Monod. It rests largely upon the implications of the following genetic evidence. Organisms which can produce an inducible enzyme may mutate to give cells which make the enzyme *constitutively*, i.e. the enzyme is produced in large amounts whether the substrate is present or not. Furthermore, in diploids it is found that the inducible character is dominant over the constitutive. Accordingly the inducible character is represented as i^+ and the constitutive as i^-. It is proposed that the i^+ strain produces an active repressor molecule, R, which may be a protein and which specifically prevents the transcription of the relevant structural gene into messenger-RNA. If the substrate is present it interacts with the repressor and makes it inactive, thus allowing free transcription and consequent synthesis of the enzyme. The constitutive strain, i^-, makes either defective molecules of R or none at all. In these cells, therefore, transcription of the structural gene into mRNA and synthesis of the enzyme go on all the time. The model is illustrated in Figure 1 and is discussed more critically in Chapter 8.

Enzymes of Class I are also subject to a second type of control, known as the 'glucose effect'. That is, when the cells are growing in glucose they will not form inducible enzymes, even when the inducers for these enzymes are present in the medium. A culture grown in the presence of glucose and some other substrate such as lactose will then grow in two phases—diauxic growth—(see Figure 4, p. 24), separated by a lag. In the first phase the glucose is used exclusively as the carbon source. Then the cells, having exhausted the glucose, stop growing altogether. During the lag period they synthesize the requisite enzyme. A second phase of growth then begins at the expense of the alternative substrate. Manifestly, glucose by its presence prevents the formation of the induced enzyme,

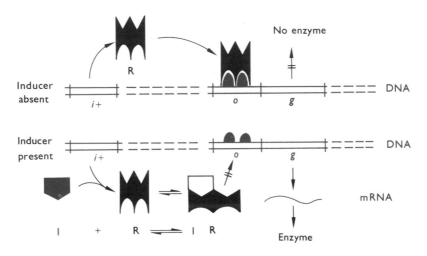

Figure I. Jacob and Monod's model of regulation of enzyme synthesis by repression. It is suggested that the gene controlling inducibility (i) codes for a repressor molecule (R) which interacts with the DNA and thus blocks transcription of the structural gene (g) for the enzyme. When inducer I (usually the substrate) is added it combines with R converting it to an inactive form. Transcription of the structural gene into mRNA and hence formation of the enzyme can now take place.

even though the substrate is present all the time. The glucose effect is also exerted by other 'good' growth substrates—good in this context meaning that a high rate of growth is observed when these compounds are present. The mechanism of the glucose effect will be discussed in some detail later (Chapter 8).

   The dual control of Class I enzymes by induction *and* the glucose effect means that (a) inducible enzymes are not formed unless the substrates are present; (b) even if they are present, the enzymes will not be formed if the cell is already supplied with glucose or some other good source of carbon.

Regulation of Class II enzymes

   Control by repression. Enzymes of this class synthesize the building blocks for the macromolecules (amino acids, nucleotides, etc.) and many of them are regulated both by alteration in the amount of enzyme synthesized and by inhibition of existing enzyme. We shall first consider the control of enzyme synthesis. Suppose a series of reactions leads from a precursor S_1 to a metabolite M_1 which might be an amino acid. S_2, etc. are intermediary metabolites, and E_1, etc. are the corresponding enzymes. An example is the formation of arginine from glutamate by a series of reactions:

$$S_1 \xrightarrow{E_1} S_2 \xrightarrow{E_2} S_3 \longrightarrow \longrightarrow \longrightarrow \longrightarrow M_1$$

(S_1 is itself derived from the action of a Class I type of enzyme.)

In general it is found that if M_1 is added to the growth medium, so that the cells no longer have to synthesize it, the formation of all the enzymes E_1, E_2, etc. stops almost immediately. The process of specifically preventing the formation of an enzyme or a group of metabolically related enzymes is referred to as *repression*. The element of specificity has been emphasized because antibiotics

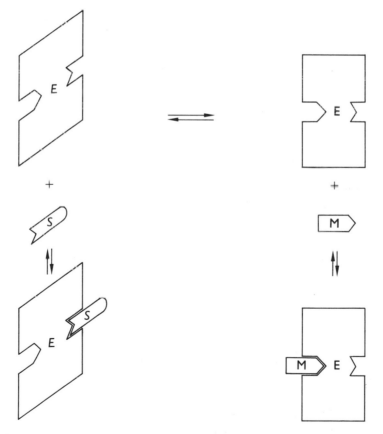

Figure 2. Schematic illustration of allosteric interaction. The enzyme molecule (E) has two recognition sites: one is specific for the substrate (S) and the other for a particular metabolic inhibitor (M). When M combines with E there is a conformational change which alters the recognition site for S.

and many other toxic substances often cause a general inhibition of enzyme formation.

Repression has been explained on the basis of the same sort of model as that used to explain induction (see Chapter 8).

Control by end-product inhibition. The allosteric effect. When M_1 is added to the growth medium it not only stops the synthesis of the enzymes E_1, E_2, etc. but *inhibits* the functioning of the first enzyme E_1. Again, the effect is highly specific and it is immediate, so that as soon as M_1 is available the cells cease to make it. This is often called feedback inhibition and it is formally analogous to a simple thermostatic regulation system.

It is important to note that the enzyme E_1 must have *two* specific recognition sites. One is the active site for its substrate, S_1, and the other, equally specific, is the site for the end-product inhibitor M_1. If S_1 and M_1 were chemically similar substances it would be reasonable to assume that they competed for the same site on the protein molecule. This would be an *isosteric* interaction. In fact they are usually so dissimilar that it is reasonable to suppose that their interaction is *allosteric* and it is assumed that when M_1 is attached to the enzyme E_1 the protein molecule is distorted so that S_1 no longer fits properly (Figure 2).

This description of end-product inhibition is over-simplified because the same precursor, e.g. glutamate or aspartate (see Figure 1, p. 30) is often the starting point for the synthesis of several end-products. In branched pathways of this sort the situation is more complex and will be discussed in Chapter 8.

The mechanism of end-product inhibition ensures that the materials of the environment will not be wasted in the manufacture of end-products which are already provided from the outside.

Regulation of Class III enzymes

These enzymes assemble the macromolecules of the cell and it is the macromolecules, i.e. polysaccharides, proteins, DNA, etc., that will differ from those of all other types of bacteria. These substances *must* be assembled by the cells; they cannot just be picked up from the environment. Every growing cell will thus have all the enzymes of Class III and, on *a priori* grounds, it seems unlikely that these enzymes will vary greatly in amount. Their functioning is more likely to be controlled by alterations in the rate of reaction. Not much is known about regulation of these enzymes but some speculations will be mentioned in Chapter 8.

To summarize, we can say that we have a fairly good idea of the way in which the information systems built into the cell regulate the degradation and the biosynthesis of small molecules and we can see that the whole arrangement is such as to give the maximum yield of bacterial protoplasm in the minimum time. What we do not yet know is how the cell regulates the synthesis of its macromolecules, nor do we understand the way in which information, presumably stored in the DNA, is translated into structure. For instance, we do not know why the organism has its characteristic size and shape, nor do we know how certain types of enzymes come to be associated with the membrane while others are found in the ribosomes, nor do we know why the cell divides when it gets to a certain critical size.

Differentiation: sporogenesis and germination

So far we have considered the co-ordination of metabolism in actively growing bacteria. However, exponential growth is unlikely to continue for very long without some nutrient becoming growth-limiting (see Section 2). When this happens a population of bacteria can react in one of two ways. In most species, the cells will make use of endogenous reserves of carbohydrate and then begin to degrade their proteins and nucleic acids and combust the products to obtain the energy necessary to maintain the integrity of the cell. Death and lysis of the cells may then ensue.

In other species deprivation of nutrients acts as a stimulus for *sporogenesis* — a sequence of events in which a spore develops within the cytoplasm and is then liberated. Briefly the process consists of the enclosing of a complete copy of the genome together with proteins and RNA within a laminated structure which is refractile. The rest of the cell disintegrates leaving the free spore which contains virtually no water, is resistant to heat and to many sorts of toxic reagents, and which can remain in a dormant state for years. When the spore is restored to a nutrient medium *germination* takes place: the refractility disappears and so do the properties of resistance to heat and to toxic chemical substances. Much of the structure of the spore breaks down, the enzymes that are needed for growth are synthesized and in due course a cell of normal shape and dimensions grows out of what is left of the spore.

The morphological changes that we have described can be regarded as a very primitive form of differentiation.

Part II

Chapter 1

The Bacterial Cell: Major Structures

The first part of this book is a brief account of the way in which bacterial cells are put together. We shall now examine the processes in more detail following the same sequence so that the succeeding chapters correspond more or less to the previous sections.

In this first chapter the anatomy of bacterial cells is considered as well as something of the chemistry of the components. The aim is to describe these constituents and to show what function they perform in the normal, healthy bacterium.

Investigations have proceeded along two main lines: firstly, the electron microscope has revealed details of surface structures and appendages in a way hitherto impossible; and secondly, biochemical investigations of the whole cell or isolated parts of it have shown us the chemical material of which the individual structures are composed. An attempt is frequently made to relate the microscopical information obtained from the study of sections with the biochemical results obtained from studies of cell fractions and so to build up a co-ordinated picture of function and structure in the intact cell.

Not all the structures revealed by the electron microscope and by other means appear to be necessary for survival of the bacterium. By comparing different kinds of bacteria it is possible to arrive at the lowest common denominator in terms of those structures that are necessary for growth and division. These 'essential' structures are depicted diagrammatically in Figure 1a, together with other structures which are found in some, but not all, bacteria (Figure 1b). Thus, with the exception of a few specialized forms, bacteria have a *wall*, which varies considerably in chemical composition and fine structure from one organism to another: this wall surrounds the *cytoplasm* which is bounded by a delicate *cytoplasmic* or *protoplast* membrane. In electron micrographs the cytoplasm has a granular appearance resulting from the presence of a large number of *ribosomes*. Within the cytoplasm the *nuclear area* (*chromatinic body, bacterial nucleoid*) can be seen in thin sections. The nucleus has a very fine fibrillar network but unlike the animal or the plant cell nucleus is not separated from the cytoplasmic contents by a nuclear membrane.

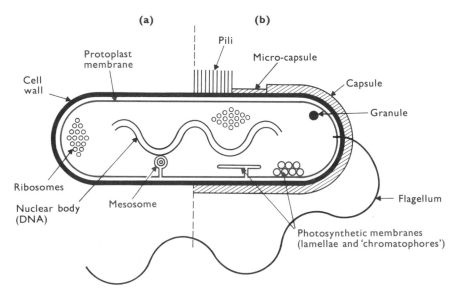

Figure 1. Cross-section of a generalized bacterial cell showing (a) essential and (b) inessential structures.

In all bacteria, the structure of which has been studied extensively, invaginations of the cytoplasmic membrane occur which are termed *mesosomes* or *chondrioids*; these are frequently found associated with the nucleus and at the sites of septation and spore formation. Various *granules* (lipid, glycogen, sulphur, metaphosphate or volutin) may be found in the cytoplasm, and in photosynthetic bacteria membranous lamellae or membrane-bounded vesicles can be seen: these are connected to the cytoplasmic membrane and are the organelles in which photosynthesis occurs.

The cell wall may be covered with thin threads termed *pili* (also known as *fimbriae*): these are of several types, one of which is involved in sexual conjugation, but little is known of the function of the remaining types. Motile bacteria have at least one *flagellum*; flagella are very long thin structures composed of protein sub-units. Many bacteria have a *capsule* external to the cell wall and this structure is sometimes many times thicker than the bacterium itself. Some bacteria produce a slime which in a few instances appears to have an ordered structure. The location of these 'inessential' structures is shown in Figure 1b.

Electron microscopy

Although bacteria are easily visible when viewed by phase-contrast microscopy little information concerning the anatomical structures within them can be gained other than by electron microscopy. Three techniques are used routinely with this instrument to investigate bacterial structure: negative staining enables the topography of the whole cell or isolated parts of it to be examined, thin sectioning reveals cellular profiles, and freeze-etching permits the examination of various surfaces in the bacterium and of cross-sectional views of the cell wall, protoplast membrane and any organelles present.

Negative staining involves mixing the material to be examined with an electron-dense material such as sodium phosphotungstate and drying down a small quantity of the mixture on a grid. Regions of the cell which the stain cannot penetrate are electron-transparent and are viewed against the electron-dense background.

The preparation of thin sections of bacteria is preceded by a number of manipulative steps which include drastic fixation, dehydration and embedding processes. Any one of these might introduce artifacts which may confuse those who attempt to interpret the resulting electron micrographs.

The relatively new technique of freeze-etching aims at overcoming some of the criticisms that can be applied to earlier methods and at providing high resolution pictures of the cell surface. The technique involves rapid freezing of a small sample to $-170°$ followed by fracture of the specimen with a cold knife under high vacuum (freeze-fracturing). The cut surface is then etched by raising the temperature for a short time to $-100°$, shadowed with a heavy metal and a carbon replica is made. The specimen is dissolved away and the replica examined in the electron microscope. The preparation is not subjected at any stage to chemical procedures such as fixation and staining. The fracture plane frequently runs along the surface of a wall or membrane component and the technique is therefore useful for the examination of surfaces of cells and of their organelles. Since viable bacteria have been recovered from specimens which have undergone the freeze-etching process but which have not been fractured it is claimed that artifacts are less likely than with thin sectioning or negative staining procedures.

The appearance of the bacterium *Escherichia coli* after subjection to these three electron microscopical techniques is shown in Figure 2. Negative staining reveals the convoluted appearance of the outer cell wall layer but little else, thin sectioning shows internal components and gives some idea of the thickness of the cell wall layers, while freeze-etching reveals the appearance of the surface of the protoplast membrane and cell wall in much greater detail than can be seen by negative staining. The various processes must affect the shape of the bacterium and Figure 3 gives an idea of how the cross-sectional shape may alter depending on the processes used. The removal of water during fixation before

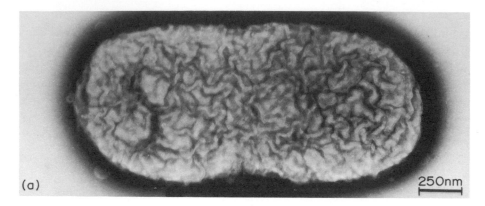

(a) 250nm

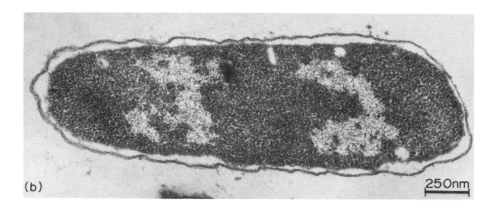

(b) 250nm

(c) 250 nm

Figure 2. Comparison of *Escherichia coli* from logarithmically-growing cultures after (a) negative staining, (b) fixation, dehydration, embedding and thin sectioning, (c) freeze etching (the intact cell surface is seen on the left, the cytoplasmic membrane is exposed in the centre and the cytoplasm on the right of the picture. Reproduced with permission of M. E. Bayer and C. C. Remsen. *J. Bacteriol.* **101** (1970) 304.

thin sectioning also results in a drastic shrinkage of the cellular material (possibly by as much as 30%). It must therefore be appreciated that an electron micrograph may be akin to an artist's impression of the cell.

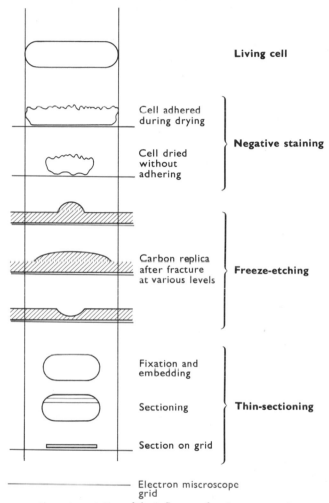

Figure 3. Diagrammatic representation of the influence of various preparative procedures on the dimensions of bacteria.

The earlier electron micrographs tended to show a limited range of bacteria which were the subjects of choice for biochemical experiments. The examination of a wider range of organisms has emphasized the diversity of structures found in bacteria—particularly the arrangement of macromolecules in the various layers of the wall. It is important not to make generalizations on the basis of what is seen in one particular electron micrograph or to assume that the structural arrangement present will be found in other bacteria of the same genus. The selection of micrographs in this chapter should illustrate this.

BACTERIAL CELL WALLS

Appearance in the electron microscope

Although there is considerable variability between genera and species, bacteria are placed in one of two main groups—referred to as Gram-positive and Gram-negative. This classification is based on the ability of some bacteria to retain the basic dye crystal violet when they are washed with ethanol: the distinction between the two groups (Gram-positive bacteria retain the stain, Gram-negative do not) is thought to be dependent on some difference in the chemistry of the surface layers.

Gram-positive bacteria

In thin sections of Gram-positive bacteria the wall appears as an amorphous structure 15–80 nm thick, apparently lacking fine structure (Figure 4). A number of layers can sometimes be distinguished on the basis of their electron-transparency (Figure 5) the thickness of each layer varying with the fixative and stain used but it is not clear whether this is due to a genuine difference in the chemistry of the layers or is an artifact of fixation or staining. When the surface of different bacteria within a single genus is observed by freeze-etching many different types are found—the wall of *Bacillus megaterium* apparently lacks fine structure

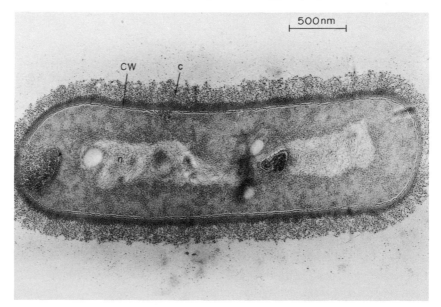

Figure 4. Thin section of *Bacillus megaterium* showing the amorphous cell wall (cw) and capsule (c). The electron-transparent nuclear material (n) is surrounded by the densely-stained ribosomes. Reproduced with permission of D. J. Ellar, D. G. Lundgren and R. A. Slepecky. *J. Bacteriol.* **94** (1967) 1189.

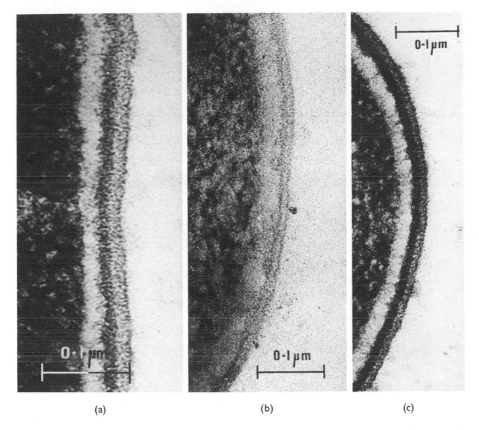

(a)   (b)   (c)

Figure 5. Cell wall profile of *Bacillus polymyxa* after fixation with (a) glutaraldehyde—post-stained with uranyl acetate and lead citrate, (b) glutaraldehyde—post-stained with 1% OsO_4 followed by lead citrate, (c) glutaraldehyde—post-stained with 4% OsO_4 for 30 min., then 1% uranyl acetate followed by lead citrate. Reproduced with permission of M. V. Nermut and R. G. E. Murray. *J. Bacteriol.* **93** (1967) 1949.

(Figure 6), an observation which is in accord with pictures obtained by shadowing isolated cell walls of the same species (Figure 7) while that of *B. anthracis* contains a number of structured layers overlying the non-structured inner wall layer (Figure 8 and inset).

Gram-negative bacteria

The walls of these organisms are extremely complex both chemically and structurally. In thin section a five-layered wall can be seen (3 electron-dense and

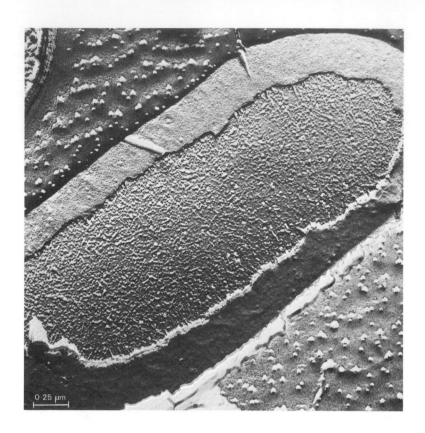

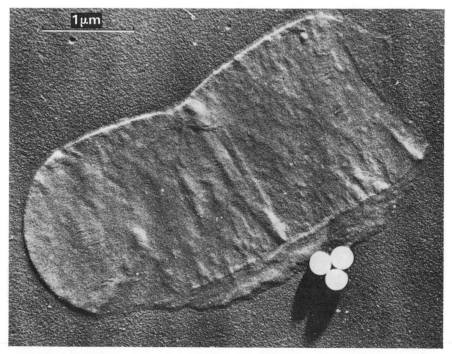

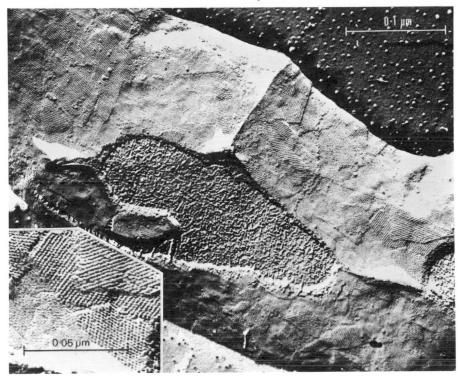

Figure 8. Freeze-etched preparation of *Bacillus anthracis*. The outer layer of the cell wall is composed of particles under which the inner layer, lacking fine structure, is visible. The inset shows that at least three structural layers are present above the inner layer. Reproduced with permission of S. C. Holt and E. R. Leadbetter. *Bact. Rev.* **33** (1969) 346.

2 electron-transparent layers) outside the three layers of the cytoplasmic membrane (Figures 9 and 10). In some organisms the innermost of the three dense layers is thickened so that no electron-transparent layer is evident between the L5 and L3 layers; it is this L5 layer whether thick or thin which disappears when organisms are treated with lysozyme and ethylenediamine tetra-acetic acid, leaving behind the outermost three layers of the wall (Figure 11). Lysozyme is an enzyme which hydrolyses bonds in the peptidoglycan, the rigid component of the wall (see page 87) so the L5 layer may be composed, at least partially, of this polymer. The outermost three-layered structure of the envelope is sometimes

Facing page top
Figure 6. Freeze-etched preparation of *Bacillus megaterium* showing an apparent lack of fine structure in the cell wall. Reproduced with permission of S. C. Holt and E. R. Leadbetter. *Bact. Rev.* **33** (1969) 346.

Facing page bottom
Figure 7. Isolated cell wall of *Bacillus megaterium*. Reproduced with permission of M. R. J. Salton and R. C. Williams. *Biochim. Biophys. Acta* **14** (1955) 455.

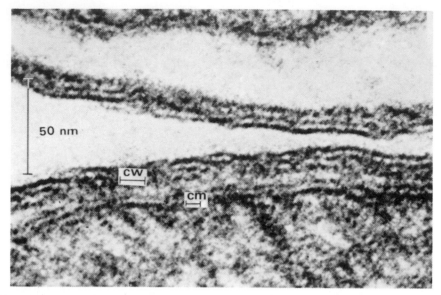

Figure 9. Thin section of cell envelope of *Escherichia coli* showing the five layers of the cell wall (cw) and three layers of the cytoplasmic membrane (cm). In the upper cell the wall has become detached from the membrane. Reproduced with permission of S. de Petris. *J. Ultrastructure Res.* **12** (1965) 247.

convoluted (Figure 12, see also Figure 2a) and can be removed with hot phenol; it has been shown to contain lipopolysaccharide and lipoprotein.

It seems probable that wall constituents extend further than the outermost electron-dense layer. In micrographs showing bacteria joined together there is a gap about 2–4 nm wide which is not found in mutants of the same strain lacking the 'O' specific chains of the lipopolysaccharide (see page 95).

Walls of many Gram-negative bacteria show a structured appearance (Figure 13) but it should not be assumed that the ordered substructure frequently seen in freeze-etched specimens (Figure 14a) is necessarily on the surface of the three-layered outer wall structure. A thin section of this organism (Figure 14b)

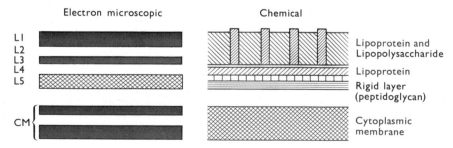

Figure 10. Diagrammatic representation of the cell envelope of *Escherichia coli* based on electron microscopic observations and chemical analysis of the isolated macromolecules.

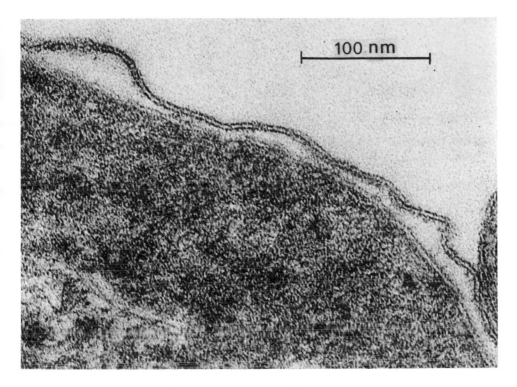

Figure II. Thin section of lysozyme-EDTA spheroplast of *Escherichia coli*. The innermost layer of the cell wall has been removed by lysozyme leaving the outer wavy membrane that is now separated by a gap from the cytoplasmic membrane. Reproduced with permission of S. de Petris. *J. Ultrastructure Res* **19** (1967) 45

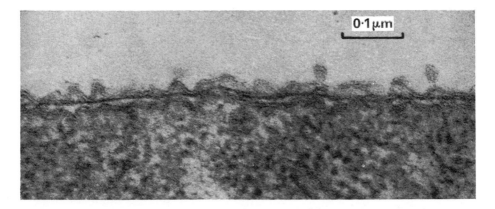

Figure 12. Thin section of *Spirillum serpens* in which portions of the outer layers of the wall are not in close contact with the rigid layer of peptidoglycan. Reproduced with permission of R. G. E. Murray. *Canad. J. Microbiology* **11** (1965) 547.

Figure 13. Cell wall of *Spirillum serpens* showing regular array of subunits. Reproduced with permission of R. G. E. Murray, in *The Bacterial Cell Wall*, by M. R. J. Salton, p. 74, Fig. 26d. Elsevier, London.

indicates that an additional component is present consisting of 'wine-glass'-shaped subunits attached to a common base layer. It is not clear whether this type of component (which is found also in some other bacteria) should be regarded as part of the wall or classed as a capsule, although 'classical' capsules never show such a high degree of ordered sub-structure.

The layered arrangement of the polymers of Gram-negative walls can also be deduced from experiments in which components of the wall are selectively removed by chemical or enzymic treatments followed by analyses of the products. This demonstrates the presence of an outermost layer of lipoprotein and lipopolysaccharide and an inner rigid layer which retains the shape of the bacterium and is composed of peptidoglycan with lipoprotein covalently attached to it. A diagrammatic representation of the arrangement of polymers based on the available evidence is given in Figure 10.

Halobacteria
These organisms form a special group in that they grow in high concentrations of salts (e.g. 4 *M* NaCl). Removal of salt results in dissolution of the wall which explains why this structure was not seen in earlier electron micrographs. The wall structure is preserved if salt is included in the fixative before sectioning (Figure 15). The wall is amorphous, rather like that of Gram-positive bacteria

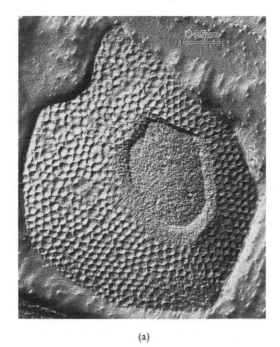

(a)

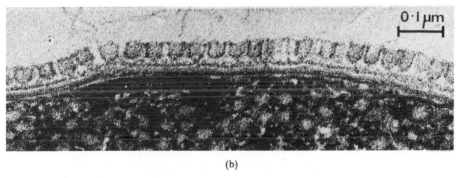

(b)

Figure 14. Cell wall of *Chromatium buderi*. (a) Freeze-etched preparation showing the outer surface composed of tightly-packed cup-shaped units. (b) Thin section showing the multi-layered nature of the cell wall with the outermost layer corresponding to 'wine-glass' shaped subunits. Reproduced with permission of C. C. Remsen, S. W. Watson and H. G. Truper. *J. Bacteriol.* **103** (1970) 254.

but it differs in that it does not contain the typical peptidoglycan constituents which make up 50% or more of the weight of the Gram-positive wall.

Isolation of cell walls

Before cell walls could be studied chemically they had to be isolated and freed from other components of the cell. The technique used most frequently is that

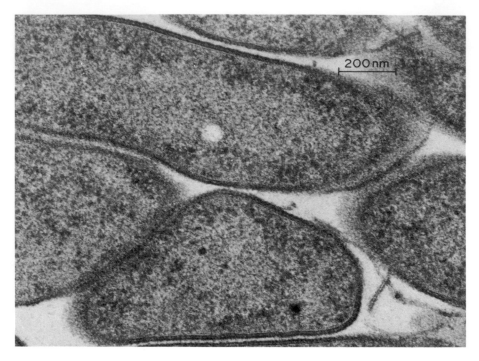

Figure 15. Thin section of *Halobacterium halobium* showing the amorphous appearance of the cell wall. Reproduced with permission of W. Stoeckenius and R. Rowen. *J. Cell. Biol.* **34** (1967) 365.

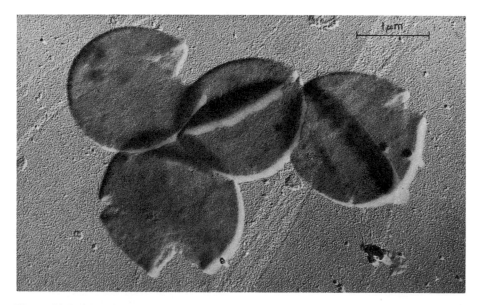

Figure 16. Isolated cell wall of *Staphylococcus lactis*. Reproduced with permission of J. Baddiley. *Proc. Roy. Soc. B* **170** (1968) 331.

of mechanical disintegration of the cells followed by several cycles of differential centrifugation to separate the walls from the heavier unbroken bacteria and the lighter ribosomes and other cytoplasmic constituents. The particular method of disintegration that is employed affects the final size of the cell wall fragments. Thus, mechanical agitation with glass beads may result in only one tear in the wall and the consequent production of one large fragment per cell, while sonic oscillation produces much smaller fragments. Cell walls often appear flattened when shadowed preparations are examined but retain the overall shape of the organism (Figures 7 and 16) suggesting that it is the wall which contains the rigid components which determine the shape of the bacterium. In general the walls of Gram-positive bacteria are easier to separate and purify than those from Gram-negative organisms which tend to remain associated with the cytoplasmic membrane: consequently preparations from Gram-negative organisms are generally referred to as the 'cell envelope fraction'.

Chemistry

Chemical analysis of isolated cell walls has revealed the complexity and heterogeneity of the individual components. In general the walls of Gram-negative bacteria have a high lipid and low amino sugar content (10–20% and 2–5% of the dry weight respectively) and contain the full range of amino acids found in proteins, while those from Gram-positive organisms have little or no lipid, have a high content of amino sugar (15–20%), but sometimes only a limited range of amino acids. Until it was realized that proteins were present in the walls of many Gram-positive bacteria purification of the isolated walls often included treatment with a proteolytic enzyme which explains why the full range of amino acids was not found in hydrolysates. Protein where present forms only a low percentage of the total weight of the wall; consequently the amino acids present in peptidoglycan are present at several times the concentration of the other

Table 1. Principal components of bacterial cell walls

Component	Gram-positive cell wall	Gram-negative cell envelope
Peptidoglycan	+	+
Teichoic acid and/or Teichuronic acid	+	−
Polysaccharide	+	+
Protein	± (not all)	+
Lipid	−	+
Lipopolysaccharide	−	+
Lipoprotein	−	+

amino acids. The principal polymers found in cell walls are shown in Table 1.

Peptidoglycan (mucopeptide, glycosaminopeptide, murein)

Structure

Although peptidoglycan is present in all bacteria living in hypotonic environments the proportion of the dry weight of the wall which it forms varies considerably. The cell wall itself may comprise 20% of the dry weight of the organism and the amount of peptidoglycan in it may range from 50–80% in most Gram-positive bacteria (e.g. Bacilli, Staphylococci) to 1–10% in Gram-negative organisms (e.g. Salmonellae, *Escherichia coli*). A characteristic pattern

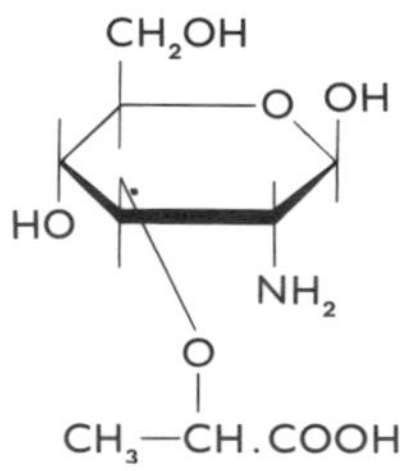

Figure 17. Muramic acid.

of amino sugars and amino acids is found in hydrolysates of purified peptidoglycan and of these *N*-acetyl muramic acid (Figure 17), diaminopimelic acid (DAP) and the D-isomers of glutamic acid and alanine are found uniquely in this structure. Consequently their presence in bacteria, blue-green algae and Streptomyces spp. has been used as evidence of the occurrence of peptidoglycan in cell walls of these micro-organisms and of their close evolutionary relationship. Structural studies on peptidoglycans from a number of bacteria indicate that the polymer is made up of a polysaccharide-type backbone to which short peptide chains are linked: some or all of these peptide chains are joined together either directly or by other short peptide chains.

Two amino sugars are present in the polysaccharide backbone, *N*-acetylglucosamine and its 3-*O*-D-lactyl derivative, *N*-acetylmuramic acid, in alternating sequence with all the linkages $\beta 1$–4. It is this portion of the polymer which is sensitive to lysozyme, an enzyme which hydrolyses the $\beta 1$–4 link between *N*-acetylmuramic acid and *N*-acetylglucosamine (Figure 18) and the backbone is therefore split into disaccharide units.

The peptide chains linked to the carboxyl group of muramic acid contain four amino acid residues and have the following sequence —L-Ala—D-Glu—R—D-Ala. Exceptions to this include the substitution of L-serine or glycine for L-alanine in some bacteria. The amino acid R can be *meso*- or LL-diaminopimelic acid, L-lysine, L-ornithine, L-diaminobutyric acid or L-homoserine; it is in-

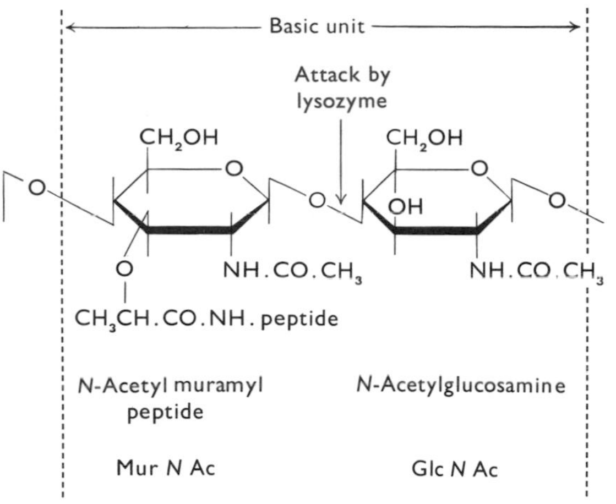
Basic unit
Attack by lysozyme
CH₂OH
O
CH₂OH
O
O
O
OH
O
O
NH.CO.CH₃
NH.CO.CH₃
CH₃CH.CO.NH.peptide
N-Acetyl muramyl peptide
N-Acetylglucosamine
Mur N Ac
Glc N Ac

Figure 18. Polysaccharide backbone of cell wall peptidoglycan showing linkage attacked by lysozyme.

variably the L-centre of the amino acid which takes part in the peptide link and with the exception of homoserine all the amino acids at this position have two amino groups. The structure of the tetrapeptide subunit is given in Figure 19. The α-COOH group of the D-glutamic acid residue is sometimes present as the amide or can be combined with glycine in some *Micrococci*.

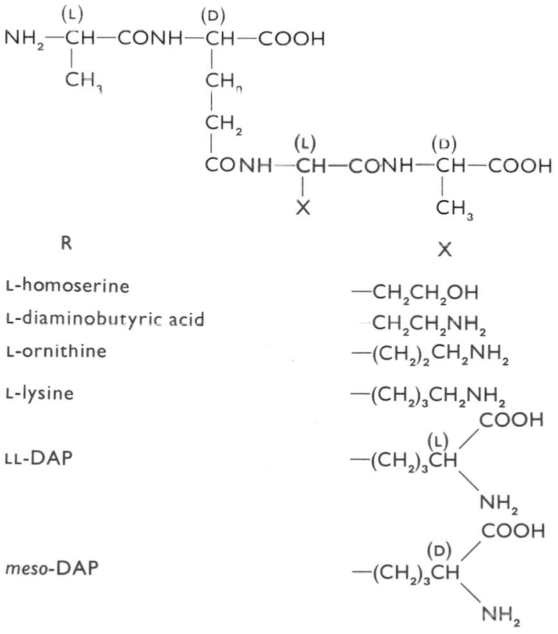
(L)
(D)
NH₂—CH—CONH—CH—COOH
CH₃
CH₂
CH₂
CONH—CH—CONH—CH—COOH
(L)
(D)
X
CH₃
R
X
L-homoserine
—CH₂CH₂OH
L-diaminobutyric acid
—CH₂CH₂NH₂
L-ornithine
—(CH₂)₂CH₂NH₂
L-lysine
—(CH₂)₃CH₂NH₂
LL-DAP
—(CH₂)₃CH
(L)
COOH
NH₂
meso-DAP
—(CH₂)₃CH
(D)
COOH
NH₂

Figure 19. General structure of the tetrapeptide subunit L-Ala—D-Glu—R—D-Ala. Redrawn from J. M. Ghuysen and M. Leyh-Bouille. *FEBS Symposium* **20** (1970) 59.

The variation in structure of peptidoglycans from different species arises from the manner and degree of cross-linking the tetrapeptide chains. There are thought to be four types of cross-linking but the terminal D-alanine of one chain is always involved. The first type may be common to all Gram-negative bacteria and consists of a direct linkage from D-alanine to the amino group on the D-carbon atom of *meso*-DAP in another chain (Figure 20). This type is found

Figure 20. Peptidoglycan of *Escherichia coli* (Type 1). A direct cross-link (in red) joins the tetra-peptides. Redrawn from J. M. Ghuysen and M. Leyh-Bouille. *FEBS Symposium* **20** (1970) 59.

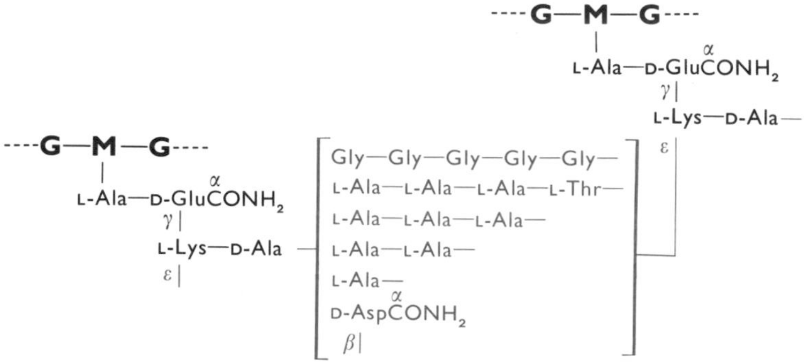

Figure 21. Peptidoglycan of many Gram-positive bacteria (Type 2). A short peptide chain or a single amino acid links the tetrapeptides. Redrawn from J. M. Ghuysen and M. Leyh-Bouille. *FEBS Symposium* **20** (1970) 59.

also in many of the Gram-positive bacilli. The second type of cross-linkage is present in the majority of Gram-positive bacteria examined and involves a short peptide or a single amino acid extending from the D-alanine of one chain to the free amino group of a diamino acid in another chain (Figure 21). The penta-glycine cross-link found in *Staphylococcus aureus*, one of the first organisms whose wall structure was carefully investigated, comes in this category. The walls of *Micrococcus lysodeikticus* have a third type of cross-link, again extend-ing from a C-terminal D-alanine to the ε-amino group of a lysine residue as in some type 2 linkages, but comprising varying amounts of a short peptide of the same composition as the peptide substituting the muramic acid (Figure 22). When no di-amino acid is present in the tetrapeptide the cross-link is between the D-alanine and the α-COOH group of glutamic acid and consists of a di-amino acid (either D-lysine or D-ornithine (Figure 23)).

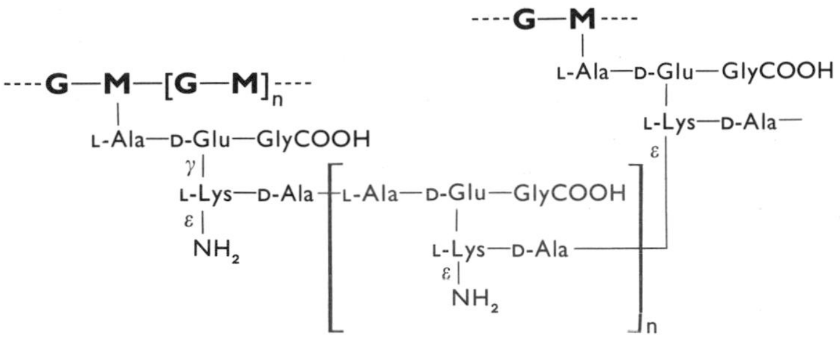

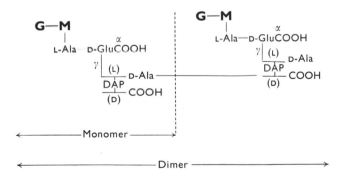

Figure 22. Peptidoglycan of *Micrococcus lysodeikticus* (Type 3). The peptide chain linking the tetrapeptides has the same composition as the tetrapeptide. Redrawn from J. M. Ghuysen and M. Leyh-Bouille. *FEBS Symposium* **20** (1970) 59.

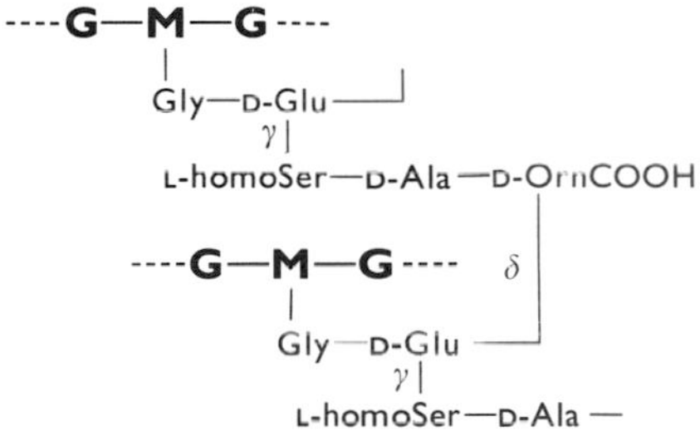

Figure 23. Peptidoglycan of *Corynebacterium poinsettiae* (Type 4). The tetrapeptide does not contain a dibasic amino acid; consequently the cross-link is between two free carboxyl groups. Redrawn from J. M. Ghuysen and M. Leyh-Bouille. *FEBS Symposium* **20** (1970) 59.

Figure 24. Monomer and dimer units obtained from the peptidoglycan of *Escherichia coli* by digestion with lysozyme.

Because the degree of cross-linking in the peptidoglycan affects the size of fragments resulting from digestion of the polymer with lysozyme, the types and proportions of such fragments give an insight into the structure of the intact polymer. A high proportion of monomers and dimers (Figure 24) is indicative of a low degree of cross-linkage as in bacilli and most Gram-negative bacteria;

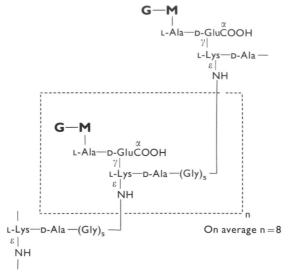

Figure 25. Fragment of Staphylococcal wall obtained after digestion with lysozyme.

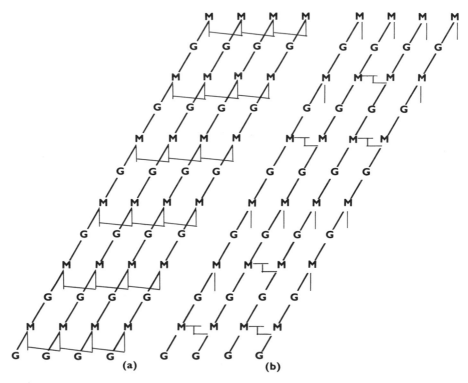

(a)   (b)

Figure 26. Diagrammatic representation of a single layer of peptidoglycan from (a) *Staphylococcus aureus* and (b) *Escherichia coli*. (a) Redrawn from J. M. Ghuysen, J. L. Strominger and D. J. Tipper, in *Comprehensive Biochemistry,* **26A** (1968) 53. Eds. M. Florkin and E. H. Stotz. Elsevier, Amsterdam, London and New York. (b) Redrawn from J. M. Ghuysen. *Bact. Rev.* **32** (1968) 425.

the larger fragments obtained from Staphylococcal walls (Figure 25) suggest that cross-linkage is almost complete. The logical conclusion from these observations is that the peptidoglycans of *Staphylococcus aureus* and *Escherichia coli* have the structures shown in Figure 26 (although only a single layer is depicted here). This diagram demonstrates the potentiality for size and rigidity of the polymer and it is possible that there is only one molecule per cell, hence the name 'murein sacculus'. Measurements of the thickness of the peptidoglycan layer from electron micrographs are consistent with the idea that Gram-negative polymers contain a single layer of peptidoglycan while the amount in Gram-positive organisms suggests that several layers are present but it is not clear how these are linked together.

Biosynthesis

The difficulty of assembling such a large structure outside a relatively impermeable cytoplasmic membrane is overcome by splitting the process into three main stages. In the first the precursors are synthesized in the cytoplasm by soluble enzymes. They are then transported through the membrane by a lipid-soluble carrier, sometimes further modified by membrane-bound enzymes, inserted at the growing point(s) of the wall, and cross-linked to other peptide chains to preserve the integrity of the overall structure. In detail the process is as follows. *N*-acetyl glucosamine-1-phosphate reacts with uridine triphosphate with the formation of UDP-*N*-acetylglucosamine, one of the two precursors. UDP-*N*-acetylmuramic acid is formed by the reaction of phosphoenolpyruvate with UDP-*N*-acetylglucosamine and reduction of the UDP-*N*-acetyl-3-*O*-enol-pyruvylglucosamine so formed. The peptide chain attached to the lactic acid residue of UDP-*N*-acetylmuramic acid is built up by the sequential addition of amino acids. Each amino acid is added by a specific enzyme, with the exception that the terminal dipeptide, D-alanyl-D-alanine, is added as a single unit (Figure 27); the enzymic reactions are dependent on Mn^{2+} and use ATP as an energy source. Organisms containing DAP have an enzyme which is inactive if lysine replaces DAP in the cell-free assay system: similarly, the enzyme from cells having lysine in the wall will not add DAP to UDP-Mur*N*Ac-L-ala-D-glu. The specificity of the D-alanyl-D-alanine ligase is not absolute but the preponderance of the true substrate under normal conditions is presumably sufficient to ensure that few mistakes are made in the synthesis of either the dipeptide or the UDP-Mur*N*Ac-pentapeptide.

The stepwise build-up of peptidoglycan has been studied in cell-free systems from a number of species and the postulated pathway is shown in Figure 28. The enzyme preparations consist of pieces of disintegrated membrane but may contain some wall material. In some preparations ribosomes are present but are considered unnecessary since treatment with RNA-ase does not affect the activity of the membrane fragments in peptidoglycan synthesis. Phosphoryl-

MurNAc-pentapeptide is first transferred with the release of UMP to an acceptor present in the membrane fragments. This has been identified as the monophosphate derivative of a C-55 polyisoprenoid alcohol (Figure 29). A transglycosylation reaction in which N-acetylglucosamine is transferred from UDP-GlcNAc results in the production of a disaccharide-pentapeptide derivative of the polyisoprenoid alcohol pyrophosphate. It is clear that this material (in some cases slightly modified) represents the subunit structure of the peptidoglycan. Modifications, such as the amidation of the α-COOH group of glutamic acid or the insertion of amino acids which are found in the cross-linking peptide generally occur at this stage. It is interesting that these amino acids are generally inserted from the corresponding amino acyl-tRNA although ribosomes and messenger-RNA are not involved. It is assumed that the acceptor for the final

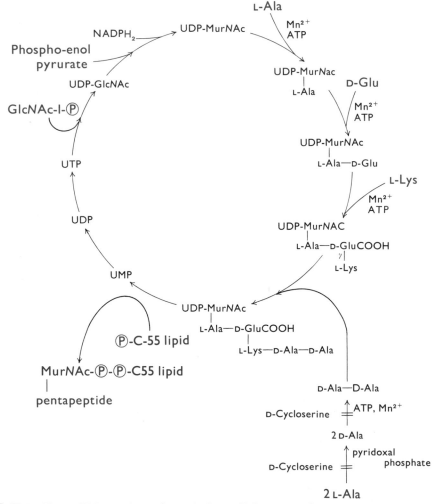

Figure 27. Biosynthesis of peptidoglycan. (I) Formation of nucleotide precursors.

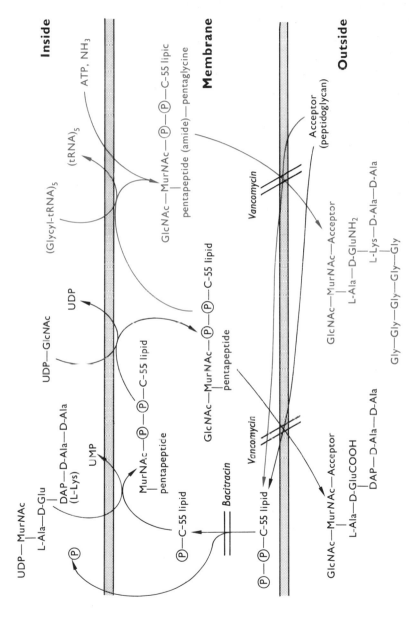

Figure 28. Biosynthesis of peptidoglycan. (II) Incorporation of precursors into linear glycan chains by membrane-bound enzymes. The pathway printed in black is believed to be common to all bacteria. The reactions in red indicate modifications to the pentapeptide chain which occur in some organisms.

polymerization reaction is the growing point of the peptidoglycan although pure membrane components catalyse this reaction *in vitro*. This reaction presumably takes place on the outside surface of the cytoplasmic membrane and it seems that the C-55 phospholipid transports the precursors across the membrane. The same C-55 phospholipid or one closely related to it has been shown to be involved in the synthesis of the 'O' specific chains of lipopolysaccharides (p. 98), a mannan polymer in *Micrococcus lysodeikticus*, a capsular polysaccharide, and

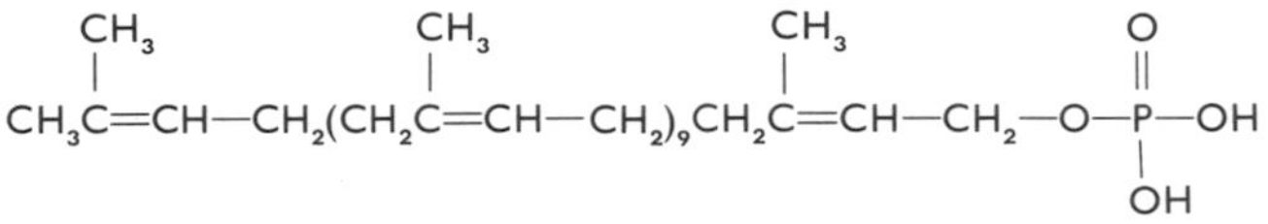

Figure 29. C-55-polyisoprenoid alcohol.

a wall teichoic acid from a staphylococcus (p. 92). These observations emphasize the importance of this compound in the synthesis of polymers outside the cytoplasmic membrane. The polymerization reaction releases the C-55 polyisoprenoid alcohol pyrophosphate which is dephosphorylated to complete the cycle.

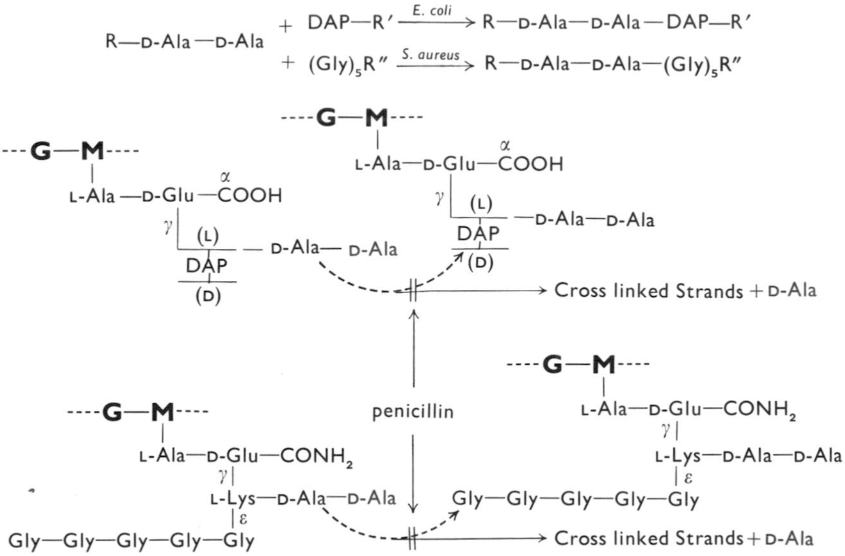

Figure 30. Biosynthesis of peptidoglycan. (III) The bridge closure reaction in (a) *Escherichia coli* and (b) *Staphylococcus aureus*.

The formation of the peptide bond in the cross-link must occur outside the membrane where ATP is not available. The nascent peptidoglycan chains at the end of stage 2 in the biosynthesis contain two D-alanine residues at the C-terminal ends of the peptides and it is believed that the terminal D-alanines are removed by an enzyme which is able to utilize the bond energy to effect the cross-link (Figure 30). This type of reaction gives rise to the structure shown in Figure 20 (*E. coli*) or Figure 21 (*S. aureus*).

Importance of peptidoglycan

The fact that the cell walls of all bacteria (except halobacteria and mycoplasmas) contain peptidoglycan and that it is often the major component of the wall, amounting to as much as 80% of the dry weight, is strong circumstantial evidence that this component is virtually indispensable. The semi-permeable cytoplasmic membrane lying immediately beneath the wall is so fragile that it has to be protected by the cell wall against the osmotic forces exerted within the protoplast. The osmotic pressure may be as much as 25 atmospheres in a Gram-positive coccus and about 5 in a Gram-negative rod. The wall must be rigid and there is abundant evidence that the peptidoglycan component confers this property on the cell wall as a whole.

Enzymic attack by lysozyme: formation of protoplasts

Lysozyme, obtained from egg-white, hydrolyses $\beta1$–4 links between *N*-acetylmuramic acid and *N*-acetylglucosamine. Certain bacteria (e.g. *Bacillus megaterium*) are lysed by this enzyme unless the treatment is carried out in hypertonic media. Under these conditions protoplasts are formed (Figure 31) which lyse

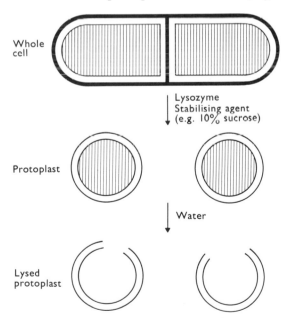

Figure 31. Formation of protoplasts from *Bacillus megaterium* by treatment with lysozyme. The cell wall surrounding the protoplast membrane has been removed by the enzymic digestion. Redrawn from McQuillen. *The Bacteria*, Volume I, p. 249 (p. 262).

on dilution with water. Degradation of the peptidoglycan therefore results in the disappearance of the cell wall and the loss of rigidity. The cytoplasm, still surrounded by the membrane, becomes spherical and is referred to as a *proto-plast*. These structures do not have any of the wall components adhering to them. As far as can be ascertained they are capable of carrying out most of the biosynthetic processes occurring in the intact bacteria including the synthesis of nascent peptidoglycan chains but only under very special circumstances do they revert to the parental shape. Good preparations of protoplasts incorporate radioactive precursors into nucleic acid and protein at rates similar to those of intact cells and can synthesize inducible enzymes. They support the growth of bacteriophage but are not infected by phage since they do not possess the bacteriophage receptors.

Antibiotics affecting cell wall synthesis

A chemotherapeutic agent which selectively inhibits synthesis of the peptido-glycan is potentially useful in the treatment of bacterial infections since this component of the wall is found uniquely in bacteria and closely related micro-organisms. Many antibiotics were recognized as cell wall inhibitors when it was shown that they inhibited the incorporation of typical cell wall amino acids and caused the concomitant accumulation of UDP-Mur$N$Ac-peptides, D-cycloserine bears a striking similarity to D-alanine and it competitively inhibits two enzymes which convert L-alanine to a racemic mixture and then synthesize the D-alanyl-D-alanine dipeptide. Since D-cycloserine binds considerably more strongly to the enzyme than does D-alanine it is a very effective inhibitor at relatively low concentrations.

Penicillin, bacitracin and vancomycin affect peptidoglycan synthesis at some stage after the formation of UDP-Mur$N$Ac-pentapeptide. Vancomycin inhibits the transfer of disaccharide-pentapeptide units from the C-55 polyisoprenoid alcohol phosphate to the growing point of the wall while bacitracin prevents the regeneration of the C-55 phospholipid from its pyrophosphate derivative by inhibiting the dephosphorylation reaction. Therefore both antibiotics bring the cycle of reactions shown in Figure 28 to a halt. Whether the bacteria are lysed under these circumstances depends on the activity of autolytic enzymes. Penicillin is believed to inhibit the cross-linking reaction outlined in Figure 30 by acting as an analogue of acyl-D-alanyl-D-alanine so that nascent uncross-linked peptidoglycan continues to be made and a structurally-weak polymer is found. The continued activity of autolytic enzymes which are themselves im-portant for growth of the peptidoglycan results in lysis of the growing organisms.

If an antibiotic affects only peptidoglycan synthesis the organism should not be unduly damaged if the treatment is carried out in protective hypertonic media: this has been found to be the case. Rod-shaped organisms treated with penicillin in the presence of a stabilizing concentration of sucrose lose the shape conferred on them by the rigid cell wall and tend to become spherical (Figure

32). Part of the cell wall remains adhering to the protoplast membrane and because of this the resulting structure is termed a *spheroplast* (*cf.* protoplast). If penicillin is then removed from the medium some of the spheroplasts revert to rod-shaped organisms. A similar sequence of events occurs if a typical peptidoglycan amino acid that cannot be synthesized by the organisms (e.g. DAP in a DAP-requiring organism) is omitted from the medium in which the cells are cultured.

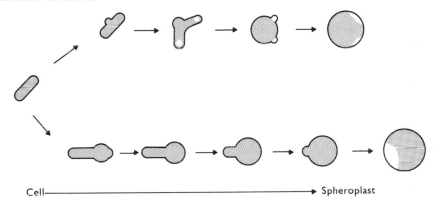

Cell————————————————————————→ Spheroplast

Figure 32. Formation of spheroplasts. These forms can be induced by growth of organisms in the presence of penicillin or by growing DAP-requiring cells in the absence of the amino acid. The cells may bulge either in the centre or terminally. Redrawn from McQuillen, in *The Bacteria*, Volume I (1960), p. 274. Eds. R.Y. Stanier and I.C. Gunsalus. Academic Press, New York and London.

L-forms

L-forms of bacteria are generally regarded as protoplasmic elements not having a defined shape and lacking the rigid component of the cell wall. The loss of shape can be correlated with the loss of peptidoglycan constituents. L-form cultures of many bacterial species have been obtained by growing organisms on agar plates in the presence of pencillin. Although the majority of the parent bacteria are killed, a few survive and after several sub-cultures in the presence of penicillin may not revert to bacteria when they are grown in its absence. They are then termed *stable L-forms*. The appearance of the cells of L-forms is similar irrespective of the parental bacterial shape.

Growth of the peptidoglycan polymer

The peptidoglycan is thought to be mainly if not totally responsible for the rigid properties of bacterial cell walls and its role can be likened to that of wire-netting in reinforced concrete. If the bacterium is to increase in length or diameter it follows that some linkages in the rigid polymer must be broken in order to insert new material. Cutting some of the strands in wire-netting does not lead to collapse of the entire framework. Similarly, controlled hydrolysis of some of the bonds either in the glycan backbone or in the peptide cross-linking chains will not result in much loss of rigidity if bonds are reformed following the addition of new material. Consequently growth of the polymer involves the controlled activity of a number of autolytic enzymes together with the insertion of new subunits. It is not known how these processes are regulated but there are indications that both synthetic and autolytic processes are localized and do not occur all over the bacterial surface.

Teichoic acids

The name 'teichoic acids' has been applied to a class of compounds, rich in phosphorus, that can be extracted from cells or isolated cell walls of Gram-positive bacteria by prolonged treatment with cold 5% trichloroacetic acid or by enzymic dissolution of the walls. They are present in the cell wall and also either in the periplasmic space between the wall and the cytoplasmic membrane or attached to the membrane. The 'membrane' teichoic acids are invariably polymers of glycerol phosphate but many different types of compounds rich in phosphorus (and therefore classed as teichoic acids) can be extracted from the walls of different organisms. The majority of them contain either glycerol or

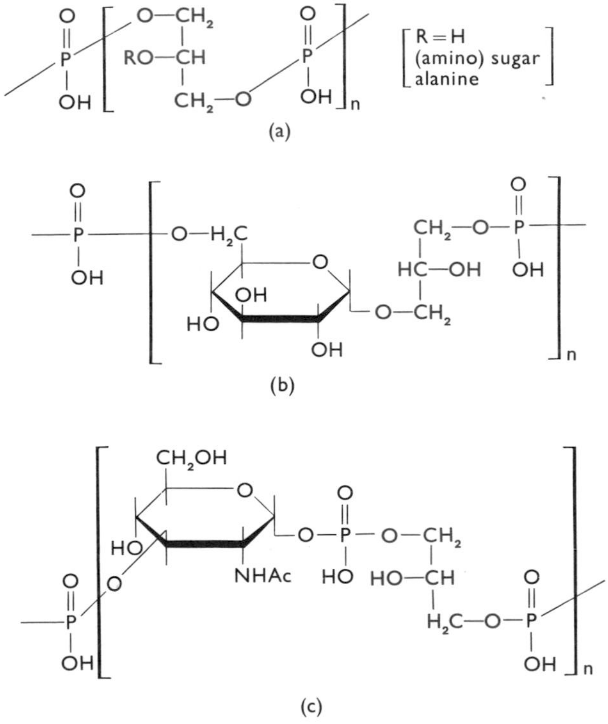

Figure 33. Glycerol teichoic acids. (a) The repeating unit is glycerol phosphate and the sugar or alanine substituent of the glycerol is not in the backbone chain. (b) The repeating unit is glucosyl-glycerol phosphate. (c) The repeating unit is N-acetylglucosamine-1-phosphate-glycerol phosphate. In (b) and (c) the sugar molecules are present in the backbone chain.

ribitol. The main types of glycerol-containing teichoic acids are illustrated in Figure 33. Variations of the type shown in Figure 33a include glycerol phosphate residues linked 1–2 rather than 1–3 and in this case the R substituent is on position 3. Polymers from the same organism are found which differ in the substituent and in the degree of substitution—the structures found are so variable that the only common factor seems to be the presence of phosphate. If a (amino)sugar is present in the polymer it can be found either as a substituent of the glycerol (Figure 33a) or as a component of the backbone chain (Figure 33b, c), the difference between these two structures being in the ratio of glycerol to phosphate in the polymer chain. The presence of a sugar residue in the chain as opposed to being attached to the glycerol has a significant effect on the biosynthetic pathway.

The general structure of the other main type of wall teichoic acid (ribitol teichoic acid) is given in Figure 34; this is typical of various bacilli. The position

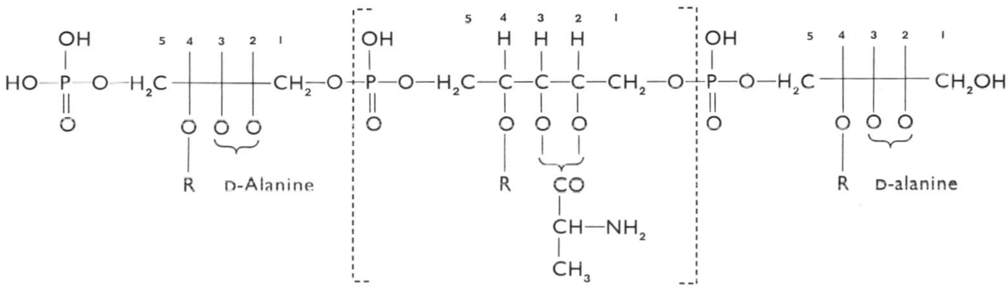

R = H or sugar or amino-sugar [linked α or β]

Figure 34. Ribitol teichoic acid. The repeating unit is ribitol phosphate and the sugar and alanine substituents are not in the backbone chain.

of D-alanine is still not certain though one ribitol teichoic acid contains glucose substituents on C-2 and C-3 which suggests that D-alanine is present on C-4. As with the wall glycerol teichoic acids, considerable variations may occur within a single species in respect to degree and type of substitution and the actual linkage of the sugar residues (α or β). More complicated teichoic acids are found in the walls of pneumococci.

The length of teichoic acid chains which tend to be short is determined chemically, by periodate oxidation of the polymer freed from alanine. The polymers from *Bacillus subtilis*, *Lactobacillus arabinosus* and *Staphylococcus aureus* are all less than 10 subunits in length.

Biosynthesis

The discovery of teichoic acids was preceded by the isolation of cytidine diphosphate ribitol (CDP-ribitol) and cytidine diphosphate glycerol (CDP-glycerol) from *Lactobacillus arabinosus*. The biosynthetic role of these nucleotide derivatives was suspected before the corresponding polymers were themselves isolated and characterized, but a direct demonstration of polymer synthesis was not achieved until much later. Particulate enzyme preparations have been obtained which catalyse the synthesis of a polyglycerophosphate polymer in the following manner:

$$\text{L-}\alpha\text{-glycerophosphate} + \text{CTP} \longrightarrow \text{CDP-glycerol} + ⓟ—ⓟ$$
$$\text{CDP-glycerol} + \text{(glycerophosphate)} \longrightarrow \text{CMP} + \text{(glycerophosphate)}_{n+1}$$

Similar preparations composed of membrane fragments catalyse the synthesis of polyribitol phosphate using CDP-ribitol as a precursor. The monosaccharide and alanine substituents are added to the polyol-phosphate backbone chain, the (amino)sugar components being transferred from the corresponding nucleotide precursor.

The more complex polymers involve the participation of more than one nucleotide derivative. The structure shown in Figure 33b can be synthesized in a cell-free system containing fragmented cytoplasmic membrane from UDP-glucose and CDP-glycerol while the polymer in Figure 33c requires UDP-*N*-acetylglucosamine and CDP-glycerol. It is claimed that a C-55 polyisoprenoid alcohol phosphate similar to the one involved in peptidoglycan synthesis is involved in the synthesis of this teichoic acid.

Function

The only factor common to the structure of teichoic acids is the presence of phosphate and this, together with the effect of growth conditions on the amount of teichoic acid present in the wall, provides an insight into the possible function of these polymers. When *Bacillus subtilis* is grown under conditions of magnesium limitation the amount of teichoic acid in the wall may be as high as 50%. If the magnesium in the medium is increased and the phosphate decreased practically all the phosphate is incorporated into nucleic acid and phospholipid and no teichoic acid is found in the wall. Instead another negatively charged polymer (teichuronic acid) containing *N*-acetylgalactosamine and glucuronic acid (for structure see Figure 35) appears in the wall. Other species react similarly and it is possible that a negatively charged polymer is necessary in the wall to provide an appropriate environment for membrane-bound enzymes that require magnesium or other cations. In Gram-negative bacteria this function may be carried out by the lipopolysaccharide (p.100).

It has been shown that teichoic acids are antigenic, i.e. when injected into animals they cause the production of antibodies. The type of teichoic acid, particularly the sugar substituents and the linkage to the backbone chain determines which antibodies are produced.

Linkage to other wall components

The attachment of teichoic acids to other polymers can be studied by treating isolated walls with autolytic enzymes and subjecting the fragments produced to

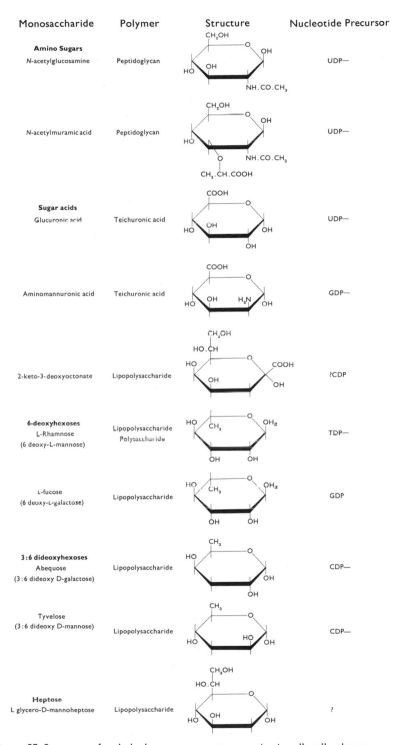

Monosaccharide	Polymer	Structure	Nucleotide Precursor

Figure 35. Structure of carbohydrate components occurring in cell wall polymers.

detailed chemical analysis. It has proved difficult to establish the linkage beyond question but fragments containing teichoic acid linked to peptidoglycan have been isolated and it seems probable that teichoic acids are linked through phospho-diester linkages to muramic acid. The occurrence of muramic acid-6-Ⓟ in the hydrolysates of some peptidoglycans suggests the 6-position of muramic acid as a possible linkage point (Figure 36).

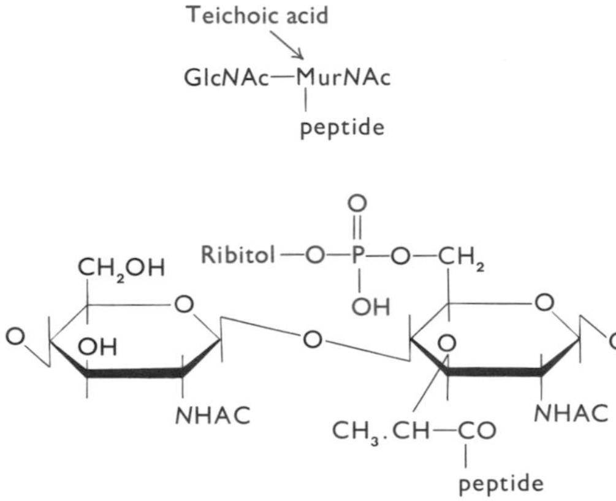

Teichoic acid

GlcNAc—MurNAc
          |
       peptide

Ribitol—O—P—O—CH₂

Figure 36. Probable linkage of teichoic acid to peptidoglycan.

Polysaccharides and amino-sugar polymers

The presence of sugars and amino-sugars in teichoic acids complicates studies of polysaccharide polymers: nevertheless it has been shown that glucose is present not only in the teichoic acid of *Micrococcus lysodeikticus*, but also in a glucose polymer to which are linked amino-mannuronic acid residues. This polymer may be similar to the teichuronic acid polymer of *Bacillus subtilis* which contains equimolar amounts of glucuronic acid and *N*-acetylgalactos-amine. The possible function of these anionic polymers has been discussed in the section dealing with the function of teichoic acids (p. 92).

The isolated walls of many Gram-positive and -negative bacteria contain sugars and amino-sugars that are not associated with peptidoglycan or teichoic acid. Some of the polymers are easily extracted, others are probably covalently bound to peptidoglycan. Their function has not been elucidated. The majority of polysaccharides present in the walls of Gram-negative bacteria are in the form of lipopolysaccharides. These substances are antigenic and will be discussed in a later section.

Proteins

The cell walls of many Gram-positive bacteria (e.g. staphylococci) when isolated by mechanical disintegration contain no protein at all. Others (streptococci) contain an immunologically type-specific protein that can be released by trypsin digestion leaving the structure of the wall unaltered, at least as seen in electron micrographs. Apparently, then, this protein does not contribute to the structural properties of the wall although it may be linked covalently to the other macromolecules. In many bacilli a structural array of protein is seen on the outermost layer of the wall (Figure 8, p. 71); the function of this layer has yet to be ascertained.

Isolated envelope preparations of Gram-negative organisms contain part or all of the protoplast membrane so it is not surprising to find that all of the commonly occurring amino acids are found in these preparations. Some of the protein is found in association with lipid while part is free or linked to peptidoglycan. The latter is a lipoprotein, present in large amounts with one molecule bound for every ten disaccharide subunits in the murein. The linkage is from an *N*-terminal lysine residue in the lipoprotein to a DAP residue in the peptidoglycan. This is the globular protein layer shown between the L5 and L3 layers in Figure 10 (p. 72).

Lipids

A conspicuous difference in lipid content is found between cell walls isolated from Gram-positive and -negative bacteria. Gram-positive organisms have very little extractable lipid in the wall with the exception of the mycobacteria which have as much as 60% of the dry weight of the wall composed of wax, phosphatides and bound lipids. As it is difficult to separate the cytoplasmic membrane from the cell wall of Gram-negative bacteria, it is not clear whether the extractable lipids (predominantly phospholipid) are derived from both components or not.

Lipopolysaccharides

The monosaccharides in the cell walls of Gram-negative bacteria such as *Salmonella typhimurium* and *Escherichia coli* bear a striking similarity to those found in the immunologically specific lipopolysaccharides isolated by extraction of whole cells with phenol/water. The resulting aqueous phase contains nucleic acids and polysaccharides in addition to lipopolysaccharide which can be sedimented by centrifugation at $100,000 \times g$. Thin-section microscopy shows that bacteria that have undergone this extraction procedure have lost the outer

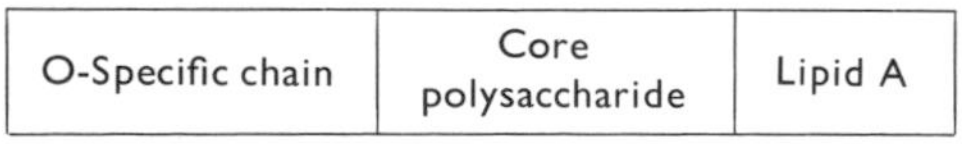

O-Specific chain	Core polysaccharide	Lipid A

Figure 37. Sequence of the three structural regions of lipopolysaccharides. Redrawn from O. Luderitz. *Angewandte Chemie* **9** (1970) 649.

3-layered membrane of the cell envelope but still retain the cytoplasmic membrane and the rigid layer of the wall.

Chemical investigation of a large number of lipopolysaccharides indicates the presence of the general structure shown in Figure 37. Lipid A is found in all lipopolysaccharides and the core polysaccharide region is attached to it. There are a number of mutant strains which lack part of this region and it is through analysis of the 'deficient' lipopolysaccharides in these strains that the detailed chemical structure has been elucidated. The structure of the core is thought to be similar if not identical in closely related strains. The outermost region of the lipopolysaccharide comprises the 'O' specific chains, the structures of which are highly species specific and thousands of different lipopolysaccharides exist.

Lipopolysaccharides from all wild-type strains of salmonellae contain glucose, galactose, N-acetylglucosamine, 2-keto-3-deoxyoctonate (KDO) and L-glycero-D-mannoheptose (core region) together with a number of other sugars which may include hexoses (galactose, mannose), pentoses (ribose, xylose), 6-deoxyhexoses (rhamnose, fucose) and 2 : 6-dideoxyhexoses (abequose, tyvelose etc.). These sugars are found in the 'O' specific chains. One such chain from *Salmonella typhimurium* contains galactose, mannose, rhamnose and abequose and its structure is shown in Figure 38. It contains several repeating units of a specific sequence of sugars.

O-Specific chain

Figure 38. Structure of the O-specific chains in the lipopolysaccharide of *Salmonella typhimurium*. The repeating unit imparts antigenic activity to the lipopolysaccharide; the sugar composition is strain specific. Redrawn from O. Luderitz. *Angewandte Chemie* **9** (1970) 649.

The order of sugar residues within the polysaccharide chain of the core region has been determined by the use of mutants blocked in the synthesis of, or transfer from, nucleotide-sugar precursors or in the transfer of phosphate residues. The lipopolysaccharides formed under such conditions lack certain sugar residues but this, apparently, has no effect on the ability of the mutants to survive under laboratory conditions. The compositions of the lipopolysaccharides from a number of mutant classes are shown in Table 2 and the composite structure derived from these and other more detailed chemical studies is given in Figure 39. It is thought that all closely related strains possess this structure. The 'O' specific chains are believed to be attached to the glucose II residue.

The core polysaccharide is in turn linked through one of the KDO residues to lipid A the structure of which has been determined recently (Figure 40). It contains two glucosamine residues, fully substituted with long chain fatty acids, β-hydroxymyristic acid and phosphate.

Table 2. Structure of the lipopolysaccharides of *Salmonella* mutants unable to synthesize the complete core region

Mutant type	Deficiency	Components present
1	Addition of heptose	[KDO]$_3$→lipid A
2	Addition of second heptose	Hep→[KDO]$_3$→lipid A
3	Transfer of phosphate	Hep→Hep→[KDO]$_3$→lipid A
4	Synthesis of UDP-Glc	Hep→Hep→[KDO]$_3$→lipid A Ⓟ, Ethanolamine
5	Synthesis of UDP-Gal	Glc→Hep→Hep→[KDO]$_3$→lipid A Ⓟ, Ethanolamine
6	Synthesis of UDP-GlcNAc	Glc→Gal→Glc→Hep→Hep→[KDO]$_3$→lipid A Gal Ⓟ, Ethanolamine
7	'Rough' mutants synthesis of 'O' specific chain	GlcNAc→Glc→Gal→Glc→Hep→Hep→[KDO]$_3$→lipid A Gal Ⓟ, Ethanolamine

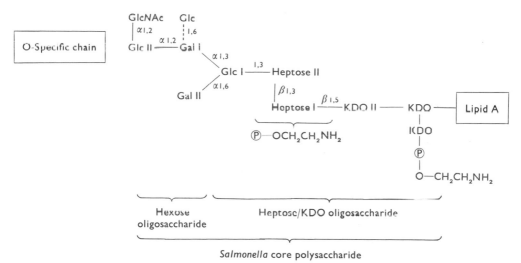

Salmonella core polysaccharide

Figure 39. Structure of the 'core' polysaccharide. The structure has been deduced from the analysis of partial acid hydrolysates of lipopolysaccharides isolated from normal strains and from mutants unable to synthesize the complete structure. Redrawn from O. Luderitz. *Angewandte Chemie* **9** (1970) 649.

It is apparent that the molecular weight of the complex, lipid A—core polysaccharide—'O' specific chain, is several thousand and the availability of phosphate residues in the core polysaccharide and lipid A may result in phosphodiester linkages between individual subunits with increase in the size of the polymer. These large molecules may be important constituents of the outer membrane of the cell envelope because they contain both hydrophobic and hydrophilic residues.

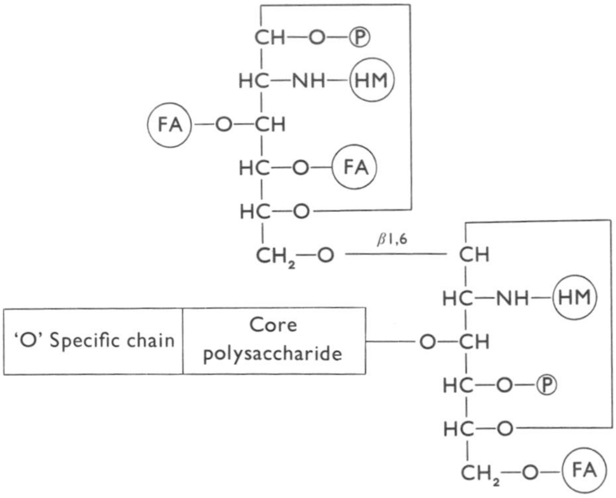

Figure 40. Structure of a lipid A unit with an attached KDO oligosaccharide. FA = long chain fatty acid. HM = β-hydroxymyristic acid. Redrawn from O. Luderitz. *Angewandte Chemie* **9** (1970) 649.

Biosynthesis

The 'O' specific chains are synthesized in a manner analogous to peptidoglycans —again an extracellular polymeric component being made from small precursors. The repeating unit is first assembled as an oligosaccharide linked to a C-55 polyisoprenoid alcohol phosphate (Figure 29) each sugar being added from its nucleotide precursor. In *Salmonella typhimurium* this involves four different bases in the nucleotides, UDP-galactose, TDP-rhamnose, GDP-mannose and CDP-abequose (Figure 41). The completed oligosaccharide is then polymerized to form the 'O' specific chain still attached to the phospholipid in the cytoplasmic membrane. The completed chain is then transferred to the core of the lipopolysaccharide.

The core polysaccharide is built up sequentially by the addition of sugars to the incomplete core. Little is known of how the heptose residues are transferred since the nucleotide precursor has not been identified. The addition of glucose, galactose and N-acetylglucosamine has been studied using cell envelope preparations from mutant organisms as a source of enzymes and as the receptor. In addition to the particulate enzyme preparations, soluble enzymes have been isolated which catalyse the addition of glucose I and galactose I to boiled cell envelope preparations from suitable mutant organisms. The product of the first reaction is used as the substrate for the second as is illustrated in the scheme for the synthesis of part of the core polysaccharide (Figure 42). The cell envelope

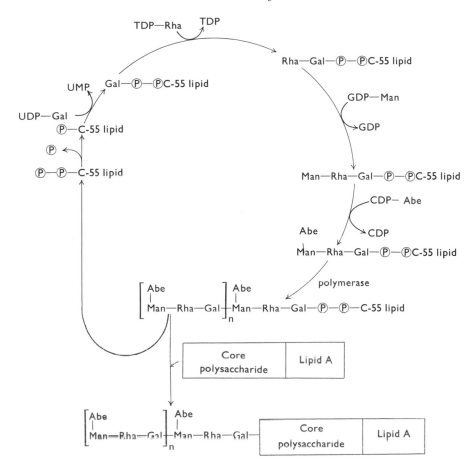

Figure 41. Biosynthesis of O-specific chain.

has been fractionated in an attempt to find the material which accepts the monosaccharides. The active material is apparently a complex of lipopolysaccharide and lipid since the purified lipopolysaccharide alone does not function as an acceptor, while a mixture of lipopolysaccharide and lipid is active if it is prepared under specified conditions of heating and slow cooling. A complex containing lipid, lipopolysaccharide and the soluble enzyme which catalyses the addition of galactose to the incomplete lipopolysaccharide has been obtained. Thus it seems that the lipid in the wall of Gram-negative organisms may be important in the stabilization of lipopolysaccharide acceptor-enzyme complexes.

Function

Since mutants containing only the lipid A and part of the core polysaccharide regions of the lipopolysaccharide grow in the laboratory just as well as wild-type strains it is clear that the 'O' specific chains and monosaccharides found in the core structures are dispensable. However, only wild-type strains containing the complete 'O' specific chain are pathogenic so these components appear to be

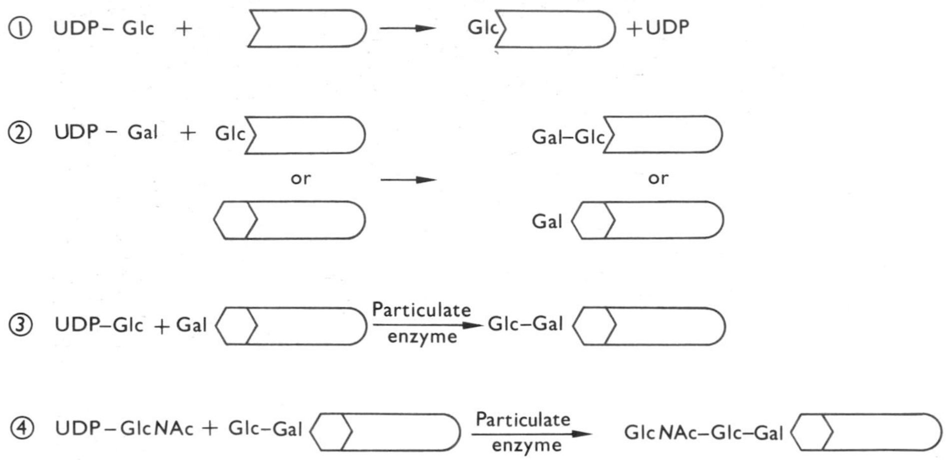

Figure 42. Biosynthesis of core of lipopolysaccharide. This has been elucidated using mutants which are unable to synthesize either UDP-glucose or UDP-galactose. The former mutant contains an incomplete lipopolysaccharide to which glucose can be added, the second, one to which galactose can be added. Further additions, catalysed by particulate enzyme preparations, are then possible.

essential for successful invasion of the host. Further the presence of charged phosphate groups may permit the polymer to function in the binding of ions and possibly to maintain a defined ionic environment in this area of the wall (*cf.* the function of teichoic acids in Gram-positive organisms, page 92).

Site of cell wall formation

In order to distinguish newly synthesized cell wall it is important to mark or label the cell wall that is already in existence: the label must be visible and should not interfere with cell growth or division. The technique of labelling with fluorescent antibody has yielded valuable information and the procedure is as follows. Whole cells or cell walls of the organism to be studied are injected into rabbits and after several weeks antibodies are prepared from the serum. The antibody molecules are treated with fluorescein isothiocyanate to give a fluorescent derivative which still reacts with cell wall material. Bacteria are grown in the presence of this fluorescent antibody for a certain length of time: the antibody is then removed and the bacteria are reincubated in fresh medium. Samples are examined microscopically by ultra-violet illumination: old wall is brilliantly fluorescent while new wall is not. The reverse technique is also of

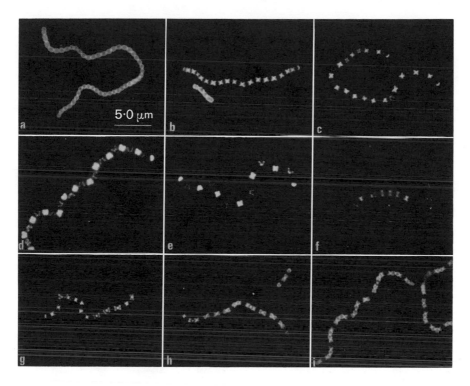

Figure 43. Ultra-violet photomicrographs of *Streptococcus pyogenes* showing cell wall immuno-fluorescence patterns. In photographs a–e, taken at 15 min intervals, new wall appears as dark patches between the light semi-circles. In photographs f–i, also taken at 15 min intervals, newly-synthesized wall is labelled with the fluorescent antibody. Reproduced with permission of R. M. Cole. *Science N.Y.* **135** (1962) 722.

great importance; organisms are exposed initially to unlabelled antibody followed by growth in fresh medium. Bacterial smears taken from the re-incubated culture can be stained on slides with fluorescent antibody to show the formation of new wall, or alternatively, labelled antibody may be included in the incubation medium. Results obtained by the use of this method must be interpreted with caution since it has not been established that all wall components are replicated simultaneously and at the same sites: the antibody is specific only for the antigenic surface components of the wall, and the pictures obtained by this method do not necessarily show where peptidoglycan is synthesized. It is, however, encouraging that pictures of *Streptococcus pyogenes* stained with antibodies specific for two different wall components (protein and polysaccharide) give essentially the same results. Although only a few species of organisms have been examined there are indications that at least two types of cell wall growth occur, one represented characteristically by *Streptococcus pyogenes* and the other by *Salmonella typhosa*. In the former, cell wall growth is initiated equatorially and synthesis of the cross-wall and peripheral wall occur simultaneously. In this manner the new halves of the daughter cocci are formed back-to-back. The 'old' ends of the cocci are preserved intact but are gradually pushed further apart as shown in Figures 43 and 44. Recent results with a Gram-positive bacillus using a fluorescent antibody reacting with peptidoglycan indicate a similar type of cell wall growth. In *Salmonella typhosa* on the other hand the fluorescence of the wall fades gradually as incubation in antibody-free

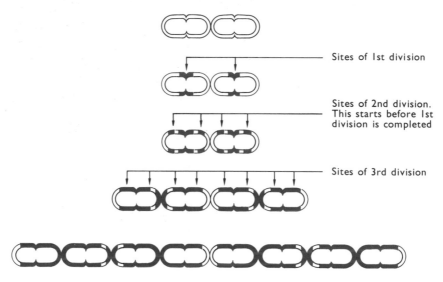

Sites of 1st division

Sites of 2nd division. This starts before 1st division is completed

Sites of 3rd division

Figure 44. Diagrammatic representation of Fig. 43 illustrating the deposition of new wall material in *Streptococcus pyogenes* = old wall, — new wall. Redrawn—reproduced with permission of R. M. Cole. *Science N.Y.* **135** (1962) 722.

medium is prolonged (Figure 45). The insertion of new wall material apparently occurs along the length of the wall though it is possible that new molecules are inserted at so many new sites that discrete gaps in the fluorescent labelling pattern cannot be resolved microscopically. This observation is not in agreement with current suggestions that peripheral wall grows outward from the site of septation. However, it must be emphasized that the antibody used in experiments such as that illustrated in Figure 45 does not reveal sites of peptidoglycan deposition.

The formation of the cross wall and subsequent cell division in Gram-positive and -negative organisms is thought to occur by similar processes although it has proved difficult to obtain reliable electron micrographs with Gram-negative bacteria. The formation of the septum is preceded by an involution of the cytoplasmic membrane which is thought to control the process. In Gram-positive bacteria (Figure 46) the thickness of the mature cross wall is double that of the peripheral wall and cell division apparently occurs by its being split presumably by autolytic enzymes. A double layer of the rigid

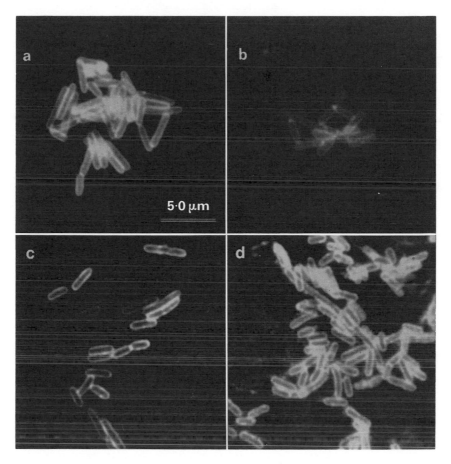

Figure 45. Growth of the cell wall of *Salmonella typhosa*. Ultra-violet micrographs of the direct (a–b) and reverse (c–d) methods of labelling with fluorescent antibody (see text) demonstrate the absence of localisation of new cell wall material. Photographs a and b, and c and d were taken at 60 min intervals. Reproduced with permission of R. M. Cole. *Science N.Y.* **143** (1964) 820.

peptidoglycan is present in the septum of the Gram-negative bacterium shown in Figure 47 and it is envisaged that separation of the two daughter cells is achieved by the ingrowth of the outer cell wall layers which are not present in the septum at early stages of its formation.

CAPSULES

Capsules are manufactured by certain Gram-positive and Gram-negative bacteria and form the outermost layer of the cell. They are not indispensible structural elements and their occurrence is subject to cultural conditions. The

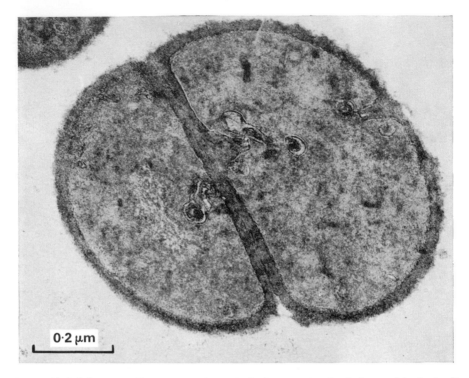

Figure 46. Cell division in Gram-positive bacteria. A septum, twice the thickness of the final wall, is formed across the cell before cell division occurs. Reproduced with permission of W. Van Iterson. Unpublished.

capsule is not shown up easily in electron micrographs (Figure 4, p. 68) but its presence can be demonstrated immunologically or by dispersing the cells in Indian ink (Figure 48).

Before it was known that some capsules were immunologically active, it was believed that they contained homogeneous accumulations of amorphous material surrounding the cell wall. However, many of the capsules are heterogeneous and the presence of polypeptide and polysaccharide material has been demonstrated in the capsule of *Bacillus megaterium* strain M after exposure of organisms to antibodies reacting with these two components. Striated structures within a capsular matrix have been detected in *Escherichia coli* strain Lisbonne, while the capsule of other organisms may not be of even thickness over the whole surface of the organism.

Many organisms do not produce a well-defined capsule but secrete a loose slime or extracellular gum composed of polysaccharide. This material may be loosely adsorbed on the cell surface and can frequently be removed by washing. Occasionally it may be laid down in an ordered manner (Figure 49).

Sugars and amino sugars are found most frequently in capsules: uronic acids also occur and distinguish the capsular polysaccharide from cell wall polysaccharides in which they are seldom detected. The capsules of a number of bacilli contain polypeptide material in addition to polysaccharide: the peptide is composed of D-glutamic acid units linked together through the α-amino and

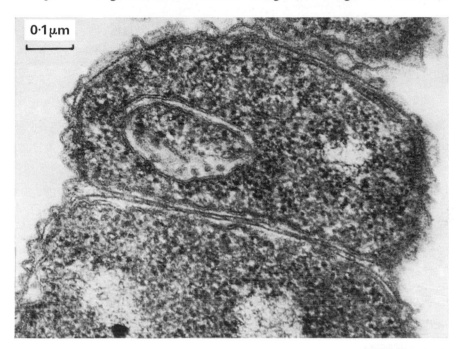

0·1 μm

Figure 47. Cell division in Gram-negative bacteria. In some electron micrographs division apparently proceeds by constriction and no septum can be distinguished. In other micrographs (as here) a septum, containing only the inner layer of the cell wall and the protoplast membrane on each side, divides the cell into two compartments before division proceeds. Reproduced with permission of P. Steed and R. G. E. Murray. *Canad. J. Microbiol.* **12** (1966) 263.

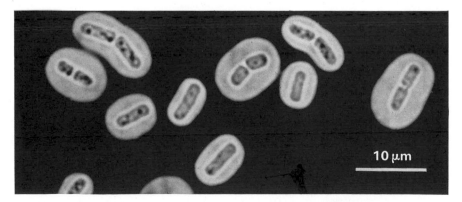

10 μm

Figure 48. Capsule of *Bacillus megaterium* demonstrated by dispersing the cells in Indian ink. Reproduced with permission of C. F. Robinow. Unpublished.

γ-carboxyl groups. The α-carboxyl groups are apparently in the amide form—reminiscent of the D-glutamyl residues of the cell wall peptidoglycan.

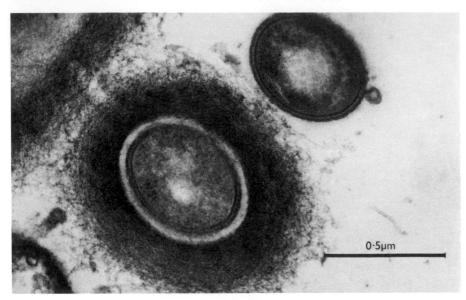

Figure 49. Polysaccharide slime layer, stained with ruthenium red, showing a concentric arrangement of the strands. Reproduced with permission of H. C. Jones, I. L. Roth and W. M. Sanders. *J. Bacteriol.* **99** (1969) 316.

It is not clear what forces or bonds hold the capsule in position. The fact that material chemically related to wall polymers has been found in some capsules (e.g. teichoic acid in pneumococci; peptidoglycan components in *Bacillus anthracis*) suggests that some wall substances may protrude from the wall, and it is just conceivable that covalent bonds may link these protrusions with the true capsular materials.

SURFACE APPENDAGES

Pili (Fimbriae)

Two types of filamentous appendages can be detected attached to the bacterial surface: pili and flagella. Pili (hair-like structures) have been found in Gram-negative bacteria only: they are invariably shorter and thinner than flagella being approximately 7 nm in diameter and having an axial hole 2–5 nm in diameter. They are thought to contain protein subunits polymerized as a right-handed helix. Six types of pili have been shown to exist (e.g. type I, type F) but the function of most of them remains in doubt. Type I pili occur fairly evenly over the surface of the cell (Figure 50): their length is variable and they tend to

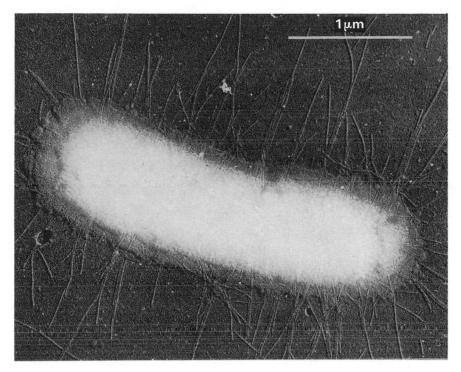

Figure 50. *Escherichia coli* with attached pili. Reproduced with permission of C. Brinton. Unpublished.

be relatively straight, rigid structures. Type F pili are involved in the transfer of nucleic acid between mating cells of *Escherichia coli*. There are generally only one or two F pili per cell and their length varies from a short stub to several times that of the bacterium—most are 2 μm long. The presence of F pili enables the bacteria to adsorb a male-specific RNA phage (Figure 51) and this serves to distinguish them from the more numerous type I pili. The protein (F pilin) of which the F pilus is composed has at least three functions: it polymerizes to form pili, it must trigger the release of RNA from the male phage M-12 which adsorbs to the pilus, and it conducts DNA from the donor to the host cell in a mating pair.

Flagella

The majority of motile bacteria have flagella which enable them to travel at speeds as great as 50 μm (i.e. fifty times their length) per second. Flagella are several times as long as the bacterium and their diameter varies from 12 nm (*Proteus vulgaris*) to 30 nm (*Vibrio metchnikovii*). Bacteria may have a single flagellum, or a tuft of flagella arising from one or both poles of the cell, or flagella arranged over the whole surface (Figure 52). Purified preparations of flagella consist almost entirely of the protein flagellin which is characterized by a high content of acidic amino acids, a low content of aromatic amino acids, and, in many instances, the absence of cysteine. Some preparations contain carbohydrate or RNA or both. The flagellum of *Salmonella typhimurium* contains the globular sub-units of flagellin arranged in rows along the axis (Figure 53) though in some micrographs the sub-units appear to be in helices: the nature of the linkages between the flagellin molecules is not known. The number

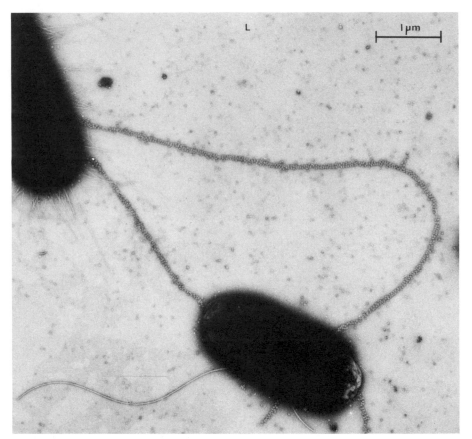

Figure 51. F pili, 'stained' with MS–2 phage, forming a link between an Hfr cell and a F⁻ cell of *Escherichia coli*. Reproduced with permission of R. Curtiss, L. G. Caro, D. P. Allison and D. R. Stallions. *J. Bacteriol* **100** (1969) 1091.

of rows of flagellin molecules is variable and depends both on the size of the protein molecule and on the diameter of the flagellum. Flagella arise either from or within the cytoplasmic membrane as is indicated by the fact that they are retained when the walls of lysozyme-sensitive organisms are digested. The flagella of many types of organism have at their base an organelle consisting of two basal discs (Figure 54a) which appear to be derived from the cell wall and cytoplasmic membrane: these may be concerned with the synthesis or functioning of the flagellum. The 'basal organelle' seen adhering to the base of the flagellum in autolysing cells (Figure 54b) represents a fragment of cytoplasmic membrane that has pulled away with the flagellum.

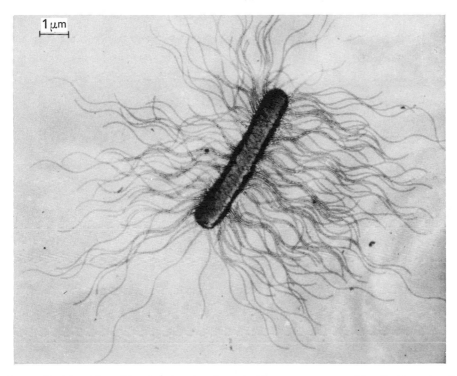

Figure 52. *Salmonella typhimurium* showing the peritrichate arrangement of flagella. Reproduced with permission of J. Hoeniger. *J. Gen. Microbiol.* **40** (1965) 29.

Figure 53. Preparation of isolated flagella demonstrating the presence of a regular array of subunits. Reproduced with permission of A. M. Glauert. *Laboratory Investigation* **14**, 331.

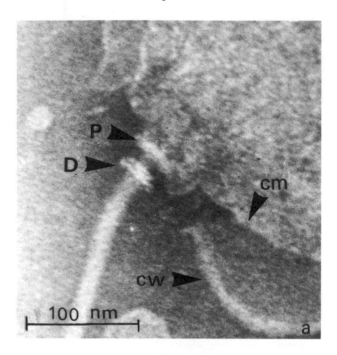

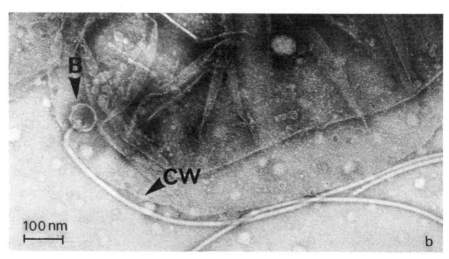

Figure 54. Negatively-stained, autolysing *Vibrio metchnikovii* showing the structures at the base of the flagellum. In (a) two discs are evident, one of which (P) appears to be continuous with the cytoplasmic membrane (cm), the other (D) is separated from the cell wall (cw). In (b) the 'basal bulb' (B) consists of a fragment of cytoplasmic membrane. The base of the flagellum with the membrane fragment attached is still within the cell wall (cw). Reproduced with permission of Z. Vaituzis and R. N. Doetsch. *J. Bacteriol.* **100** (1969) 512.

BACTERIAL CYTOPLASM

In general bacterial cytoplasm appears to lack the membrane systems that are such a feature of plant and animal cells but has a high overall density resulting from the presence of large numbers of ribosomes. Improved fixation and thin-sectioning techniques have revealed details of the fine structure which will be presented in this section.

MEMBRANES

Cytoplasmic membrane (protoplast membrane, plasma membrane)

Structure

The first direct indication that protoplast membranes and cell walls were separate structures resulted from the isolation of protoplasts of *Bacillus megaterium*: a limiting semi-permeable membrane surrounds these spherical bodies and is preserved intact when the protoplasts are maintained in a hypertonic medium. If such protoplast suspensions are diluted with water the protoplasts lyse and the membrane fraction can be isolated by differential centrifugation for chemical characterization. Treatment of Gram-negative bacteria with lysozyme leaves the outer membrane intact (Figure 11, p. 73); lysis of the resulting spheroplasts and centrifugation of the particulate material gives rise to a mixture of cytoplasmic and outer membranes which is difficult to separate into its two components. It is impossible to state whether any membrane preparation is 'pure' since other cell components may be strongly adsorbed to it after breakage; conversely, some 'true' membrane components may be removed by the stringent washing procedures. The preparations that are obtained consist almost entirely of lipid and protein and account for approximately 10% of the dry weight of the bacteria: other substances found associated with the membrane include carbohydrate, RNA and DNA.

Characterisation

The main class of lipid present in bacterial cytoplasmic membranes is phospholipid: Gram-positive bacteria contain several types within this class while Gram-negative organisms possess mainly phosphatidylethanolamine (Table 3). The types of phospholipid found and the amount of each present depend on the bacterial strain and on the growth conditions, particularly the pH value of the medium: for this reason the percentage composition cannot be included in Table 3. The structure of the various phospholipids is shown in Figure 55. Other classes of lipid found in bacterial membranes include free fatty acids and glycolipids in which glucose or mannose frequently appear.

Gram-positive and -negative bacteria differ markedly in the fatty acid composition of the membrane lipids. Branched-chain fatty acids are characteristic of Gram-positive organisms (e.g. *Micrococcus lysodeikticus* contains 90%

Table 3. Principal classes of phospholipids found in the membranes of Gram-positive bacteria
and the cell envelopes of Gram-negative organisms

Gram-positive

Bacillus megaterium	PE, PG, di-PG, lys-PG
Bacillus subtilis	PE, PG, di-PG, αα-PG
Bacillus licheniformis	PE, PG
Streptococcus faecalis	PA, PG, di-PG, αα-PG
Staphylococcus aureus	PA, PG, di-PG, αα-PG
Micrococcus lysodeikticus	PG, di-PG, PI

Gram-negative

Azotobacter spp.	PE
Escherichia coli	PE (PG, PS, αα-PG)
Halobacteria spp.	PG, di-PG

Abbreviations

PE	Phosphatidylethanolamine
PG	Phosphatidylglycerol
di-PG	Di-phosphatidylglycerol (cardiolipin)
lys-PG	Lysyl-phosphatidylglycerol
αα-PG	Aminoacyl-phosphatidylglycerol
PA	Phosphatidic acid
PI	Phosphatidylinositoi
PS	Phosphatidylserine

of its fatty acid residues as branched-chain C-15) and virtually no unsaturated fatty acids are found. Gram-negative bacteria contain a mixture of saturated and unsaturated fatty acids in which chain lengths of C-16 and C-18 predominate, and also cyclopropane acids. The ratio of saturated to unsaturated fatty acids is dependent on the growth temperature. Thus in cultures growing at low temperatures (e.g. 20°) unsaturated fatty acids predominate but at higher temperatures (e.g. 37°) some of the unsaturated fatty acids are replaced by saturated derivatives and the ratio of the two types approaches unity. This effect is even more marked at the two extremes of the range of growth temperature. Thus in psychrophilic organisms growing at 2°, practically all of the fatty acids are unsaturated whereas in thermophilic bacteria growing at 60° they are all saturated. It appears that this response to temperature is important in maintaining the fluidity of the lipids in the membrane.

The major component of the cytoplasmic membrane is protein (60–80%) but it seems unlikely that any one protein is present in large amounts. Polyacrylamide gel electrophoresis of the proteins that have been made soluble by dissolving the membrane in the detergent, sodium dodecyl sulphate, indicates that at least twenty protein components are normally present, some of them being glycoproteins. Some are easily removed by simple aqueous washing and are apparently loosely bound to the surface (epi-proteins) while others are only

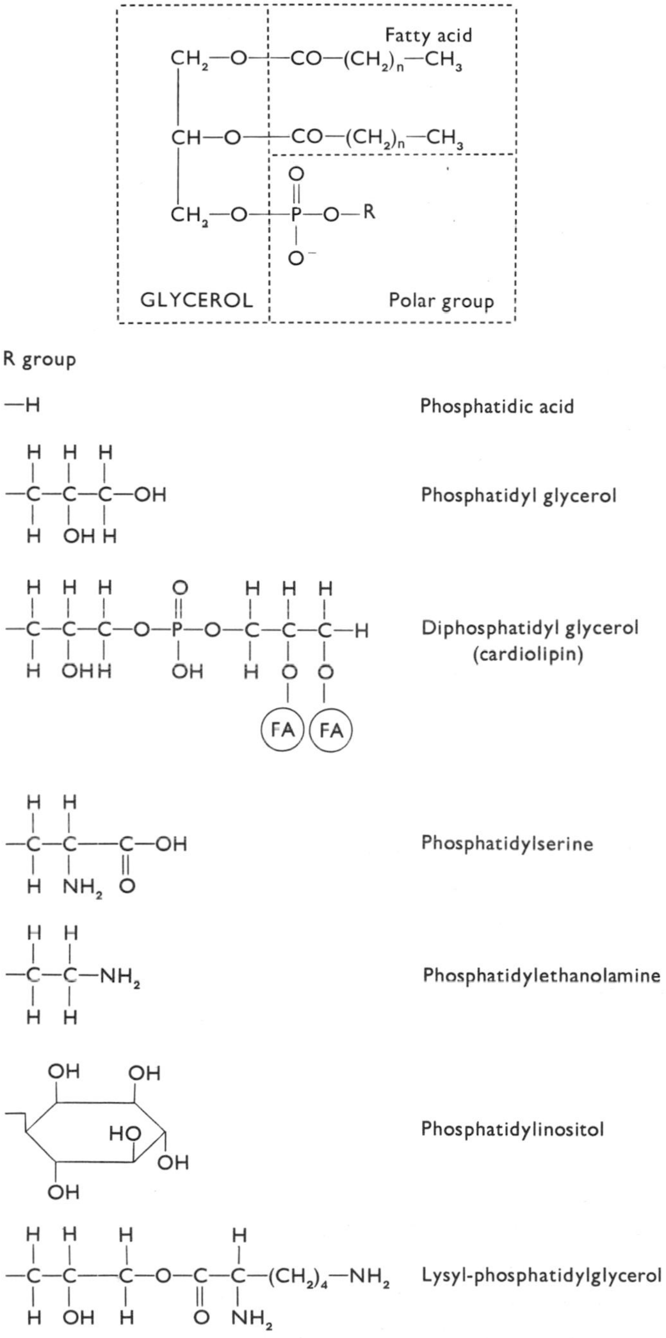

Figure 55. Phospholipids found in bacteria.

extracted by organic solvents and are presumed to penetrate the hydrophobic core of the membrane (endo-proteins).

Models of membrane structure

In thin sections the cytoplasmic membrane appears as a three-layered structure (Figure 9, p. 72) with two electron-dense layers separated by an electron-transparent zone. This observation, together with X-ray diffraction data led Robertson to propose the 'unit-membrane' hypothesis which is an extension of

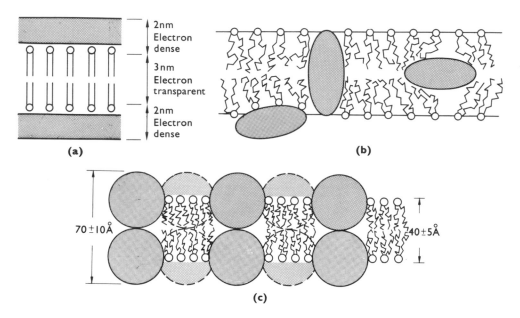

Figure 56. Diagrammatic representation of cross-section through the cytoplasmic membrane. (a) Bimolecular lipid leaflet. (b) Lipid bilayer, showing penetration by protein molecules. (c) Repeating lipoprotein subunit in which a region of lipid bilayer fills the pores between the double layer of protein molecules. The dashed circles represent protein molecules behind the plane of the section which are in contact with the protein molecules denoted by full circles.

the ideas elaborated by Danielli and Davson from physical studies involving measurements of permeability, surface tension and conductivity. The 'unit membrane' is rigidly defined as a bimolecular lipid leaflet in which the hydrocarbon chains of the phospholipids are close-packed and oriented perpendicular to the surface of the leaflet (Figure 56a). The 'core' of the membrane, therefore, has hydrophobic properties and it is this region which is considered to be electron-transparent in micrographs of thin sections. The polar groups of the phospholipids are on the outside of the leaflet and are linked electrostatically to charged groups of the proteins or glyco-proteins. These hydrophilic regions were considered to be equivalent to the electron-dense areas on both sides of the membrane and it was further proposed that the proteins were in the extended β-configuration. This model explains many of the properties of cytoplasmic membranes, particularly their relative impermeability to many simple substances such as amino acids, sugars and even ions, and various electrical properties such as resistance and capacitance. However, the rigid definition of the model as presented here is not in accord with detailed experimental investigation.

Firstly, the image in the electron microscope is of material stained with osmium tetroxide. There is uncertainty as to the groups which bind osmium but one possibility is that it becomes attached across the double bond of an unsaturated fatty acid. However it is the polar and not the apolar portion of the membrane which is stained. In addition, membranes from which the lipid has been removed by solvent extraction still give the triple layer image and some bacterial membranes which do not contain any unsaturated fatty acids give the same picture. Consequently it is unwise to try to relate electron density and chemical structure when considering membranes.

Secondly, although a cursory inspection of thin sections of membranes from different sources reveals an apparent identity in appearance, detailed examination shows that membranes vary in width from 5–13 nm. If the fatty acid side-chains are close-packed and are arranged perpendicular to the surface, the hydrophobic part of the membranes from different sources should be relatively constant whereas, in fact, the electron-transparent zone varies from 3–10 nm (the length of two C-18 chains end to end is approximately 3·2 nm).

Thirdly, the hypothesis was based on investigations carried out with myelin, a highly specialized membrane which is atypical in that it has no enzymic activity. Furthermore the lipid composition is abnormal because there are several species of poly-unsaturated fatty acids, and the protein/lipid ratio is so low that approximately half of the lipid could not be covered with protein if the lipid were present in a bimolecular leaflet. Since bacterial membranes have a protein/lipid ratio close to 5 in terms of area it could be argued that the structure of myelin has very little bearing on bacterial membrane structure. However, the important consideration is not the ratio of protein area to lipid area but the total amount of lipid and whether it is sufficient to cover the surface of the bacterium in the form of a bimolecular lipid leaflet. Unfortunately such calculations have not been made for bacterial cells.

Fourthly, physical determinations indicate that hydrophobic bonding exists between protein and protein and between protein and lipid, and there is further evidence that the protein is not present in the extended form but in a more globular form with a considerable amount of α-helix structure.

Most physical determinations using modern techniques (circular dichroism, optical rotatory dispersion, electron spin resonance, differential scanning colorimetry) support a more liberal interpretation of the unit membrane hypothesis.

This suggests that membranes consist largely of a lipid bilayer in which the fatty acid side-chains are not arranged in a regular manner but are in constant motion. The *Mycoplasma* are particularly useful in these investigations since these organisms incorporate into membrane lipids whatever fatty acids are present in the growth medium. Investigations are carried out under conditions in which the fatty acid composition is varied and the effect of the changes on such properties as permeability and lipid mobility is then measured. Since membranes can be prepared very easily from this group of organisms (they have no wall components) it is possible to carry out investigations with intact cells, with membranes derived from them, and with model lipid bilayer systems in which the same phospholipids are present as in the cytoplasmic membrane. The three different systems yield almost identical results which support the existence of a lipid bilayer over 80–90% of the area covered by the cytoplasmic membrane.

Proteins are believed to penetrate this fairly fluid bilayer to a lesser or greater extent as shown in Figure 56b, causing limited disturbance of the fatty acid side-chains in the phospholipids. In extreme situations the protein molecule may pass right through the membrane or may exist in a completely hydrophobic environment in the middle of the bilayer. Electron micrographs of membranes that have been subjected to freeze-etching support the idea that proteins are located within the hydrophobic core of the membrane. In general the bacterial cytoplasmic membrane would be expected to have much more protein on both its surfaces than is indicated in Figure 56b in view of the high protein/lipid ratio and this is borne out by electron micrographs of negatively-stained membrane preparations and also by micrographs of freeze-etched bacteria.

Another suggestion is that the membrane consists of repeating lipoprotein subunits having functional and structural properties (Figure 56c). Repeating units have been demonstrated in electron micrographs of positively- and negatively-stained animal and plant material and have also been reported in bacterial membranes. A structure of this sort would require much less genetic information for its specification than would a lipid bilayer and it would be more simply synthesized. A limited number of different lipoprotein subunits would be expected to exist but there is no evidence as to how many different types of subunit exist in the membrane of a single species. Support for the subunit theory comes from studies with isolated membranes of *Halobacterium halobium* and *Mycoplasma laidlawii*. These can be dissociated by special treatments into small lipoprotein complexes (these have been resolved further) which can be re-aggregated to form vesicular membranes resembling the intact membrane when examined by electron microscopy. However, it is difficult to demonstrate that these complexes are true subunits; they might represent small pieces of membrane that have been fragmented by treatment with a detergent and which reassociate spontaneously on its removal. Until a *homogeneous* preparation of lipoprotein subunits has been demonstrated to aggregate to form a functionally active membrane much of this work must be regarded with caution. If such sub-units exist it is presumably some chemical property which enables them to associate in such a way that a functional membrane is formed just as viral coat proteins associate or molecules of flagellin come together to form the bacterial flagellum.

Function

Membranes are involved in the maintenance of a permeability barrier, the transport of a variety of small and large molecules, energy production, and also as a site for the localization of enzymes, possibly including those for the synthesis of proteins, RNA and DNA. The membranes are semi-permeable and do not allow free passage of small molecules such as ions, amino acids and sugars. The direction of flow of small molecules is oriented and the cell is therefore able to concentrate substances inside the cell and to maintain concentration gradients. Some of the enzymes and carriers that are involved in these transport processes have been released from intact cells by a procedure involving osmotic shock and are thought to be derived from the membrane. In one instance a protein known to be involved in transport of β-galactosides has been specifically labelled and the label shown to be present exclusively in the membrane fraction. The ability to concentrate molecules enables the cell to maintain a constant intracellular environment, the energy for this being supplied by adenosine triphosphate. However, not all substances are transported actively into bacterial cells—some enter by diffusion.

A number of the enzymes and carriers concerned with the terminal stages of biological oxidations are present in purified membrane preparations. These include various dehydrogenases, the complete cytochrome system, and ATP-ase, an enzyme known to function in ATP synthesis. In this respect the bacterial cytoplasmic membrane is apparently analogous to the inner mitochondrial membrane which contains similar enzymes and electron carriers.

Another group of membrane-bound enzymes is intimately involved in the synthesis of extracellular polymers. These enzymes function as multi-enzyme systems and the loss or dislocation of one enzyme may result in the entire sequence being halted. These syntheses involve a C-55 polyisoprenoid alcohol phosphate in addition to enzymes and the reactions are considered in detail on pages 83 and 98.

The role of the membrane in the synthesis of DNA, RNA and protein is not clear. It is difficult to prove whether substances adhering to the membrane after several washings are contaminants or are membrane components. The converse is also true—some truly membrane components are easily washed off. Nevertheless it seems likely that membrane-bound polysomes are active in protein synthesis though it has not been established whether they are concerned only in synthesis of certain classes of proteins (e.g. extracellular proteins). It appears likely that DNA is attached to the cytoplasmic membrane either directly or through the mesosomal membranes (p. 122) since *membrane* fractions have been obtained containing a small percentage of the total membrane together with most of the DNA (free DNA would not be obtained in this fraction). It is possible that membranes are connected in some way with DNA metabolism either in respect to DNA synthesis or to the separation of the replicating strands.

There are vast gaps in our knowledge concerning the location of various enzyme systems in bacteria but it is likely that many such systems will be traced to the protoplast membrane. As yet nothing is known either of the relative amounts of enzymic and structural proteins present or of the molecular orientation of the proteins in the membrane.

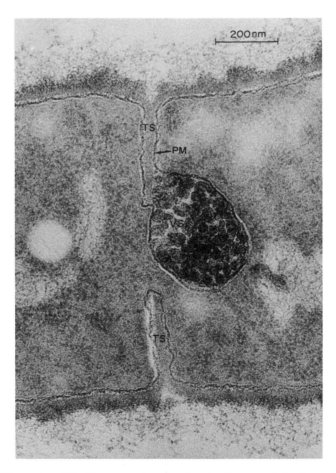

Figure 57. Thin section of *Bacillus megaterium* showing the close association between the transverse septum (TS), the plasma membrane (PM) and the mesosomal vesicles (VS). Reproduced with permission of D. J. Ellar, D. G. Lundgren and R. A. Slepecky. *J. Bacteriol.* **94** (1967) 1189.

Mesosomes

Structure

The majority of Gram-positive and Gram-negative bacteria contain membranous organelles formed by involution of the protoplast membrane and located in the region of the bacterial DNA or at the site of cell division or spore formation. The terms *mesosomes*, *chondrioids* or *peripheral bodies* have been applied to these structures. The mesosome shown in Figure 57 is associated with an early stage of septum formation and thin sections showing later stages of the process demonstrate that mesosomes are present at this site throughout its entire construction (Figure 58). Different arrangements of the membranes in mesosomes have been observed; vesicles, probably interconnected, are seen in

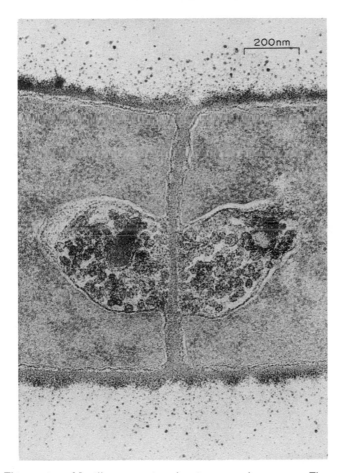

Figure 58. Thin section of *Bacillus megaterium* showing a complete septum. The septum is only half the thickness of the finished wall in the zone where the mesosomes are present, suggesting that the mesosome is involved in synthesis of the septum. Reproduced with permission of D. J. Ellar, D. G. Lundgren and R. A. Slepecky. *J. Bacteriol.* **94** (1967) 1189.

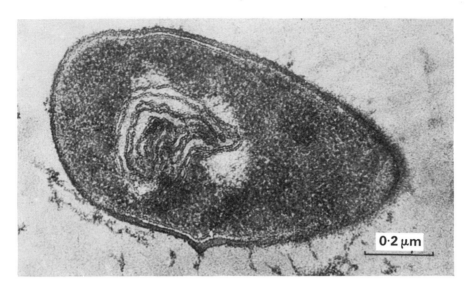

Figure 59. Mesosome of *Bacillus subtilis* sectioned in the region of the nucleus. The internal structure takes the form of whorls of membranes. Reproduced with permission of W. Van Iterson. *Bact. Rev.* **29** (1965) 299.

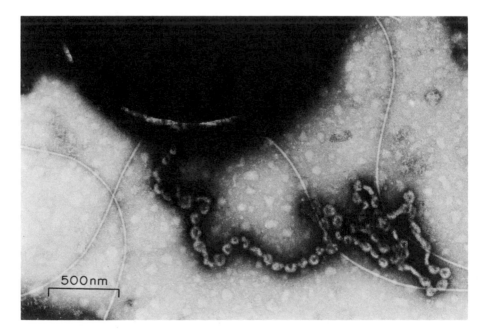

Figure 60. Negatively-stained protoplast of *Bacillus subtilis* to which is attached a long appendage consisting of a string of small vesicles. This appendage is thought to be an extruded mesosome. Reproduced with permission of A. Ryter. *Bact. Rev.* **32** (1968) 39.

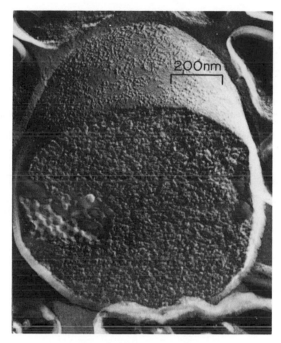

Figure 61. Freeze-fractured *Bacillus subtilis* showing a mesosome in close contact with the septum. Reproduced with permission of N. Nanninga. *J. Cell. Biol.* **39** (1968) 251.

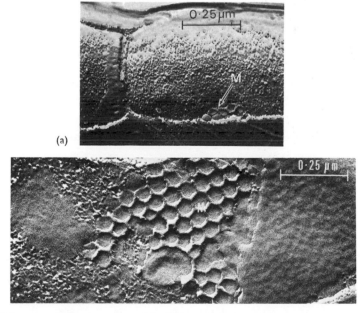

Figure 62. Freeze-etched preparations of *Bacillus cereus* showing mesosomes in close contact with the protoplast membrane. Reproduced with permission of S. C. Holt and E. R. Leadbetter. *Bact. Rev.* **33** (1969) 346.

most sections (Figure 57) while whorls of membranes are seen in others (Figure 59). It has been suggested that the appearance may be dependent on the method of fixation prior to sectioning. When protoplasts are obtained from the parent organism the contents of mesosomes are extruded and the 'pearl-string' appearance of the material seen in Figure 60 is remarkably similar to that of the interconnected vesicles of intact mesosomal structures. Vesicles may fuse to give

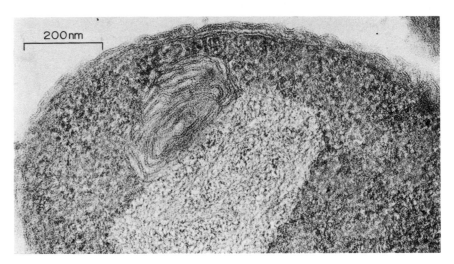

Figure 63. Thin section of *Escherichia coli* showing a mesosome in contact with the nuclear material. Reproduced with permission of A. Ryter. *Bact. Rev.* **32** (1968) 39.

lamellar-type membrane structures and if this does occur it suggests that the mesosomal contents are unstable in structure and can change from one form to the other. Micrographs obtained by the more reliable freeze-etching procedure emphasize the close association between the mesosomes and the cytoplasmic membrane and give the impression that the internal membranes are present as vesicles (Figures 61 and 62). Mesosomes of Gram-negative bacteria are much less prominent than those of Gram-positive organisms: they are present mainly as small infoldings of the cytoplasmic membrane and are frequently associated closely with the nuclear material (Figure 63).

Function
It has been suggested that mesosomes may act as a site for cell respiration and energy production, a specific site for cross-wall formation in Gram-positive bacteria, a control centre for orderly cell division, an organ of attachment for the bacterial nucleus during replication, and a site for DNA uptake during transformation. Methods for the separation of cytoplasmic membrane and the internal membranes of mesosomes have been developed which enable some of these hypotheses to be tested. As shown in Figure 60 the conversion of intact

cells to protoplasts results in the extrusion of mesosomal vesicular membranes and these can be separated by lowering the Mg^{2+} concentration. The intact protoplasts and mesosomal membranes can then be recovered separately by differential centrifugation. It should be emphasized that the mesosomal membrane fraction may not contain the bounding membrane of the mesosome which probably becomes part of the protoplast membrane: therefore a negative result in respect to an enzyme under investigation does not prove categorically that the mesosome is not involved in a certain process. The mesosomal membranes are practically devoid of many enzymes and carriers involved in respiration and are not specific sites for the synthesis of phospholipid or peptidoglycan. The location of mesosomes at the site of septum formation provides strong support for some role in its synthesis—whether as a site of assembly of the building blocks or by directing and controlling autolytic enzymes that are necessarily involved in cell division has yet to be proved. Furthermore, mesosomes are re-formed in cells that have been rapidly chilled (and consequently have lost the mesosomes) before septum formation and cell division take place. It is also pertinent that protoplasts do not contain mesosomes and do not divide when incubated in liquid media. Another suggested function concerns anchorage of the chromatinic body during replication of the bacterial nucleus: this is based on analysis of thin sections which show that contact is maintained between the mesosome and the replicating DNA.

Photosynthetic membranes

Purple and green bacteria have the ability to carry out photophosphorylation, i.e. they are photosynthetic organisms. The purple bacteria contain membrane-bounded structures arranged as lamellae or as vesicles. The vesicles or 'chromatophores' of *Rhodospirillum rubrum* (Figure 64) can be obtained by differential centrifugation of extracts of photosynthetically-grown organisms that have been mechanically disrupted. The particles are approximately 60 nm in diameter, they are chemically similar to membrane preparations, and they catalyse a light-dependent process resulting in the production of energy. The cellular content of vesicles is paralleled by the chlorophyll concentration: as the light intensity under which the bacteria are grown is increased so both the cellular content of 'chromatophores' and the amount of chlorophyll decrease and *vice versa*. Because the membrane fraction of *Rhodospirillum rubrum* prepared from osmotically-lysed organisms contains the entire content of chromatophores it may be that the vesicles are attached to or arise from the protoplast membrane. Thin sections of lysed organisms (Figure 65) demonstrate that the protoplast membrane is continuous with the membrane surrounding some of the vesicles or lamellar structures. It is believed that the vesicular structures obtained by physical fractionation techniques are artifacts resulting from detachment of the photosynthetic membranes from the cytoplasmic membrane and the subsequent formation of vesicles. Many other purple bacteria contain lamellar structures arranged in stacks near the periphery of the cell (Figure 66) and attached to the cytoplasmic membrane. Although it has not been definitely

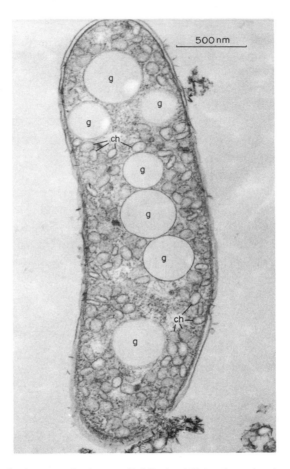

Figure 64. Longitudinal section of an intact cell of *Rhodospirillum rubrum* showing chromatophores (ch) and granules of poly-β-hydroxybutyrate (g). Reproduced with permission of E. S. Boatman. *J. Cell Biol.* **20** (1964) 297.

established that the lamellae are the sites of photosynthesis the circumstantial evidence is strong; the amounts of membrane and of chlorophyll alter in parallel on changing light intensity, temperature or oxygen tension.

Until recently it was believed that the chlorophyll pigments in Green bacteria were not located in any special structure. However improved fixation and embedding techniques have revealed large vesicles (30×100 nm) lying immediately beneath the cytoplasmic membrane (Figure 67). These vesicles have been isolated and shown to be enriched in relation to the photosynthetic pigments.

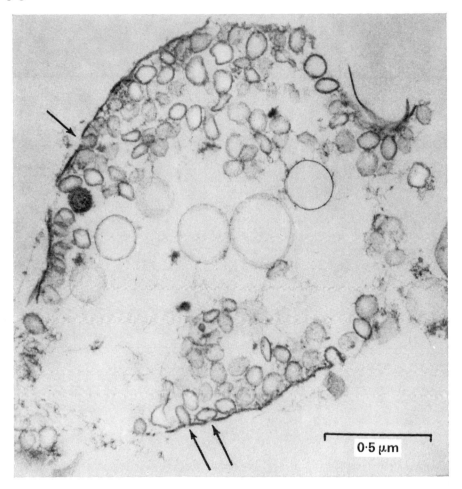

Figure 65. Section of spheroplast of *Rhodospirillum rubrum* demonstrating continuity of the protoplast membrane with the membrane surrounding the chromatophores (arrows). Reproduced with permission of E. S. Boatman. *J. Cell Biol.* **20** (1964) 297.

Gas vacuoles

Gas vacuoles are found in many non-motile bacteria living in aquatic environments. They are present in green and purple bacteria and in some non-photosynthetic organisms (e.g. halobacteria). Gas vesicles may be arranged side by side to form a compound gas vacuole (Figure 67) or they may occur singly as in some halobacteria (Figure 68). The vesicles are of varying shapes (Figure 69), are bounded by a membrane (not a 'unit membrane') and appear regularly

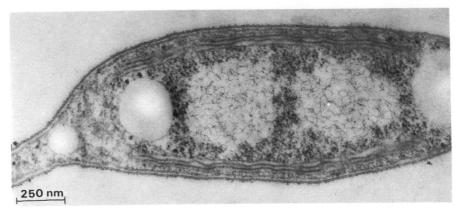

Figure 66. Transverse section of *Rhodomicrobium vannielii* showing the symmetrical stacking of lamellar membranes on either side of the cell. Reproduced with permission of W. C. Trentini and M. P. Starr. *J. Bacteriol.* **93** (1967) 1699.

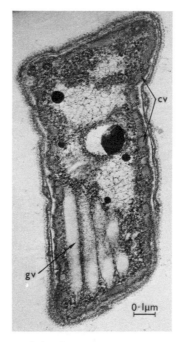

Figure 67. Longitudinal section of the Green bacterium *Pelodictyon clathratiforme* showing a single gas vacuole composed of four gas vesicles (gv), and a cortical array of the photopigment-bearing chlorobium vesicles (cv). Reproduced with permission of G. Cohen-Bazire, R. Kunisawa and N. Pfennig. *J. Bacteriol.* **100** (1969) 1049.

Figure 68. Longitudinal section of *Halobacterium* showing an irregular distribution of short gas vesicles throughout the cell. Reproduced with permission of G. Cohen-Bazire, R. Kunisawa and N. Pfennig. *J. Bacteriol.* **100** (1969) 1049.

striated when viewed by negative-staining. It is possible that they function as buoyancy tanks since they can be deflated by a sudden increase in pressure which causes loss of buoyancy. It is thought that they are freely permeable to gases.

BACTERIAL NUCLEI
(CHROMATINIC BODIES, DNA-PLASM)

The bacterial nucleus is not surrounded by a membrane, has no defined shape and is seen in electron micrographs as an irregularly-shaped, electron-transparent area within the cytoplasm. It has been calculated that the genetic information of an *Escherichia coli* cell is contained in approximately 1.6×10^7 nucleotides (*c.* 3% of the dry weight of the cell). DNA preparations from lysed spheroplasts spread on thin protein films support the view that one or very few molecular strands are present (Figure 70) and only a few free ends can be seen. The amount of DNA per cell varies considerably depending upon the rate at which the cell is growing. This difference is reflected in the number of nuclei per cell; rapidly dividing cells contain 2–4 nuclei while slow-growing cells possess 1–2 nuclei. At high growth rates DNA replication occupies approximately 80% of the division time and cytological observations suggest that nuclear separation occurs almost immediately after DNA replication is completed.

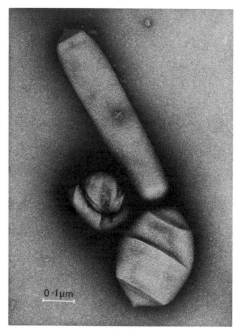

0·1μm

Figure 69. Isolated gas vesicles of *Halobacterium*. Reproduced with permission of G. Cohen-Bazire, R. Kunisawa and N. Pfennig. *J. Bacteriol.* **100** (1969) 1049.

It is obvious that the double strand of DNA, 2000 μm long, has to be folded several hundred times in order to be compressed into a bacterium perhaps 2 μm long. The state of the DNA in the intact nucleus is seen in thin sections of bacteria that have been fixed by the Ryter-Kellenberger (R-K) method. In this procedure fixation with osmium tetroxide is carried out in the presence of calcium ions and amino acids followed by treatment with an aqueous solution of uranyl ions. This technique prevents shrinkage during the subsequent dehydration step and results in a homogeneous appearance of the nucleus (Figure 71). Normal fixation methods not involving the use of Ca^{2+} and amino acids result in side-to-side aggregation of the DNA fibrils (Figure 72). The main disadvantage of the R-K technique is the difficulty in recognizing the orientation

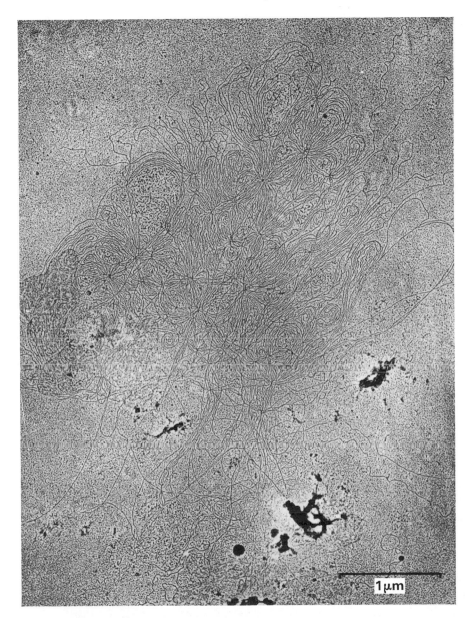

Figure 70. Nuclear DNA from *Micrococcus lysodeikticus* obtained from lysing protoplasts by the technique of Kleinschmidt. The protoplasts are lysed gently in a solution of cytochrome at pH 5 and spread on a solution of cytochrome. The apparent thickness of the DNA strand is obtained by circular shadowing. Few free ends are visible. Reproduced with permission of A. Kleinschmidt in the chapter by E. Kellenberger and A. Ryter in *Modern Developments in Electron Microscopy* (1962), Fig. 12, p. 367, edited by B. M. Siegel, Academic Press, New York.

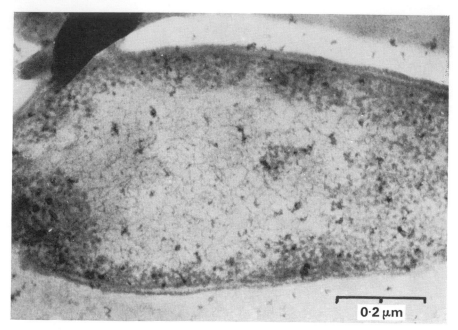

Figure 71. Section of *Escherichia coli* after R-K fixation (see text) showing homogeneous appearance of the nucleus. The black deposits in the nucleus are artifacts of staining. Reproduced with permission of G. Wolfgang Fuhs. *Bact. Rev.* **29** (1965) 277 and Pergamon Press, Oxford.

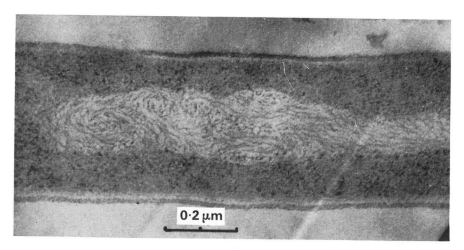

Figure 72. Thin section of *Bacillus subtilis* after osmic fixation and uranyl treatment. The fibrous appearance results from the random side-to-side aggregation of parallel-oriented DNA helices. Reproduced with permission of G. Wolfgang Fuhs. *Bact. Rev.* **29** (1965) 277 and Pergamon Press, Oxford.

of the DNA fibres but the micrographs obtained using this method show that the DNA-plasm is present in the hydrated state. It is possible that the high degree of hydration is particularly favourable for the penetration of enzymes and DNA-precursors between the strands during the replication process but this has not been determined experimentally.

The process of DNA replication (involving uncoiling of the Watson–Crick double helix) must be very complex unless large numbers of single strand nicks are present in the continuous molecule of DNA comprising each nucleus. The mechanism by which separation of the two daughter nuclei is effected in a dividing cell is not known, though mesosomes may be involved (see p. 122).

RIBONUCLEOPROTEIN PARTICLES (RIBOSOMES)

The cytoplasm of rapidly growing bacteria is filled with darkly staining granules approximately 10–20 nm in diameter and grouped in clusters. These granules are equivalent to the ribosomes which can be obtained by high-speed centrifugation of disrupted bacteria. If cytoplasmic material is released from protoplasts by osmotic shock, fixed preparations can be obtained in which the lightly stained substructure seems to be continuous over the whole micrograph while the densely-staining areas are ribosomes (Figure 73); those that are associated in clusters are presumably equivalent to polyribosomes although it is not

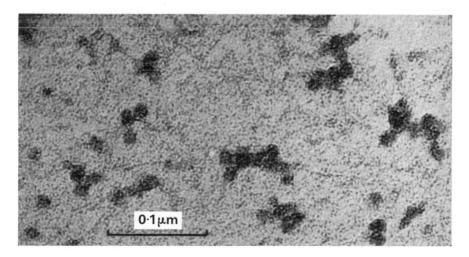

Figure 73. Unpurified ribosomal preparation of *Bacillus subtilis* obtained by fixation of cytoplasmic material released by osmotic shock. The dense material probably consists of polyribosomes (polysomes) though the strands of messenger RNA are not visible. Reproduced with permission of W. Van Iterson. *Bact. Rev.* **29** (1965) 299.

possible to distinguish the strand of messenger RNA. It is difficult to investigate the state in which ribosomes exist *in vivo*, but the available data suggest a high degree of organization within the cytoplasm, a view that is consistent with the great metabolic activity of bacteria.

CYTOPLASMIC INCLUSIONS

Reserve materials may accumulate in the cytoplasm when certain bacteria are incubated under specific conditions: among these products can be included glycogen, lipid droplets, polymerized inorganic metaphosphate and sulphur. Of these the first three are generally absent during active growth of the organism while sulphur granules seem to be essential as an energy reserve for sulphur bacteria.

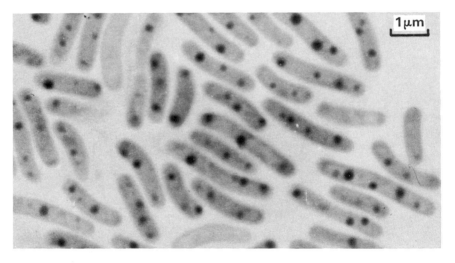

Figure 74. Metachromatic granules of *Spirillum* revealed by staining with toluidine blue. Photographed by Dr. J. P. Truant and reproduced with permission of R. G. E. Murray. See *The Bacteria*, Volume I (1960), Plate I, Fig. 6, p. 41. Eds. R. Y. Stanier and I. C. Gunsalus, Academic Press. New York and London.

Glycogen granules (granulose)

These appear as electron-transparent spheres without a limiting membrane; they may account for as much as 50% of the dry weight of the organism and tend to accumulate under conditions of nitrogen starvation in *Escherichia coli* when the organisms are provided with a carbon source. If nitrogen is supplied the accumulated glycogen is broken down and used as a carbon source in the same way as glucose.

Lipid granules

Poly-β-hydroxybutyric acid is accumulated in large granules—easily visible by phase contrast microscopy—by a wide variety of bacteria including bacilli and *Rhodospirillum rubrum*. It is a storage product and may occupy a considerable proportion of the cytoplasm in cells of old cultures (Figure 64). Large numbers of lipid droplets may be found in bacteria: these can be stained with lipid-soluble dyes such as Sudan Black.

Metachromatic granules (volutin granules, polymetaphosphate granules)

In the majority of organisms which contain these granules their appearance and the concomitant accumulation of polyphosphate tends to occur towards the end

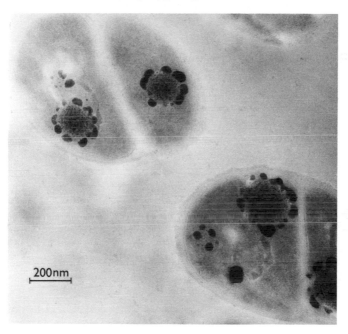

Figure 75. Thin section of *Micrococcus lysodeikticus* showing the electron-dense particles containing inorganic polyphosphate. Reproduced with permission of I. Friedberg and G. Avigad. *J. Bacteriol.* **96** (1968) 544.

of the growth cycle. They are composed in part of inorganic metaphosphate in a polymer of high molecular weight and are readily stained with toluidine blue (Figure 74), hence the name metachromatic. The granules from *Micrococcus lysodeikticus* are atypical in that they accumulate during the log phase of growth and disappear in the stationary phase. The main constituents of the granules are lipid and protein in addition to the phosphate which may represent more than 50% of the total phosphate in the cell. The electron-dense phosphate is present in a rosette arrangement around the periphery of the granule (Figure 75).

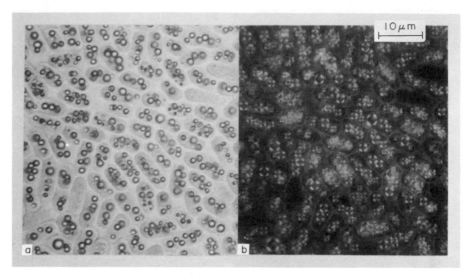

Figure 76. Photomicrograph of *Chromatium okenii* containing sulphur granules. (a) Bright-field microscopy, (b) polarising microscopy. Reproduced with permission of G. J. Hageage, E. D. Eanes and R. L. Gherna. *J. Bacteriol.* **101** (1970) 464.

Sulphur droplets

Prominent deposits of sulphur (Figure 76) are found in the large purple sulphur bacteria growing in the presence of sulphide. The sulphur arises from the oxidation of sulphide during photosynthesis and is in turn oxidized to sulphate when the cells have been depleted of sulphide, and so the droplets disappear. The sulphur is probably in an unstable form since the stable orthorhombic allotopic form does not exist as wet spherical droplets. The small purple sulphur bacteria and green sulphur bacteria do not store sulphur within the cytoplasm but deposit it outside the cell.

FURTHER READING

A Cell walls

General reviews
1 Ghuysen J-M., Strominger J. L. and Tipper D. J. (1968) Bacterial cell walls. *Comprehensive Biochemistry* **26A**, 53.
2 Osborne M. J. (1969) Structure and biosynthesis of the bacterial cell wall. *Ann. Rev. Biochem.* **38**, 501.

The Cell surface
3 Glauert A. M. and Thornley M. J. (1969) The topography of the bacterial cell wall. *Ann. Rev. Microbiol.* **23**, 159.
4 Holt S. C. and Leadbetter E. R. (1969) Comparative ultrastructure of selected aerobic spore-forming bacteria: a freeze-etching study. *Bact. Rev.* **33**, 346.

Peptidoglycan
5 Ghuysen J-M. and Leyh-Bouille M. (1970) Biochemistry of the bacterial wall peptidoglycan in relation to the membrane. *FEBS Symposium* **20**, 59.
6 Ghuysen J-M. (1968) Use of bacteriolytic enzymes in determination of wall structure and their role in metabolism. *Bact. Rev.* **32**, 425.

Teichoic acids
7 Baddiley J. (1968) Teichoic acids and the molecular structure of bacterial walls. *Proc. Roy. Soc. ser. B* **170**, 331.
8 Archibald A. R., Baddiley J. and Blumson N. L. (1968) The teichoic acids. *Adv. Enzymol.* **30**, 223.

Lipopolysaccharides
9 Luderitz O., Jann K. and Wheat R. (1968) Somatic and capsular antigens of Gram-negative bacteria. *Comprehensive Biochemistry* **26A**, 105.
10 Luderitz O. (1970) Recent results on the biochemistry of the cell wall lipopolysaccharides of *Salmonella* bacteria. *Angewandte Chemie* **9**, 649.

B Membranes
11 Stoeckenius W. and Engleman D. (1969) Current models for the structure of biological membranes. *J. Cell Biol.* **42**, 613.
12 Korn E. D. (1969) Cell membranes: structure and synthesis. *Ann. Rev. Biochem.* **38**, 263.
13 Salton M. R. J. (1967) Structure and function of bacterial cell membranes. *Ann. Rev. Microbiol.* **21**, 417.

Photosynthetic membranes
14 Cohen-Bazire G. and Sistrom W. R. (1966) The prokaryotic photosynthetic apparatus. In *The Chlorophylls*, p. 313. Vernon L. P. and Seely G. R. (eds.), Academic Press, New York.
15 Lascelles J. (1968) The bacterial photosynthetic apparatus. *Advances in Microbial Physiology* **2**, 1.

C Books and reviews of general interest
16 Salton M. R. J. (1964) *The bacterial cell wall*. Elsevier, London.
17 Rogers H. J. and Perkins H. R. (1968) *Cell walls and membranes*. E. and F. N. Spon Ltd., London.
18 Pollock M. R. and Richmond M. H. (eds.) (1965) *Function and structure in Micro-organisms*. Fifteenth Symposium of the Society for General Microbiology. Cambridge University Press.

19   Chapman D. (ed.) (1968) *Biological membranes, physical fact and function*. Academic Press,
     London and New York.
20   Cole R. M. (ed.) (1965) Symposium on the fine structure and replication of bacteria and their
     parts. *Bact. Rev.* **29**, 277. (Contributions on nuclei, cytoplasm and cell walls.)
21   Rogers H. J. (1970) Bacterial growth and the cell envelope. *Bact. Rev.* **34**, 194.

Chapter 2
Growth: Cells and Populations

THE GROWTH OF POPULATIONS

Introduction

Unlike the higher forms of life, bacteria do not have obligatory life cycles and can reproduce continually as vegetative, undifferentiated cells. In adequate culture media they grow bigger and eventually divide into two, often very rapidly. Indeed, bacteria can be kept from growing only by making the environment inhospitable or fallow. This may result from the growth of the organisms since, when a certain density is reached, the environment will be depleted of essential nutrients or toxic products of metabolism may have accumulated. Bacteria differ from many differentiated metazoans because even if they are not actually growing, they are poised to do so.

Free life and unicellularity also require a high degree of adaptability. Changes in the environment cannot be met by homeostasis since this demands the interaction of specialized cells. Rather, bacteria deal with physical or chemical alterations of their surroundings by exhibiting a different set of metabolic activities. In other words, they are capable of existing in a variety of physiological states which can be quite different from one another. In addition, bacteria shift from one such physiological state to another in a rapid and efficient manner.

Balanced growth

When bacterial cells are placed in a nutritionally compatible environment they will grow and, in due time, divide. If the environment is not altered the resulting daughter cells will similarly grow and divide. This will go on at a rate which is characteristic of the particular strain and of the particular culture medium. When the population reaches a certain level the environment will be perturbed due to exhaustion of nutrients, lessened availability of oxygen, or the accumulation of toxic products. But until this occurs the bacteria grow in an unhindered manner, and each cell will grow, on the average, at the same rate as its predecessors. If we measure growth in any one of many ways, for instance by determining the mass, we will find that the *rate of increment* is constant. This condition has been called *balanced growth*, since all cell constituents increase proportionally over the same period of time. It is not likely that such steady states exist for long in natural situations since even small environmental changes may perturb the growth of bacteria. In the laboratory, however, conditions of approximately balanced growth can be obtained by the use of experimental procedures designed to keep the environment constant.

Working with populations of bacteria in balanced growth has several practical advantages. Samples taken at different times from the same culture vary only by the increment in growth between those times. For instance, if a sample of a culture is taken at a given time and another sample of the same size is taken after the population has doubled the *content of any cell constituent* in the two samples will vary by the same factor, namely two. Moreover if growth was balanced over the interval between the two samplings the average physiological properties of individual cells will be unchanged. If an investigator maintained a culture in his laboratory in balanced growth, he would have physiologically comparable populations 'on tap'. Another advantage in working with cultures whose growth is balanced is that the same *rate constant* defines not only the overall growth of the population but also the synthesis of each cell component. If a culture doubles in 30 minutes, the rate constants of the syntheses of DNA, RNA, proteins and cell envelopes, will all be one doubling per 30 minutes. The advantage lies in the fact that kinetics of the system can be totally defined *in relative terms* without resorting to involved analytical procedures. It is sufficient to determine the rate constant of any easily measurable parameter.

Balanced growth can be described mathematically as follows. Let M be the mass of bacteria, t the time, then

$$dM/dt = kM$$

or, in words, the change in the amount of mass with regard to time is proportional to the mass present. The expression $(dM/dt)(1/M)$, the change in mass with regard to time per unit of mass, is equal to k, the *growth rate constant*. As we said before, mass can be replaced in the equation by any measurable cell constituent or by the number of cells. Note that this is the equation for a first order chemical reaction. It states what should be intuitively apparent, that the more bacteria are present, the more bacteria will be made, as long as growth is not perturbed.

The equation can be integrated to give

$$M_2/M_1 = e^{k(t_2 - t_1)},$$

where M_1 is the bacterial mass at time t_1 and M_2 that present at time t_2. Taking logarithms of both sides,

$$\ln M_2/M_1 = k(t_2 - t_1),$$

which shows that a plot of the logarithm of the mass versus the time of measurement gives a straight line with slope k. This equation differs in the units from one that describes the increase in the number of bacteria as a result of binary fission. If at each division one bacterium becomes two and these become four, the series $2, 4, 8, 16, \ldots, 2^n$ describes the increase in bacterial numbers. Thus, starting with a single cell,

$$N = 2^n,$$

where N is the number of bacteria and n the number of generations. If the number of cells at time t_1 is N_1 and at time t_2, N_2, then

$$\ln N_2/N_1 = n \ln 2.$$

Since the relative increment of cell mass and cell members is the same as long as growth is balanced

$$\frac{N_2}{N_1} = \frac{M_2}{M_1} \quad \text{and} \quad k = \frac{n \ln 2}{t_2 - t_1}$$

thus relating mathematically what can readily be said in words, namely that the *growth rate constant* is a function of the number of times the cells will, on the average, divide in a given time period. The growth rate constant can then be expressed as the *mean generation time*, the interval of time required for a doubling in number, mass, etc. of a bacterial population. As will be discussed later (p. 156), the generation times of individual cells may vary; therefore, the mean generation time refers to populations and not to individual cells.

The measurement of growth

In the laboratory growth is defined operationally in a variety of ways, depending on what is measured. For instance, growth may be represented by the increment in cell mass, cell numbers, or any cell constituent. One can also measure, and relate to growth, the utilization of nutrients or the accumulation of metabolic products. If cultures are in balanced growth these measurements will be equivalent in the sense that they have the same rate constant. For this reason it is not unexpected that the growth parameters most often used are those that can be easily and accurately measured, e.g. the turbidity of a liquid culture which depends on the amount of light that is scattered by the organisms and is related to the bacterial mass.

The *total* number of organisms can be determined by direct count under a microscope or by the use of electronic particle counters. These measurements do not differentiate between dead and living bacteria and this is a serious handicap in some experimental situations. However, under conditions of balanced growth, cultures of most of the commonly used species consist largely of living cells. The number of *living* bacteria is usually determined by colony count, that

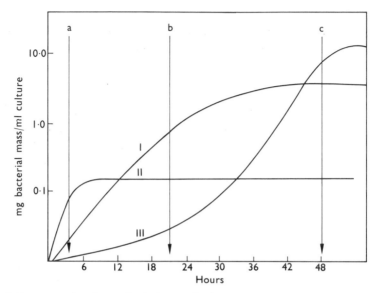

Figure 1. Various modes of growth of a bacterial culture.

This graph shows in a schematic fashion how bacteria may grow in different environments. Culture I grew exponentially until it reached a density of about 1 mg/ml bacterial dry weight, then slowed down. Culture II grew faster but its growth was terminated more abruptly and at a lower density. Culture III showed a long lag in the initiation of growth but achieved a higher density than culture I.

If growth were measured after 3, 20, or 48 hours, different answers would be obtained. After 3 hours, the highest amount of growth would be in culture II, after 20 hours in culture I, and after 48 hours in culture III. This indicates that a single measurement may give a spurious answer about the growth response to a particular environment.

is by placing an appropriate dilution on or in a solid medium. Since colonies arise from single living cells, the number of colonies multiplied by the dilution factor equals the number of living bacteria originally present.

Growth is often measured as the response of bacteria to environmental changes. For instance, the most general question that can be asked about the action of a drug on bacteria is how it affects their growth. Evidently there are many experimental approaches to such questions. The simplest experiment that can be done is to expose flasks or tubes of freshly inoculated cultures to various conditions and to measure the effect on growth. It should be apparent, however, that if the yield of bacteria is measured only at one time, the information obtained may be misleading. In one flask growth may have ceased before the measurement was made but in another growth may only have been retarded and the organisms be still growing at the time of the measurement. The yield of organisms would then not be a reliable index of the effect of the environment. An example of this is shown in Figure 1. The pitfalls of this method can be avoided if, instead of determining the bacterial yield at one time, the rate of growth is measured over a suitable interval of time. It should be evident that of all the measurable characteristics, the rate constant reflects the whole phenomenon of growth in the most direct and comprehensive manner.

Different rates of balanced growth

Probably more experiments have been carried out with *Escherichia coli* than with all other bacterial species. There are many reasons for this, one of which is that it grows well in chemically defined media. Some of these media are very simple, containing only mineral salts and a single organic component as the source of carbon and energy. When glucose is the organic component the bacterial mass may double every 45–60 minutes. Growth is slower when other substances such as acetate or succinate are used as the carbon source. The growth rate is increased considerably by the addition of mixtures of amino acids, purines and pyrimidines, B vitamins, etc.; but of all chemically defined substances, the amino acids have the greatest effect on the growth rate. However, the fastest growth is achieved when a complex mixture of nutrients is present. Such mixtures, which are found in meat infusions and in the water-soluble extracts of meat or yeast, are called *infusion* or *nutrient broths*. In addition to most small molecular weight building blocks, these media contain peptides and other complex substances, and, of course, are undefined in composition.

In defined and in complex media, cultures of organisms like *Escherichia coli* grow until they reach densities of several hundred micrograms of bacterial dry weight per millilitre. Actual quantities vary with the strain, the culture medium employed, and the efficiency of aeration. One reason for concern about the upper limit of balanced growth is that biochemical determinations often require rather large samples and the investigator needs to harvest his culture at the highest cell density consistent with his purpose.

Continuous culture

Balanced growth can be prolonged simply by diluting the culture at regular intervals with fresh medium. The least degree of perturbation will be introduced when this is carried out frequently, for instance, every time the culture doubles in mass. This can be done over a long period of time, producing a saw-tooth pattern of growth (Figure 2, p. 23).

For many bacterial species in common use in the laboratory a variety of growth rates can be obtained simply by choosing the appropriate medium. With *Escherichia coli* at 37° these rates range from doubling times of 18–20 minutes in a rich meat infusion broth to about 2 hours in media where the sole carbon and energy source is one of several sugars which are metabolized inefficiently. Even slower rates of balanced growth at 37° can be obtained by special manipulations. Thus a culture may be allowed to grow until something in the medium becomes limiting. Fresh medium can then be run continuously into a vessel at a rate which doubles the volume, say every 3 hours. Clearly,

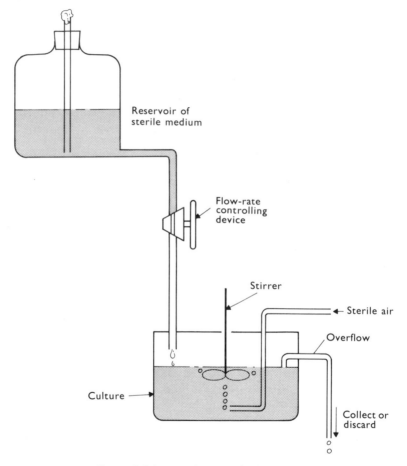

Reservoir of
sterile medium

Flow-rate
controlling
device

Stirrer

← Sterile air

Overflow

Culture ──

Collect or
discard

Figure 2. Schematic diagram of a chemostat.

growth will be limited by the flow rate. To maintain this rate at a constant value an amount of culture equivalent to that flowing in must be withdrawn simultaneously. If the process is carried out over a sufficient time to achieve equilibrium the culture will have not only a constant volume but also a constant cell density since a constant number of organisms is withdrawn per unit of time. This method is called *continuous cultivation*. In practical terms it is carried out in an apparatus that can vary considerably in outward appearance and which is called a 'chemostat' or 'bactogen' (Figure 2). Any chemostat consists of a culture vessel which can be maintained at a proper temperature and which can be aerated. This is usually done by introducing filtered air through a device that ensures a suitably small size of the air bubbles through the medium. The vessel is provided with a syphon or overflow line which maintains the liquid level constant. The fresh culture medium is introduced from a reservoir through a system which regulates the flow rate. These systems vary in different laboratories and may be a capillary or a stopcock that makes suitable resistance to flow, a set of suitably timed solenoids, or some kind of an adjustable delivery pump. The rate of flow of the fresh medium into the culture vessel determines the rate of growth.

In chemostats special consideration must be given to the culture medium, for the following reason. If the usual culture media are used, growth will proceed until the medium becomes exhausted or toxic. The bacterial density would then be very great and beyond the limits which permit balanced growth. For this reason it is customary to limit the density of the population by the simple device of making the concentration of any one nutrient limiting. Thus, in a defined medium one of several essential elements can be added at a 'concentration below that which supports the maximum yield of bacterial mass. The limiting factor may be carbon, nitrogen, phosphorus, sulphur, magnesium or one of a number of other essential ingredients of the medium. This permits one to choose not only the bacterial density (by the concentration of the limiting nutrient) but also one of a continuum of growth rates, from the maximum afforded by the medium to very slow ones (by adjusting the flow rate). Indeed, doubling times of 24 hours or longer can be obtained with bacteria that would otherwise double in less than one hour.

Response of growth to environmental changes

So far we have assumed that the bacteria were placed in reasonably favourable circumstances and that the only variable was the kind of nutrient or the rate of its addition. But how does bacterial growth respond to other physical and chemical variables? This question can only be answered in very general terms since the number of these variables is great and there are large numbers of bacterial species which exhibit different responses.

The rate of bacterial growth has an optimum range of temperature and pH, being slower on either side of it, but there are species that can grow at values of pH or temperature that would be extremely unfavourable for higher forms of life. Whereas many of the micro-organisms commonly studied in the laboratory grow well at about ph 7, some bacteria grow at pH 2, others at pH 12. However, it is unlikely that these extreme pH values of cultivation are the same as the intracellular pH and it would be of great interest to determine how such differences are maintained. Unlike the pH value, the temperature inside the cell must be the same as that of the environment. Therefore, the range of temperatures over which an organism can grow must also be the range over which its enzymic machinery can function. Our understanding of the relationship between growth and enzymic action has been advanced in recent years by the study of temperature-sensitive mutants of *Escherichia coli.* Two types of mutant have been found: those that cannot grow at the high temperatures of the usual range of this organism and those that cannot grow at lower temperatures. In some mutants of the first class certain enzymes have become more sensitive to heat than the corresponding ones in the parent organism. On the other hand, in mutants that require high temperatures, the enzymes probably operate at a rate which is too slow at lower temperatures for the metabolic needs of growth.

The structure and composition of bacteria is related to the growth rate

At any temperature, the structure and chemical composition of bacteria vary with the growth rate. In general, when bacteria grow faster they are larger and contain a high amount of RNA. In several species, this variation has been shown to be related to the rate of growth in a simple and systematic manner. In the experiment shown in Figure 3, cultures of the same species of organism, *Salmonella typhimurium*, were grown in a variety of media that permitted different

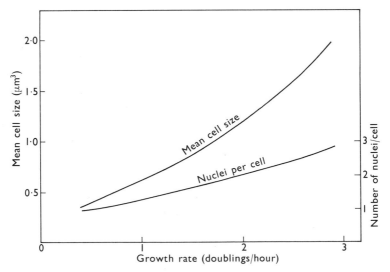

Figure 3. The relationship of the mean cell size and the number of nuclei per cell to the rate of growth of *Salmonella typhimurium.*

Cultures of *Salmonella typhimurium* were grown in a variety of culture media that supported different rates of balanced growth. The mean cell size and the mean number of nuclei per cell were estimated and plotted against the growth rate. As shown in the diagram, there is a smooth and continuous relationship between the rate of growth and these cytological properties.

rates of balanced growth. The variation in cell size was quite large, extending over a six-fold range. In other species even greater variation occurs, suggesting that with regard to structure and composition bacteria can undergo phenotypic changes of great magnitude. Obviously, the cell size of bacteria is not a species characteristic and, if used in taxonomy, must be determined under specified conditions.

There are situations where the cell dimensions are not related in a simple way to the growth rate. For instance, cell division may be prevented without substantially inhibiting the synthesis of cellular material by low levels of many physical or chemical agents which at higher levels are bacteriostatic, e.g., ultraviolet light, penicillin. This leads to the formation of very long filaments which grow at nearly normal rates. In other instances the cell wall is weakened, leading to cells that are globular or irregular in shape. It is therefore important to keep in mind that the relatively simple dependence of cell size on growth rate holds only for situations where growth is unhampered or uninhibited and, consequently, balanced.

Although the great variability in the chemical composition and size of bacterial cells is obviously a complex phenomenon it is possible to consider it on different levels. On a cytological level it is striking that the average number of nuclear bodies per bacterium is greater at faster growth rates (Figure 3). The reason for this is not known but it is likely that it is related in some manner to the process of cell division. Bacteria growing slowly have a higher content (per unit mass) of cell membrane and cell wall than bacteria growing faster. Consequently, at fast growth rates synthesis of these cell envelopes might be limiting. The cells could then adjust to the relatively smaller amount of envelope material by an increase in volume and hence by a decrease in the ratio of surface to volume. At a subcellular level, the cell mass *per nucleus* (regarded as the amount of cytoplasm directly controlled by one individual genome) varies mainly because of variations in the ribosome content (Figure 4). The amount of these particles varies with the growth rate much more than does that of any other cell component, from a very low value in slow-growing cultures to one-third or more of the dry weight of organisms growing fast. It has been found that the number of ribosomes is directly proportional to the rate of growth which, in balanced growth, means also to the rate of protein synthesis. Thus, ribosomes function at a unique level of efficiency which appears to be the same in bacteria growing at different rates. Changes in the rate of protein synthesis require changes in the amount of participating ribosomes.

Bacteria growing at different temperatures in the same medium do not show substantial changes in their size and composition even though their rates of growth may be quite different. This suggests that it is predominantly the nutritional composition of the environment which determines the size and composition of the cells.

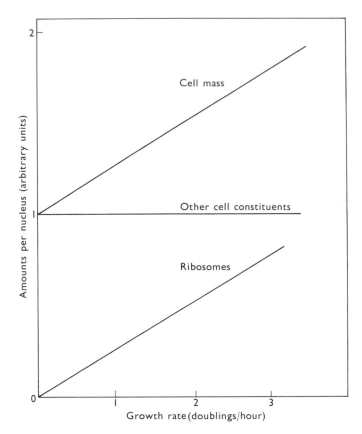

Figure 4. The relationship between the amount of various cell constituents per nucleus and the growth rate of *Salmonella typhimurium*.

The amount of cytoplasm *per nucleus* increases with the rate of balanced growth. As shown in this figure, this increment is due to the larger content of ribosomes in the faster growing cells.

Unbalanced growth

Natural environments vary rapidly and extensively and it would not be expected that bacteria would remain in a state of balanced growth for long periods of time. Obviously cultures do not grow indefinitely and, as growth ceases, a large number of gradual and interrelated events ensue. These vary with the organisms and the conditions of cultivation.

It seems easier to begin the discussion of these complex situations with a consideration of some of the events which take place in a controllable situation, namely when cultures are shifted artificially from one steady state to another. Cultures in balanced growth can be readily transferred from a medium giving one rate of growth to a medium giving another. The study of the transition is helped by the fact that the cells are in defined physiological states at the start and at the end of the experiment. There are two major types of transition which can be carried out in the nutritional sense. One consists of shifting the cells from a slow to a fast condition of growth. This can be done by adding richer nutrients or, in the case of a chemostat, by increasing the rate of inflow of nutrients into the culture vessel. This has been called the *shift-up*. The converse is the slowing down of growth by depleting the medium of some of its nutrients. This operation, called the *shift-down*, can be carried out by washing cells grown in a rich medium and transferring them into a nutritionally poorer medium. Thus, the addition of nutrient broth to a culture of *Escherichia coli* growing in a synthetic minimal medium constitutes a *shift-up*, whereas the washing of a culture grown in nutrient broth and its resuspension in the synthetic medium is a *shift-down*.

Since cells growing at the faster rate are larger and richer in RNA, it is pertinent to ask how, during a *shift-up*, small cells relatively poor in RNA acquire

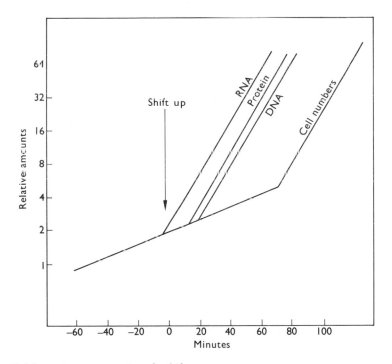

Figure 5. Schematic representation of a shift-up.

A culture of *Escherichia coli* growing at 37° in a synthetic medium at a rate of 1·2 doublings per hour is shifted at the time zero to nutrient broth, where the rate is three doublings per hour. Since growth is balanced, the semilogarithmic plot of the data before the shift includes in a single line the *relative* increment of the various components. After the shift the synthesis of RNA, protein, and DNA change to the new, faster rate in that order. Subsequently the rate of cell division increases to the same new value. The size and composition characteristic of the new medium is reached only 70 minutes after the time of the shift.

the characteristics of a fast-growing culture. As shown schematically in Figure 5, the rate of RNA synthesis speeds up almost immediately, is followed by an increase in the rate of protein synthesis and, a few minutes later, by an increase in the rate of DNA synthesis. Cell division continues at the original rate until the cells have acquired the large size characteristic of fast growth. Thus the composition characteristic of the new medium is established very rapidly. In general terms this sequence of events during a *shift-up* follows what would be expected from the earlier discussion. If the rate of protein synthesis is proportional to the ribosome content, this rate cannot be increased before the number of ribosomes becomes larger. What is perhaps surprising is the rapid and seemingly abrupt manner in which these changes in rate take place. This indicates that the syntheses of various types of macromolecules are not only related sequentially but are also subject to control mechanisms which, broadly speaking, are rapid and efficient.

Cells growing in a rich medium are provided with a variety of utilizable nutrients. The presence of these substances can be expected to result in widespread repression of many enzyme systems concerned with the synthesis of these compounds. Phenotypically, bacteria growing in a nutrient broth are devoid of the capacity for synthesis of amino acids, purines, pyrimidines, etc. Indeed, this sparing action by environmental components is probably the reason why bacteria grow faster in rich media. When such bacteria are 'shifted-down' to a poorer medium, such as one where a single sugar constitutes the only source of carbon and energy, they find themselves without many of the enzymes necessary for growth. The result, which can be readily observed, is a long lag in the synthesis of bacterial mass. However, DNA synthesis and cell division continue for some time, resulting in small uni-nucleated cells, which are poor in RNA (Figure 6). During this period the cells synthesize the enzymes they lack. Since there is no net synthesis of RNA or protein, this must be done by reorganizing the existing cell material (see Chapter 8). Gradually, net protein synthesis is resumed and after many hours the cells will have acquired the characteristics expected after growth in the poorer medium.

Transition situations of this type, where bacteria are removed from a particular condition of balanced growth, have been collectively called *unbalanced growth*. This refers not only to nutritional changes such as *shifts-up* and *-down* but also to physical or chemical disturbances created in the culture. Unbalanced growth can obviously be produced in a large number of ways. In some cases the changes are reversible while others result in the death of the cell. Many situations of unbalanced growth have been studied with bacteria and it is found that within limits and to varying extents certain synthetic processes can proceed while others are inhibited. Thus it has been possible to show that synthesis of the principal types of macromolecules can occur, at least in a limited way, in the absence of synthesis of others (see Chapter 8). Examples of typical manipulations which lead to the uncoupling of synthetic activities are shown in Table 1.

The synthesis of DNA is particularly easy to dissociate from other synthetic activities. This is true in both senses: DNA synthesis can be prevented without affecting other synthetic processes for considerable times, as, for instance, when

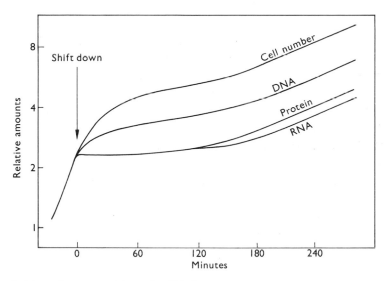

Figure 6. Schematic representation of a shift-down.

A culture of *Escherichia coli* growing in nutrient broth at a rate of three doublings per hour is washed by centrifugation and, at zero time, placed in a synthetic medium where the growth rate is 1·2 doublings per hour. As in Figure 5, a single line denotes the relative increment of various cell components before the shift. After the shift, cell division and DNA synthesis proceed for some time, resulting rapidly in small cells. Net protein and RNA synthesis, however, are resumed only gradually and the size and composition characteristic of the new medium are only obtained after three to four hours. During the initial lag, extensive turnover of RNA takes place.

thymine is removed from a mutant that requires it; conversely, DNA synthesis may continue in the absence of RNA or protein synthesis. This is observed in a *shift-down* or under certain conditions of starvation, such as when amino acids are removed from mutants requiring them (Table 1). Sometimes the amount of DNA made corresponds to that required for the completion of molecules which, at the onset of starvation, were in the process of synthesis. DNA synthesis is a sequential process, a cycle, and in a population that is not dividing synchronously, individual cells will be in a different stage of replication. During starvation each cell can complete its cycle of DNA synthesis but cannot initiate new cycles. This finding is of interest because it yields cells which are equal with regard to their cycle of DNA synthesis, namely, they have all completed it (Table 1).

The syntheses of RNA and of proteins, although readily dissociated from the synthesis of DNA, are closely connected to each other. Thus, if RNA synthesis is inhibited, for instance by the addition of the antibiotic actinomycin D, protein synthesis will proceed at a gradually diminishing rate which parallels the rate of decay of messenger RNA.

Table I. Examples and consequences of unbalanced growth

Cause of inhibition	Synthesis inhibited	Residual synthesis				Some consequences of unbalanced growth
		DNA	Messenger RNA	Ribosomal RNA	Protein	
Removal of thymine from requiring mutants	DNA	0	+	+	+	Cells die, sometimes due to induction of prophages
Addition of low levels of actinomycin D	RNA	+	0	0	+	Enzymes continue to be made on pre-existing messenger RNA
Removal of amino acids from 'stringent' mutants	Protein and ribosomal RNA	+	+	?	0	Each chromosome finishes its round of replication; messenger RNA synthesis is balanced by its turnover, thus, no net RNA accumulates
Removal of amino acids from 'relaxed' mutants	Protein	+	+	+	0	Cells accumulate protein-poor ribosomal particles
Addition of penicillin	Mucopeptide	+	+	+	+	Cells lyse due to continued cytoplasmic synthesis and lack of cell wall synthesis

When protein synthesis is inhibited by the addition of a variety of antibiotics (chloramphenicol, the tetracyclines, etc.), the synthesis of RNA continues at a fast rate and proceeds until a two or threefold increment has occurred. Depriving cells of a single amino acid causes not only immediate cessation of protein synthesis, but, in the case of some mutants, of ribosomal RNA synthesis as well. These are called 'stringent', as opposed to others, called 'relaxed' which continue to synthesize ribosomal RNA as well as messenger RNA, in the absence of the required amino acids. This suggests that the syntheses of messenger RNA and ribosomal RNA are controlled by different regulatory mechanisms. These operate on the level of RNA polymerase. Recent work has shown that this enzyme requires a specific protein factor for the transcription of ribosomal RNA cistrons (see p. 312).

Unbalanced growth can also arise as the result of inhibition of the synthesis of components of cell envelopes, notably cell wall peptidoglycan. This can be caused by the addition of penicillin at suitable concentrations or by the removal of diaminopimelic acid from mutants that require it. Under these conditions cell wall peptidoglycan is not synthesized but other cell components such as nucleic acids and proteins continue to be made for many hours. Under normal laboratory conditions this leads to the weakening of the cell envelopes and to extrusion of the protoplasm (Chapter 1).

The growth cycle of bacterial cultures

If an organism such as *Escherichia coli* is inoculated into a suitable growth medium in the evening, by the following morning there will be signs of considerable growth. If the culture medium was of a common type, and the incubation temperature 37°, and if the culture had been aerated by shaking or bubbling air through it, it is likely that bacterial density in this culture will be around 1 mg of cell mass per ml. In this condition the bacteria are not growing and have reached what is called the *stationary phase*. In order to obtain growing bacteria, a portion of this overnight culture must be sub-cultured into fresh medium. Growth will not seem to begin immediately but will start slowly and reach the maximum rate of growth in a gradual manner. This period is called the *lag phase*. The duration of this period depends on many considerations: the organism, the medium, the conditions of previous cultivation, the degree of aeration, and the way in which bacterial growth is measured. The latter point is particularly important because an overnight culture often consists of a mixture of living and dead organisms. Thus, estimates of live bacteria will give different answers from determinations of total bacterial mass (Figure 7). If for example, growth is followed by measuring the turbidity of the culture, the figure obtained will include the contribution of the dead organisms. The result will show a false lag in the resumption of growth.

If the behaviour of the live organisms alone is followed, cellular mass is seen to increase before the cell number increases, as in a slow-growing culture that had been transferred to a rich medium. In other words, the living cells act as if they had been made to undergo a *shift-up*, where small, RNA-poor cells became large and rich in RNA before reaching a steady state. Under certain circumstances some species of bacteria show a time lag of many hours when old cultures are transferred to new media. The nature of this lag depends on the organisms and the culture conditions and, in general, is poorly understood.

At the end of the lag phase (real or apparent) the cells are in a physiologically steady state, i.e. they undergo balanced growth. This condition, commonly called *exponential growth*, is terminated more or less gradually when the cells begin to exhaust essential nutrients or when they accumulate toxic products. Again, the details of these processes are entirely dependent on the species

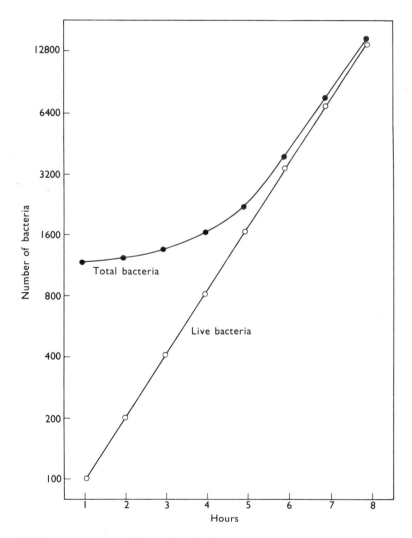

Figure 7. The measurement of live and total cells.

If in a population consisting initially of 100 living cells and 900 dead cells growth is measured by estimating the total number of cells, an apparent lag will be seen. However, if the living cells only are measured, there is no lag. This is what may happen when an old culture (24 hours or more) of *Escherichia coli* is transferred to fresh medium.

of organism and the actual conditions of cultivation. It is as difficult to generalize about the onset of the *stationary phase* as it is about the lag phase. When either the carbon or the nitrogen source is limiting in a synthetic medium, growth of *Escherichia coli* ends very abruptly. But cell division and DNA synthesis continue for some time, resulting in very small cells. This situation is a drastic *shift-down*, where cells go from a given growth condition to one where the growth rate is slower, in this case zero. Cultures in nutrient broth cease growing gradually,

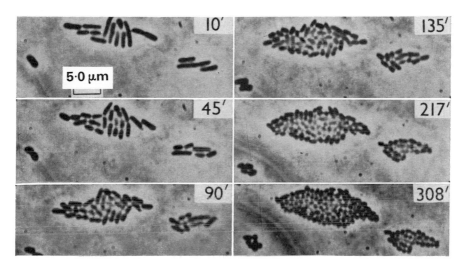

Figure 8. Morphological events during starvation.

A culture of *Escherichia coli* previously growing in a nutrient broth was placed at zero time on the surface of an agar block containing phosphate buffer and other salts but no organic compounds. The cells were observed with a phase contrast microscope and photomicrographs were taken at the times shown. The progressive decrease in size of the organisms as the result of division without growth can be clearly observed.

as if depleting many different nutrients in sequence. This may be thought of as a series of gradual *shifts-down* since these cells also decrease in size and RNA content. Figure 8 shows an example of the morphological changes which occur when the stationary phase is simulated by starvation of a culture. It must be borne in mind that in natural media, the nutritional aspects of the *shift-down* can be obscured by the accumulation of acids or other toxic metabolites. The end of the exponential phase usually occurs when cultures of *Escherichia coli* reach densities of about 10^8 cells per ml and when the culture contains approximately 10^9 bacteria per ml, it comes to a standstill. Under these conditions there is cell death and extensive cell turn-over may occur in some species of bacteria. That is to say some cells die and others grow by utilizing the products of disintegration of the dead cells. Therefore, a measurement of the bacterial mass in a culture in the stationary phase gives no indication of the complex events that take place at the level of individual cells.

Under ordinary conditions the whole cycle of growth will have taken place during a working day. Perhaps because all this can be so readily observed in the laboratory, and often yields interesting-looking S-shaped growth curves, there has been a tendency in the past to suggest that bacteria undergo life-cycles. In the light of present knowledge this no longer seems true since this growth cycle can readily be interrupted and is certainly not obligatory for the reproduction of the organisms. Many of the fundamental aspects of this growth curve can be better studied by examining the characteristics during balanced growth and during the transition between defined physiological states.

It must be evident that this discussion of the behaviour of bacterial populations has been framed by reference to somewhat idealized experiments. There are variations on this theme which must be interpreted from totally different points of view. For instance, the formation of bacterial spores during the stationary phase of some organisms introduces a different quality in growth physiology because spores respond to environmental changes in ways that cannot be predicted from knowing how the vegetative cells react (see Chapter 9).

THE GROWTH OF SINGLE CELLS

The life of an individual bacterium

One of the advantages of working with bacteria lies in the fact that their populations may often be regarded as being homogeneous. In many physiological senses a given cell of a population undergoing balanced growth is very similar to all other cells of the population. It is thus possible to carry out a measurement on a relatively large sample of the population and to consider it characteristic of individual cells. This is tacitly implied in the overwhelming number of investigations on bacterial physiology. However, the cells of a growing bacterial population are not all alike; they differ with regard to the time when they divide. If we call a cell which has newly emerged after division 'young' and one which will divide within a short time 'old', we expect that a bacterial population consists of cells whose age follows a certain distribution. For many types of experiment this heterogeneity does not matter because the behaviour of the *average cell* is reflected in the culture as a whole. However, if physiological processes are to be understood at the level of the *individual cell*, it must be realized that the effective life of an organism is the time between two successive divisions. To understand bacterial growth it is then necessary to know how biochemical events are interrelated throughout the cell cycle. This is a relatively new field of investigation and our knowledge of it is scant. In large part this is due to practical difficulties, since, obviously, it is not easy to make measurements on individual living bacterial cells throughout the cell cycle. However, a number of ingenious ways of doing this have been developed.

Methods for studying the bacterial cell cycle

The growth of individual bacteria can be observed under a microscope if the organisms are placed on a thin agar block and covered with a cover glass. This permits measurement of a few things such as the increment in the cells' dimensions or mass, but, even with refined techniques, few cytological measurements or observations are possible. Under suitable conditions of phase contrast microscopy the nuclear bodies of living bacteria can be made visible and followed through their division. Disappointingly little in the way of cytological details is seen since these bodies lack a mitotic apparatus and are seen simply to pull apart, often with a number of apparently uninterpretable contortions. Direct microscopy yields no more useful information about the behaviour of other cellular elements throughout the division cycle but ultra-thin sections of bacteria examined with the electron microscope have revealed considerable morphological details related to the division process. For instance, the close proximity of the cell's nuclear bodies to the invaginations of the cell membrane known as mesosomes has suggested that a primitive form of mitotic division may exist. The mesosomes might act as anchor points for the dividing nuclear bodies, leading to their separation as a consequence of elongation of the cell membrane.

There exist a number of ultramicrochemical techniques which permit measurements on a single growing bacterium but they involve difficult methods and, by and large, have not been applied to the study of the bacterial cell cycle. Our knowledge of the cycle is obtained through other approaches, principally that of synchronizing the division of cells so that large samples of the population reflect the behaviour of individuals. The study of *synchronous growth*, as this form of multiplication has come to be called, would be impeccable were it not that the condition has to be artificially imposed. Bacterial cultures usually consist of individuals which do not divide at the same time but are entirely unsynchronized. Division synchrony may be induced by one of several manipulations. For instance, in some species synchronous divisions take place after shifting the temperature of cultivation or after restoring nutrients to a culture that has been starved. These procedures often yield interesting results but must be interpreted with caution. The observed results may be spurious and due to the stimulus used to induce synchrony, without necessarily reflecting what takes place in the normal division cycle. This difficulty has been overcome, with various degrees of success, by avoiding 'shock' treatments to induce synchrony. Thus, synchronously dividing cultures are obtained by selecting cells of *similar size* from populations in balanced growth. For this purpose bacterial suspensions are passed through filters of pore size such that only the smallest cells are allowed to go through. Alternatively, a different pore size may be used for the selective retention of the largest cells, which can then be harvested. By using an analogous principle, the heaviest or the lightest cells can be selected by centrifugation through density gradients. In these methods it is assumed that cells of *similar size* or *density* are of *similar age*, which is usually a good approximation. Recently, a technique has been developed that takes advantage of the fact that cells of certain strains adsorb to nitrocellulose surfaces and can grow attached to them. For unknown reasons, the 'young' cells which emerge by division do not stick and will float off. Such free cells harvested over a short period of time are nearly the same age and, if permitted to grow, will divide synchronously. This experimental set-up has been called, appropriately, the 'baby machine'. Usually synchrony does not persist and, after a few divisions, the cells are again out of step with one another.

In addition to synchronous growth there are a number of more indirect ways of usefully studying the cell division cycle of bacteria. One method is to establish whether among cells in a non-synchronous population all or only a proportion are synthesizing particular cell constituents. The experimental plan may be as follows: a radioactive precursor of DNA, tritium-labelled thymidine, is added to the culture for a short fraction of one mean generation time, say 10%. If all the cells become labelled, it must be concluded that DNA synthesis occurs substantially throughout the whole cell cycle. Conversely, if only a proportion of the cells is labelled, DNA is synthesized only at certain times during the cycle and not at others. The presence of label in individual bacteria may be determined by radio-autography, a technique in which labelled cells are placed on a microscope slide and are covered with a photographic emulsion. They are then kept for a suitable period of time to allow some of the isotope to decay. This results in sensitization of the photographic grains which, after suitable development, are seen as black dots in the vicinity of the cells. Homogeneity of the population is indicated by a random distribution of grains about each bacterium.

Are all cell cycles equal?

In some species, bacterial multiplication produces exceedingly regular arrays of cells. Thus the chains of some *Streptococci* are composed of cocci in identical stages of division. One can observe cells arranged in square sheets in *Lampropedia* and in perfect cubes in *Sarcina*. Such order must arise from regularity in the individual cell divisions and, where it takes place, all cells have nearly the same life span. On the other hand in *Escherichia coli* individual cells may take quite different periods of time to divide. This can be determined by timing individual cell division under a microscope. The variability in division times has been the subject of considerable speculation since it has been thought to reflect the 'individual variation' which many people have attributed to biological phenomena. On the other hand, it is possible that the variation is not in the process of a cell dividing into two physiologically distinct daughter cells, but rather in the final separation of two cells still held together by portions of their cell envelopes. This explanation makes the partition of cytoplasm a more regular phenomenon than is observed but does not change the fact that the separation into distinct cells is variable. Since studies on the cell cycle deal with cells as physically separate entities, they must take into account the statistical variability encountered in some species.

Macromolecular activities during the cell division cycle

In animal or plant cells a complicated and well-timed sequence of events occurs during the cell cycle. DNA is synthesized during the interphase, often long before the mitotic division of chromosomes. In many cells mitosis occupies a short fraction of the cell cycle and is closely followed by cell division. In those bacteria where suitable studies have been carried out, division of the nuclear bodies takes place about half-way through the division cycle. Bacteria also differ from animal or plant cells with regard to their DNA synthesis because this process is not restricted to a portion of the growth cycle of the bacteria. Continuous DNA synthesis may be a consequence of fast growth and the short period of time between divisions. Thus, the whole of the division cycle may have to be used to make the required complement of DNA. This implies that there is a maximum rate of DNA synthesis and that this process cannot be speeded up beyond this rate. This notion is supported by the following observations. It has been demonstrated that bacterial DNA replicates sequentially, from one end of the molecule to the other. Consequently, the enzymic apparatus that performs this synthesis must travel along the length of the molecule and its position at any time constitutes what might be called a *growing point*. In *Escherichia coli* growing in a synthetic medium with glucose as the sole carbon source (about one doubling per hour at 37°), there is only one growing point per DNA molecule. When the organisms grow faster the number of growing points increases. This implies that the maximum rate of synthesis per growing point has been exceeded and that, in order to double the complement of DNA during this shorter division cycle, the number of such growing points has been increased. This is represented in Figure 9. This observation illustrates a principle proposed by Maaløe and Kjeldgaard, that the *output* of macromolecules is controlled principally by the frequency with which their synthesis is initiated and not by changes in the rate of 'growth' of individual molecules.

Not much is known about the behaviour of other macromolecules during the division cycle. In a gross sense, the syntheses of RNA, proteins, and cell wall components appear to take place continuously between divisions. However, certain enzymes are synthesized at definite times during the cycle in bacteria as well as in yeasts. At first glance this might be interpreted as indicating that transcription of DNA into messenger RNA resembles DNA replication in being

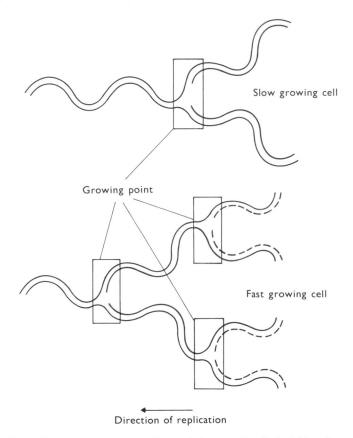

Figure 9. The replicating chromosome of fast and slow growing *Escherichia coli*.

Replication of the chromosome of *Escherichia coli* is sequential, i.e. it proceeds in order from one end to the other. In slow-growing cells there is only one place in the chromosome where this takes place at any given time. In fast-growing cells a new replication or 'growing' point may be formed before the first round is completed. Note that the rate of synthesis per chromosome is three times faster in the fast-growing cells than in the slow-growing ones, assuming that the rate at each growing point is constant.

sequential and, at any given time, taking place on a distinct portion of the DNA only. However, synthesis of new enzymes after induction takes place at the same rate in cells which are in different stages of the division cycle. This indicates that the transcription process is not sequential along the DNA and leaves the finding of the 'bursts' of enzyme synthesis without a direct interpretation at this time.

The cell cycle of bacteria emerges as rather undifferentiated and perhaps less refined than that of higher cells. It is likely, however, that this simplicity only reflects our primitive stage of experimentation. The complicated events that constitute growth, bacterial or otherwise, of a cell must be harmoniously inter-related and knowledge of the way in which this regulation is effected is still on a primitive level.

FURTHER READING

1　Maaløe O. and Kjeldgaard N. O. (1966) *Control of Macromolecular Synthesis*. Benjamin, New York.

2　Helmstetter C. E. (1969) Sequence of bacterial replication. *Ann. Rev. Microbiol.* **23**, 1527.

3　Geiduschek E. P. (1970) Synthesis of DNA and its transcription as RNA. In *Aspects of Frotein Biosynthesis* (Part A), C. B. Anfinsen, Jr., Editor. Academic Press, New York and London.

4　Pritchard R. H., Barth P. T. and Collins J. (1969) Control of DNA synthesis in bacteria. In *Microbiol Growth, Symp. Soc. for Gen. Microbiol.* **19**, 263.

Chapter 3
Class I Reactions: Supply of
Carbon Skeletons

Most bacteria can use any of a number of compounds as carbon sources. The pseudomonads for instance can grow on various hydrocarbons, phenols, aliphatic amides, etc. but most other types of bacteria are more restricted in their biochemical potentialities. However, almost all can grow on glucose and it will be useful to consider the ways in which this substance is metabolized to secure carbon skeletons and energy for biosynthesis.

By a series of enzymically catalysed reactions the hexose molecule is degraded, usually after the introduction of inorganic phosphate, to smaller molecules which can then be utilized as building blocks for the synthesis of new cell material. In the course of these reactions energy in the form of ATP is produced for utilization in the subsequent energy-requiring synthetic reactions. At present, four major metabolic pathways for glucose catabolism have been discovered and, as we shall see, these are usually related to the particular mode of life of the organism concerned. The differences in these metabolic sequences reside in the presence or absence of specific enzymes catalysing particular reaction sequences and, in fact, all four pathways utilize an identical series of enzymes for converting triose phosphate to pyruvate. It is possible, therefore, to visualize a central area of metabolism common to all pathways of glucose metabolism, with characteristic differences lying outside this zone.

According to species, bacteria are able to live either in the presence or absence of oxygen; some are intolerant of one or other of these conditions and are classified as strict aerobes or anaerobes, but many bacteria can tolerate both conditions and are referred to as facultative anaerobes. The pathway for glucose metabolism which operates in bacteria must therefore be related to the conditions of life under which the organism can exist. Under aerobic conditions oxygen will be the ultimate hydrogen acceptor but under anaerobiosis, products of glucose metabolism must serve as hydrogen acceptors in a series of balanced oxidation-reduction reactions if glucose degradation and growth are to occur.

These aspects of glucose metabolism will become apparent in the following discussions of metabolic pathways.

The four routes of glucose catabolism currently recognized are: (1) Embden–Meyerhof glycolysis, (2) pentose phosphate cycle, (3) Entner–Doudoroff pathway, and (4) phosphoketolase pathway. The distinguishing features of these will now be discussed.

PATHWAYS OF GLUCOSE CATABOLISM

The Embden–Meyerhof glycolytic pathway

The Embden–Meyerhof glycolytic pathway (Figure 1) consists of some ten enzymes which effect the conversion of glucose (C_6) to pyruvate (C_3). The

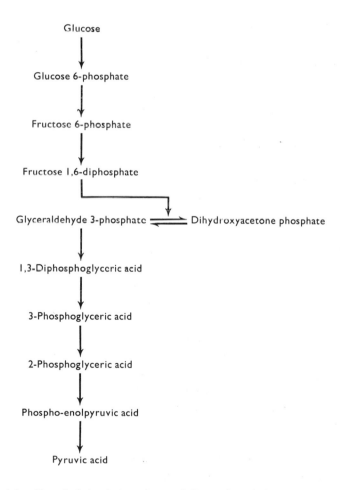

Figure 1. Embden–Meyerhof glycolytic pathway of glucose degradation.

system can operate under both aerobic and anaerobic conditions. Aerobically it usually functions in conjunction with the tricarboxylic acid cycle which can oxidize pyruvate to CO_2 and H_2O. Anaerobically, pyruvate, or products of its further (anaerobic) metabolism, must be reduced, e.g. with the formation of lactate or ethanol. The details of this sequence of reactions will now be considered.

Glucose must first be phosphorylated in the 6-position by the enzyme *hexokinase* which requires Mg^{2+} for activity. This essentially irreversible reaction yields glucose 6-phosphate and ADP.

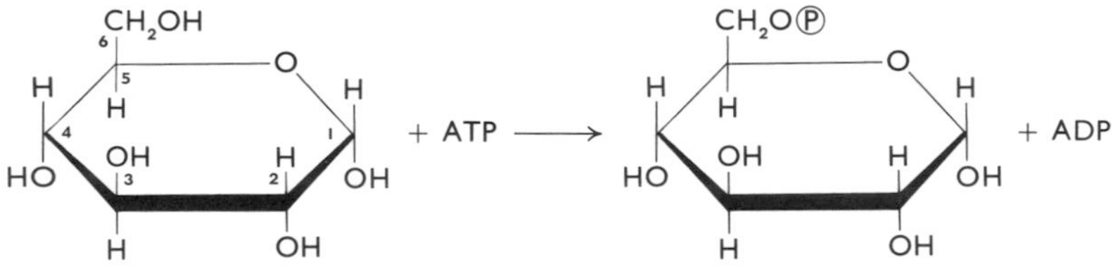

Glucose   Glucose 6-phosphate

Next follows an isomerization of glucose 6-phosphate to fructose 6-phosphate catalysed by *phosphohexose isomerase*, which has no cofactor requirement. At equilibrium some 70% of the glucose form is present.

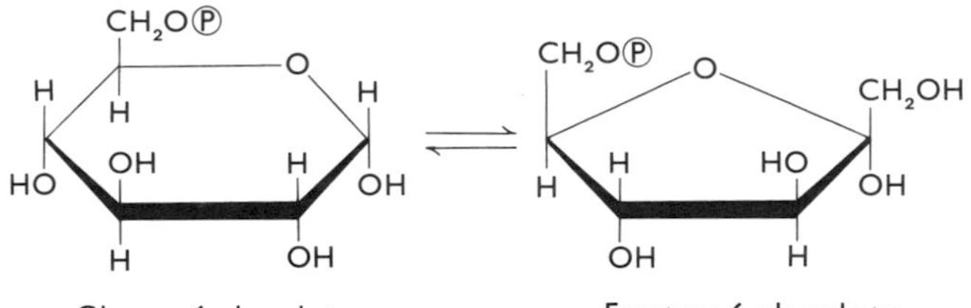

Glucose 6-phosphate   Fructose 6-phosphate

A second phosphate group is now introduced into the 1-position of the fructose by the action of *phosphofructokinase*, an enzyme characteristic of this pathway of glucose metabolism, and the presence of which in a given type of bacterium is taken as good evidence for the operation of glycolysis. Like hexokinase, it requires ATP and Mg^{2+} and is irreversible.

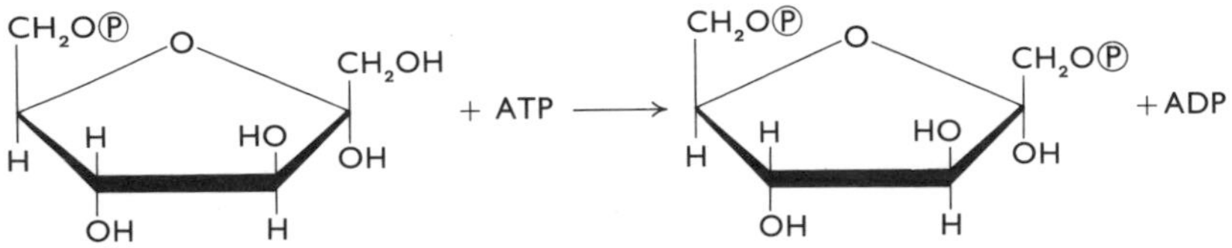

Fructose 6-phosphate   Fructose 1,6-diphosphate

The hexose molecule is now split by *fructose diphosphate aldolase* into two C_3 units—the triose phosphates, glyceraldehyde 3-phosphate and dihydroxyacetone phosphate [4].

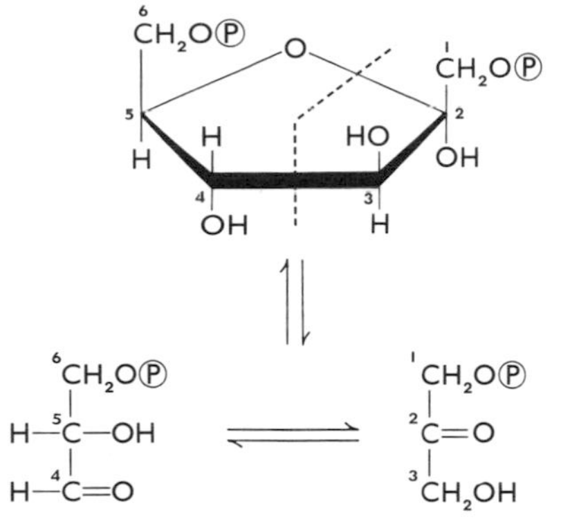

Glyceraldehyde 3-phosphate   Dihydroxyacetone phosphate

The triose phosphates are produced in equivalent amounts but due to the action of *triose phosphate isomerase*, which catalyses their interconversion, they are brought to an equilibrium which is about 95% in favour of dihydroxyacetone phosphate. On account of this, dihydroxyacetone phosphate was originally believed to be the sole product of aldolase action. Dihydroxyacetone phosphate has been found to inhibit fructose diphosphate aldolase competitively. The fission of the hexose occurs in such a way that dihydroxyacetone phosphate is derived from carbon atoms 1, 2 and 3 and glyceraldehyde 3-phosphate from atoms 4, 5 and 6 of the fructose molecule.

The triose phosphate stage of glycolysis marks the point of linkage with fat metabolism because of the possibility of interconversion of dihydroxyacetone phosphate and glycerol. In some bacteria glycerol is first phosphorylated by a kinase and ATP (in *Escherichia coli* this reaction is the pacemaker for glycerol metabolism), and then the α-glycerophosphate formed undergoes oxidation to dihydroxyacetone phosphate catalysed by α-glycerophosphate dehydrogenase.

α-Glycerophosphate   Dihydroxyacetone phosphate

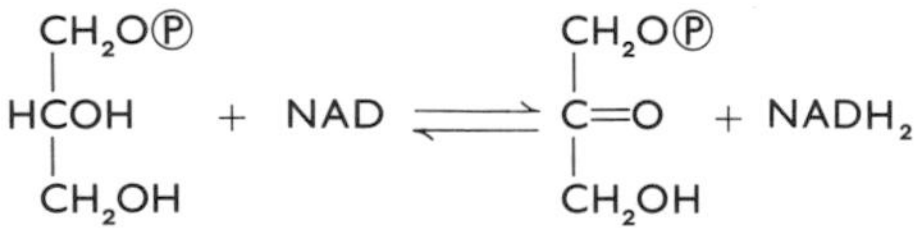

Despite the fact that the equilibrium of triose phosphate isomerase greatly favours dihydroxyacetone phosphate, glyceraldehyde 3-phosphate is the substrate for the next step in glycolysis—an oxidation to 1,3-diphosphoglyceric acid. The enzyme responsible, *triose phosphate dehydrogenase*, requires NAD as coenzyme, and inorganic phosphate, which is incorporated into the product, is also necessary for the reaction to occur:

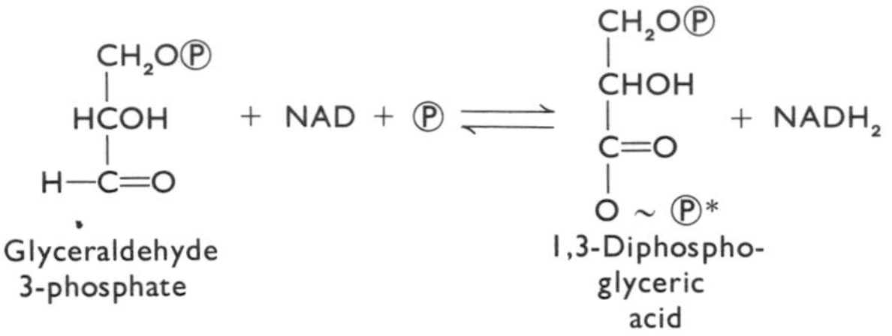

$$
\begin{array}{l}
CH_2O\textcircled{P} \\
| \\
HCOH \\
| \\
H-C=O
\end{array}
\quad + \text{ NAD } + \textcircled{P} \rightleftharpoons
\quad
\begin{array}{l}
CH_2O\textcircled{P} \\
| \\
CHOH \\
| \\
C=O \\
| \\
O \sim \textcircled{P}*
\end{array}
\quad + \text{ NADH}_2
$$

Glyceraldehyde   1,3-Diphospho-
3-phosphate   glyceric
                                       acid

* $\sim$ Indicates a high-energy phosphate bond (cf. p. 41).

This reaction represents the first and only oxidation in the sequence of glycolysis leading to pyruvic acid. The $NADH_2$ must be reoxidized for the reactions to continue. Aerobically this occurs via the cytochrome system (see Appendix B). Anaerobically it is coupled with reduction of an organic compound—often one produced later in the glycolysis pathway, namely pyruvate. Thus pyruvate oxidizes $NADH_2$ and is itself reduced to lactate.

It should be noted that the product 1,3-diphosphoglycerate contains a high-energy bond (the acyl phosphate in the 1-position). The next reaction is the transfer of this phosphate group to ADP by the action of *phosphoglycerate kinase* leaving 3-phosphoglycerate as the other product. The energy made available by the dehydrogenation process is thus coupled to ATP synthesis. Mg^{2+} is a cofactor for this enzyme, as with other kinases.

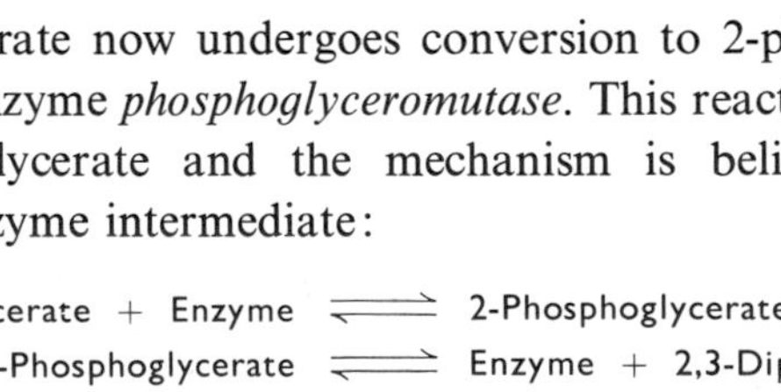

$$
\begin{array}{l}
CH_2O\textcircled{P} \\
| \\
CHOH \\
| \\
C=O \\
| \\
O \sim \textcircled{P}
\end{array}
\quad + \text{ ADP } \rightleftharpoons
\quad
\begin{array}{l}
CH_2O\textcircled{P} \\
| \\
CHOH \\
| \\
COOH
\end{array}
\quad + \text{ ATP}
$$

1,3-Diphosphoglyceric   3-Phosphoglyceric
         acid   acid

3-Phosphoglycerate now undergoes conversion to 2-phosphoglycerate by the action of the enzyme *phosphoglyceromutase*. This reaction requires a trace of 2,3-diphosphoglycerate and the mechanism is believed to involve a phosphorylated enzyme intermediate:

2,3-Diphosphoglycerate + Enzyme $\rightleftharpoons$ 2-Phosphoglycerate + Enzyme $\textcircled{P}$

Enzyme $\textcircled{P}$ + 3-Phosphoglycerate $\rightleftharpoons$ Enzyme + 2,3-Diphosphoglycerate

$$
\begin{array}{l}
CH_2O\textcircled{P} \\
| \\
CHOH \\
| \\
COOH
\end{array}
+
\begin{array}{l}
CH_2O\textcircled{P} \\
| \\
CHO\textcircled{P} \\
| \\
COOH
\end{array}
\rightleftharpoons
\begin{array}{l}
CH_2O\textcircled{P} \\
| \\
CHO\textcircled{P} \\
| \\
COOH
\end{array}
+
\begin{array}{l}
CH_2OH \\
| \\
CHO\textcircled{P} \\
| \\
COOH
\end{array}
$$

3-Phospho-   2,3-Diphospho-   2,3-Diphospho-   2-Phospho-
glycerate   glycerate   glycerate   glycerate

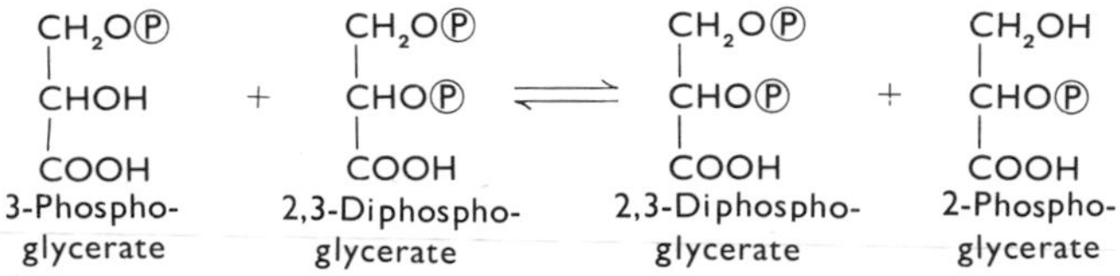

It will be noticed that the donor molecule, by transfer of its phosphate group in the 3-position, becomes the product of the reaction and the original substrate appears as the diphosphate which can then in turn act as a donor molecule.

Water is now removed to yield phospho-enolpyruvic acid by the action of *enolase* which requires a divalent metal ion such as Mg^{2+}, Mn^{2+} or Zn^{2+}:

$$
\begin{array}{ccc}
CH_2OH & & CH_2 \\
| & & \| \\
CHO℗ & \rightleftharpoons & CO\sim℗ + H_2O \\
| & & | \\
COOH & & COOH
\end{array}
$$

2-Phospho-   Phospho-
glyceric   enolpyruvic
acid   acid

Enolases from organisms which require Mg^{2+} are susceptible to inhibition by fluoride, due to the formation of magnesium fluorophosphate, whereas those requiring Mn^{2+} are not because the manganese salt, unlike the magnesium salt, is soluble. Inhibition by fluoride results in the accumulation of phospho-glycerate and for many years the detection of phosphoglycerate under these conditions was taken as evidence for the operation of glycolysis. However it is now known that enolase is part of the central area of glucose metabolism common to the different pathways.

The effect of dehydration is to produce another high-energy phosphate bond, this time an enol-phosphate, which is now transferred to ADP by a further enzyme, *pyruvate kinase*. Thus the energy yielded by the dehydration step is coupled to the synthesis of another molecule of ATP. The equilibrium of the kinase reaction greatly favours ATP formation and the reverse reaction is extremely difficult to demonstrate:

$$
\begin{array}{ccc}
CH_2 & & CH_3 \\
\| & & | \\
C-O\sim℗ + ADP & \longrightarrow & C=O + ATP \\
| & & | \\
COOH & & COOH
\end{array}
$$

Phospho-enolpyruvic   Pyruvic acid
acid

Fermentation products

Bacteria differ from yeast in that the majority of them give rise to a diversity of products by the further metabolism of pyruvate. These reactions permit an overall oxidation-reduction balance to be preserved under anaerobic conditions and the products formed are characteristic of the particular species and may be used as a biochemical aid to identification. Some of these reactions can give rise to additional energy, for example the conversion of an acyl coenzyme A

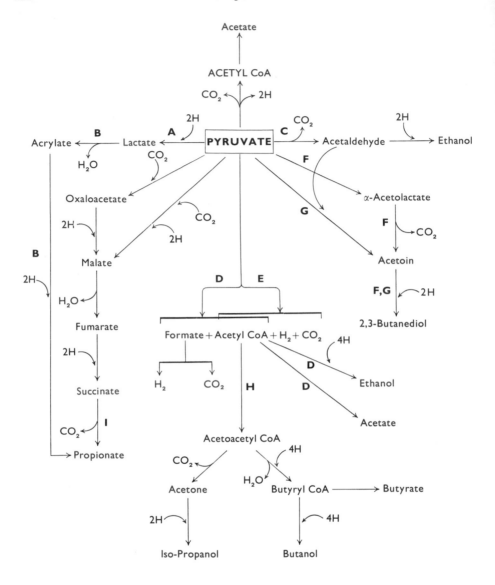

Figure 2. Bacterial fermentation products of pyruvate. Pyruvate formed by the catabolism of glucose is further metabolized by pathways which are characteristic of particular organisms and which serve as a biochemical aid to identification. End products of fermentations are indicated in red.

A Lactic acid bacteria (*Streptococcus,*
   Lactobacillus)
B *Clostridium propionicum*
C Yeast, Acetobacter, Zymomonas,
   Sarcina ventriculi, Erwinia amylovora.
D Enterobacteriaceae (coli-aerogenes)

E Clostridia
F Aerobacter
G Yeast
H Clostridia (butyric, butylic organisms)
I Propionic acid bacteria

derivative to the free acid

$$\text{Acetyl CoA} + \text{ADP} \rightleftharpoons \text{Acetate} + \text{ATP} + \text{CoA}$$

$$\text{Butyryl CoA} + \text{ADP} \rightleftharpoons \text{Butyrate} + \text{ATP} + \text{CoA,}$$

but, in general, the energy yield is poorer than that derived from the conversion of glucose to pyruvate. The fermentation reactions of pyruvate carried out by various micro-organisms are summarized in Figure 2.

The full sequence of reactions from glucose to pyruvate has now been described and it will be noted that, since each fructose 1,6-diphosphate molecule gives rise to two triose phosphates, four moles of ATP are produced per mole of fructose diphosphate fermented. As one mole of ATP is used for the phosphorylation of glucose and a second for the phosphorylation of fructose 6-phosphate, the net yield is two moles of ATP per mole of glucose fermented.

The fate of pyruvate produced by bacteria varies tremendously, depending on the species of organism and environmental conditions, such as pH, so that products of pyruvate metabolism can be used as an aid to classification. Homolactic acid bacteria (i.e. those which produce lactate as the major product of glucose fermentation, as opposed to heterolactic bacteria which produce substantial amounts of other products in addition to lactate) resemble muscle in yielding two moles of lactate per mole of glucose fermented. The enzyme involved in the reduction of pyruvate, *lactate dehydrogenase*, utilizes the $NADH_2$ produced by the triose phosphate dehydrogenase reaction, as previously discussed, thus permitting anaerobic oxidation of glyceraldehyde 3-phosphate. The linked reactions are:

$$\text{Glyceraldehyde 3-phosphate} + \text{NAD} \rightleftharpoons \text{3-Phosphoglycerate} + \text{NADH}_2$$

$$\underset{\text{Pyruvic acid}}{CH_3CO.COOH + NADH_2} \rightleftharpoons \underset{\text{Lactic acid}}{CH_3CHOH.COOH + NAD}$$

A feature of the Embden–Meyerhof mechanism is that fission of fructose 1,6-diphosphate occurs in such a way that the carboxyl groups of the resulting pyruvate molecules are derived from carbon atoms 3 and 4 of the original glucose molecule. Consequently, in the lactic fermentation the carboxyl groups of lactate are similarly derived from 3-C and 4-C. In the alcohol fermentation, however, these carbon atoms are eliminated as carbon dioxide and the acetaldehyde and ethanol are derived from 1-C, 2-C and 5-C, 6-C as shown at the top of the next page.

Using specifically labelled [^{14}C]glucose the distribution of the carbon atoms of glucose in intermediates and products of different metabolic pathways can be established and this enables deductions to be made concerning the occurrence and quantitative significance of these pathways.

$$\overset{1}{C}-\overset{2}{C}-\overset{3}{C}-\overset{4}{C}-\overset{5}{C}-\overset{6}{C}$$

$$\overset{1}{CH_3}\overset{2}{CHOH}.\overset{3}{COOH} \longleftarrow \overset{1}{CH_3}\overset{2}{CO}.\overset{3}{COOH} \qquad \overset{4}{HOOC}.\overset{5}{CO}.\overset{6}{CH_3} \longrightarrow \overset{4}{HOOC}.\overset{5}{CHOH}.\overset{6}{CH_3}$$

$$\overset{1}{CH_3}\overset{2}{CHO} \qquad \overset{3}{CO_2} \qquad\qquad \overset{4}{CO_2} \qquad \overset{5}{OHC}.\overset{6}{CH_3}$$

$$\overset{1}{CH_3}\overset{2}{CH_2}OH \qquad\qquad\qquad\qquad HO.\overset{5}{H_2C}.\overset{6}{CH_3}$$

Control of glycolysis

The principal functions of glycolysis are two-fold, to provide energy (directly and via intermediates which are oxidized in the tricarboxylic acid cycle) and to furnish carbon skeletons for the biosynthesis of cellular components. The term *amphibolic* has been applied to such dual function pathways, to distinguish them from the strictly catabolic and anabolic sequences of metabolism (see also Chapter 8, p. 423). The ATP produced in the exergonic reactions is utilized in the endergonic synthetic pathways with the generation of ADP or AMP and, in consequence, the energy state of a bacterial cell depends on the balance between the concentrations of the adenine nucleotides ATP, ADP and AMP. When all the adenylate is present as ATP the cell will have a maximum energy level and conversely when all is present as AMP the minimum level will be manifest. The reactions involved are:

$$
\begin{array}{lll}
(1) & ATP + H_2O \rightleftharpoons ADP + ⑨ \\
(2) & ADP + H_2O \rightleftharpoons AMP + ⑨ \\
(3) & ATP + 2H_2O \rightleftharpoons AMP + 2⑨
\end{array}
$$

To express quantitatively the energy state of a cell Atkinson has introduced the concept of the *energy charge* of the adenylate system, somewhat analogous to the charge of a storage battery. Reaction (3) formally represents energy acceptance and donation by the adenylate system and the degree of charge is proportional to the amount of phosphate added to AMP; the addition of 2 moles of phosphate per mole of adenylate fully charges the system which then exists solely as ATP. To enable the charge to range from 0 to 1 instead of 0 to 2, Atkinson defines energy charge as half the number of anhydride-bound phosphate groups per adenine moiety, i.e. in terms of concentrations

$$\text{Energy charge} = \frac{(ATP) + 0.5\,(ADP)}{(ATP) + (ADP) + (AMP)}$$

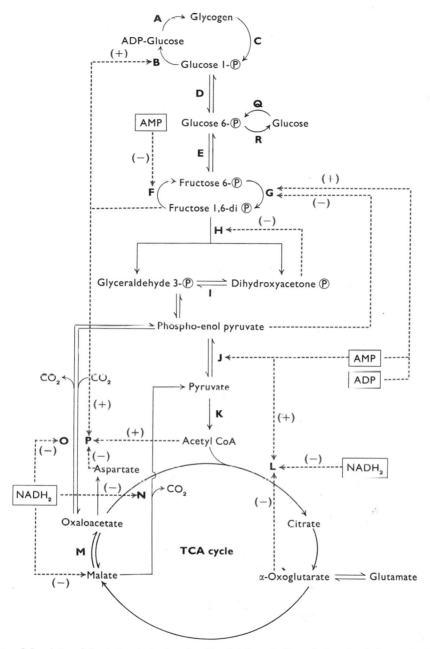

Figure 3. Regulation of glycolysis and tricarboxylic acid cycle in bacteria. Plus and minus signs indicate activation and inhibition respectively of the appropriate enzymes by various metabolites. Note that the controls indicated do not all necessarily apply to a single species of organism.

A   Glycogen synthetase
B   ADP-glucose pyrophosphorylase
C   Phosphorylase
D   Phosphoglucomutase
E   Phosphohexose isomerase
F   Fructose 1,6-diphosphatase
G   Phosphofructokinase
H   Fructose diphosphate aldolase
I   Triosephosphate isomerase
J   Pyruvate kinase (there are two pyruvate kinases

in enteric bacteria, one activated by fructose 1,6-diphosphate and the other by AMP)
K   Pyruvate dehydrogenase
L   Citrate synthase
M   Malate dehydrogenase
N   Malate enzyme
O   Phospho-enolpyruvate carboxykinase
P   Phospho-enolpyruvate carboxylase
Q   Hexokinase
R   Glucose 6-phosphatase

This single parameter enables the energy level of the cell to be defined and is of great assistance when the energy-linked control of metabolic systems is considered, on account of the heterogeneity of response to the individual adenine nucleotides encountered with certain regulatory enzymes, e.g. some respond primarily to the individual concentration of ATP, ADP or AMP, whereas others respond either to the ratio of (ATP)/(ADP) or (ATP)/(AMP). The concept of the energy charge thus unifies the treatment of these cases.

When the energy charge of a bacterial cell is low there must be some means of increasing the rate of glycolysis and, conversely, when the energy charge is high of decreasing the rate of ATP production. This is achieved by modulating the rates of two key reactions in glycolysis, the phosphorylation of fructose 6-phosphate to fructose 1,6-diphosphate catalysed by phosphofructokinase, and the conversion of phospho-enolpyruvate to pyruvate effected by pyruvate kinase. Phosphofructokinase is activated by ADP or AMP and inhibited by phospho-enolpyruvate, while pyruvate kinase is activated by AMP and fructose 1,6-diphosphate. Energy-linked control of this type is clearly of great value for the regulation of catabolic sequences which, in general, are not subject to end-product or negative feedback control of the type encountered in biosynthetic pathways (see Chapter 4, pp. 202–207). However, in the case of glycolysis the dual function of energy generation and production of intermediates for biosynthesis imposes upon it a form of regulation rather more complicated than that encountered with a strictly catabolic pathway involving only energy-linked control. An additional control feature, apparently unique to amphibolic pathways, is *precursor activation* which, in a sense, is the opposite of feedback control where the last metabolite of a pathway inhibits the first enzyme of the sequence. In precursor control the first metabolite of the sequence activates the last enzyme of that sequence. It is not possible here to enter into a detailed discussion of precursor activation and one example must suffice. Fructose 1,6-diphosphate may be regarded as the first metabolite of a metabolic sequence in glycolysis which leads to pyruvic acid, the final enzymic step being the conversion of phospho-enolpyruvate to pyruvate, catalysed by a pyruvate kinase. As already mentioned, fructose 1,6-diphosphate activates this pyruvate kinase thus functioning in a positive 'feedforward' manner.

Figure 3 illustrates the regulatory mechanisms currently known to operate in glycolysis and the tricarboxylic acid cycle.

Glucogenesis

When bacteria grow on pyruvate or other C_3 or C_4 compounds as the sole source of carbon it is essential that they synthesize glucose from these substrates, a process termed glucogenesis (or gluconeogenesis). As three reactions of glycolysis are virtually irreversible on account of their highly exergonic nature, namely those catalysed by hexokinase, phosphofructokinase and pyruvate kinase, it is not possible for the cell simply to reverse the glycolytic sequence in order to synthesize glucose, and alternative means have to be provided to circumvent these irreversible steps.

In *Escherichia coli* it has been found that two enzymes which enable the organism to grow on C_4 compounds such as succinate or malate are phospho-enolpyruvate carboxykinase and malate enzyme

$$HOOC.CH_2CO.COOH + ATP \xrightarrow{\text{PEP carboxy kinase}} CH_2:CO\sim\textcircled{P}.COOH + CO_2 + ADP$$

$$HOOC.CH_2CHOH.COOH + NADP \xrightarrow{\text{malate enzyme}} CH_3CO.COOH + CO_2 + NADPH_2$$

Loss of the carboxykinase by mutation completely prevents growth on C_4 compounds and acetate, yet such mutants are still able to grow on C_3 compounds such as pyruvate or lactate. This observation was explained by the discovery of an enzyme *phospho-enolpyruvate synthase*, induced during growth on C_3 compounds, which requires Mg^{2+} and effects the reaction:

$$CH_3COCOOH + ATP \longrightarrow CH_2{=}CO\sim\textcircled{P}.COOH + AMP + \textcircled{P}$$

In this way phospho-enolpyruvate is synthesized from pyruvate at the expense of two of the high energy bonds of ATP, one of which is preserved in the phospho-enolpyruvate formed. In passing, it should be noted that growth on C_3 compounds demands that the organism be able to synthesize C_4 compounds from C_3 and CO_2 in order for the tricarboxylic acid cycle to operate to produce both energy and intermediates for biosynthesis (see p. 188). Studies with mutants of the Enterobacteriaceae, unable to grow on glucose, glycerol, pyruvate or their precursors unless the medium is supplemented with utilizable intermediates of the tricarboxylic acid cycle, have revealed that the key enzyme is *phospho-enolpyruvate carboxylase*, which catalyses the reaction

$$CH_2{=}CO\sim\textcircled{P}.COOH + CO_2 \longrightarrow HOOC.CH_2CO.COOH + H_3PO_4$$

This enzyme therefore fulfils a replenishing role by permitting net synthesis of C_4 compounds and belongs to the category of anaplerotic enzymes (see p. 190).

The irreversibility of the phosphofructokinase reaction is overcome by the enzyme fructose diphosphatase which hydrolyses phosphate from the 1-position, thus

$$\text{Fructose 1,6-diphosphate} + H_2O \longrightarrow \text{Fructose 6-phosphate} + \textcircled{P}$$

The necessity for this reaction in glucogenesis is illustrated by the fact that mutants of *Escherichia coli* devoid of the enzyme are unable to grow on substrates such as acetate, succinate or glycerol and have an absolute requirement for hexoses.

By utilizing these reactions in conjunction with the reversible steps of glycolysis the bacterial cell is able to synthesize glucose 6-phosphate from C_3

and C_4 compounds. Studies with mutants of *Escherichia coli* lacking triose-phosphate isomerase have demonstrated that this enzyme is essential for glucogenesis, thus lending support to the pathway proposed.

Glycogenesis

It has been mentioned that bacteria such as *Escherichia coli* accumulate substantial amounts of glycogen under conditions where the nitrogen of the medium becomes exhausted in the presence of excess carbon sources. Glycogen synthesis requires glucose 1-phosphate as its substrate and this is formed from glucose 6-phosphate by the enzyme *phosphoglucomutase*, which undergoes phosphorylation and dephosphorylation during the reaction and involves glucose 1,6-diphosphate as an intermediate.

Glucose 6-phosphate		Glucose 1,6-diphosphate		Glucose 1-phosphate
+	$\rightleftharpoons$	+	$\rightleftharpoons$	+
Phosphoenzyme		Dephosphoenzyme		Phosphoenzyme

Catalytic amounts of glucose 1,6-diphosphate are essential for the reaction to proceed and these are formed by the action of phosphoglucokinase on glucose 1-phosphate and ATP.

Glucose 1-phosphate reacts with ATP under the influence of ADP-glucose pyrophosphorylase to yield ADP-glucose and inorganic pyrophosphate

$$\text{Glucose 1-phosphate} + \text{ATP} \longrightarrow \text{ADP-glucose} + Ⓟ \sim Ⓟ$$

ADP-glucose then transfers its glucosyl moiety to a glycogen primer (having $\alpha[1 \rightarrow 4]$ bonds) in a reaction catalysed by glycogen synthetase

$$\text{ADP-glucose} + \text{(Glucose)}_n \longrightarrow \text{ADP} + \text{(Glucose)}_{n+1}$$

Glycogen breakdown to glucose 1-phosphate occurs by a different route, involving the enzyme *phosphorylase*

$$\text{(Glucose)}_n + Ⓟ \longrightarrow \text{(Glucose)}_{n-1} + \text{Glucose 1-}Ⓟ.$$

These differences of synthesis and degradation permit the regulation of glycogen synthesis which, in bacteria, occurs by modulation of the activity of ADP-glucose pyrophosphorylase. This enzyme is strongly activated by fructose 1,6-diphosphate and $NADPH_2$ and, to a lesser extent, by glyceraldehyde 3-phosphate and phospho-enolpyruvate. It would seem that the signals for glycogen synthesis are the concentrations in the intracellular pools of these glycolytic intermediates and the accumulation of $NADPH_2$ in the cell under circumstances where they are not being utilized for biosynthetic reactions.

Mammalian glycogen synthesis differs from that in bacteria and plants in employing UDP-glucose as the glucosyl carrier and by being regulated at the levels of glycogen synthetase and phosphorylase. To date, no allosteric control of bacterial phosphorylase has been observed.

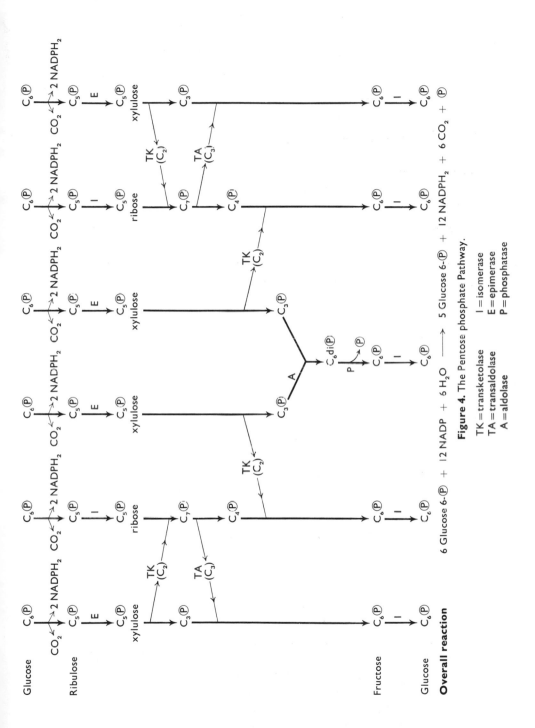

Figure 4. The Pentose phosphate Pathway.

$$6 \text{ Glucose } 6\text{-}\textcircled{P} + 12 \text{ NADP} + 6 \text{ H}_2\text{O} \longrightarrow 5 \text{ Glucose } 6\text{-}\textcircled{P} + 12 \text{ NADPH}_2 + 6 \text{ CO}_2 + \textcircled{P}$$

TK = transketolase   I = isomerase
TA = transaldolase   E = epimerase
A = aldolase   P = phosphatase

The pentose phosphate cycle

For many years Embden–Meyerhof glycolysis was considered to be the sole pathway for glucose catabolism, but the process referred to as the *pentose phosphate cycle* has now been shown to be of major importance to living organisms, including many bacterial species. The overall effect of the cycle may be summarized as:

6 Glucose 6-$\textcircled{P}$ + 12 NADP + 6H$_2$O $\longrightarrow$ 5 Glucose 6-$\textcircled{P}$ + 6CO$_2$ + 12 NADPH$_2$ + $\textcircled{P}$

This is illustrated diagrammatically in Figure 4 and it will be appreciated that the cyclic involvement of six molecules of glucose 6-phosphate results in the complete oxidation of one of them to CO_2 and water, with the regeneration of five molecules of glucose 6-phosphate. The sequence involves the oxidative decarboxylation of a hexose to a pentose followed by anaerobic rearrangements of the carbon skeletons of pentose so produced. The mechanism of these reactions will now be considered.

The pentose phosphate pathway diverges immediately from Embden–Meyerhof glycolysis in that oxidation of glucose 6-phosphate occurs to yield 6-phosphogluconolactone, catalysed by *glucose 6-phosphate dehydrogenase*. This lactone is then hydrolysed by a *lactonase* to yield 6-phosphogluconate which is subsequently oxidatively decarboxylated to give the pentose phosphate, ribulose 5-phosphate. Experiments with isotopes have clearly shown that it is the 1-C atom of glucose which is eliminated as CO_2 (Figure 5).

Ribulose 5-phosphate, under the influence of the enzymes *pentose phosphate isomerase* and *xyluloepimerase*, forms an equilibrium mixture of ribose 5-phosphate, ribulose 5-phosphate and xylulose 5-phosphate. So far, then, the hexose monophosphate, by oxidation and by loss of CO_2, has been converted to a mixture of isomeric pentose monophosphates. Now rearrangements of the carbon skeletons of these pentose monophosphates occur under the influence of the enzymes *transketolase* and *transaldolase*. The intermediate formation of C_7 and C_4 sugars occurs in a series of reactions which does not require the presence of oxygen. First, one molecule each of the ribose 5-phosphate and xylulose 5-phosphate react to form the seven-carbon sugar, sedoheptulose 7-phosphate and the triose, glyceraldehyde 3-phosphate; the reaction is catalysed by transketolase which requires thiamine pyrophosphate (TPP) as cofactor. In keeping with the known mechanism of action of TPP (see Appendix B) it is believed that an 'active' glycolaldehyde complex is formed with the enzyme-coenzyme by fission of xylulose 5-phosphate into glyceraldehyde 3-phosphate and a two-carbon fragment. The active glycolaldehyde then becomes attached to ribose 5-phosphate forming sedoheptulose 7-phosphate, thus:

```
      CH₂OH                    ⎡ CH₂OH ⎤  TPP-Enzyme
       |                       ⎢   |   ⎥
      C=O                      ⎢ H—C=O ⎥
       |          + Enzyme-TPP ⎣   +   ⎦
  HO—C—H          ───────────→   O=C—H
       |                           |
   H—C—OH                      H—C—OH
       |                           |
      CH₂O℗                      CH₂O℗
     Xylulose                 Glyceraldehyde
    5-phosphate                3-phosphate
```

```
   ⎡ CH₂OH ⎤  TPP-Enzyme          CH₂OH
   ⎢   |   ⎥                       |
   ⎣ H—C=O ⎦                      C=O
       +                           |
     O=C—H                    HO—C—H
       |                           |
   H—C—OH                      H—C—OH
       |                           |
   H—C—OH        ──────→       H—C—OH  + Enzyme-TPP
       |                           |
   H—C—OH                      H—C—OH
       |                           |
     CH₂O℗                      CH₂O℗
  Ribose 5-phosphate        Sedoheptulose 7-phosphate
```

A further transfer reaction between sedoheptulose 7-phosphate and glyceraldehyde 3-phosphate is catalysed by the enzyme transaldolase, and by its action fructose 6-phosphate and the four-carbon sugar erythrose 4-phosphate are formed. The reaction resembles that catalysed by transketolase but in this case the active moiety transferred is dihydroxyacetone and no cofactor requirement has been demonstrated:

```
   CH₂OH                                                        CH₂OH
    |                                                            |
   C=O                                                          C=O
    |                                                            |
  HO—C—H                                                    HO—C—H
    |                                                            |
  H—C—OH                          O=C—H      HO—C—H          H—C—OH
    |           H—C=O               |          |               |
  H—C—OH  +   H—C—OH   ──────→   H—C—OH  +  H—C—OH           H—C—OH
    |           |                   |          |               |
   CH₂O℗      CH₂O℗              CH₂O℗      CH₂O℗           CH₂O℗
 Sedoheptulose Glyceraldehyde   Erythrose   Fructose
 7-phosphate    3-phosphate     4-phosphate 6-phosphate
```

Transketolase now catalyses a reaction between erythrose 4-phosphate and xylulose 5-phosphate such that another mole of fructose 6-phosphate and one of glyceraldehyde 3-phosphate are formed (see top of next page).

The fructose 6-phosphate molecules, by the action of phosphohexose isomerase, are converted to glucose 6-phosphate and may then re-enter the cycle. The overall effect of the cycle at this stage is therefore:

6 Glucose 6-phosphate $\longrightarrow$

4 Glucose 6-phosphate + 2 Glyceraldehyde 3-phosphate + $6CO_2$

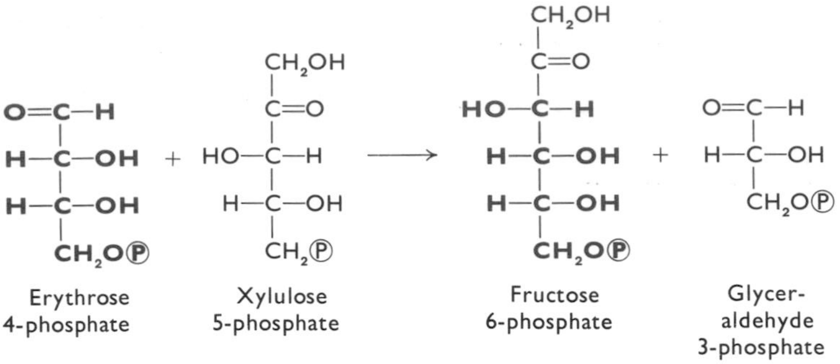

Chapter 3

$$
\begin{array}{ccccccc}
\text{O}=\text{C}-\text{H} & & \text{CH}_2\text{OH} & & \text{CH}_2\text{OH} & & \\
\text{H}-\text{C}-\text{OH} & + & \text{C}=\text{O} & & \text{C}=\text{O} & & \text{O}=\text{C}-\text{H} \\
\text{H}-\text{C}-\text{OH} & & \text{HO}-\text{C}-\text{H} & \longrightarrow & \text{HO}-\text{C}-\text{H} & + & \text{H}-\text{C}-\text{OH} \\
\text{CH}_2\text{O}\text{\textcircled{P}} & & \text{H}-\text{C}-\text{OH} & & \text{H}-\text{C}-\text{OH} & & \text{CH}_2\text{O}\text{\textcircled{P}} \\
& & \text{CH}_2\text{\textcircled{P}} & & \text{H}-\text{C}-\text{OH} & & \\
& & & & \text{CH}_2\text{O}\text{\textcircled{P}} & &
\end{array}
$$

Erythrose 4-phosphate + Xylulose 5-phosphate → Fructose 6-phosphate + Glyceraldehyde 3-phosphate

(See previous page.)

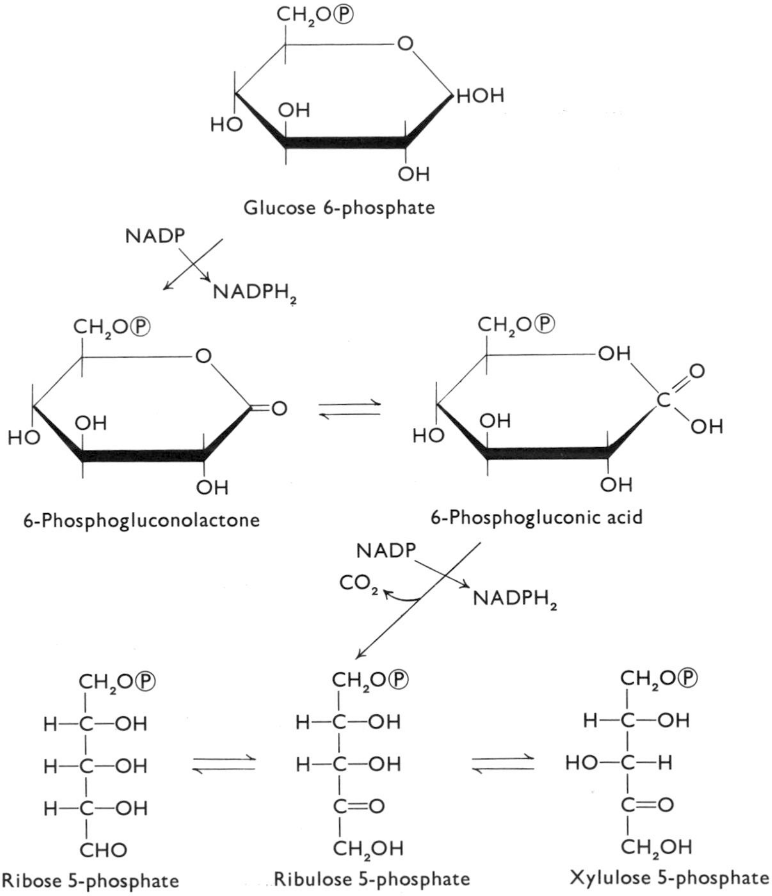

Figure 5. Oxidative decarboxylation of glucose 6-phosphate.

Triose phosphate isomerase converts one molecule of glyceraldehyde 3-phosphate to dihydroxyacetone phosphate and this is condensed with another molecule to yield fructose 1,6-diphosphate by a reversal of the aldolase reaction:

Glyceraldehyde 3-$\textcircled{P}$

$\updownarrow$

Dihydroxyacetone $\textcircled{P}$

$+$ $\rightleftharpoons$ Fructose 1,6-di $\textcircled{P}$

Glyceraldehyde 3-$\textcircled{P}$

The hexose diphosphate is dephosphorylated and isomerized to give glucose 6-phosphate. This yields a fifth molecule of glucose 6-phosphate, i.e. six molecules of hexose have been converted to five and six molecules of carbon dioxide have been formed.

It will be seen in Figure 5 that two reactions result in formation of $NADPH_2$ so that 12 molecules have to be reoxidized for each turn of the cycle, since six molecules of hexose are involved. Appendix B indicates that 3 ATP's can be formed from the oxidation of each $NADPH_2$. The ATP production is thus 36 molecules but one is expended in the phosphorylation of glucose to glucose 6-$\textcircled{P}$, hence the overall reactions may be written as:

$$Glucose + 6 O_2 \longrightarrow 6 CO_2 + 6 H_2O + Energy\ [=35\ ATP]$$

These reactions illustrate how the pentose cycle yields energy in effect by oxidizing glucose completely to carbon dioxide and water. But it is also evident that $NADPH_2$ formed in the pentose cycle serves as an important, and probably the major, source of reducing power for fatty acid biosynthesis in a series of reactions which effects the condensation and reduction of acetate units to yield long chain fatty acids, as described in Chapter 4, pp. 242–245. In this sense, then, $NADPH_2$ may be regarded as an alternative form of energy and one which is utilized for fatty acid synthesis. However, it must be appreciated that, besides producing energy for biosynthetic reactions and cellular maintenance, the pentose cycle can also serve to furnish the intermediates required for biosynthesis, and compounds such as pentose and triose phosphates will be drained from the cycle for this purpose.

It should be noted that since the anaerobic rearrangements of the carbon skeletons in the pentose cycle are reversible, it is possible for five molecules of hexose phosphate to give rise to six molecules of pentose phosphate by non-oxidative reactions, as opposed to their direct oxidation to pentose phosphate and carbon dioxide.

The Entner–Doudoroff pathway

A new pathway of glucose metabolism was discovered by Entner and Doudoroff who observed that *Pseudomonas saccharophila* preferentially released $^{14}CO_2$

from [1-^{14}C]glucose and that the pyruvate produced was labelled in a manner different from that characteristic of the Embden–Meyerhof route; the carboxyl of one pyruvate was derived from 1-C and the other from 4-C:

1 COOH
  |
2 CO
  |
3 CH$_3$

4 COOH
  |
5 CO
  |
6 CH$_3$

The isolation of a new intermediate, 2-oxo-3-deoxy-6-phosphogluconate and the discovery of two new enzymes, one a dehydratase for 6-phosphogluconate and the other an aldolase specific for the new intermediate, enabled them to formulate the metabolic sequence shown in Figure 6.

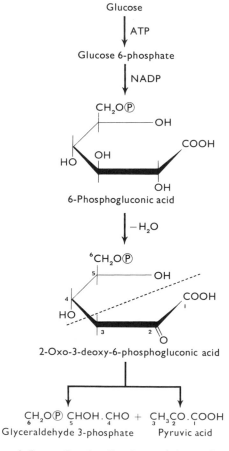

Figure 6. Entner–Doudoroff pathway of glucose degradation.

The dehydratase removes the elements of water from 6-phosphogluconate and requires Fe^{2+} ions and glutathione for maximal activity. No cofactors have been demonstrated for the aldolase which splits the resulting 2-oxo-3-deoxy-6-phosphogluconate to glyceraldehyde 3-phosphate and pyruvate; the former can be converted to a second molecule of pyruvate by the reactions common to glycolysis.

This sequence of reactions, now commonly referred to as the Entner–Doudoroff pathway, has been found to occur in a number of Gram-negative species, particularly among pseudomonads, but it can also be induced by growing the Gram-positive, lactate-producing organism *Streptococcus faecalis* on gluconate. Studies with labelled substrates have revealed that glucose is metabolized exclusively by the Entner–Doudoroff pathway in *Pseudomonas saccharophila*, *Zymomonas mobilis* (*Pseudomonas lindneri*) and *Zymomonas anaerobia* and is the major pathway in other pseudomonads.

Energetically, the anaerobic operation of the Entner–Doudoroff pathway is only half as efficient as anaerobic glycolysis, for the net yield of ATP per mole of glucose is 1 mole instead of 2, because only 1 mole of triose phosphate is produced and oxidized. This gives 2 moles of ATP but the net yield is only one because of the need to phosphorylate glucose.

Phosphoketolase pathway

Some kinds of bacteria carry out a fermentation of glucose which yields lactate together with other major products such as CO_2, acetate or ethanol. They also ferment pentoses with the formation of lactate and acetate. An example is *Leuconostoc mesenteroides* which ferments glucose according to the equation:

$$\text{Glucose} \longrightarrow \text{Lactate} + \text{Ethanol} + CO_2$$

The enzymes phosphofructokinase, aldolase and triose phosphate isomerase are absent, indicating a departure from Embden–Meyerhof glycolysis. This was confirmed by the use of $[1\text{-}^{14}C]\text{-}$ and $[3,4\text{-}^{14}C_2]\text{-glucose}$ which indicated the following derivations:

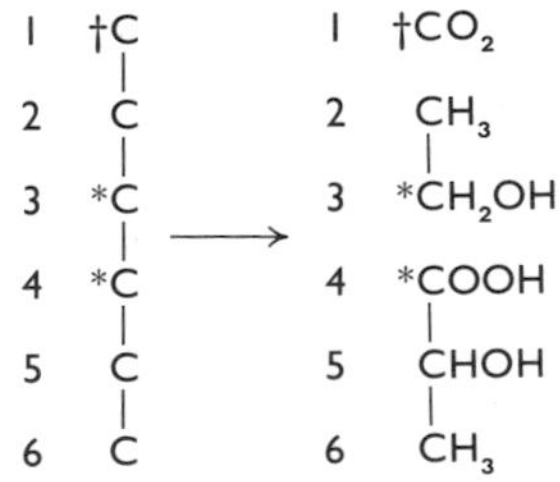

```
1   †C              1   †CO2
    |
2   C               2   CH3
    |                   |
3   *C      ------>  3   *CH2OH
    |                   |
4   *C              4   *COOH
    |                   |
5   C               5   CHOH
    |                   |
6   C               6   CH3
```

These results suggest that the hexose is decarboxylated to a pentose which then undergoes fission, a conclusion supported by the presence of glucose 6-phosphate dehydrogenase and 6-phosphogluconate dehydrogenase in the organism. These enzymes bring about the oxidative decarboxylation of glucose 6-phosphate to ribulose 5-phosphate and are NAD- rather than NADP-dependent.

The substrate for the fission is xylulose 5-phosphate and *phosphoketolase* cleaves this pentose phosphate to acetyl phosphate and glyceraldehyde 3-phosphate. The enzyme requires TPP, inorganic phosphate, Mg^{2+} and a thiol compound for activity:

$$
\begin{array}{c}
CH_2OH \\
| \\
C{=}O \\
| \\
HO{-}C{-}H \quad + \quad \text{\textcircled{P}} \longrightarrow \\
| \\
H{-}C{-}OH \\
| \\
CH_2O\text{\textcircled{P}}
\end{array}
\qquad
\begin{array}{c}
CH_3 \\
| \\
\text{\textcircled{P}}{\sim}O{-}C{=}O \\
\text{Acetyl phosphate} \\
\\
O{=}C{-}H \\
| \\
H{-}C{-}OH \\
| \\
CH_2O\text{\textcircled{P}}
\end{array}
$$

Xylulose   Glyceraldehyde
5-phosphate   3-phosphate

The purified enzyme is specific for xylulose 5-phosphate and does not attack fructose 6-phosphate or sedoheptulose 7-phosphate although a similar enzyme from *Acetobacter xylinum* does split fructose 6-phosphate to yield erythrose 4-phosphate and acetyl phosphate. In the presence of ADP, cell extracts catalyse the formation of glyceraldehyde 3-phosphate, acetate and ATP so that the cleavage yields energy to the organism. Again in this fermentation the triose phosphate is converted to pyruvate by the glycolytic enzymes and the pyruvate is reduced to lactate. The other ultimate product of the fission, i.e. acetate or ethanol, depends on the oxidation-reduction balance of the system which, in turn, depends on whether the substrate is a hexose or a pentose. Figure 7 reveals that conversion of glucose to xylulose 5-phosphate involves two oxidation steps, with the formation of $2NADH_2$, or the equivalent of 4H. Metabolism of glyceraldehyde 3-phosphate to pyruvate yields a further 2H making 6H in all. Two of these are utilized for the reduction of pyruvate to lactate and, to balance the overall reaction, the remaining 4H are used to reduce acetyl phosphate to ethanol, via acetyl coenzyme A and acetaldehyde. However, when pentose is the substrate the two initial oxidation steps are eliminated and the reducing power formed is equivalent to only 2H, which are utilized for lactate formation, leaving acetyl phosphate as the other product.

This difference in products is also reflected in the energetics of the fermentations for whereas there is a net yield of 2 moles of ATP per mole of pentose fermented, the yield from glucose is only one as a consequence of the loss of the high energy bond of acetyl phosphate by its reduction to ethanol. This observation underlines the general principle that alcohol formation by any route other than pyruvate decarboxylation and acetaldehyde reduction is energetically wasteful to the organism.

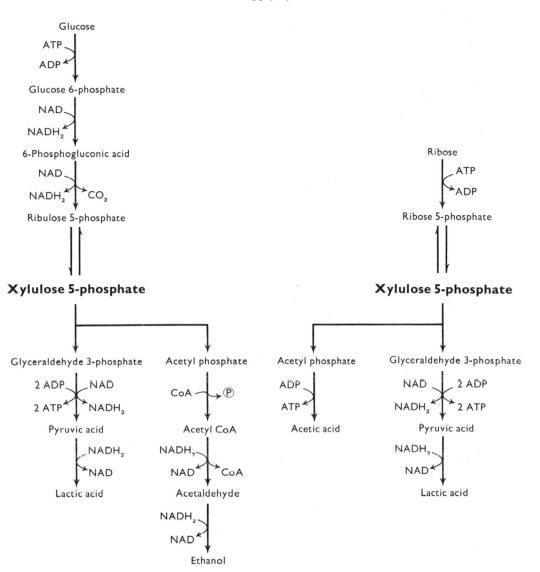

Figure 7. Phosphoketolase pathway. Fermentation of glucose to lactic acid, ethanol and CO_2 and of ribose to lactic acid and acetic acid.

Glucose $\longrightarrow$ Lactic acid + ethanol + CO_2 + ATP
Ribose $\longrightarrow$ Lactic acid + acetic acid + 2 ATP

Catabolism of lipids

Two of the most important kinds of lipids are triglycerides and phospholipids. The former are long-chain fatty acyl esters:

$$CH_2O.OC(CH_2)_pCH_3$$
$$CHO.OC(CH_2)_qCH_3$$
$$CH_2O.OC(CH_2)_rCH_3$$

where p, q and r may be the same or different, are usually even numbers, and are frequently about 16. Phospholipids have the third acyl group replaced by phosphate linked to ethanolamine, serine or inositol. Lipids occur as components of membranes and can also serve as reserves of carbon and energy. When they are catabolized they are first hydrolysed to yield the free fatty acids. As shown in Figure 8 these are then degraded stepwise to acetyl coenzyme A which is further metabolized in the tricarboxylic acid cycle.

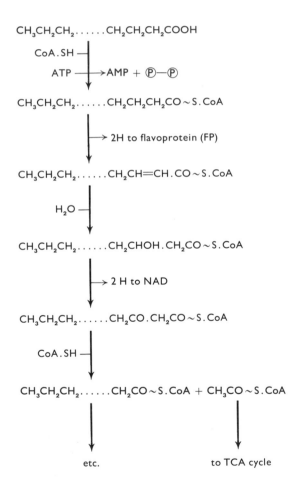

Figure 8. *β*-Oxidation of fatty acids.

THE TRICARBOXYLIC ACID CYCLE

We have seen how the various pathways of glucose metabolism lead to the formation of pyruvate. Under aerobic conditions pyruvate is then oxidized by a cyclic process termed the *tricarboxylic acid cycle*, entry to which is gained after pyruvate has been converted to a C_2-unit by loss of carbon dioxide (Figure 9). The tricarboxylic acid cycle effects the oxidation of two-carbon units to carbon dioxide and water and constitutes the most important single mechanism for the generation of ATP in aerobic organisms. Additionally, it is of importance for the production of carbon skeletons for synthetic reactions, particularly those leading to the synthesis of amino acids, e.g. aspartic and glutamic acids. The relative importance of these two functional roles of the cycle depends to a large extent on whether or not the cells are growing. This aspect of metabolism is considered subsequently on page 188. Before entering the tricarboxylic acid cycle, pyruvate is converted to acetyl coenzyme A. This oxidative decarboxylation is brought about by a multi-enzyme complex which comprises three enzymes, five different cofactors and requires Mg^{2+} ions. The cofactors include thiamine pyrophosphate (TPP), coenzyme A (CoA), lipoic acid, flavin adenine dinucleotide (FAD) and NAD (see Appendix B). Coenzyme A combines with substrates via its thiol group; for this reason it is usually denoted as CoA.SH. Combination with acyl groups, e.g. acyl-S.CoA, gives rise to C—S bonds which have a high free energy of hydrolysis and such compounds therefore belong to the class of high-energy compounds.

Lipoic acid (thioctic acid) is a dithiol compound and the open chain form can be reversibly oxidized to a disulphide, five-membered ring form; these are denoted by $lip(SH)_2$ and $lipS_2$ respectively:

Reduced   Oxidized

Lipoic acid

FAD is a hydrogen carrier which undergoes reversible reduction and oxidation by accepting and relinquishing two hydrogen atoms (see Appendix B):

$$FAD + 2H \rightleftharpoons FADH_2$$

The first step in the oxidative decarboxylation of pyruvate is catalysed by pyruvate dehydrogenase which requires TPP and Mg^{2+} as cofactors. An enzyme-bound 'active acetaldehyde'-TPP complex, which has been shown to be 2-hydroxyethyl-2-thiamine pyrophosphate, is produced and this is then transferred to one of the sulphur atoms of the disulphide form of lipoic acid, being oxidized to acetyl lipoate.

$$CH_3CO.COOH + TPP \longrightarrow [CH_3CHO]TPP + CO_2$$

Pyruvate   Active acetaldehyde

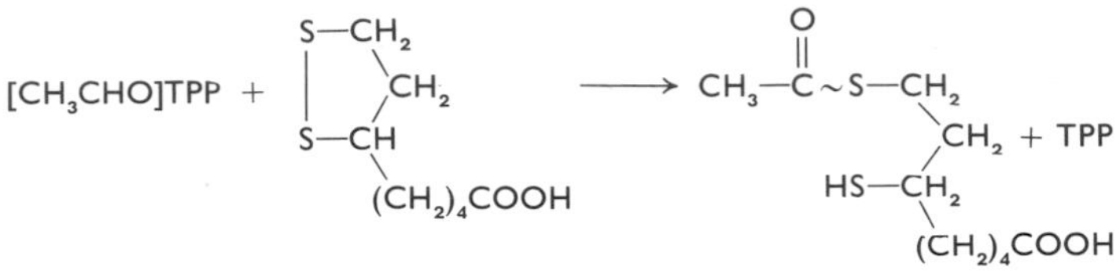

$$[CH_3CHO]TPP +$$
$$\begin{array}{c} S-CH_2 \\ | \quad CH_2 \\ S-CH \\ \quad (CH_2)_4COOH \end{array}$$
$$\longrightarrow$$
$$\begin{array}{c} O \\ \| \\ CH_3-C\sim S-CH_2 \\ \quad CH_2 + TPP \\ HS-CH_2 \\ \quad (CH_2)_4COOH \end{array}$$

Lipoic acid   Acetyl lipoic acid
(disulphide form)

The lipoic acid is covalently bound to the second enzyme of the complex, *dihydrolipoyl transacetylase*. The C $\sim$ S bond in the acyl lipoates resembles that in acyl CoA compounds and has a high free energy of hydrolysis, i.e. acyl lipoates are high-energy compounds. The TPP is liberated and is thus able to act catalytically. The acetyl lipoate now serves as an acyl donor for coenzyme A, resulting in the formation of free acetyl coenzyme A and reduced lipoic acid bound to the transacetylase enzyme.

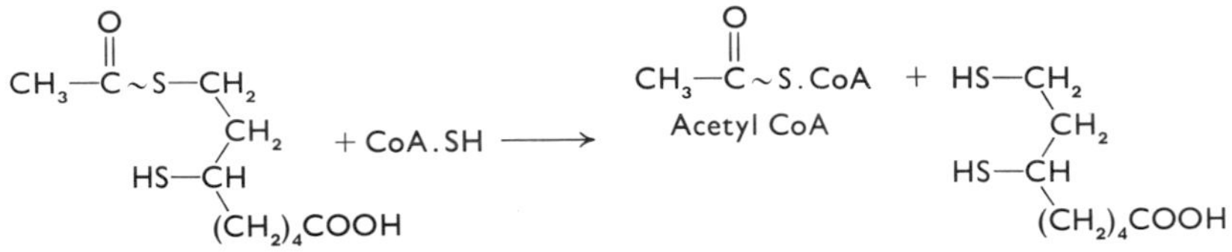

$$\begin{array}{c} O \\ \| \\ CH_3-C\sim S-CH_2 \\ \quad CH_2 \\ HS-CH \\ \quad (CH_2)_4COOH \end{array} + CoA.SH \longrightarrow \begin{array}{c} O \\ \| \\ CH_3-C\sim S.CoA \\ \text{Acetyl CoA} \end{array} + \begin{array}{c} HS-CH_2 \\ \quad CH_2 \\ HS-CH \\ \quad (CH_2)_4COOH \end{array}$$

Acetyl lipoic acid   Reduced lipoic acid

Reduced lipoyl transacetylase is then re-oxidized by the third enzyme of the complex, *dihydrolipoyl dehydrogenase*, which contains tightly bound FAD; this becomes reduced and is in turn re-oxidized in a mechanism involving NAD.

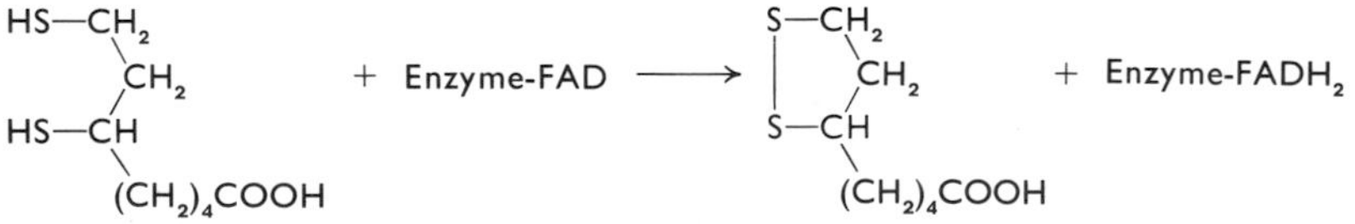

$$\begin{array}{c} HS-CH_2 \\ \quad CH_2 \\ HS-CH \\ \quad (CH_2)_4COOH \end{array} + \text{Enzyme-FAD} \longrightarrow \begin{array}{c} S-CH_2 \\ | \quad CH_2 \\ S-CH \\ \quad (CH_2)_4COOH \end{array} + \text{Enzyme-FADH}_2$$

$$\text{Enzyme-FADH}_2 + \text{NAD} \longrightarrow \text{Enzyme-FAD} + \text{NADH}_2$$

The overall reaction for the oxidation of pyruvate may be written as:

$$\text{Pyruvate} + \text{CoA.SH} + \text{NAD} \longrightarrow \text{Acetyl} \sim \text{S.CoA} + \text{NADH}_2 + \text{CO}_2$$

The pyruvate dehydrogenase complex has been isolated from *Escherichia coli* and found to have a particle weight of four million. It consists of 24 molecules of pyruvate dehydrogenase each binding one molecule of TPP, one molecule of dihydrolipoyl transacetylase containing 24 polypeptide chains each possessing one molecule of lipoic acid, and 12 molecules of dihydrolipoyl dehydrogenase each containing one molecule of FAD.

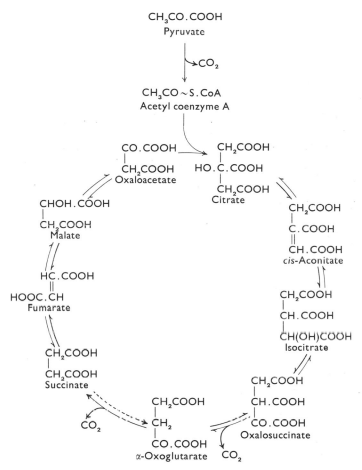

Figure 9. The tricarboxylic acid cycle.

Under certain circumstances energy may be made available by the conversion of acetyl CoA to acetate and CoA with concomitant formation of a mole of ATP.

Quantitatively, however, the most important reaction which acetyl CoA undergoes in metabolism is condensation with oxaloacetate to form citrate and thus to gain entry to the tricarboxylic acid cycle. The reaction is catalysed by *citrate synthase*.

$$
\begin{array}{c}
\text{Acetyl CoA} \\
CH_3CO \sim S.CoA \\
+ \\
O{=}C{-}COOH \\
| \\
CH_2COOH \\
\text{Oxaloacetic acid}
\end{array}
\rightleftharpoons
\begin{array}{c}
CH_2COOH \\
| \\
HOC{-}COOH + CoA.SH \\
| \\
CH_2COOH \\
\text{Citric acid}
\end{array}
$$

The formation of citrate is followed by isomerization to D-isocitrate via *cis*-aconitate. This is catalysed by the enzyme *aconitate hydratase* (*aconitase*) which is specific for *cis*-aconitate and D-isocitrate and which, therefore, catalyses two distinct dehydration reactions, one involving a hydroxyl group attached to a tertiary carbon atom and the other a secondary hydroxyl attached to a secondary carbon atom. This seemed so remarkable that it was originally assumed that two different aconitases existed, one for each reaction. All attempts to resolve the enzyme into two such components have failed and the relative activities of the two reactions remain constant throughout purification:

$$
\begin{array}{c}
CH_2COOH \\
| \\
HO.C.COOH \\
| \\
CH_2COOH \\
\text{Citric acid}
\end{array}
\underset{\pm H_2O}{\overset{\text{Aconitase}}{\rightleftharpoons}}
\begin{array}{c}
CH.COOH \\
\| \\
C.COOH \\
| \\
CH_2COOH \\
\text{cis-Aconitic acid*}
\end{array}
\underset{\pm H_2O}{\overset{\text{Aconitase}}{\rightleftharpoons}}
\begin{array}{c}
CH(OH)COOH \\
| \\
HC.COOH \\
| \\
CH_2COOH \\
\text{Isocitric acid}
\end{array}
$$

*There is some uncertainty concerning the role of *cis*-aconitate as an obligatory intermediate. It has been suggested that an enzyme-bound carbonium ion is the true intermediate and *cis*-aconitate is produced from this ion as a side product.

In the next step of the cycle isocitrate undergoes oxidative decarboxylation to α-oxoglutarate. This is catalysed by *isocitrate dehydrogenase* for which both NAD and NADP requirements have been established, depending upon the source of the enzyme:

$$
\begin{array}{c}
HOCH.COOH \\
| \\
CH.COOH + NADP \\
| \\
CH_2COOH \\
\text{Isocitric acid}
\end{array}
\rightleftharpoons
\begin{array}{c}
O{=}C{-}COOH \\
| \\
CH_2 \\
| \\
CH_2COOH \\
\text{α-Oxoglutaric acid}
\end{array}
+ NADPH_2 + CO_2
$$

Oxalosuccinic acid has been postulated as an intermediate in this reaction and it has been shown that highly purified isocitrate dehydrogenase will catalyse the decarboxylation of this compound.

The overall oxidative decarboxylation reaction is reversible and, therefore, enables carbon dioxide fixation to occur. It also is one of the energy-yielding steps of the cycle since reduction of nicotinamide nucleotide occurs.

α-Oxoglutarate is now oxidized to succinate and carbon dioxide by the action of *α-oxoglutarate oxidase*, a reaction similar in many respects to pyruvate oxidation which has already been discussed:

$$O=C-COOH \atop CH_2 \atop CH_2 \atop COOH \quad + \; CoA.SH \; + \; NAD \longrightarrow \quad O=C\sim S.CoA \atop CH_2 \atop CH_2 \atop COOH \quad + \; CO_2 \; + \; NADH_2$$

α-Oxoglutaric acid   Succinyl CoA

Succinyl CoA is then cleaved to succinic acid and CoA with the concomitant formation of a mole of ATP.

Succinate next undergoes oxidation to fumarate by *succinate dehydrogenase*:

$$CH_2COOH \atop CH_2COOH \quad + \; FP \; \xrightarrow[\text{dehydrogenase}]{\text{Succinate}} \quad HC-COOH \atop HOOC-CH \quad + \; FPH_2$$

Succinic acid   Fumaric acid

As this enzyme is associated with insoluble particles which contain also the necessary enzymes for transferring electrons to molecular oxygen, it is often referred to as the succinoxidase system. Succinate dehydrogenase is a flavoprotein (FP) which is oxidized via the cytochrome system.

Fumarate is reversibly hydrated to malate by the enzyme *fumarate hydratase* (*fumarase*) which therefore carries out a similar type of reaction to aconitase. Only the L-isomer of malate is formed and only fumarate, and not maleate, can serve as the substrate:

$$HC-COOH \atop HOOC-CH \quad + \; H_2O \; \xrightarrow{\text{Fumarase}} \quad HOCH-COOH \atop CH_2COOH$$

Fumaric acid   L-Malic acid

The last step of the cycle is the oxidation of L-malate to regenerate oxaloacetate by a nicotinamide nucleotide-requiring *malate dehydrogenase*. At neutral pH values the equilibrium is greatly in favour of malate formation. However, under the normal conditions of operation of the tricarboxylic acid cycle, oxaloacetate is removed at a rapid rate and therefore malate oxidation proceeds readily:

$$HOCH-COOH \atop CH_2COOH \quad + \; NAD \; \xrightarrow[\text{dehydrogenase}]{\text{Malate}} \quad O=C-COOH \atop CH_2COOH \quad + \; NADH_2$$

L-Malic acid   Oxaloacetic acid

In passing, it may be noted that another enzyme for the oxidation of malate is known, namely *malate dehydrogenase* (*decarboxylating*). This so-called '*malate enzyme*' catalyses an oxidative decarboxylation to pyruvate and CO_2 using NADP as cofactor:

$$HOCH-COOH \atop CH_2COOH \quad + \; NADP \; \rightleftharpoons \quad O=C-COOH \atop CH_3 \quad + \; CO_2 \; + \; NADPH_2$$

L-Malic acid   Pyruvic acid

There is no evidence that oxaloacetate is an intermediate and since this compound must be formed to permit the cycle to function, the malate enzyme is not believed to play a role in the tricarboxylic acid cycle. Present views assign to this enzyme the function of producing pyruvate for glucose synthesis (glucogenesis) when growth occurs on dicarboxylic acids such as succinate and malate, and the formation of $NADPH_2$ for reductive biosynthesis. Enzymes which catalyse the decarboxylation of oxaloacetate to pyruvate and CO_2 and require only a divalent metal ion as cofactor are also known. Again, these enzymes are not strictly relevant to the present discussion of the cycle.

We have now seen how a two-carbon fragment produced in metabolism is condensed with oxaloacetate and, by undergoing a series of reactions, regenerates oxaloacetate and is itself oxidized to two molecules of carbon dioxide and two molecules of water. Oxaloacetate is thus enabled to combine with another molecule of acetyl CoA and participate in another cycle. It must be emphasized that the carbon atoms of the regenerated oxaloacetate are not identical with those of the molecule of oxaloacetate which initiated the cycle.

Significance of the tricarboxylic acid cycle

The tricarboxylic acid cycle fulfils two major roles, the production of energy and the provision of intermediates for biosynthesis, and their relative importance depends on whether or not the cells are growing. By isotopic experiments it has been shown that during exponential growth of *Escherichia coli* in glucose ammonium salts medium the principal function of the tricarboxylic acid cycle is to provide intermediates while glycolysis furnishes energy. When growth ceases, a switch of roles occurs, the cycle now generates energy and glucose is used for glycogenesis, leading to a deposition of glycogen in the cells.

From our previous discussion it is apparent that the tricarboxylic acid cycle is important in the total oxidation of carbohydrates. However, it is also important in the oxidation of many other substances. For instance, fatty acids are degraded stepwise to acetyl-CoA, many amino acids are converted to the keto acids, pyruvic, α-oxoglutaric and oxaloacetic acids, and indeed intermediates of the cycle such as citric and succinic acids may be directly oxidized by this pathway.

Furthermore, as outlined in Section 4, these same substances may be precursors in the biosynthesis of amino acids and nucleotides.

Regulation of the tricarboxylic acid cycle

Recent work has shown that control of the tricarboxylic acid cycle is exerted principally at the citrate synthase stage, i.e. on the first enzyme of the cycle. Weitzman has found that the citrate synthases of 18 genera of Gram-negative bacteria (including Azotobacter and Pseudomonas) were inhibited by $NADH_2$ whereas the enzyme from aerobic Gram-positive bacteria was unaffected. Some of these enzymes are, however, inhibited by ATP. Within the $NADH_2$—susceptible group, subgroups could be distinguished on the basis of the ability of AMP to reverse the $NADH_2$ inhibition. It has been suggested that only those organisms with a relatively simple mesosomal structure are susceptible to inhibition by $NADH_2$. Thus under conditions which result in an accumulation of reducing power and ATP, e.g. when biosynthesis is curtailed, the overall operation of the tricarboxylic acid cycle will be inhibited in susceptible bacteria, thus conserving energy and carbon. The citrate synthase of *Escherichia coli* has also been shown to be inhibited by α-oxoglutarate.

Although the oxidation of pyruvate to acetyl CoA is not part of the cycle it does furnish one of the substrates for citrate synthase and is itself also subject to regulation. Acetyl CoA, the product of the oxidation, inhibits pyruvate dehydrogenase, the first enzyme of the complex, by negative feedback inhibition (see Chapter 4, p. 202). This enzyme is activated by AMP and inhibited when the energy charge is high.

Energy yields as a result of glucose catabolism

The generation of energy in the tricarboxylic acid cycle occurs in the oxidative steps, namely the oxidation of pyruvate, isocitrate, α-oxoglutarate, succinate and malate. We can consider the total yield of energy in terms of molecules of ATP produced from pyruvate in the course of one turn of the tricarboxylic acid cycle, i.e. in the oxidation of pyruvate to carbon dioxide and water. The energy yielding steps are as follows:

Pyruvate + NAD ⟶ Acetyl CoA + $NADH_2$	3 ATP
Isocitrate + NADP ⟶ α-Oxoglutarate + $NADPH_2$	3 ATP
α-Oxoglutarate + NAD ⟶ Succinyl CoA + $NADH_2$	3 ATP
Succinyl CoA ⟶ Succinate + CoA	1 ATP
Succinate + FP ⟶ Fumarate + FPH_2	2 ATP
Malate + NAD ⟶ Oxaloacetate + $NADH_2$	3 ATP
	Total 15 ATP

Hence the oxidation of pyruvate by the cycle yields 15 molecules of ATP. Thus when glucose is metabolized by the Embden–Meyerhof pathway and the tricarboxylic acid cycle there is a net yield of 38 molecules of ATP (Figure 10). This may be compared with anaerobic glycolysis or yeast fermentation:

Glucose ⟶ 2 Lactate + 2 ATP	
Glucose ⟶ 2 Ethanol + $2 CO_2$ + 2 ATP	

when the much greater efficiency of the aerobic process in terms of energy is apparent.

The energy which is biologically useful to an organism is that which is obtained in the form of ATP, and is not necessarily identical with the free energy change of the process. When the energy yield is measured in terms of the molar growth yield (grams dry weight of organism produced per mole of glucose utilized as the energy source) the values reflect the yield of ATP. The measurement of molar growth yields thus represents a valuable experimental technique for the assessment of the energy (ATP) derived from a specified carbon source by a given type of bacterium, and can be used in deducing the metabolic pathways involved. For example, the anaerobic fermentation of glucose by the Embden–Meyerhof, Entner–Doudoroff and phosphoketolase pathways yields respectively 2, 1 and 1 moles of ATP per mole of glucose fermented; consequently the amount of growth supported per mole of glucose by the Embden–Meyerhof pathway will be double that obtained with either of the other two types of metabolism.

THE GLYOXYLATE CYCLE

As mentioned above, intermediates of the tricarboxylic acid cycle may be continuously drawn off and used as precursors of amino acids, etc. They must be replenished if the cycle is to go on functioning. This usually happens by carboxylation of pyruvate or phospho-enolpyruvate (PEP) to yield oxaloacetate. The following reactions bring this about:

$$CH_3CO.COOH + CO_2 + ATP \xrightarrow{\text{pyruvate carboxylase}} HOOC.CH_2CO.COOH + ADP + \textcircled{P}$$

$$CH_2:CO\sim\textcircled{P}.COOH + CO_2 + H_2O \xrightarrow{\text{PEP-carboxylase}} HOOC.CH_2CO.COOH + \textcircled{P}$$

The carboxylation of PEP by PEP-carboxylase appears to be the essential reaction for *E. coli in vivo* since only mutants which lack this enzyme fail to grow on pyruvate or its precursors unless tricarboxylic acid cycle intermediates are added to the growth medium. PEP-carboxylase is absent, however, from pseudomonads and Arthrobacter which employ the ATP-dependent pyruvate carboxylase to produce oxaloacetate.

However many micro-organisms are able to utilize acetate as the sole carbon source and if under these conditions the tricarboxylic acid cycle is to continue to provide energy and intermediates for biosynthesis, there must be an alternative method of replenishing its intermediates from C_2-units. A pathway which does this is the *glyoxylate cycle* (Figure 11) and pathways having this function are sometimes referred to as 'anaplerotic' (from the Greek for 'filling up'), since they replenish intermediates drained off for biosynthesis. Effectively the net result of this cycle is the condensation of two molecules of acetate to give one of succinate. This is made possible by two additional enzymes that we have not so far discussed. One of these is *isocitrate lyase* and the other is *malate*

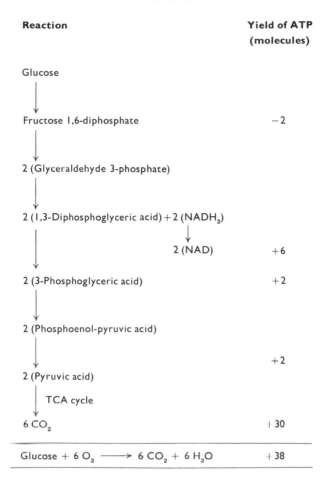

Reaction   **Yield of ATP**
                                                    (molecules)

Glucose

Fructose 1,6-diphosphate -2

2 (Glyceraldehyde 3-phosphate)

2 (1,3-Diphosphoglyceric acid) + 2 (NADH$_2$)

                          2 (NAD) $+6$

2 (3-Phosphoglyceric acid) $+2$

2 (Phosphoenol-pyruvic acid)

                                                        $+2$

2 (Pyruvic acid)

    TCA cycle

6 CO$_2$ $+30$

Glucose + 6 O$_2$ $\longrightarrow$ 6 CO$_2$ + 6 H$_2$O $+38$

Figure 10. Yield of ATP as a result of aerobic catabolism of glucose via Embden–Meyerhof pathway and tricarboxylic acid cycle.

synthase and they are found in greatly increased amounts in cells grown on acetate. The reactions then are as follows:

(1)   Acetyl CoA + Oxaloacetate + H$_2$O $\longrightarrow$ Citrate + CoA
(2)   Citrate $\rightleftharpoons$ Isocitrate
(3)   Isocitrate $\rightleftharpoons$ Succinate + Glyoxylate
(4)   Acetyl CoA + Glyoxylate + H$_2$O $\longrightarrow$ Malate + CoA
(5)   Malate + $\frac{1}{2}$O$_2$ $\longrightarrow$ Oxaloacetate + H$_2$O

2 Acetyl CoA + $\frac{1}{2}$O$_2$ + H$_2$O $\longrightarrow$ Succinate + 2 CoA

It will thus be seen that the glyoxylate cycle interlocks with the tricarboxylic acid cycle and that some enzymes and intermediates are common to both. The tricarboxylic acid cycle can then operate normally for the oxidation of acetate and the production of biosynthetic intermediates, being replenished by the succinate produced via the glyoxylate cycle.

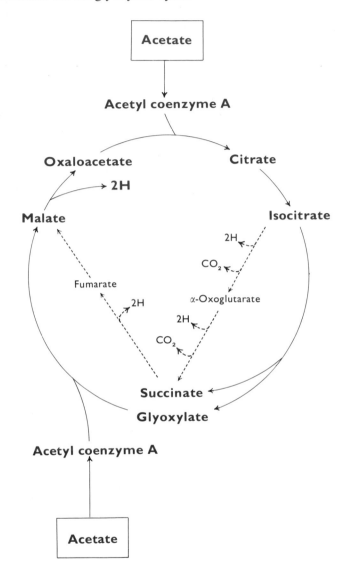

Figure 11. The glyoxylate cycle. The dotted lines indicate reactions of the tricarboxylic acid cycle.

The essential requirement of isocitrate lyase for growth on acetate was demonstrated by the inability of mutants devoid of the enzyme to grow on this substrate, although they were still able to oxidize it. Such mutants oxidized acetate to completion, as opposed to the wild type organism which oxidized acetate with the simultaneous assimilation of 20–30% of the carbon, thus demonstrating both the role of the anaplerotic enzymes in biosyntheses from acetate and also the independence of these enzymes from those concerned with the provision of energy.

The glyoxylate cycle appears to be regulated by inhibition of isocitrate lyase. In *Escherichia coli* this enzyme is powerfully inhibited by phospho-enolpyruvate in a non-competitive manner, which may be regarded as a feedback mechanism. Accumulation of phospho-enolpyruvate would thus inhibit the key enzyme of the anaplerotic sequence leading to phospho-enolpyruvate formation. The operation of this mechanism has been verified by the use of mutants of *Escherichia coli* which lack phospho-enolpyruvate carboxylase (p. 190) and consequently, although able to form phospho-enolpyruvate from pyruvate via the synthase reaction (p. 171), are unable to remove it. The addition of pyruvate to cultures of the mutant arrests growth on acetate because it causes accumulation of phospho-enolpyruvate.

In other micro-organisms compounds such as succinate, glycollate and pyruvate have also been found in inhibit isocitrate lyase.

AUTOTROPHS AND HETEROTROPHS

Most species of bacteria use organic carbon compounds and from them make their own cellular components. They derive energy from exergonic catabolic reactions of these organic substrates and are called *heterotrophs*. The *autotrophs* can use carbon dioxide as their sole source of carbon. The energy they require to do this may be supplied by light, as in *photosynthesis* or from exergonic inorganic oxidation reactions as occurs in the *chemosynthetic* bacteria. Because photosynthetic organisms can utilize organic compounds other than carbon dioxide, a modified terminology has now been adopted somewhat replacing the older autotroph/heterotroph and photosynthetic/chemosynthetic categories. The *lithotrophs* use carbon dioxide and reduce it to organic compounds. This requires energy and reducing power. The latter is supplied by inorganic substances, the former by radiation in the *photolithotrophs* and by inorganic oxidation in the *chemolithotrophs*. The organotrophs use organic substrates and may be *photo-organotrophs* using light energy for assimilation of carbon dioxide and organic substances or may be *chemo-organotrophs* using oxidation or fermentation of organic compounds. The reducing power needed in the photo-organotrophs is again supplied by organic material.

Bacterial photosynthesis closely resembles the process found in plants and blue-green algae but differs in utilizing somewhat different pigment systems and in not producing oxygen.

Phototrophic bacteria

Photosynthesis comprises an extremely complex series of reactions and it is only within recent years that the nature of these has been elucidated. There are two essential, closely integrated yet quite distinct processes involved, namely the absorption of light energy by the photosynthetic pigments and its conversion to the chemical bond energy of ATP, and the biosynthetic reactions leading from CO_2 or organic compounds to cell materials and utilizing the chemical bond energy derived from the light.

Whereas the first process is dependent on light and unique to photosynthetic organisms, the reactions leading to synthesis of cell material can occur in the dark and are also found in many non-photosynthetic bacteria. Even the key reactions by which CO_2 fixation is achieved in photosynthetic bacteria are found in chemolithotrophic organisms.

The main difference between plant and bacterial photosynthesis lies in the nature of the ultimate electron donor. Plants utilize water as the electron donor and release oxygen; isotopic studies with water enriched with ^{18}O have revealed that the oxygen is derived from the water, so that the overall series of reactions may be designated as:

$$CO_2 \ + \ 2H_2^{18}O \ \xrightarrow{\text{Light}} \ (CH_2O) \ + \ ^{18}O_2 \ + \ H_2O.$$

Photosynthetic bacteria do not use water but rather reduced sulphur compounds or organic compounds or even molecular hydrogen as electron donors, the particular donor depending on the species. The reaction is anaerobic.

Photosynthetic bacteria fall into two principal groups, referred to as the green and purple bacteria and distinguished by their photosynthetic pigments. The green *Chlorobium* genus contains one of two different chlorophylls together with alicyclic carotenoids, while the purple bacteria always contain one particular type of chlorophyll, bacteriochlorophyll, together with various aliphatic carotenoids. In consequence, organisms of the two groups absorb light of different wavelengths for the process of photosynthesis.

The green bacteria are usually photolithotrophic and strictly anaerobic organisms which utilize CO_2 as the carbon source and H_2S as an electron donor. The purple bacteria can be classified on a physiological basis as purple sulphur bacteria (Thiorhodaceae) or non-sulphur purple bacteria (Athiorhodaceae). The purple sulphur bacteria resemble the green bacteria in being anaerobic and photolithotrophic whereas the non-sulphur purple bacteria prefer organic compounds as the source of carbon and reducing power for photosynthesis, although they can also reduce CO_2. Unlike most photosynthetic bacteria, some of the non-sulphur purple organisms are able to tolerate oxygen and can grow in the dark, deriving their energy by oxidation of organic compounds and using oxygen as the final electron acceptor, i.e. as chemo-organotrophs.

The green sulphur bacteria and the purple sulphur bacteria use hydrogen sulphide as an exogenous electron donor for cellular synthesis from CO_2. Hydrogen sulphide is oxidized to sulphate in two stages, the first of which is analogous to the overall equation for plant photosynthesis:

$$CO_2 + 2\,H_2S \xrightarrow{\text{Light}} (CH_2O) + H_2O + 2\,S \qquad \text{(Bacteria)}$$

$$\text{cf.} \quad CO_2 + 2\,H_2O \xrightarrow{\text{Light}} (CH_2O) + H_2O + O_2 \qquad \text{(Plants)}$$

Elemental sulphur frequently accumulates in the purple sulphur bacteria and then, when the exogenous source of hydrogen sulphide is exhausted, disappears as it is further oxidized to sulphate:

$$3\,CO_2 + 2\,S + 5\,H_2O \xrightarrow{\text{Light}} 3\,(CH_2O) + 2\,H_2SO_4$$

Some purple bacteria, both sulphur and non-sulphur, can use molecular hydrogen and their photosynthetic reaction can be formulated:

$$CO_2 + 2\,H_2 \xrightarrow{\text{Light}} (CH_2O) + H_2O$$

The non-sulphur purple bacteria can use organic substances as electron donors:

$$CO_2 + 2\,CH_3CHOH.CH_3 \xrightarrow{\text{Light}} (CH_2O) + H_2O + 2\,CH_3COCH_3$$

but usually can also assimilate the organic substrate. Thus many strains use lower fatty acids. Acetate, for instance, can be assimilated to form the reserve material poly-β-hydroxybutyrate $[(C_4H_6O_2)_n]$. This anaerobic process requires ATP (generated from the light) and reducing power, derived by anaerobic break-down of some acetate *via* the tricarboxylic acid cycle:

$$CH_3COOH + 2\,H_2O \longrightarrow 2\,CO_2 + 8\,H$$

Pairs of acetate molecules are combined, reduced and polymerized:

$$2\,CH_3COOH + 2\,H \longrightarrow [CH_3CH.CH_2\overset{|}{C}{=}O] + 2\,H_2O$$
$$\underset{\underset{[C_4H_6O_2]}{|}}{\overset{|}{O}}$$

The overall reaction is thus:

$$CH_3COOH + 2\,H_2O \longrightarrow 2\,CO_2 + 8\,H$$
$$3\,CH_3COOH + 8\,H \longrightarrow 4\,(C_4H_6O_2) + 8H_2O$$
$$\overline{}$$
$$9\,CH_3COOH \xrightarrow{\text{Light}} 4\,(C_4H_6O_2) + 2\,CO_2 + 6\,H_2O$$

In general, however, photosynthesis results in endergonic reduction of carbon dioxide to carbohydrate.

Fixation of carbon dioxide: photolithotrophs and chemolithotrophs

Experiments with $^{14}CO_2$ show that in extremely short time periods the first labelled compound formed is glycerate 3-phosphate with the isotope in its carboxyl group. The detection also of labelled sedoheptulose 7-phosphate and ribulose 1,5-diphosphate in the early stages of photosynthesis suggested that some of the reactions of the pentose phosphate cycle might be involved in the process. It is now known that of the fifteen or more reactions involved in photosynthesis only two are specific to photosynthetic and chemolithotrophic organisms. The rest are reactions common to glycolysis, the pentose phosphate cycle and the formation of carbohydrate from non-carbohydrate precursors, i.e. reactions found in non-photosynthetic organisms.

The two specific reactions are the phosphorylation of ribulose 5-phosphate to ribulose 1,5-diphosphate under the influence of the enzyme *phosphoribulokinase*, and the fission of ribulose 1,5-diphosphate by carbon dioxide and water, to yield two molecules of glycerate 3-phosphate, catalysed by the enzyme *carboxydismutase* (*ribulose diphosphate carboxylase*). Phosphoribulokinase is analogous in its action to the phosphofructokinase of the Embden–Meyerhof sequence, although differing in its specificity:

$$
\begin{array}{ccc}
\text{CH}_2\text{OH} & & \text{CH}_2\text{O}\circledP \\
| & & | \\
\text{C}=\text{O} & & \text{C}=\text{O} \\
| & & | \\
\text{H}-\text{C}-\text{OH} \;+\; \text{ATP} \;\longrightarrow\; & & \text{H}-\text{C}-\text{OH} \;+\; \text{ADP} \\
| & & | \\
\text{H}-\text{C}-\text{OH} & & \text{H}-\text{C}-\text{OH} \\
| & & | \\
\text{CH}_2\text{O}\circledP & & \text{CH}_2\text{O}\circledP \\
\text{Ribulose} & & \text{Ribulose} \\
\text{5-phosphate} & & \text{1,5-diphosphate}
\end{array}
$$

The carboxydismutase reaction is a complex one in which ribulose 1,5-diphosphate reacts with carbon dioxide to give two molecules of glycerate 3-phosphate:

$$
\begin{array}{ccc}
& & \text{CH}_2\text{O}\circledP \\
& & | \\
\text{CH}_2\text{O}\circledP & & \text{H}-\text{C}-\text{OH} \\
| & & | \\
\text{CO} & & \text{COOH} \\
| & & \\
\text{CO}_2 + \text{H}-\text{C}-\text{OH} \;+\; \text{H}_2\text{O} \;\longrightarrow\; & & + \\
| & & \\
\text{H}-\text{C}-\text{OH} & & \text{COOH} \\
| & & | \\
\text{CH}_2\text{O}\circledP & & \text{H}-\text{C}-\text{OH} \\
& & | \\
\text{Ribulose 1,5-diphosphate} & & \text{CH}_2\text{O}\circledP \\
& & \text{Glycerate} \\
& & \text{3-phosphate}
\end{array}
$$

The subsequent reactions of glycerate 3-phosphate are common to glycolysis although operating in the reverse direction and employing NADP rather than NAD. Glycerate 3-phosphate is thus phosphorylated by ATP in the presence of *phosphoglycerate kinase* to glycerate 1,3-diphosphate and then reduced to glyceraldehyde 3-phosphate by triose phosphate dehydrogenase and $NADPH_2$. An equilibrium mixture of glyceraldehyde 3-phosphate and dihydroxyacetone phosphate is produced by the action of triose phosphate isomerase, and one molecule of each then combines with the other under the influence of aldolase to form fructose 1,6-diphosphate. The phosphate group in the 1-position is then removed by a specific phosphatase and the resulting fructose 6-phosphate undergoes reactions of the pentose phosphate cycle catalysed by transketolase and transaldolase.

Fructose 6-phosphate and glyceraldehyde 3-phosphate are converted to xylulose 5-phosphate and erythrose 4-phosphate by transketolase. Erythrose 4-phosphate and glyceraldehyde 3-phosphate yield sedoheptulose 1,7-diphosphate in a reaction catalysed by transaldolase, and the latter compound loses a phosphate group to become sedoheptulose 7-phosphate. This reacts with glyceraldehyde 3-phosphate to yield one molecule each of xylulose 5-phosphate and ribose 5-phosphate, again catalysed by transketolase. Pentose phosphate isomerase and xyluloepimerase convert ribose 5-phosphate and xylulose 5-phosphate to ribulose 5-phosphate and phosphorylation of this yields ribulose 1,5-diphosphate. This series of reactions can be denoted as follows:

2 Glyceraldehyde 3-phosphate $\longrightarrow$ Fructose 6-phosphate + Ⓟ

Fructose 6-phosphate + Glyceraldehyde 3-phosphate $\longrightarrow$
$\qquad\qquad\qquad$ Xylulose 5-phosphate + Erythrose 4-phosphate

Erythrose 4-phosphate + Glyceraldehyde 3-phosphate $\longrightarrow$ Sedoheptulose 7-phosphate + Ⓟ

Sedoheptulose 7-phosphate + Glyceraldehyde 3-phosphate $\longrightarrow$
$\qquad\qquad\qquad$ Ribose 5-phosphate + Xylulose 5-phosphate

Ribose 5-phosphate + Xylulose 5-phosphate $\longrightarrow$ 2 Ribulose 5-phosphate

Xylulose 5-phosphate $\rightleftharpoons$ Ribulose 5-phosphate

---

$\qquad$ 5 Glyceraldehyde 3-phosphate $\longrightarrow$ 3 Ribulose 5-phosphate + 2Ⓟ

$\qquad$ 3-Ribulose 5-phosphate + 3ATP $\longrightarrow$ 3 Ribulose 1,5-diphosphate + 3ADP

The fixation of each molecule of CO_2 in photosynthesis thus requires the expenditure of 3 molecules of ATP, two to phosphorylate the two molecules of glycerate 3-phosphate formed and one to phosphorylate ribulose 5-phosphate produced in the regeneration cycle. The overall process leading to hexose synthesis may be represented in the following way:

Fixation $\quad$ 6 Ribulose 1,5-diphosphate + 6 CO_2 + 6 H_2O $\longrightarrow$ 12 Glycerate 3-phosphate

Reduction $\quad$ 12 Glycerate 3-phosphate + 12 ATP + 12 $NADPH_2$ $\longrightarrow$
$\qquad$ 12 Glyceraldehyde 3-phosphate + 12 ADP + 12 Ⓟ + 12 NADP + 12 H_2O

Regeneration 12 Glyceraldehyde 3-phosphate + 6 ATP $\longrightarrow$
$\qquad$ 6 Ribulose 1,5-diphosphate + 6 ADP + 5 Ⓟ + Fructose 6-phosphate

---

Sum $\qquad$ 6 CO_2 + 6 H_2O + 18 ATP + 12 $NADPH_2$ $\longrightarrow$
$\qquad\qquad$ Fructose 6-phosphate + 18 ADP + 12 NADP + 17 Ⓟ

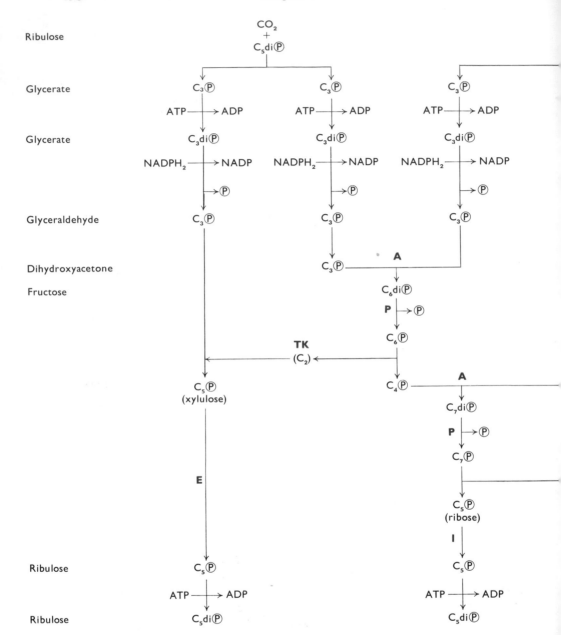

Overall reaction

$$3\,CO_2 \;+\; 3\,H_2O \;+\; 9\,ATP \;+\; 6\,NADPH_2 \;\longrightarrow\; \tfrac{1}{2}\,Fructose\ 6\text{-}\textcircled{P} \;+\; 9\,ADP \;+\; 6\,NADP \;+\; 8\tfrac{1}{2}\,\textcircled{P}$$

Figure 12. The photosynthetic or Calvin cycle.

TK = transketolase, A = aldolase, E = epimerase, I = isomerase, P = phosphatases.

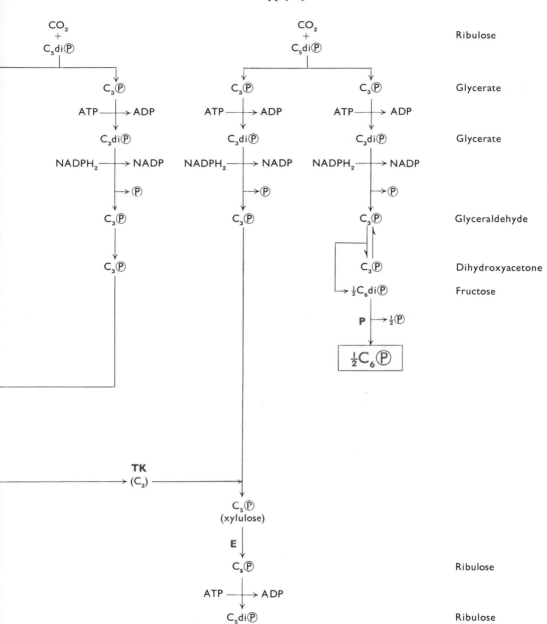

The reactions of the photosynthetic or Calvin cycle are depicted schematically in Figure 12. For purposes of clarity only half the number of molecules involved are represented leading to the formation of a 'half molecule' of hexose.

It must be emphasized that the photosynthetic cycle provides also the intermediary carbon compounds for biosynthetic reactions associated with growth, in addition to producing hexose. Thus compounds such as triose phosphates and pentoses will be drained from the cycle to fulfil these requirements.

The same pathway of CO_2 fixation as occurs in photosynthetic bacteria has now been demonstrated in many of the chemolithotrophic bacteria, the difference residing solely in the nature of the energy-yielding mechanisms. Thus the key enzymes carboxydismutase and phosphoribulokinase are present and following exposure to $^{14}CO_2$ the label appears rapidly in glycerate 3-phosphate, hexose phosphates, sedoheptulose phosphate and ribulose phosphate.

The exergonic reactions in different organisms may include oxidation of ammonia to nitrite and then to nitrate, of inorganic sulphur compounds (H_2S, S, $SO_3{}^{2-}$, etc.), of ferrous to ferric compounds and of molecular hydrogen to water. Usually the oxidizing agent is oxygen but some species are anaerobes using inorganic nitrate as oxidant. Thus *Thiobacillus denitrificans* oxidizes H_2S to sulphate while nitrate is reduced to molecular nitrogen:

$$5\,H_2S \;+\; 8\,KNO_3 \;\longrightarrow\; 3\,K_2SO_4 \;+\; 2\,KHSO_4 \;+\; 4\,N_2 \;+\; 4\,H_2O$$

This exergonic reaction is coupled to the formation of ATP from ADP and inorganic phosphate; the electron transport chain involves flavoproteins and cytochromes as in aerobic oxidations (see Appendix B).

Control of autotrophic carbon dioxide fixation
Present evidence indicates that autotrophic CO_2 fixation is regulated by modulation of the activity of phosphoribulokinase. From the preceding discussion it will be apparent that the overall reactions leading to hexose synthesis make heavy demands on the ATP and reducing power ($NADPH_2$) of the organism. Phosphoribulokinase has been shown to be sensitive to the energy charge of the cell and the enzyme from several types of chemolithotrophic bacteria is inhibited by AMP. It has also been found recently that $NADH_2$ activates the phosphoribulokinase of *Hydrogenomonas* and *Rhodopseudomonas*. Thus regeneration of the CO_2-acceptor appears to be controlled by the energy charge of the autotrophic cell and in such a way that fixation and reduction occur only when the cell is well-endowed with energy.

FURTHER READING

1 Elsden S. R. (1962) Photosynthesis and lithotrophic carbon dioxide fixation. In *The Bacteria*, Volume 3: Biosynthesis (eds. Gunsalus I. C. and Stanier R. Y.), p. 1. Academic Press, New York.

2 Krampitz L. O. (1961) Cyclic mechanisms of terminal oxidation. In *The Bacteria*, Volume 2: Metabolism (eds. Gunsalus I. C. and Stanier R. Y.), p. 209. Academic Press, New York.

3 Wood W. A. (1961) Fermentation of carbohydrates and related compounds. In *The Bacteria*, Volume 2: Metabolism (eds. Gunsalus I. C. and Stanier R. Y.), p. 59. Academic Press, New York.

4 Kornberg H. L. (1966) Anaplerotic sequences and their role in metabolism. In *Essays in Biochemistry*, Volume 2 (eds. Campbell P. N. and Greville G. D.), p. 1. Academic Press, London.

5 White A., Handler P. and Smith E. L. (1964) *Principles of Biochemistry*. McGraw-Hill, New York.

6 Mahler H. R. and Cordes E. H. (1966) *Biological Chemistry*, Harper and Row, New York, Evanston and London.

7 Lehninger A. L. (1970) *Biochemistry*. Worth, New York.

8 Sanwal B. D. (1970). Allosteric controls of amphibolic pathways in bacteria. *Bacteriological Reviews* **34**, 20.

Chapter 4
Class II Reactions: Synthesis of
Small Molecules

BIOSYNTHESIS OF AMINO ACIDS

As outlined in Section 4 and Figure 1, p. 30, the 20 amino acids commonly found in proteins are related to one another in a series of *families*, according to their biosynthetic origin. The families comprise

(a) the aromatic family with three amino acids
(b) the aspartate family with six amino acids
(c) the glutamate family with four amino acids
(d) the serine family with three amino acids
(e) the pyruvate family with three amino acids
(f) histidine, which is not related to any of the other amino acids, but the synthesis of which is closely related to that of the purines.

We shall discuss these in turn. Each family is illustrated by a detailed metabolic scheme with structural formulae. The compounds are numbered and the numbers will be referred to in the text. The enzymes involved are indicated on the schemes by letters which are used when the control of these pathways is being discussed.

Control of biosynthetic reactions is necessary in order to ensure that the cell does not waste energy and carbon synthesizing metabolites which are already in the medium. This topic is discussed further in Chapter 8. Generally the same biosynthetic pathway may be subjected to two kinds of control, represented respectively by alteration of the rate of enzyme synthesis and by modulation of enzyme activity. It will be convenient to consider the two mechanisms together and to recapitulate briefly some of the points made in Section 8.

Reduction in the differential rate of enzyme *synthesis* (see Chapter 8, p. 431) is termed *repression*; reduction in enzyme *activity* is often caused by negative *feedback inhibition*. Usually it is one or more terminal products of a biosynthetic pathway which are the active agents in repression and feedback inhibition (hence the name of the latter phenomenon, which is also called *retro-inhibition* or *end-product inhibition*). The main impression from recent work on control mechanisms, especially of feedback inhibition, is of the variety of different mechanisms observed in different bacterial species. Consequently, only an outline of some of the better authenticated examples can be given here.

Consider a branched biosynthetic pathway for two metabolites Q and Y in which there is a series of common intermediates:

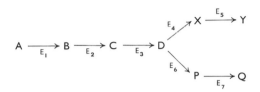

It is obvious that if Y has the property of inhibiting any of the enzymes before the branch point, its presence in the medium will arrest further synthesis of Q as well as its own synthesis. At least six different patterns of feedback inhibition have evolved, which surmount this problem in different ways. These may be briefly summarized as follows:

(i) *Concerted or multivalent feedback inhibition.* In this mechanism Q and Y separately have no effect on enzyme E_1, but together they are potent inhibitors.

(ii) *Co-operative feedback inhibition.* Here Q and Y separately are weakly inhibitory to E_1, but together they exert an effect that is more than the sum of their individual inhibitory effects.

(iii) *Cumulative feedback inhibition.* In this mechanism a given end-product inhibits E_1 by a given percentage irrespective of the presence of other inhibitors. The presence of a second product increases the total inhibition. If all the products are present, the activity of E_1 is completely inhibited.

(iv) *Compensatory antagonism of feedback inhibition.* In this situation in which Q might totally inhibit E_1 in a system in which Y is required to react with an intermediate, Z, from some other pathway, then Z is able to decrease the inhibition produced by Q, so preventing a situation in which Z accumulates because Q has shut off the synthesis of Y.

(v) *Sequential feedback inhibition.* Here Y inhibits enzyme E_4 only and Q inhibits enzyme E_6 only. However this inhibition causes the accumulation of D in either case and D is an inhibitor of E_1.

(vi) *Multiple enzymes with specific regulatory effectors.* In this case there are different but *isofunctional* enzymes catalysing the conversion of A to B and each is inhibited by one of the end-products. Thus if Q inhibits one form of E_1, Y can still be formed by the action of another.

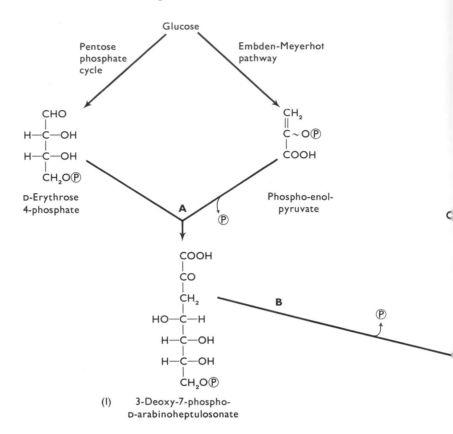

Figure Ia. The family of aromatic amino acids: enzymes **A–G**.

        A Phospho-2-keto-3-deoxyheptonate aldolase
        B 5-Dehydroquinate synthase
        C 5-Dehydroquinate dehydratase
        D Shikimate dehydrogenase

    There are also at least four patterns of repression of enzyme synthesis.

(a) In *simple end-product repression* all the enzymes of a pathway are repressed by the presence of the end-product.

(b) In other cases, the end-product represses only the first enzyme, while the others are absent unless induced by the product of the first reaction.

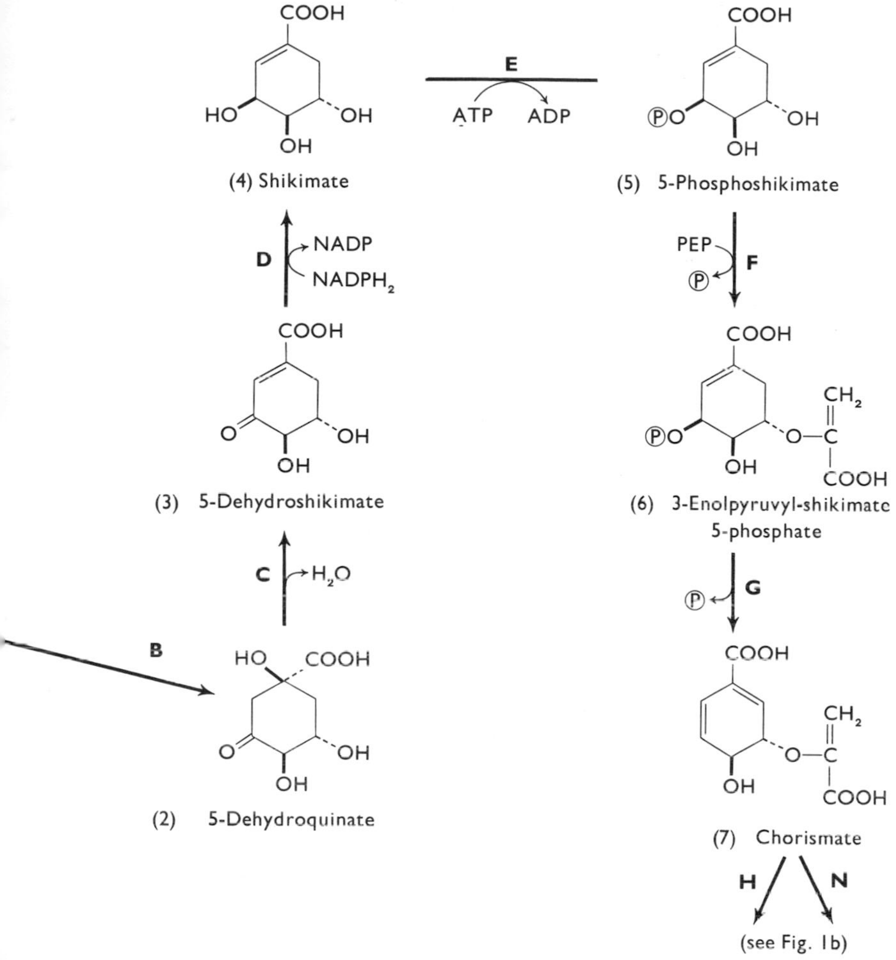

COOH

HO ''OH

OH

(4) Shikimate

$$\xrightarrow[\text{ATP} \quad \text{ADP}]{\text{E}}$$

COOH

PO ''OH

OH

(5) 5-Phosphoshikimate

D NADP / NADPH$_2$

COOH

O ''OH

OH

(3) 5-Dehydroshikimate

C H$_2$O

HO COOH

O ''OH

OH

(2) 5-Dehydroquinate

B

PEP / P **F**

COOH

PO ''OH

OH O—C CH$_2$ / COOH

(6) 3-Enolpyruvyl-shikimate 5-phosphate

P **G**

COOH

OH O—C CH$_2$ / COOH

(7) Chorismate

H / **N**

(see Fig. 1b)

E Shikimate kinase
F 3-Enolpyruvyl-shikimate 5-phosphate aldolase
G Chorismate synthase

(c) In branched pathways *multivalent repression* of all enzymes by the concerted action of all the end-products may be observed. If any one of the end-products is not present, no repression is observed.

(d) Alternatively there may be *cumulative repression* in which there are several different E$_1$ enzymes each repressed by the presence of one particular end-product.

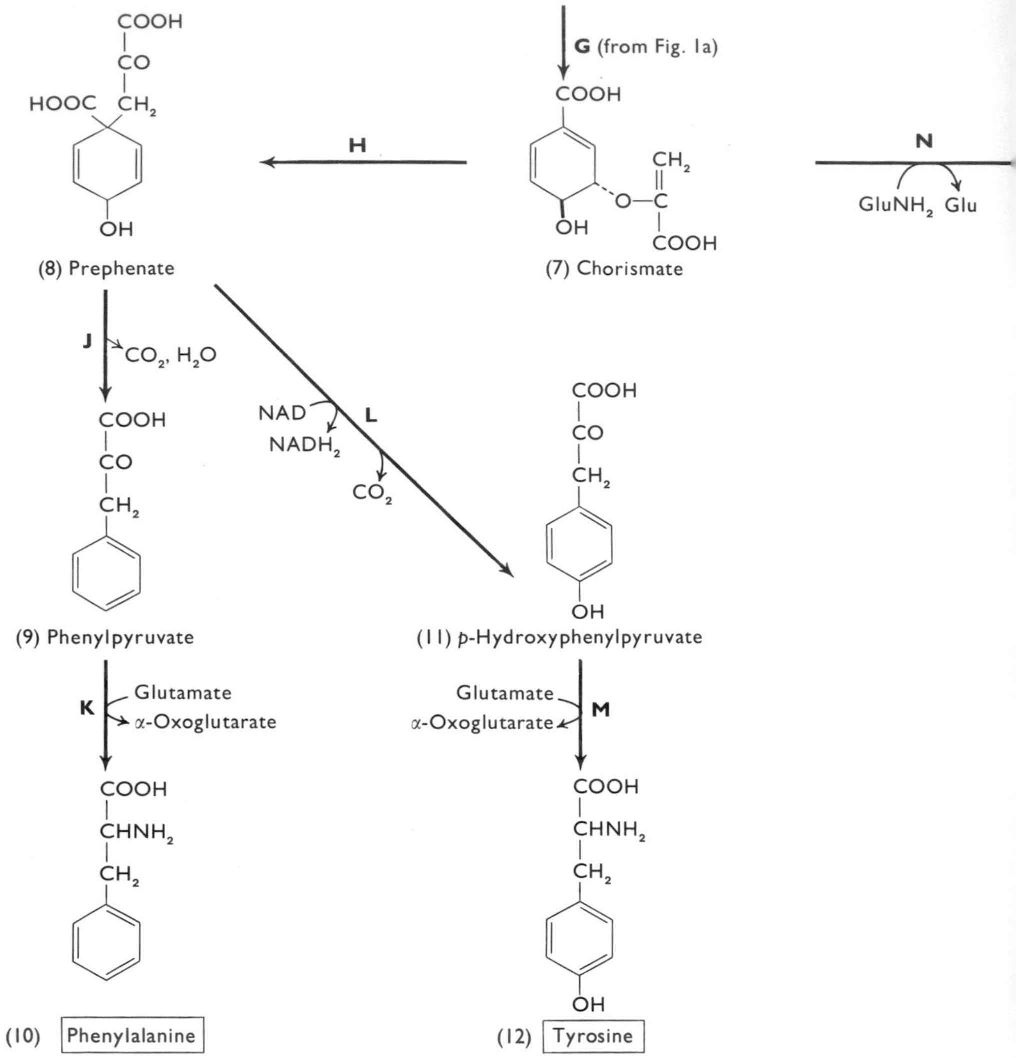

Prephenate (8), Chorismate (7), Phenylpyruvate (9), p-Hydroxyphenylpyruvate (11), Phenylalanine (10), Tyrosine (12) metabolic pathway with enzyme steps G, H, N, J, L, K, M.

(8) Prephenate

(7) Chorismate — G (from Fig. 1a)

N — $GluNH_2$ Glu

J — CO_2, H_2O

NAD / $NADH_2$ — L — CO_2

(9) Phenylpyruvate

(11) p-Hydroxyphenylpyruvate

K — Glutamate / α-Oxoglutarate

M — Glutamate / α-Oxoglutarate

(10) Phenylalanine

(12) Tyrosine

Repression of a group of enzymes controlled by a single operator is usually *co-ordinate*, all the enzymes being repressed in a strictly proportional way. If repression is not proportional, this is good indirect evidence that the structural genes are not closely linked (although the converse is not necessarily true).

Feedback inhibition is a control mechanism that works rapidly and prevents existing enzymes from wasting energy and carbon in making a metabolite already available. It operates within seconds of the addition of a controlling metabolite and on the disappearance of the exogenous metabolite the inhibition is just as quickly relieved.

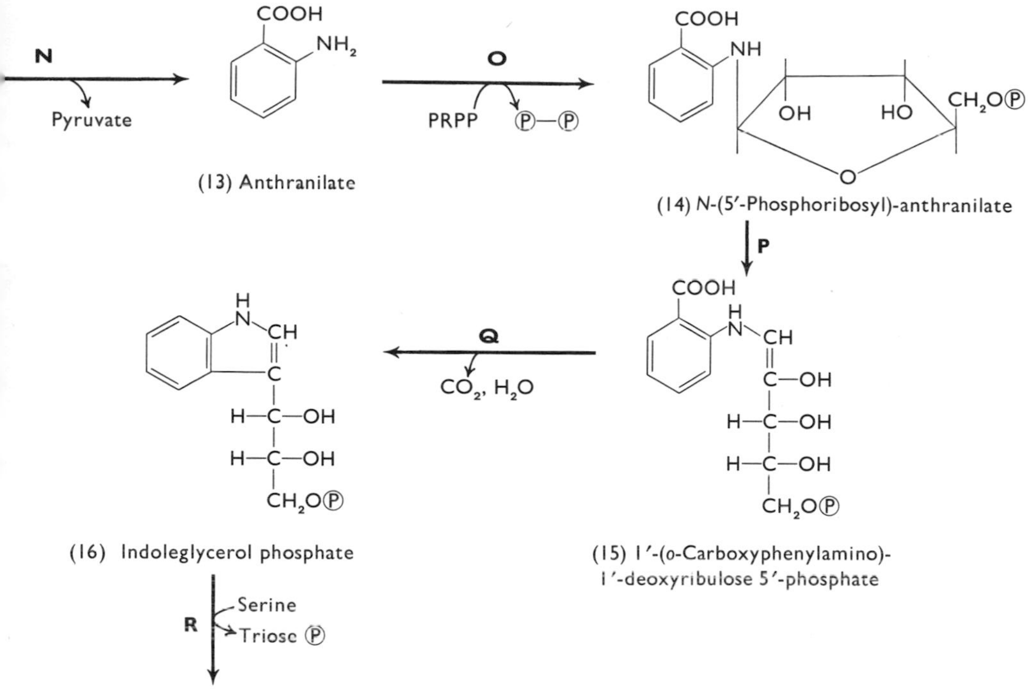

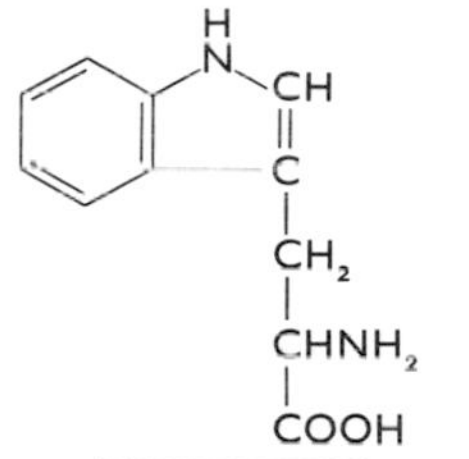

(13) Anthranilate

(14) N-(5'-Phosphoribosyl)-anthranilate

(16) Indoleglycerol phosphate

(15) 1'-(o-Carboxyphenylamino)-1'-deoxyribulose 5'-phosphate

(17) Tryptophan

Figure 1b. The family of aromatic amino acids: enzymes **H–R**.

H Prephenate synthase (chorismate mutase)
J Prephenate dehydratase
K Phenylalanine aminotransferase
L Prephenate dehydrogenase
M Tyrosine aminotransferase
N Anthranilate synthase
O Anthranilate phosphoribosyltransferase
P Phosphoribosyl-anthranilate isomerase
Q Indoleglycerol phosphate synthase
R Tryptophan synthase

Repression on the other hand is a much more slowly acting control both in its imposition and in its removal. Thus although addition of a controlling metabolite rapidly prevents further synthesis, the existing enzyme is not usually destroyed, but merely diluted out by growth, the specific activity of the enzyme falling by a factor of 2 per generation time. Similarly, when the repressing metabolite is removed, some time will have to elapse (about 2 generation times) before the specific activity of the enzyme rises to near its fully derepressed level.

The family of aromatic amino acids (Figures 1a and 1b)

The three aromatic amino acids phenylalanine, tyrosine and tryptophan are derived from a seven-carbon straight chain compound 3-deoxy-7-phospho-D-arabinoheptulosonic acid (1). This arises by condensation of a C_4 compound D-erythrose 4-phosphate (which comes from glucose via the pentose phosphate cycle) with the C_3 compound phospho-enolpyruvate (a derivative of glycolysis) with the elimination of a single molecule of orthophosphate. This reaction is mediated by *phospho-2-keto-3-deoxyheptonate aldolase* (A).

This straight chain compound then cyclizes with the elimination of another molecule of orthophosphate to give 5-dehydroquinic acid (2). This is catalysed by *dehydroquinate synthase* (B). The remaining biosynthetic steps common to the three aromatic amino acids may be regarded as the successive introduction of further double bonds until aromatization is complete.

Chorismic acid (7) is so called because it represents a branch point in the pathway, where the pathways to phenylalanine (10) and tyrosine (12) on the one hand, and tryptophan (17) on the other hand, diverge.

One reaction of chorismic acid is the rearrangement of the C_3 ether side chain with the formation of a carbon–carbon bond giving prephenic acid (8). The enzyme for this is *prephenate synthase (chorismate mutase)* (H). Prephenic acid is another branch point, since phenylalanine and tyrosine arise from it by separate pathways. Both the steps in which prephenate is concerned involve the loss of its C-1 carboxyl group.

Chorismic acid also gives rise, by an aromatization reaction in which glutamine is an amino donor, to anthranilic acid (13) with the concomitant formation of glutamate and pyruvate. This reaction is catalysed by *anthranilate synthase* (N). Anthranilic acid is the precursor of tryptophan. The final reaction is the exchange of the C_3 side chain of indoleglycerol phosphate with serine to give glyceraldehyde 3-phosphate and tryptophan (17). The enzyme catalysing this exchange is *tryptophan synthase* (R).

Thus, of the three aromatic amino acids, phenylalanine and tyrosine derive their aromatic rings from 3-deoxy-7-phospho-arabinoheptulosonate with the loss of one C atom as CO_2, and their C_3 side chains from phospho-enolpyruvate. Tryptophan derives its aromatic ring from the same source, its heterocyclic N atom from glutamine, its remaining ring C atoms from C-1 and C-2 of ribose, and its C_3 side chain from serine.

Control

At least six different control patterns have been recognized in the aromatic family of amino acids. Only the mechanisms which have been studied in detail will be mentioned here.

In *Bacillus subtilis*, the end-products respectively inhibit the first enzyme in their terminal branch of the pathway. Thus tyrosine (12) inhibits prephenate dehydrogenase (L), phenylalanine (10) inhibits prephenate dehydratase (J) and tryptophan (17) inhibits anthranilate synthase (N). These inhibitions cause accumulation of the substrates of the enzymes, i.e. prephenate (8) and chorismate (7), and these compounds in turn inhibit the first enzyme of the whole pathway, phospho-2-keto-3-deoxy-heptonate aldolase (A). This is an example of sequential feedback.

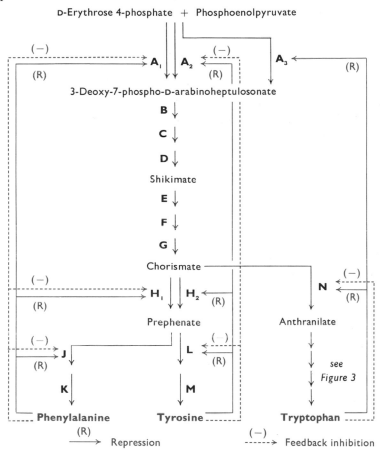

Figure 2. Control of aromatic amino acid biosynthesis in *Escherichia coli*.

In *Escherichia coli* and *Salmonella typhimurium*, in contrast, the control by feedback inhibition is quite different (Figure 2). The first reaction in the pathway is catalysed by three separate *isofunctional* phospho-2-keto-3-deoxyheptonate aldolases (A). One is repressed and inhibited by tyrosine, one by phenylalanine and the third is repressed but not inhibited by tryptophan. In addition, there are two separate chorismate mutases (enzyme H). One of these is physically associated with prephenate dehydrogenase (L) and both enzymes of the complex are repressed by tyrosine which also inhibits enzyme L. The other chorismate mutase is associated with prephenate dehydratase (J) and these enzymes are both repressed and inhibited by phenylalanine.

Although less work has been done with the aerobic pseudomonads, it appears that enzyme A is inhibited only by tyrosine. This may however only indicate that there are other isoenzymes which are either very labile or repressed.

The genes specifying the sequence of enzymes N to R in Figure 1b (chorismate to tryptophan) constitute an operon (Figure 3) in most of the bacteria studied. Again variations in control patterns exist in different bacterial species,

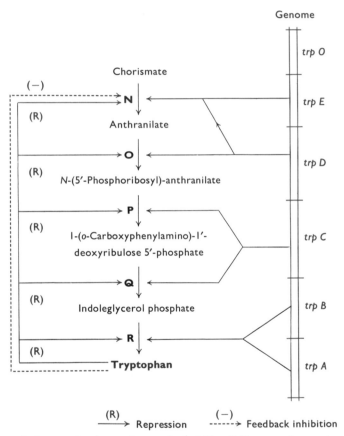

Figure 3. Control of tryptophan biosynthesis in *Escherichia coli*. The genes form a cluster and the enzymes are co-ordinately repressed by tryptophan. Enzyme N is also feedback inhibited by tryptophan.

but *Escherichia coli* and *Salmonella typhimurium* have been the most extensively examined. Reactions P and Q in *E. coli* (but not in *Pseudomonas putida*) are catalysed by the same enzyme, which is a single polypeptide chain specified by the *trpC* gene. Tryptophan synthase (R) is composed of two kinds of polypeptide chain, one specified by the *trpB* gene (the B protein) the other by the *trpA* gene (the A protein). The polypeptide chain of the enzyme anthranilate phospho-ribosyltransferase (O) specified by the *trpD* gene not only catalyses this reaction, but is also a component part of enzyme N (anthranilate synthase), the other peptide chain of which is specified by the *trpE* gene. Anthranilate synthase is feedback inhibited by tryptophan, and the site of tryptophan binding is to the product of the *trpE* gene. All four enzymes of the operon are repressed co-ordinately by tryptophan. The close association of the enzymes formed (as well as of the genes of the operon) suggests that many of the intermediates recorded in Figures 1b and 3 do not occur in the free state.

The aspartate family (Figures 4a and 4b)

Aspartic acid (1) arises by transamination between glutamate and oxalo-acetate, which is an intermediate of the tricarboxylic acid cycle. This is catalysed by *aspartate aminotransferase*. Aspartate in turn gives rise to five amino acids which are protein constituents, in addition to diaminopimelic acid, which is found in the cell wall mucopeptides but not in protein.

Asparagine (2), the β-amide of aspartic acid and an important protein constituent, is formed from aspartate by the enzyme *asparagine synthase* (A) which catalyses the reaction of ammonia with aspartate and ATP with the concomitant formation of AMP and pyrophosphate.

The four remaining amino acids arise from the β-semialdehyde of aspartic acid (4). Aspartate β-semialdehyde (4) is a branch point where the biosynthetic pathway to lysine diverges from the main sequence. The semialdehyde con-denses with pyruvate to give 2,3-dihydrodipicolinic acid (17), under the in-fluence of *dihydrodipicolinate synthase* (Q) and this ultimately gives rise to lysine (23).

After aspartate β-semialdehyde, the next compound on the main pathway is homoserine (5). This is formed by reduction (by either $NADPH_2$ or $NADH_2$) of aspartate β-semialdehyde catalysed by *homoserine dehydrogenase* (D). Homoserine is another branch point, at which the pathway to methionine diverges from the pathway to threonine and isoleucine.

Methionine is formed from homoserine via cystathionine (7) as shown in Figure 4b. The final reaction of this sequence, the methylation of homocysteine to give methionine (9), is more complex. Two alternative pathways exist. The more usual one in glucose-ammonium salt medium is the transfer of a methyl group from a conjugated form of methyl-FH_4 (see Appendix B) to homo-cysteine to give methionine, catalysed by enzyme H.

The other diverging pathway from homoserine is the formation of threonine (11). This takes place by the phosphorylation of homoserine by ATP in the presence of *homoserine kinase* (J), to give *O*-phosphohomoserine (10). This then isomerizes with the loss of a molecule of orthophosphate to give threonine (11) under the influence of *threonine synthase* (K).

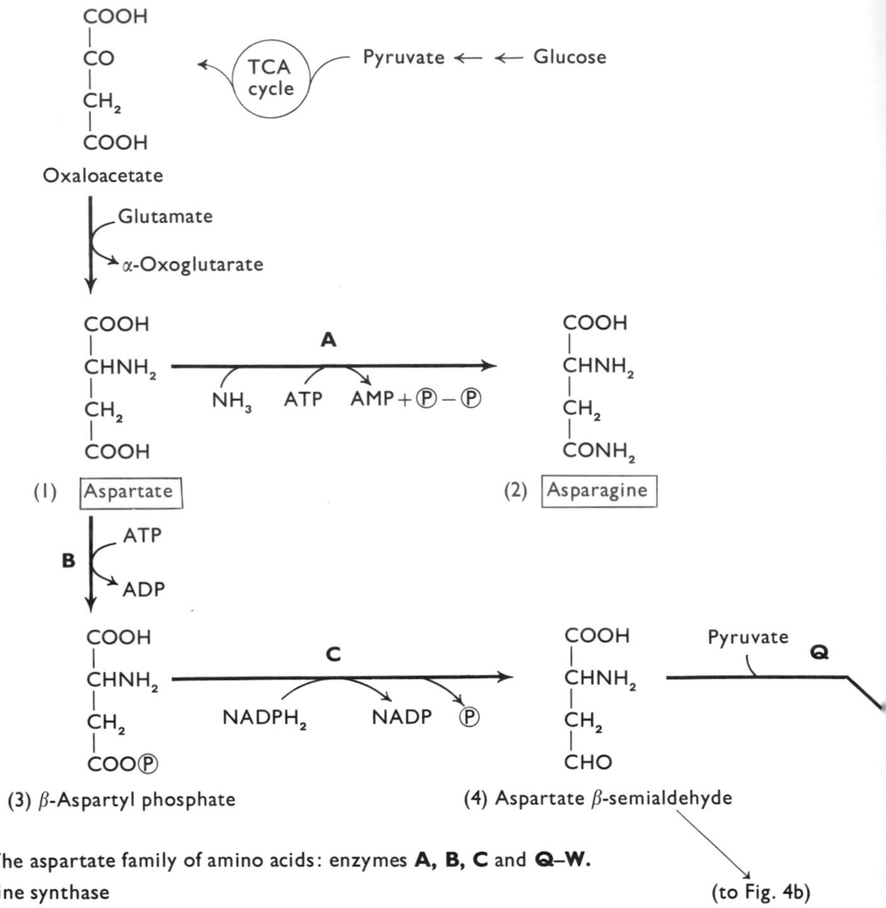

Figure 4a. The aspartate family of amino acids: enzymes **A, B, C** and **Q–W**.

A   Asparagine synthase   (to Fig. 4b)
B   β-Aspartokinase
C   Aspartate semialdehyde dehydrogenase
Q   Dihydrodipicolinate synthase
R   Dihydrodipicolinate reductase
S   Succinyloxoaminopimelate synthase
T   Succinyldiaminopimelate aminotransferase
U   Succinyldiaminopimelate desuccinylase
V   Diaminopimelate epimerase
W   *meso*-Diaminopimelate decarboxylase

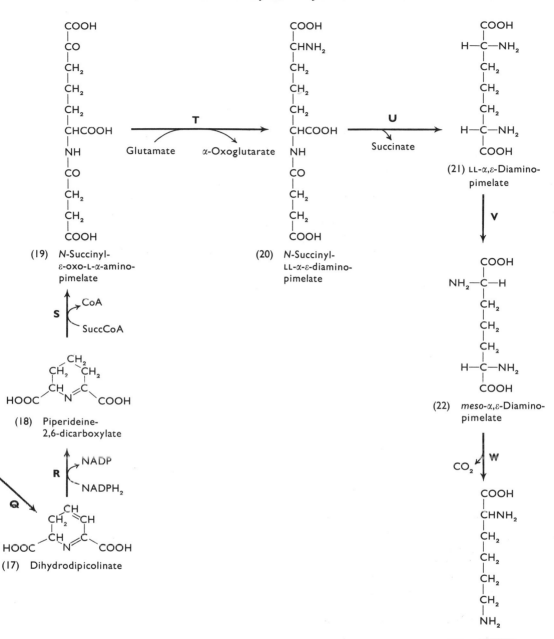

(19) N-Succinyl-ε-oxo-L-α-amino-pimelate

(20) N-Succinyl-LL-α-ε-diamino-pimelate

(21) LL-α,ε-Diamino-pimelate

(18) Piperideine-2,6-dicarboxylate

(22) *meso*-α,ε-Diamino-pimelate

(17) Dihydrodipicolinate

(23) Lysine

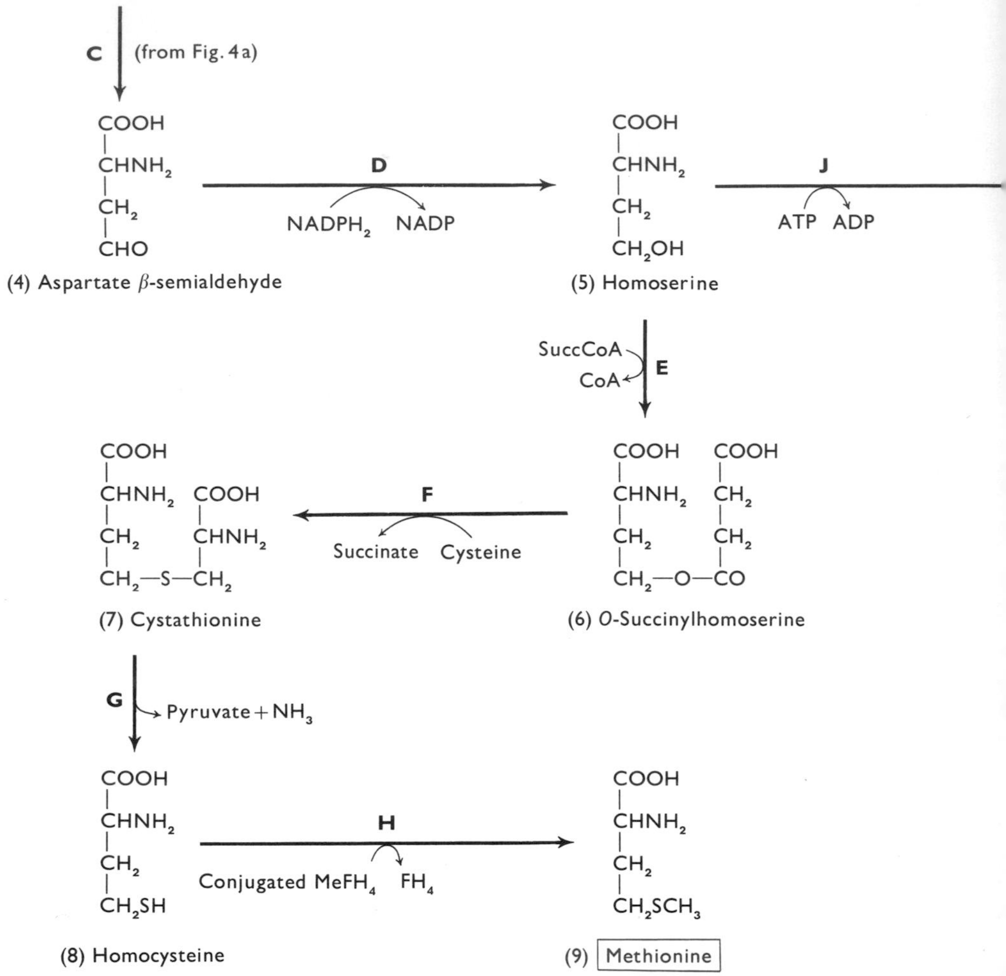

C | (from Fig. 4a)

$$
\begin{array}{c}
\text{COOH} \\
|\\
\text{CHNH}_2 \\
|\\
\text{CH}_2 \\
|\\
\text{CHO}
\end{array}
\qquad \xrightarrow[\ \ \text{NADPH}_2 \quad \text{NADP}\ \ ]{\text{D}} \qquad
\begin{array}{c}
\text{COOH} \\
|\\
\text{CHNH}_2 \\
|\\
\text{CH}_2 \\
|\\
\text{CH}_2\text{OH}
\end{array}
\qquad \xrightarrow[\ \ \text{ATP} \quad \text{ADP}\ \ ]{\text{J}}
$$

(4) Aspartate β-semialdehyde   (5) Homoserine

SuccCoA ⟍
          | E
CoA ⟋

$$
\begin{array}{c}
\text{COOH} \\
|\\
\text{CHNH}_2 \quad \text{COOH} \\
|\qquad\qquad |\\
\text{CH}_2 \qquad \text{CHNH}_2 \\
|\qquad\qquad\\
\text{CH}_2\!-\!\text{S}\!-\!\text{CH}_2
\end{array}
\qquad \xleftarrow[\ \ \text{Succinate} \quad \text{Cysteine}\ \ ]{\text{F}} \qquad
\begin{array}{c}
\text{COOH} \quad \text{COOH} \\
|\qquad\quad |\\
\text{CHNH}_2 \quad \text{CH}_2 \\
|\qquad\quad |\\
\text{CH}_2 \qquad \text{CH}_2 \\
|\qquad\quad |\\
\text{CH}_2\!-\!\text{O}\!-\!\text{CO}
\end{array}
$$

(7) Cystathionine   (6) O-Succinylhomoserine

G ⟍ Pyruvate + NH₃

$$
\begin{array}{c}
\text{COOH} \\
|\\
\text{CHNH}_2 \\
|\\
\text{CH}_2 \\
|\\
\text{CH}_2\text{SH}
\end{array}
\qquad \xrightarrow[\ \ \text{Conjugated MeFH}_4 \quad \text{FH}_4\ \ ]{\text{H}} \qquad
\begin{array}{c}
\text{COOH} \\
|\\
\text{CHNH}_2 \\
|\\
\text{CH}_2 \\
|\\
\text{CH}_2\text{SCH}_3
\end{array}
$$

(8) Homocysteine   (9) | Methionine |

Figure 4b. The aspartate family of amino acids: enzymes **D–P**.

D   Homoserine dehydrogenase
E   Homoserine O-succinyltransferase
F   Cystathionine γ-synthase
G   Cystathionase II
H   Homocysteine methyltransferase
J   Homoserine kinase
K   Threonine synthase
L   Threonine dehydratase
M   Acetohydroxyacid synthase
N   Dihydroxyacid synthase
O   Dihydroxyacid dehydratase
P   α-Oxo acid aminotransferase

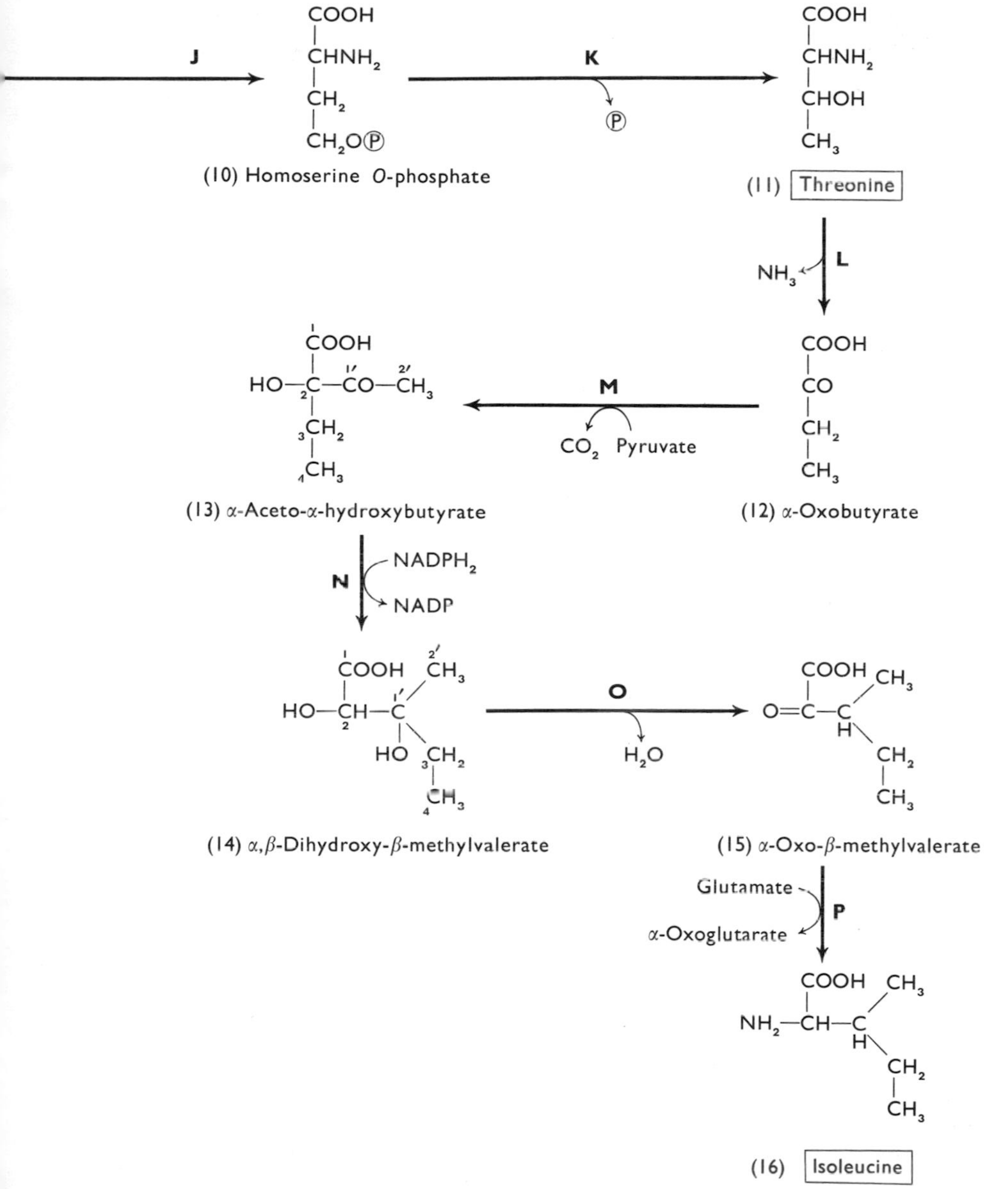

J →

COOH
|
CHNH$_2$
|
CH$_2$
|
CH$_2$O(P)

(10) Homoserine *O*-phosphate

K → (P)

COOH
|
CHNH$_2$
|
CHOH
|
CH$_3$

(11) [Threonine]

NH$_3$ ↙ **L** ↓

COOH
|
CO
|
CH$_2$
|
CH$_3$

(12) α-Oxobutyrate

M ← CO$_2$ Pyruvate

COOH
|
HO—$_2$C—$^{1'}$CO—$^{2'}$CH$_3$
|
$_3$CH$_2$
|
$_4$CH$_3$

(13) α-Aceto-α-hydroxybutyrate

N ↓ NADPH$_2$ → NADP

COOH $^{2'}$CH$_3$
| /
HO—$_2$CH—$^{1'}$C
/ \
HO $_3$CH$_2$
|
$_4$CH$_3$

(14) α,β-Dihydroxy-β-methylvalerate

O → H$_2$O

COOH CH$_3$
| /
O=C—C
\
H CH$_2$
|
CH$_3$

(15) α-Oxo-β-methylvalerate

Glutamate ↙ **P** ↓
α-Oxoglutarate ↙

COOH CH$_3$
| /
NH$_2$—CH—C
\
H CH$_2$
|
CH$_3$

(16) [Isoleucine]

In addition to its role as a protein constituent, threonine is also a precursor of isoleucine. It is deaminated to give α-oxobutyric acid (12) by *threonine dehydratase* (L). The first step in the sequence from α-oxobutyrate to isoleucine is the condensation of α-oxobutyrate with pyruvate, with elimination of CO_2. The enzyme is *acetohydroxyacid synthase* (M) and the product of the reaction is α-aceto-α-hydroxybutyrate (13).

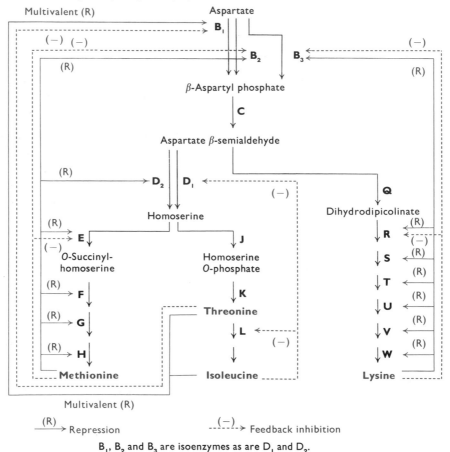

B$_1$, B$_2$ and B$_3$ are isoenzymes as are D$_1$ and D$_2$.

Figure 5. Control of aspartate-derived amino acid biosynthesis in *Escherrchia coli*.

The next step is a complex reaction in which (13) undergoes a pinacol-type rearrangement and is simultaneously reduced by $NADPH_2$ to α,β-dihydroxy-β-methylvaleric acid (14). This step is catalysed by *dihydroxyacid synthase* (*acetohydroxyacid reductoisomerase*) (N).

Compound (14) is then dehydrated to give α-oxo-β-methylvalerate (15) under the action of *dihydroxyacid dehydratase* (O). Isoleucine (16) is finally formed by transamination of (15) with glutamate catalysed by the *aminotransferase* (P). Thus both of the amino acids derived from asparate which have more than four carbon atoms derive their remaining atoms from C-2 and C-3 of pyruvate.

Control

Control of amino acid biosynthesis in the aspartate family is complicated by the large number of branches in the pathway. Accordingly a number of different mechanisms have evolved in different species for control of essentially the same biosynthetic sequence.

In *Escherichia coli* K-12 (see Figure 5), the main feature is the occurrence of a multiplicity of aspartokinases (enzyme B, Figure 4a). There are three such enzymes, two of which (aspartokinases II and III) are under simple repressive control by their respective end-products methionine and lysine. Aspartokinase I, which is inhibited by threonine, is under multivalent repression by threonine plus isoleucine. Aspartokinase II is physically associated with homoserine dehydrogenase (enzyme D) activity, both being repressed by methionine. Aspartokinase I is similarly associated with homoserine dehydrogenase I, and both activities are inhibited by threonine. Aspartokinase III is repressed and inhibited by lysine; it is not associated with enzyme D activity, since homoserine dehydrogenase is not involved in lysine formation. Mutant studies have shown that the β-aspartyl phosphate (3) produced by any of the aspartokinases can be channelled to the synthesis of the other end-products, so that there is only a single pool of compound (3) in the cell.

Lysine represses the formation of enzymes Q to W and methionine the formation of enzymes E to H. Additionally, homoserine *O*-succinyltransferase (E) is feedback inhibited by methionine and also by *S*-adenosylmethionine, and dihydrodipicolinate synthase (Q) is inhibited by lysine. In *Salmonella typhimurium* enzymes J and K, like the aspartokinase I-homoserine dehydrogenase I complex, are regulated by a multivalent repression mechanism involving both threonine and isoleucine. In contrast to this control pattern, *Bacillus polymyxa* and *Rhodopseudomonas capsulatus* each possesses a single aspartokinase which exhibits co-operative feedback inhibition. In *Rhodopseudomonas spheroides* and *Bacillus licheniformis* aspartokinase is inhibited by aspartate β-semialdehyde (4) which may possibly indicate sequential feedback inhibition.

In *Lactobacillus arabinosus*, asparagine (2) both inhibits and represses enzyme A.

Threonine dehydratase (L), which catalyses the first step in isoleucine biosynthesis, is feedback inhibited by isoleucine. This inhibition can be relieved by valine, an example of compensatory antagonism. The control of this enzyme and later enzymes in the pathway (M to P) will be discussed later, when valine biosynthesis is considered (p. 225).

The glutamate family (Figures 6a and 6b)

The amino acids glutamine, proline and arginine derive their carbon skeletons from glutamic acid. The additional carbon atom of arginine is derived from CO_2 and its two N atoms from ammonia and aspartic acid, respectively.

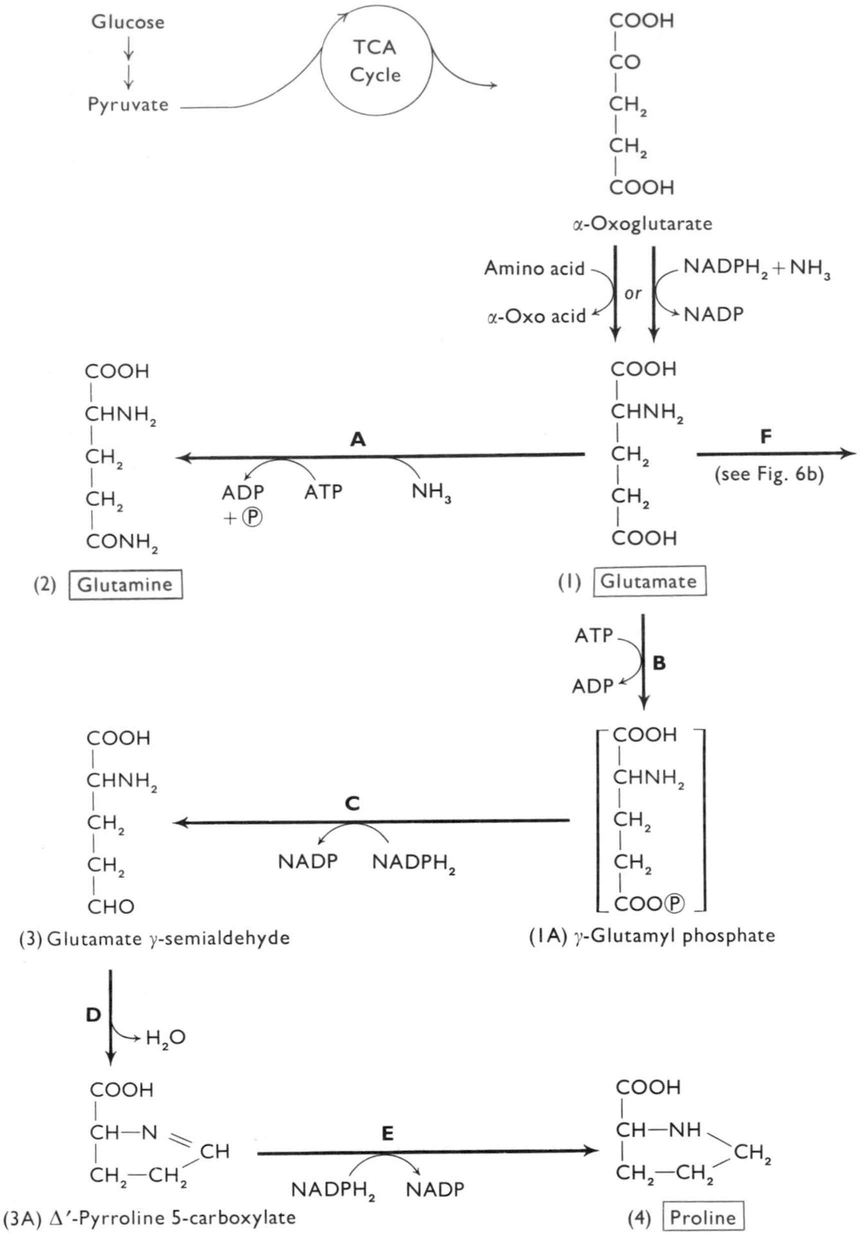

Figure 6a. The glutamate family of amino acids: enzymes **A–E**.

A   Glutamine synthase

B γ-Glutamokinase

C   Glutamate semialdehyde dehydrogenase

D   (Possibly non-enzymic)

E   Pyrroline 5-carboxylate reductase

Glutamate (1) itself is formed from α-oxoglutarate, which is an intermediate of the tricarboxylic acid cycle, either by transamination (α-oxoglutarate can act as amino acceptor for transaminases specific for almost all the L-amino acids) or else by direct reductive fixation of ammonia under the action of the $NADH_2$- or $NADPH_2$-dependent enzyme *glutamate dehydrogenase*. In most bacteria this enzyme appears to be the principal, if not the only, primary reaction for the incorporation of free ammonia into amino acids.

Glutamine (2) is formed by reaction of glutamate with ammonia. The formation of the amide bond requires a molecule of ATP, but unlike the asparagine synthase reaction, the *glutamine synthase* (A) reaction yields ADP and orthophosphate. Since this enzyme has a much higher affinity for ammonia than has glutamate dehydrogenase, under conditions of ammonia deficiency it can replace the latter enzyme as the primary port of entry of ammonia into amino acids. In these conditions, an enzyme or enzyme system catalyses the reaction of glutamine with α-oxoglutarate in the presence of $NADPH_2$ to give two molecules of glutamate.

Proline (4) is formed from glutamate by an ATP- and $NADPH_2$- requiring reduction of glutamate to glutamate γ-semialdehyde (3) (reactions (B) and (C)). The enzymology of this step has not been clarified but it seems likely that γ-glutamyl phosphate (1A) is an intermediate in this reduction.

Arginine (12) is formed from ornithine (9) which in turn arises from glutamate via a series of N-acetylated derivatives. The function of the N-acetyl groups may be to prevent the spontaneous cyclization of the ornithine precursors to give proline derivatives.

Control

Glutamine (2) is a branch-point compound with regard to its amide nitrogen atom which can be transferred to a variety of compounds in biosynthesis, e.g. histidine, tryptophan, glucosamine, carbamoyl phosphate, AMP, etc. In *Escherichia coli*, these metabolites exert a repressive effect on glutamine synthase (A) and so does ammonia, since at high concentrations, ammonia can replace glutamine in most of the reactions in which the latter acts as a nitrogen donor. Besides being subject to repression, the activity of pre-existing glutamine synthase is also under the control of cumulative feedback inhibition. This controlling effect of feedback metabolites is modulated in *E. coli* by the occurrence of enzyme A in two forms (called respectively synthases I and II), which show different susceptibilities to end-product inhibition. Glutamine synthase II is more sensitive to feedback inhibition by CTP, AMP, histidine and tryptophan. Form I, which is less sensitive, is converted into form II by an adenylylation reaction requiring ATP and a specific activating enzyme. There is also a deadenylylating enzyme which is inhibited by glutamine, and which reconverts form II to form I.

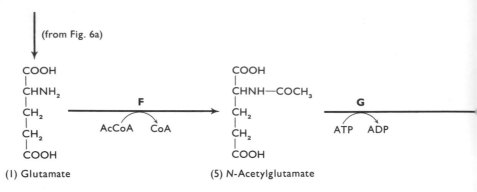

↓ (from Fig. 6a)

```
COOH                              COOH
|                                 |
CHNH₂          F                  CHNH—COCH₃        G
|         ─────────────────→      |            ─────────────
CH₂      AcCoA    CoA             CH₂           ATP    ADP
|                                 |
CH₂                               CH₂
|                                 |
COOH                              COOH
```
(1) Glutamate   (5) *N*-Acetylglutamate

```
COOH                              COOH
|              M                  |                    L
CHNH₂    ←──────────────          CHNH₂        ←──────────────────
|        AMP  ATP   Aspartate     |            Ⓟ
CH₂            +                  CH₂               Carbamoyl
|            Ⓟ—Ⓟ                 |                 phosphate  ←────
CH₂          COOH                 CH₂
|            |                    |
CH₂          CH₂                  CH₂NH—CONH₂
|                                 (10) Citrulline
NH
HN=C—NH—CH—COOH
```
(11) Argininosuccinate

```
N │
  │ ↘ Fumarate
  ↓
COOH
|
CHNH₂
|
CH₂
|
CH₂
|
CH₂
|
NH
|
HN=C—NH₂
```
(12) Arginine

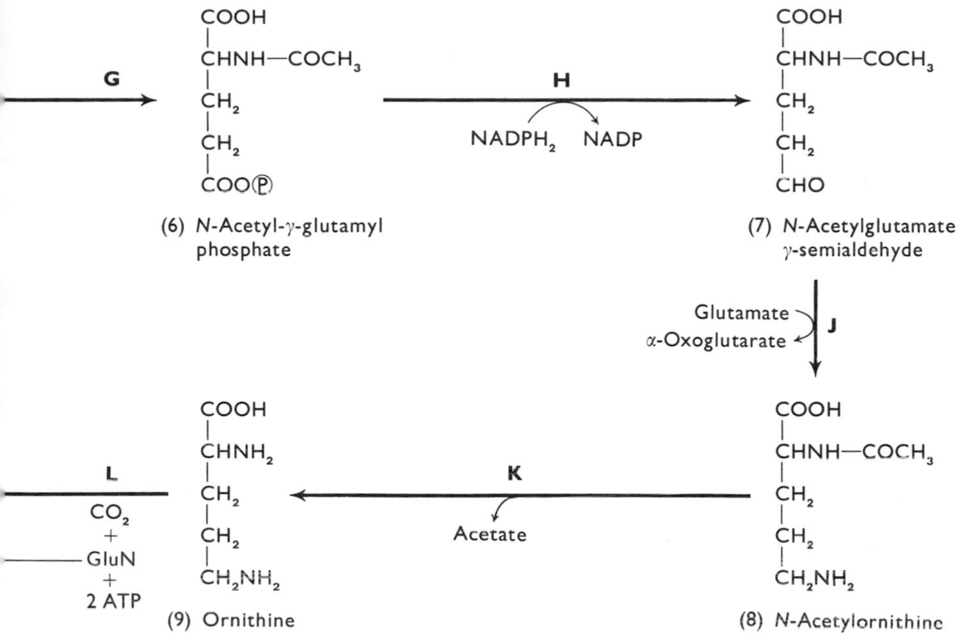

$$
\begin{array}{l}
\text{COOH} \\
|\ \\
\text{CHNH—COCH}_3 \\
|\ \\
\text{CH}_2 \\
|\ \\
\text{CH}_2 \\
|\ \\
\text{COO}\,\textcircled{P}
\end{array}
$$

(6) N-Acetyl-γ-glutamyl phosphate

G →   **H** → NADPH₂   NADP

$$
\begin{array}{l}
\text{COOH} \\
|\ \\
\text{CHNH—COCH}_3 \\
|\ \\
\text{CH}_2 \\
|\ \\
\text{CH}_2 \\
|\ \\
\text{CHO}
\end{array}
$$

(7) N-Acetylglutamate γ-semialdehyde

Glutamate ⟶ **J** ⟵ α-Oxoglutarate

$$
\begin{array}{l}
\text{COOH} \\
|\ \\
\text{CHNH}_2 \\
|\ \\
\text{CH}_2 \\
|\ \\
\text{CH}_2 \\
|\ \\
\text{CH}_2\text{NH}_2
\end{array}
$$

(9) Ornithine

L   CO₂ + GluN + 2 ATP

K ⟵ Acetate

$$
\begin{array}{l}
\text{COOH} \\
|\ \\
\text{CHNH—COCH}_3 \\
|\ \\
\text{CH}_2 \\
|\ \\
\text{CH}_2 \\
|\ \\
\text{CH}_2\text{NH}_2
\end{array}
$$

(8) N-Acetylornithine

Figure 6b. The glutamate family of amino acids: enzymes **F–N**.
F   Glutamate acetyltransferase
G   N-Acetyl-γ-glutamokinase
H   N-Acetylglutamate semialdehyde dehydrogenase
J   N-Acetylornithine δ-aminotransferase
K   N-Acetylornithinase
L   Ornithine carbamoyltransferase
M   Argininosuccinate synthase
N   Argininosuccinate lyase

Proline (4) inhibits the formation of glutamate γ-semialdehyde (3) and also can repress enzymes B to E.

The control of the biosynthesis of arginine has been studied in much detail in *E. coli*. The genes controlling arginine biosynthesis do not constitute an operon but are scattered around the genome. Nevertheless, these genes are all under the control of a single regulatory gene *argR*. In *E. coli* K-12 arginine represses all the enzymes of the biosynthetic pathway (F to N). In other strains of *E. coli*, however, the apo-repressor protein produced by the *arg R* gene does not combine so readily with arginine, so that the addition of exogenous arginine

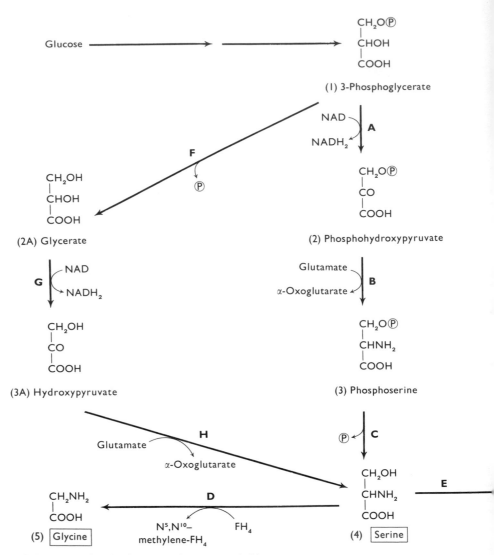

Figure 7. The serine family of amino acids: enzymes **A–N**.

A Phosphoglycerate dehydrogenase
B Phosphoserine aminotransferase
C Phosphoserine phosphatase
D Serine hydroxymethyltransferase
E Serine *O*-acetyltransferase
F Phosphoglycerate phosphatase

has a paradoxical derepression effect. Arginine also inhibits the first biosynthetic enzyme, glutamate acetyltransferase (F) by feedback inhibition. Ornithine (9) is required to make *putrescine*, a precursor of polyamines; in the

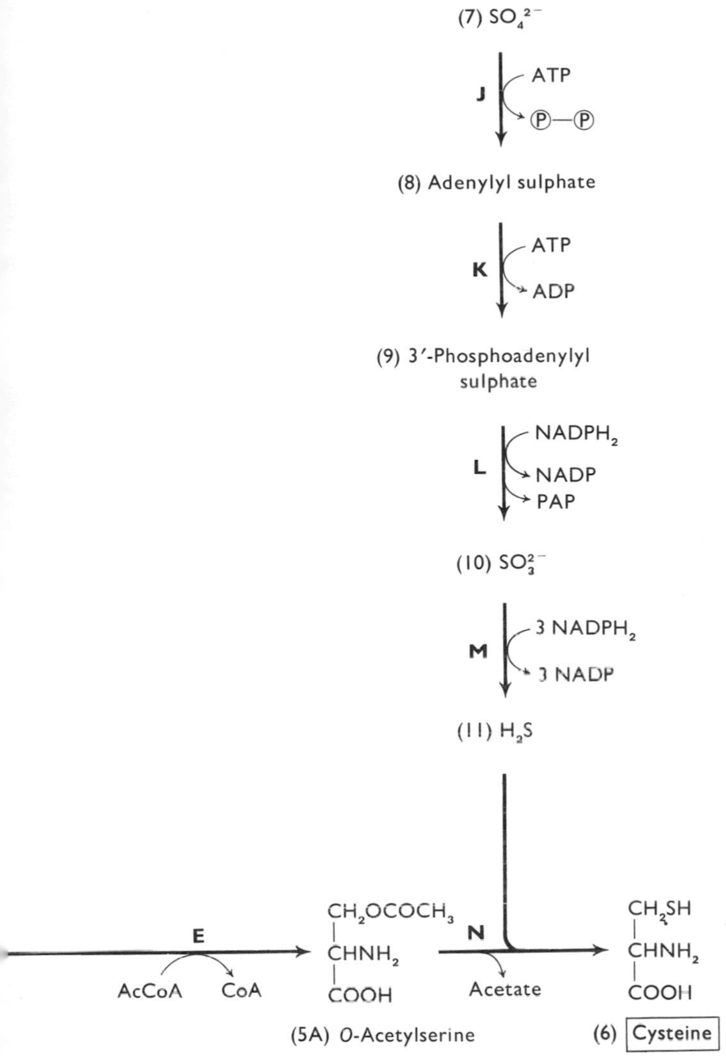

(7) SO_4^{2-}

J — ATP → $\textcircled{P}$—$\textcircled{P}$

(8) Adenylyl sulphate

K — ATP → ADP

(9) 3'-Phosphoadenylyl
sulphate

L — NADPH$_2$ → NADP → PAP

(10) SO_3^{2-}

M — 3 NADPH$_2$ → 3 NADP

(11) H_2S

$$
\begin{array}{ccc}
& CH_2OCOCH_3 & CH_2SH \\
\xrightarrow{\quad E \quad} & CHNH_2 & \xrightarrow{\quad N \quad} CHNH_2 \\
AcCoA \quad CoA & COOH & COOH \\
& \text{Acetate} & \\
(5A)\ \textit{O-Acetylserine} & & (6)\ \boxed{\text{Cysteine}}
\end{array}
$$

G   Glycerate dehydrogenase
H   Serine aminotransferase
J   Sulphate adenylyltransferase
K   Adenylyl sulphate kinase
L   Phosphoadenylyl sulphate reductase
M   Sulphite reductase
N   Acetylserine sulphhydrase

presence of excess arginine, which prevents ornithine biosynthesis, a new pathway for putrescine biosynthesis is induced involving the decarboxylation of
arginine.

The serine family (Figure 7)

There is some variation in the pathways by which serine is formed from 3-phosphoglyceric acid which is an intermediate in glucose degradation. In some organisms, a non-phosphorylated pathway (involving glyceric acid and hydroxypyruvate and the enzymes *phosphatase* (F), *glycerate dehydrogenase* (G) and *serine aminotransferase* (H)) is thought to predominate, while in others, e.g. *Escherichia coli* and *Salmonella typhimurium*, a phosphorylated pathway shown as the main sequence in Figure 7 predominates.

Serine is the precursor of glycine and the sulphur-containing amino acid cysteine. Glycine (5) is formed from serine by the transfer of a hydroxymethyl group to tetrahydrofolate (FH_4) to give N^5,N^{10}-methylene-FH_4 and glycine. This reaction is catalysed by *serine hydroxymethyltransferase* (D).

Cysteine (6) is formed in bacteria via *O*-acetylserine (5A), which is then converted to cysteine by reaction with hydrogen sulphide (11) formed by reduction of sulphate by a series of ATP- and $NADPH_2$-dependent steps, as shown in Figure 7.

Control

The serine pathway can be divided into two routes, that supplying the C_3 carbon skeleton for cysteine formation (enzymes A to C and E) and that of sulphate reduction (enzymes J to N) (see Figure 8). In *Escherichia coli*, in the first route, the initial enzyme, phosphoglycerate dehydrogenase (A) is feedback inhibited by serine. Enzyme multiplicity in this pathway is absent. No repression of enzymes A to C by serine has been observed, but since serine is a precursor of a large

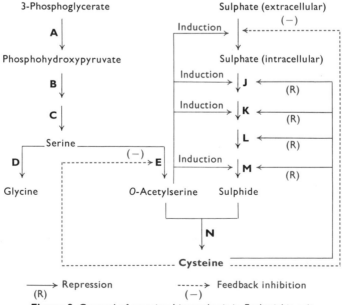

Figure 8. Control of cysteine biosynthesis in *Escherichia coli*.

number of compounds made from C_1 units (purines, thymine, methionine and histidine) as well as glycine and cysteine, repression by serine would not be likely to occur. Cysteine represses enzymes J to M in the sulphate-reduction pathway. Additionally, it feedback-inhibits serine *O*-acetyltransferase (E) and the permease catalysing uptake of sulphate by the cell. *O*-Acetylserine induces the sulphate permease and enzymes J, K and M. In the presence of cysteine both sulphate reduction and the formation of *O*-acetylserine are shut off. When the cysteine concentration falls the feedback inhibition of enzyme E is relieved and *O*-acetylserine is formed. This induces the enzymes of sulphate reduction which are also derepressed since cysteine is no longer present.

The pyruvate family (Figure 9)

Alanine (2) is formed from pyruvate (1) by transamination with glutamate (and probably other amino acids) catalysed by *alanine aminotransferase* (A). In some bacteria alanine can be formed directly by the $NADH_2$-dependent amination of pyruvate catalysed by *alanine dehydrogenase*.

Pyruvate is also the precursor of valine and leucine. The first three steps in the formation of these two amino acids from pyruvate are catalysed by enzymes B, C and D (Figure 9). These enzymes, *acetohydroxyacid synthase, dihydroxyacid synthase* and *dihydroxyacid dehydratase*, are the same as those involved in the corresponding three steps in the biosynthesis of isoleucine from α-oxobutyrate and pyruvate (enzymes M, N and O in Figure 4b). Similarly the *aminotransferase* giving rise to valine (enzyme E, Figure 9) also catalyses the formation of isoleucine (Figure 24b, enzyme P). The first step is the condensation of two molecules of pyruvate to give α-acetolactate (3) with the elimination of CO_2. Acetolactate then undergoes a reductive rearrangement to yield α,β-dihydroxy-β-methylbutyrate (4). This compound is then dehydrated to α-oxo-β-methyl-butyrate (5). This is a branch point in the sequence for valine and leucine.

One enzyme, the *aminotransferase* (E) catalyses the transamination of (5) to valine (6), while another, *2-isopropylmalate synthase* (F) catalyses the condensation of α-oxo-β-methylbutyrate with acetyl coenzyme A to 2-isopropyl-malate (7), which is a precursor of leucine (Figure 9).

Control

In *Escherichia coli* and *Salmonella typhimurium* the enzymes catalysing the conversion of compound 7 to compound 10 are under the control of four closely-linked genes, forming an operon (see Figure 10). The enzymes are co-ordinately repressed by leucine. Leucine also inhibits by feedback the first enzyme on its branch pathway (F).

The enzymes of the common isoleucine-valine biosynthetic pathway, aceto-hydroxy acid synthase (B in Figure 9, M in Figure 4b), dihydroxy acid synthase (C in Figure 9, N in Figure 4b), dihydroxy acid dehydratase (D in Figure 9,

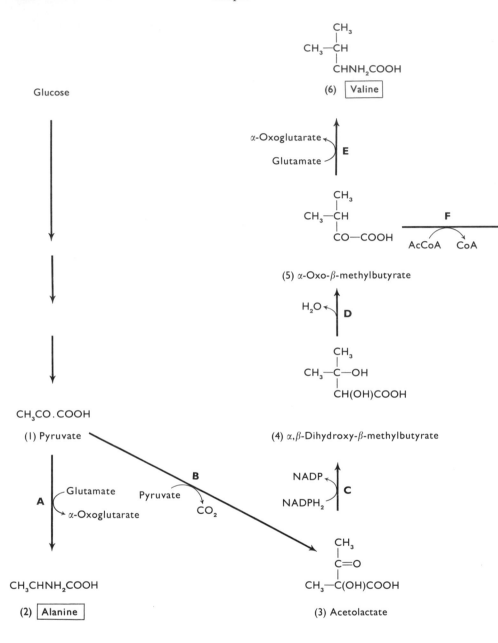

Figure 9. The pyruvate family of amino acids: enzymes **A–J**.

A   Alanine aminotransferase
B   Acetohydroxyacid synthase
C   Dihydroxyacid synthase
D   Dihydroxyacid dehydratase
E   Valine aminotransferase

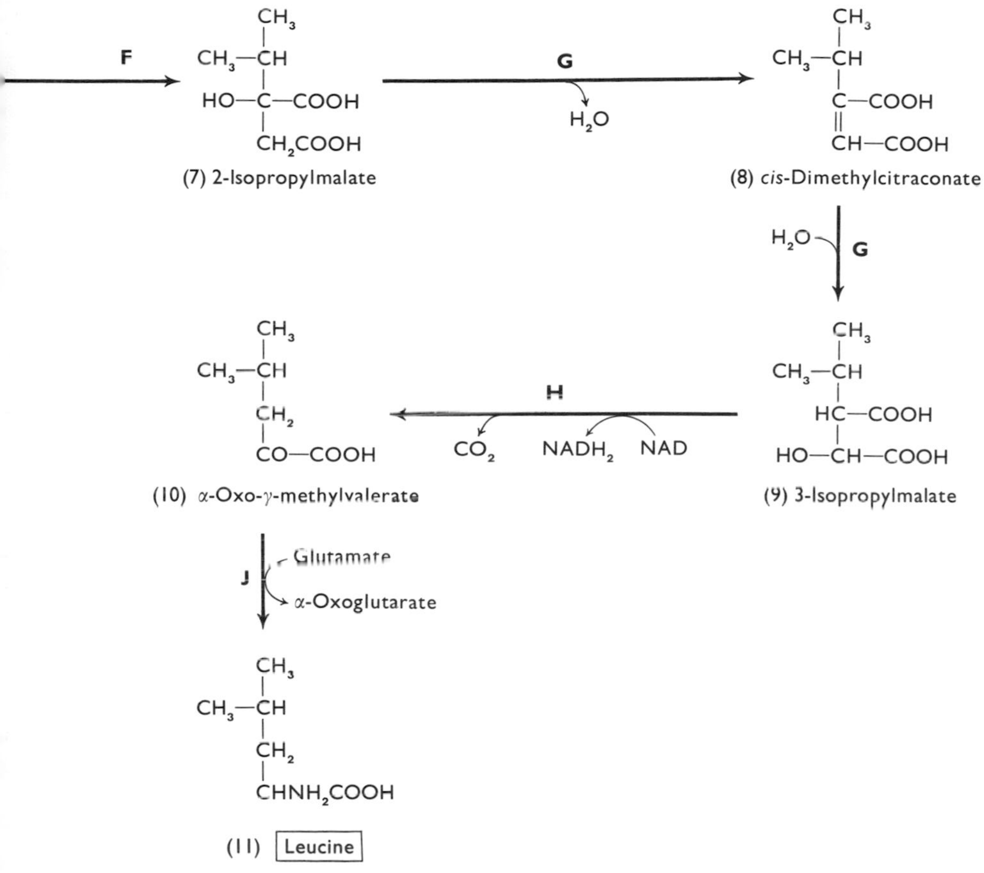

F →

CH₃
|
CH₃—CH
|
HO—C—COOH
|
CH₂COOH

(7) 2-Isopropylmalate

G → (H₂O)

CH₃
|
CH₃—CH
|
C—COOH
‖
CH—COOH

(8) *cis*-Dimethylcitraconate

H₂O ↘ **G** ↓

CH₃
|
CH₃—CH
|
HC—COOH
|
HO—CH—COOH

(9) 3-Isopropylmalate

← **H** (CO₂, NADH₂, NAD)

CH₃
|
CH₃—CH
|
CH₂
|
CO—COOH

(10) α-Oxo-γ-methylvalerate

J ↓ (Glutamate ↗, α-Oxoglutarate ↘)

CH₃
|
CH₃—CH
|
CH₂
|
CHNH₂COOH

(11) [Leucine]

F Isopropylmalate synthase
G Isopropylmalate isomerase
H Isopropylmalate dehydrogenase
J Leucine aminotransferase

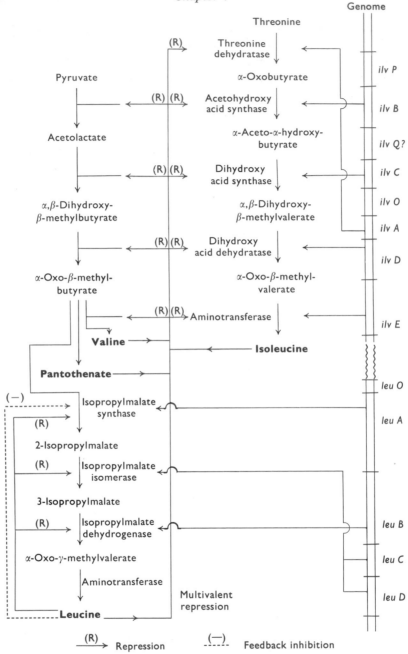

Figure 10. Regulation and gene arrangement of the enzymes of isoleucine-valine-leucine bio-synthesis. The numbers refer either to Figure 4b or Figure 9.

In *Salmonella typhimurium* multivalent repression is observed of the enzymes specified by all five *ilv* genes. In *Escherichia coli* only the *ilvADE* cluster shows multivalent repression. The two clusters of genes are separated by a large chromosomal distance (118 min. in *Salmonella* and 73 min. in *Escherichia*).

O in Figure 4b) and the aminotransferase (E in Figure 9, P in Figure 4b) are under the control of four closely-linked genes in the Enterobacteriaceae, along with the gene (*ilvA*) for threonine dehydratase (L in Figure 4b). Two, or possibly three, operator genes are also present, *ilvP* which controls structural gene *ilvB*, *ilvO* which controls genes *ilvA*, *ilvD* and *ilvE*, and possibly *ilvQ*, controlling the *ilvC* gene. The order of these and the reactions catalysed by the corresponding enzymes are shown in Figure 10.

In *Escherichia coli* K-12, the *ilvADE* cluster of genes (controlled by the *ilvO* operator) is multivalently repressed when isoleucine, valine, leucine and pantothenic acid, all of which arise by the common pathway (Figure 10) are simultaneously present in the medium. If any one of these compounds is not present in the medium, the enzymes specified by these three genes are derepressed. In *Salmonella typhimurium* all five enzymes specified by the *ilv* genes show multivalent repression.

Histidine and its relations with purine biosynthesis (Figure 11)

Extensive isotopic tracer studies have revealed that histidine derives its five-carbon backbone from ribose, its C-2 and N-3 from the same source as C-2 and N-1 of the purine ring, and its N-1 from the amide of glutamine.

Histidine and the purines arise from a common precursor, 5-phospho-α-D-ribosyl-pyrophosphate (2) which is formed from ribose 5-phosphate (1), an intermediate of the pentose phosphate cycle (p. 174), by the action of *ribose phosphate pyrophosphokinase* (A) in which the two terminal phosphates of ATP are transferred leaving AMP. Phosphoribosyl-pyrophosphate occurs at a branch point. One enzyme, *phosphoribosyl-ATP pyrophosphorylase* (B), catalyses its interaction with ATP to give *N*-1-(5′-phosphoribosyl)-ATP (3) which is the precursor of histidine. Another enzyme, *phosphoribosyl-pyrophosphate amidotransferase* (M), converts phosphoribosyl-pyrophosphate to 5-phosphoribosylamine (13), the precursor of the purines (see p. 236).

Compound (7) has not yet been characterized. It is formed by transfer of the amide group of glutamine to (6) catalysed by an *amidotransferase* (F). This intermediate is then cleaved by a *cyclase* (G) to give two compounds: D-*erythro*-imidazoleglycerol phosphate (8), which is the precursor of histidine, and 5-amino-1-(5′-(phosphoribosyl)-imidazole-4-carboxamide (20). The latter compound, which is a precursor of the purine nucleus, can be formed from phosphoribosyl pyrophosphate by an independent pathway (see p. 236). Thus, if the purine precursor is reconverted to ATP, we have a cyclic system for histidine formation as shown on p. 232.

Imidazoleglycerol phosphate then gives rise to histidine (12) as shown in Figure 11.

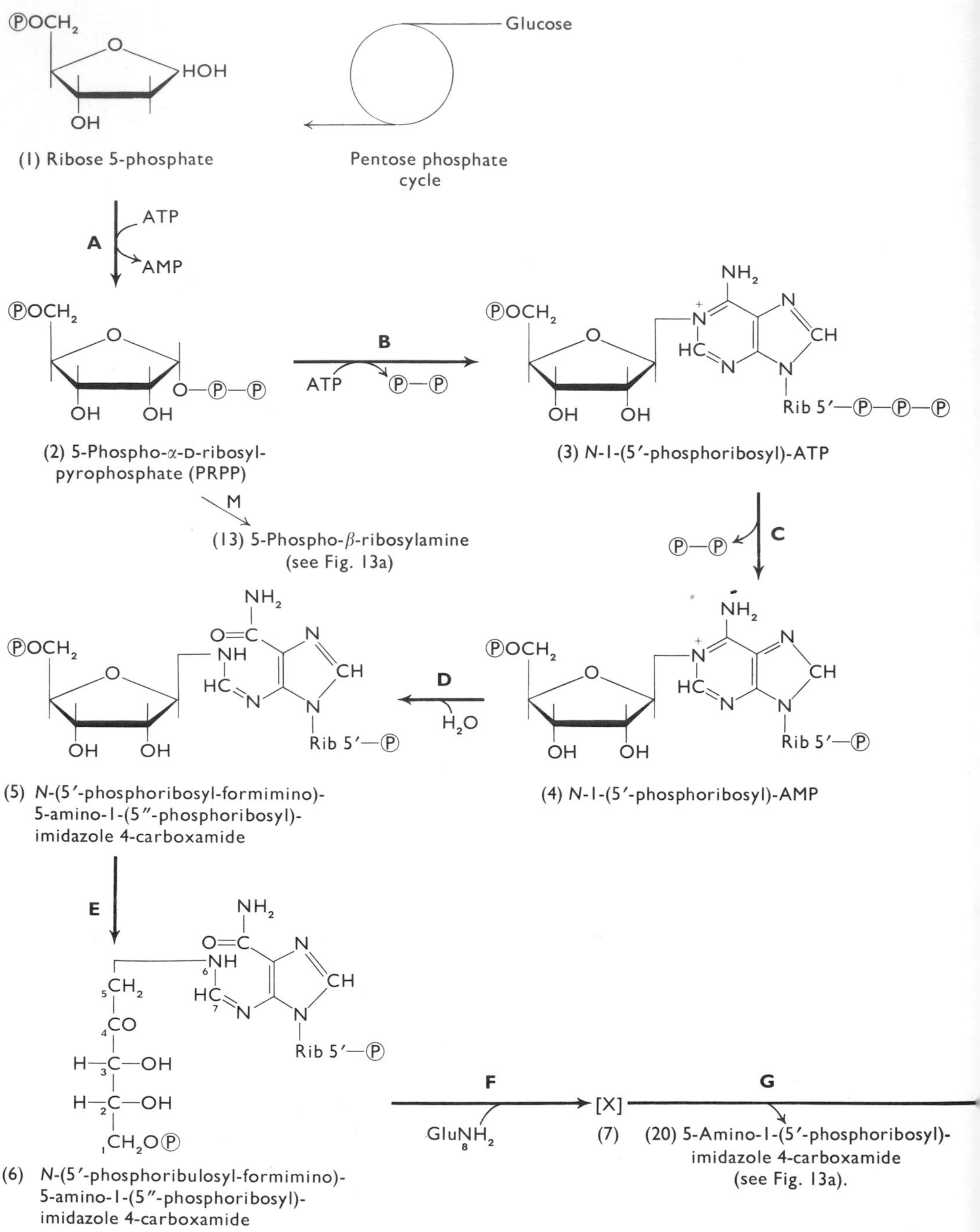

(1) Ribose 5-phosphate

Pentose phosphate
cycle

Glucose

A | ATP → AMP

(2) 5-Phospho-α-D-ribosyl-
pyrophosphate (PRPP)

B   ATP → $P-P$

(3) N-1-(5'-phosphoribosyl)-ATP

M

(13) 5-Phospho-β-ribosylamine
(see Fig. 13a)

$P-P$ ← C

(4) N-1-(5'-phosphoribosyl)-AMP

D H_2O

(5) N-(5'-phosphoribosyl-formimino)-
5-amino-1-(5″-phosphoribosyl)-
imidazole 4-carboxamide

E

(6) N-(5'-phosphoribulosyl-formimino)-
5-amino-1-(5″-phosphoribosyl)-
imidazole 4-carboxamide

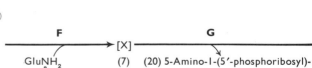

F   G
→   [X] ────────────────
GluNH$_2$ (7)   (20) 5-Amino-1-(5'-phosphoribosyl)-
                              imidazole 4-carboxamide
                              (see Fig. 13a).

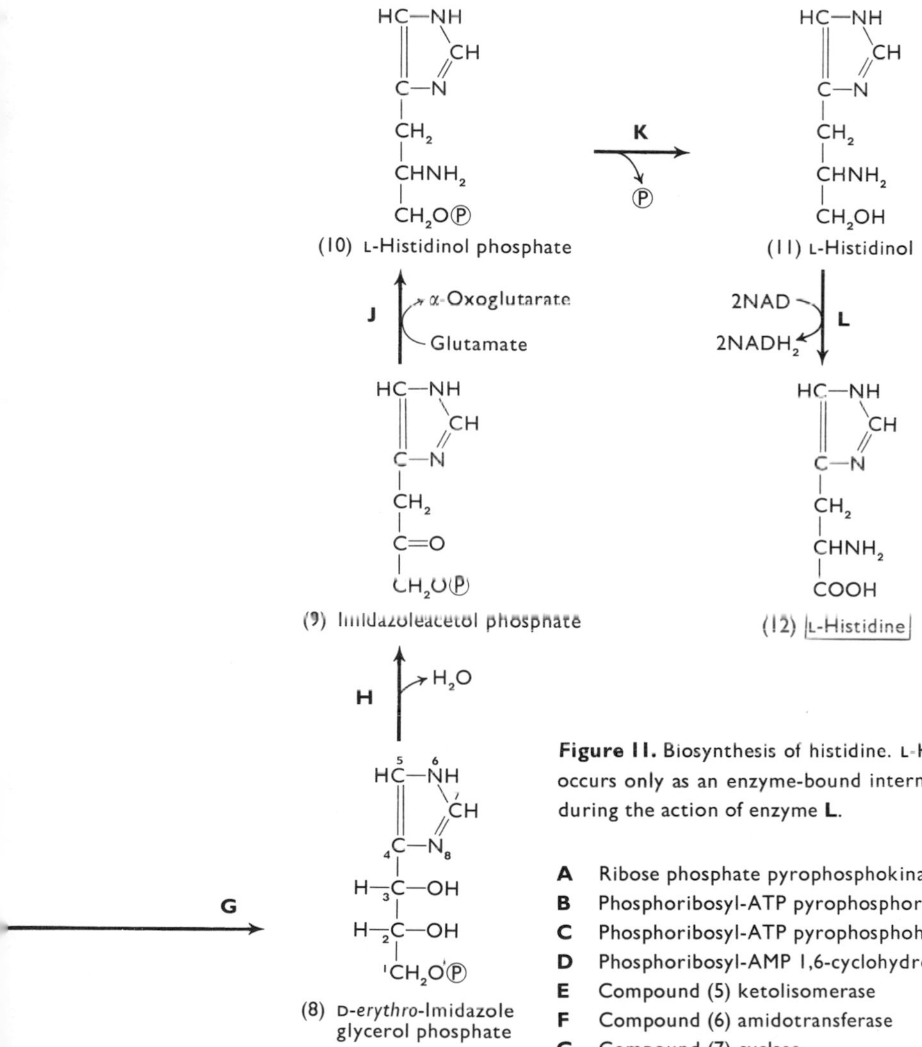

Figure 11. Biosynthesis of histidine. L-Histidinal occurs only as an enzyme-bound intermediate during the action of enzyme **L**.

A Ribose phosphate pyrophosphokinase
B Phosphoribosyl-ATP pyrophosphorylase
C Phosphoribosyl-ATP pyrophosphohydrolase
D Phosphoribosyl-AMP 1,6-cyclohydrolase
E Compound (5) ketolisomerase
F Compound (6) amidotransferase
G Compound (7) cyclase
H Imidazoleglycerol phosphate dehydratase
J Histidinol phosphate aminotransferase
K Histidinol phosphatase
L Histidinol dehydrogenase

Control

The control of histidine biosynthesis has been studied in great detail in *Salmonella typhimurium*.

The first enzyme in the sequence, phosphoribosyl-ATP pyrophosphorylase (B) is feedback-inhibited by histidine. 2-Thiazolealanine, an antagonist of histidine, inhibits growth by mimicking this feedback inhibition while being unable to replace histidine in proteins. Mutants which are resistant to 2-thiazolealanine have an altered enzyme B which is not sensitive to feedback inhibition by either histidine or 2-thiazolealanine.

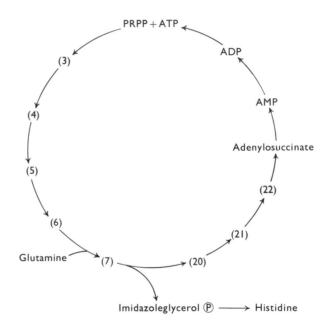

The genes controlling the structure of the enzymes of histidine biosynthesis map in a closely-linked operon (Figure 12). The letters used to denote genes should not be confused with those used to denote the enzymes. The enzymes of the operon show co-ordinate repression, and this system was the first in which this phenomenon was observed. Mutations in the histidine operon may show *polarity*, i.e. a mutation in any gene on the map may cause not only a low level of activity of the enzyme coded for by that gene, but also a low level of all the enzymes coded for by genes lying beyond it (i.e. away from the O gene) in the genetic map. Those enzymes that are coded for by genes lying between the operator and the mutation site show normal levels of activity. Co-ordinate repression still occurs, but the ratio of activities of any two of the enzymes may differ from that in the wild type. The occurrence of polarity mutants is strong evidence for the idea of a *polycistronic messenger RNA* (Chapter 6).

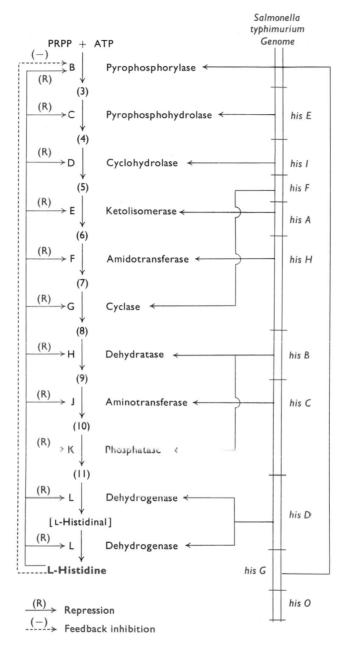

Figure 12. The histidine operon. The operon has an operator gene *hisO* and nine structural genes (*A* to *I*) which are co-ordinately controlled and which determine the structures of the 10 enzymes (the dehydratase H and phosphatase K are different functions of the same protein). The order of the genes does not follow the biochemical reaction sequence. The numbers in brackets and enzyme letters refer to Figure 11. The whole operon maps at minute 65 on the *Salmonella* chromosome. Enzyme B is feedback inhibited by histidine.

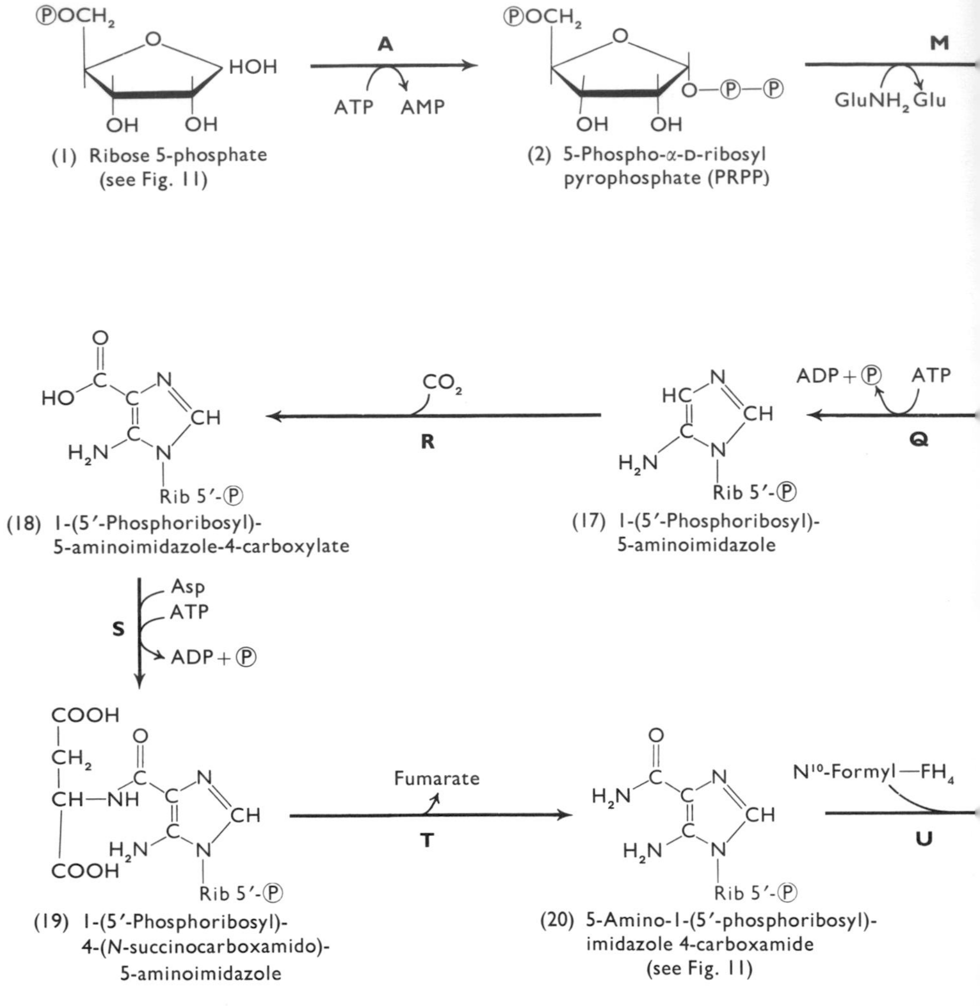

Figure 13. Biosynthesis of purine nucleotides: IMP, AMP and GMP.

A	Ribose phosphate pyrophosphokinase	**R**	Phosphoribosyl-aminoimidazole carboxylase
M	Phosphoribosyl-pyrophosphate amidotransferase	**S**	Phosphoribosyl-succinocarboxamido-aminoimidazole synthase
N	Phosphoribosyl-glycineamide synthase	**T**	Adenylosuccinate lyase
O	Phosphoribosyl-glycineamide formyltransferase	**U**	Phosphoribosyl-aminoimidazole-carboxamide formyltransferase
P	Phosphoribosyl-formylglycineamidine synthase	**V**	Adenylosuccinate synthase
Q	Phosphoribosyl-aminoimidazole synthase	**X**	Inosinate dehydrogenase
		Y	GMP synthase (XMP: ammonia ligase)

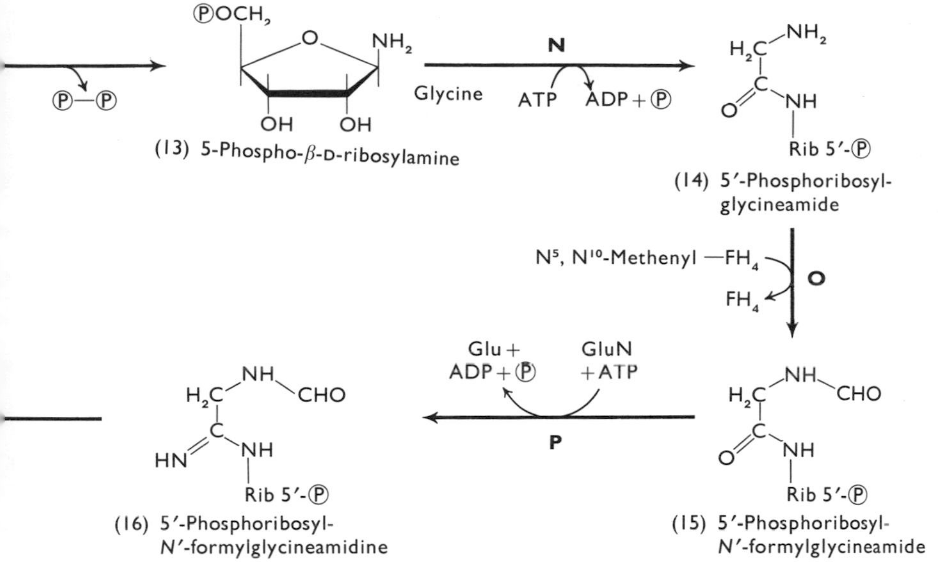

Ⓟ—Ⓟ
ⓅOCH₂
NH₂
OH OH
(13) 5-Phospho-β-D-ribosylamine
N
Glycine
ATP ADP+Ⓟ
Rib 5′-Ⓟ
(14) 5′-Phosphoribosyl-glycineamide
N⁵, N¹⁰-Methenyl —FH₄
FH₄
O
Glu+ ADP+Ⓟ
GluN +ATP
P
(16) 5′-Phosphoribosyl-N′-formylglycineamidine
(15) 5′-Phosphoribosyl-N′-formylglycineamide

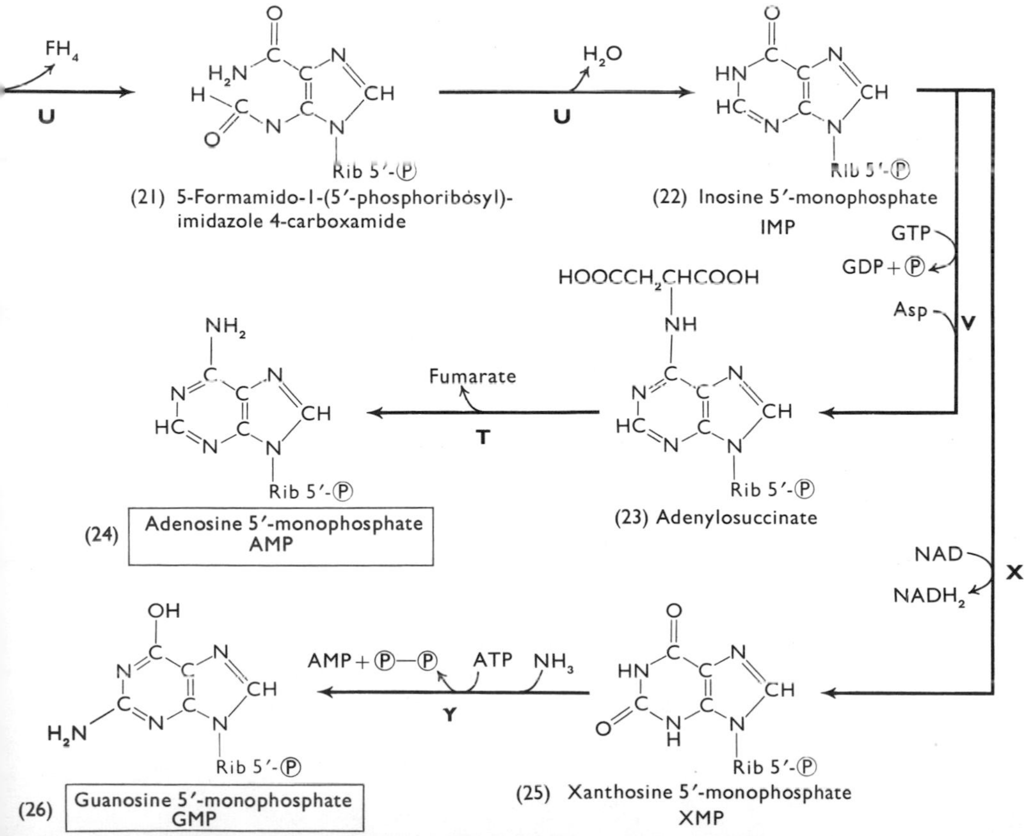

FH₄
U
(21) 5-Formamido-1-(5′-phosphoribosyl)-imidazole 4-carboxamide
H₂O
U
(22) Inosine 5′-monophosphate IMP
GTP
GDP+Ⓟ
Asp
V
HOOCCH₂CHCOOH
Fumarate
T
(23) Adenylosuccinate
(24) Adenosine 5′-monophosphate AMP
NAD
NADH₂
X
AMP+Ⓟ—Ⓟ ATP NH₃
Y
(25) Xanthosine 5′-monophosphate XMP
(26) Guanosine 5′-monophosphate GMP

2-Thiazolealanine does not mimic the action of histidine in repressing the enzymes of the *his* operon. Consequently the inhibition of growth by 2-thiazole-alanine is only transient, because the lack of histidine ultimately derepresses all the enzymes to such an extent that the increased levels overcome the inhibition by 2-thiazolealanine. This phenomenon is called *induced phenotypic resistance*.

BIOSYNTHESIS OF PURINE AND PYRIMIDINE NUCLEOTIDES

Purine nucleotides

This topic has already been mentioned in connection with histidine biosynthesis. The ultimate precursors of the individual atoms of the purine nucleotide ring have been given in Figure 3 (p. 32).

Phosphoribosyl-pyrophosphate formed from ribose 5-phosphate reacts with glutamine, under the influence of *phosphoribosyl-pyrophosphate amidotransferase* (M) (Figure 13), to yield 5-phosphoribosylamine (13), glutamate and pyrophosphate. Phosphoribosylamine then reacts with glycine and ATP under the influence of *phosphoribosylglycineamide synthase* (N) to give 5-phospho-ribosyl-glycineamide (14), ADP and orthophosphate.

The next step is the addition of a C_1 unit to yield 5-phosphoribosyl-*N*-formylglycineamide (15) which is the precursor of IMP as shown in Figure 13.

IMP (22 in Figure 13) is the precursor of both adenosine 5'-monophosphate (AMP, 24) and guanosine 5'-monophosphate (GMP, 26) and so represents a metabolic branch point.

AMP is formed from IMP by the introduction of an amino-group. The N atom is derived from aspartate and is introduced by the intermediary formation of adenylosuccinic acid (23). This compound is formed by interaction of IMP and aspartate under the influence of *adenylosuccinate synthase* (V), a reaction requiring GTP which is converted to GDP and orthophosphate. Adenylo-succinate is then cleaved by the enzyme *adenylosuccinate lyase* (T) to give fumarate and AMP. This enzyme is identical with that catalysing an earlier step in purine biosynthesis (enzyme T, Figure 13).

GMP is derived from IMP via xanthosine 5'-phosphate (XMP) (25), which is formed by the NAD-linked oxidation of IMP catalysed by *IMP dehydrogenase* (X). XMP is converted to GMP by the enzyme *XMP:ammonia ligase (GMP synthase)* (Y). In this reaction XMP, ammonia and ATP are the reactants and GMP, AMP and pyrophosphate the products.

The AMP and GMP formed by this reaction are successively converted to the 5'-di- and 5'-tri-phosphates by phosphorylation with ATP, catalysed by *nucleoside monophosphate kinase* and *nucleoside diphosphate kinase*.

Control

Study of the control of purine biosynthesis is complicated because (a) histidine biosynthesis is closely inter-related; (b) nucleotides can be formed from pre-existent bases or nucleotides present in the medium (the 'salvage pathway') and (c) inter-conversion of nucleotides can occur when they are added to crude extracts. Considerations of space prevent discussion of the 'salvage pathway' here.

The first enzyme common to both purine and histidine biosynthesis, enzyme A in Figures 11 and 13, is inhibited in the Enterobacteriaceae by ADP, CTP, GTP and UTP and also by tryptophan. The inhibition caused by the latter compound is cumulative with that due to the nucleotides. It is possible that this enzyme may be controlled by the ATP/ADP ratio in the cell. Histidine does not affect enzyme A. Enzyme M (PRPP amidotransferase) is feedback inhibited by AMP (24) and IMP (22) in *Escherichia coli* and by GMP (26) and AMP (24) in *Aerobacter aerogenes*. There is some evidence to suggest that the latter two inhibitors bind at different sites. Other enzymes between compound 13 and IMP (22) do not show feedback inhibition. These relationships are shown in Figure 14.

In Figure 13 inosinate dehydrogenase (X) is feedback-inhibited by GMP but not by adenine nucleotides, while adenylosuccinate synthase (V) is inhibited by ADP. The complicating factor in these pathways is the existence of enzymes (guanylate reductase and adenylate deaminase) catalysing the reconversion of both GMP and AMP to IMP (22). Since all these reactions are irreversible, unless all the enzymes are strictly controlled, they could catalyse a futile cycle of the following type:

$$IMP + Aspartate + GTP \longrightarrow Compound\ (23) + GDP + \textcircled{P}$$

$$Compound\ (23) \longrightarrow AMP + Fumarate$$

$$AMP + H_2O \longrightarrow IMP + NH_2$$

$$Sum: Aspartate + GTP \longrightarrow Fumarate + GDP + \textcircled{P} + NH_2 + H_2O$$

with the loss of a high energy bond. To prevent this happening, adenylate deaminase is inhibited by GTP and guanylate reductase by ATP.

Repression also occurs in this pathway. Most of the enzymes of the bio-synthetic sequences shown in Figure 13 are either totally or partially repressed when purine bases, nucleosides or nucleotides are present in the medium. The genes controlling purine biosynthesis do not show an operon-type organization but are scattered over the chromosome and repression is not co-ordinate. In the guanine branch of the pathway, however (Figure 13), the enzymes inosinate dehydrogenase (X) and GMP synthase (Y), specified by the genes *guaB* and *guaA* respectively, are co-ordinately repressed and the genes form a small operon. Gene *guaC* controlling guanylate reductase is not linked to them.

Inhibition and repression by purine nucleotide derivatives extends to some of the enzymes of one-carbon metabolism such as methylene-tetrahydrofolate dehydrogenase which produces the formyl units required by enzymes O and U in Figure 13.

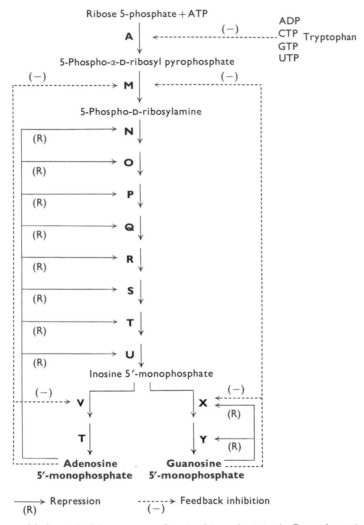

Figure 14. Control of the enzymes of purine biosynthesis in the Enterobacteriaceae.

Pyrimidine nucleotides

The precursors of the pyrimidine nucleus have been shown in Section 4 (Figure 3, p. 32). Pyrimidines are synthesized (Figure 16) from aspartic acid and carbamoyl phosphate (1). This compound is also a precursor of arginine (see p. 220). Aspartate arises from the tricarboxylic acid cycle by transamination of oxaloacetate, while carbamoyl phosphate is formed by the action of a glutamine-dependent *carbamoyl phosphate synthase* (A). Carbamoyl phosphate condenses with aspartate under the influence of *aspartate carbamoyltransferase* (B) to carbamoylaspartate (ureidosuccinate) (2) and orthophosphate. The ring is then formed by dehydration to yield dihydro-orotate (3) a reaction catalysed

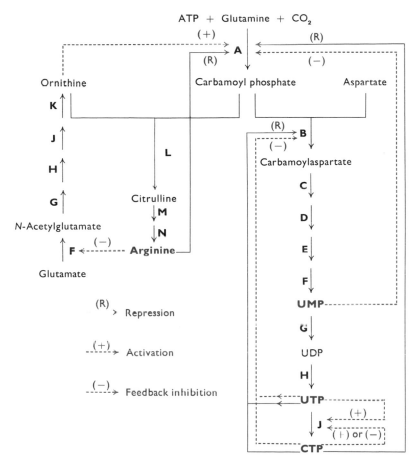

Figure 15. Control of pyrimidine biosynthesis in *Escherichia coli* and *Salmonella typhimurium* and its relation to arginine biosynthesis. Letters on the left-hand side of the diagram refer to enzymes in Figure 6b, while those on the right-hand side refer to enzymes in Figure 15.

by *dihydro-orotase* (C). Dihydro-orotic acid is oxidized to orotic acid (4) by the NAD-linked flavoprotein *dihydro-orotate dehydrogenase* (D). Orotic acid then gives rise to CTP (9) as shown in Figure 16. Cytosine nucleotides are formed from UMP via UTP. Hence UMP must be converted to UTP by *nucleoside monophosphate kinase* (G) and *nucleoside diphosphate kinase* (H) UTP then gives rise to cytidine 5′-triphosphate (9) by the action of *cytidine 5′-triphosphate*

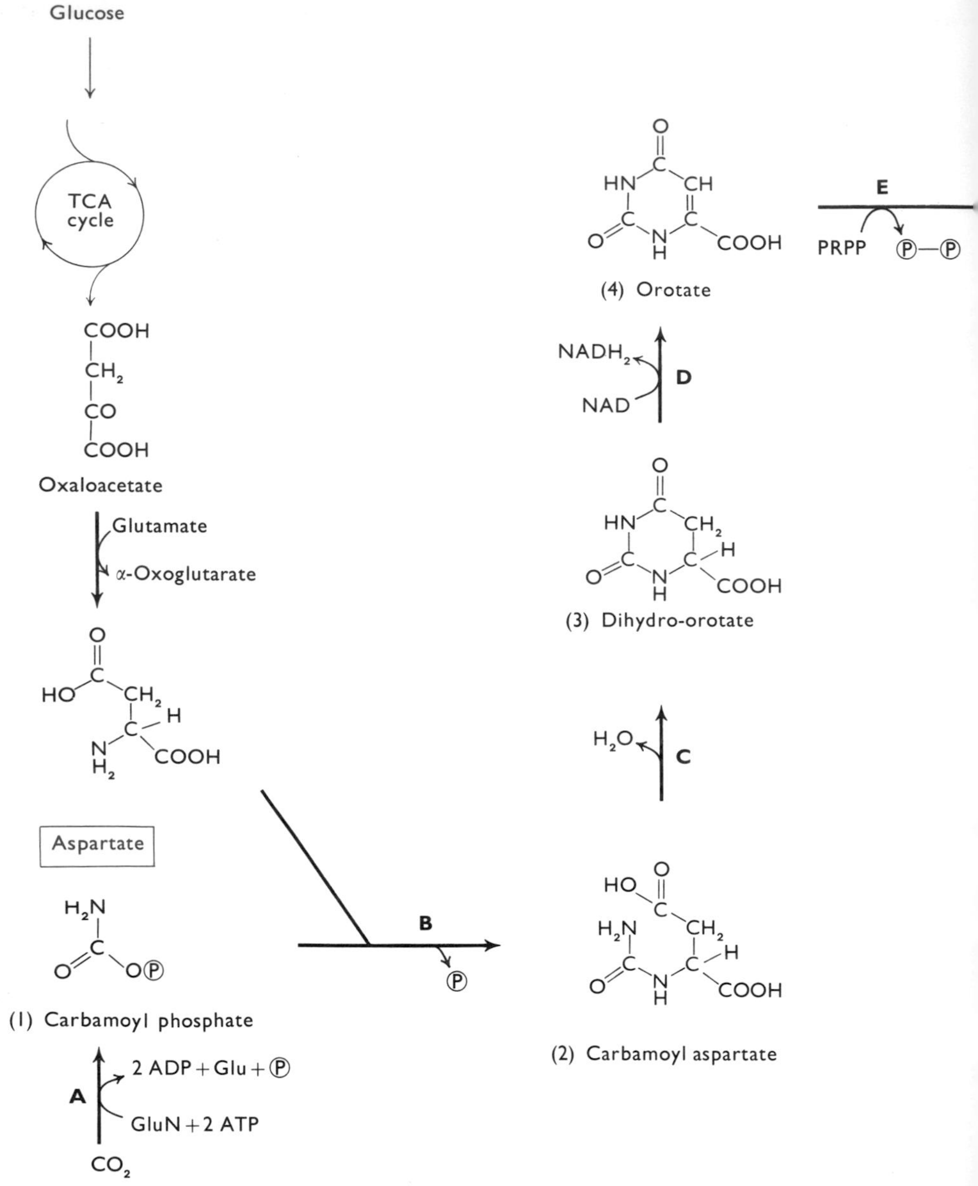
Glucose
TCA cycle
COOH
CH2
CO
COOH
Oxaloacetate
Glutamate
α-Oxoglutarate
HO
CH2
H
COOH
Aspartate
H2N
OⓅ
(1) Carbamoyl phosphate
A
2 ADP + Glu + Ⓟ
GluN + 2 ATP
CO2
B
Ⓟ
(2) Carbamoyl aspartate
H2O
C
(3) Dihydro-orotate
NADH2
NAD
D
(4) Orotate
E
PRPP
Ⓟ—Ⓟ

synthase (J) which requires ammonia and ATP. ADP and orthophosphate are formed.

Control

Some of the earliest observations on repression and feedback inhibition were those made on the biosynthesis of pyrimidine nucleotides in *Escherichia coli*,

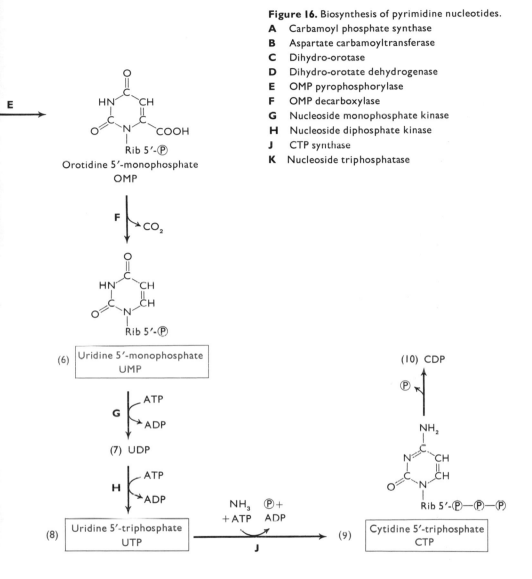

Figure 16. Biosynthesis of pyrimidine nucleotides.

A Carbamoyl phosphate synthase
B Aspartate carbamoyltransferase
C Dihydro-orotase
D Dihydro-orotate dehydrogenase
E OMP pyrophosphorylase
F OMP decarboxylase
G Nucleoside monophosphate kinase
H Nucleoside diphosphate kinase
J CTP synthase
K Nucleoside triphosphatase

and aspartate carbamoyltransferase (B) is the enzyme on which much of the early work on allosteric proteins was done (Chapter 8). It is potently inhibited by CTP (Figure 15; see also Chapter 8, p. 423) and it is activated by ATP. Other enzymes subject to feedback inhibition are carbamoyl phosphate synthase (A) and CTP synthase (J). The latter is inhibited by CTP, while GTP and UTP activate it. Carbamoyl phosphate synthase (A) is feedback-inhibited by UMP and this inhibition is antagonized by ornithine, an intermediate of arginine biosynthesis which also requires carbamoyl phosphate. In *Pseudomonas aeruginosa* the feedback inhibitor of enzyme B is UTP. In *Bacillus subtilis* and *Streptococcus faecalis* it is not subject to feedback control.

The enzymes C and F in *E. coli* are co-ordinately repressed by addition of uracil although only two genes *pyrC* and *pyrD* are at all close together on the chromosome. The pyrimidine nucleoside phosphates are probably the active repressing agents, as shown in Figure 15. Carbamoyl phosphate synthase (A) which is also concerned with arginine biosynthesis (Figure 6b) is cumulatively repressed by arginine and a cytosine derivative, while enzyme B is non-co-ordinately repressed by both uridine and cytidine nucleotides. In contrast, in *P. aeruginosa*, not only is there no linkage between any of the genes, but the enzymes of pyrimidine biosynthesis are non-repressible ('constitutive').

Deoxyribonucleotides

The deoxyribonucleotides are formed from the corresponding ribonucleotides without cleavage of the bond between base and sugar but in coliform bacteria a major fraction of dUTP is derived by deamination from dCTP. The probable intermediates are indicated in Figure 17. Ribonucleotide reductase of *Escherichia coli* is under complex feedback inhibition and activation by a variety of deoxyribonucleotides so as to provide a balanced supply of precursors for DNA synthesis. The enzyme is also repressed by deoxyribonucleotides.

Energy requirements for purine and pyrimidine biosynthesis

The biosynthesis of purine nucleotides is a highly endergonic process. From Figure 13 it can be seen that the *de novo* synthesis of one molecule of AMP or GMP requires six molecules of ATP, without taking into consideration the ATP requirements for the biosynthesis of other reactants, e.g. glutamine, or the ATP requirements for the regeneration of ATP from the AMP formed in some of the reactions. A further two molecules of ATP are necessary to bring the purine nucleotides to the level at which they can polymerize to give RNA. Reduction to deoxyribonucleotides also requires ATP. The requirements for synthesis of the pyrimidine nucleotides are very little less.

BIOSYNTHESIS OF LIPIDS

The lipid content of bacteria may vary from about 2–3% of the dry weight, most of which is associated with the lipoprotein of the cytoplasmic membrane, to more than 50% in those organisms which store lipid as a reserve material. The types of lipid differ according to the organism, and the main kinds, triglycerides and phospholipids, have been mentioned in Chapter 3.

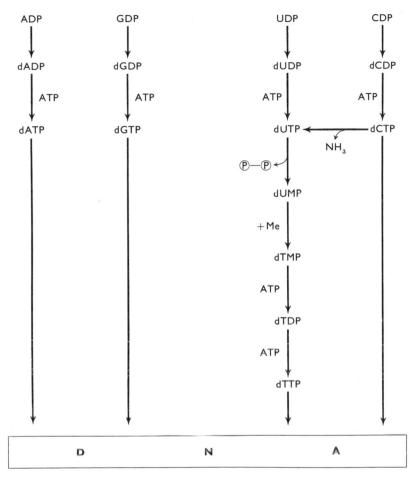

Figure 17. Formation of deoxyribonucleotides. Purine and pyrimidine ribonucleoside derivatives are reduced to the deoxy-compound without separation of the sugar from the base.

We will first consider the synthesis from glucose of glycerol and long chain fatty acids. It has been found that L-α-glycerophosphate and not free glycerol is involved in lipid synthesis. This compound is produced by reduction of dihydroxyacetone phosphate, itself a normal intermediate of glycolysis, by the action of α-*glycerophosphate dehydrogenase*, an NAD-dependent enzyme:

$$
\begin{array}{ccc}
\text{CH}_2\text{OH} & & \text{CH}_2\text{OH} \\
| & & | \\
\text{C=O} \quad + \text{ NADH}_2 & \rightleftharpoons & \text{CHOH} \; + \; \text{NAD} \\
| & & | \\
\text{CH}_2\text{O}\textcircled{P} & & \text{CH}_2\text{O}\textcircled{P}
\end{array}
$$

The synthesis of long chain fatty acids is complex and is achieved by a series of reactions which have acetyl coenzyme A as their starting point but, unlike

fatty acid oxidation, do not have thioesters of coenzyme A as their intermediates. Instead, thioesters of a low-molecular weight protein, called *acyl
carrier protein* (ACP) are involved in a manner to be described. A simplified
outline of the series of reactions is given in Figure 18. The initiating reaction is
the synthesis of malonyl coenzyme A by the carboxylation of acetyl coenzyme
A, a reaction requiring ATP and Mn^{2+} and catalysed by *acetyl CoA carboxylase*,
an enzyme which possesses a biotin (see Appendix B) prosthetic group. The
reaction is believed to occur in two stages:

$$CO_2 + ATP + Enzyme\text{-}biotin \xrightleftharpoons{Mn^{2+}} Enzyme\text{-}biotin\text{-}COOH + ADP + \circled{P} + H_2O$$

$$Enzyme\text{-}biotin\text{-}COOH + CH_3CO \sim S.CoA \rightleftharpoons Enzyme\text{-}biotin + HOOC.CH_2CO \sim S.CoA$$

$$Sum: CO_2 + ATP + CH_3CO \sim S.CoA \rightleftharpoons HOOC.CH_2CO \sim S.CoA + ADP + \circled{P} + H_2O$$

Malonyl$\sim$S.CoA and acetyl$\sim$S.CoA are then both converted to acyl carrier
protein derivatives prior to undergoing a condensation reaction in which CO_2
is eliminated, followed by reduction with $NADPH_2$. The fatty acid synthase
complex which carries out these reactions has been shown to be a soluble system
in both *Clostridium kluyveri* and *Escherichia coli* and fractionation has revealed
that seven protein fractions are necessary in the latter organism. The reactions
in *E. coli* will now be discussed in detail.

Following the formation of malonyl$\sim$S.CoA, the acyl groups of this compound and of acetyl$\sim$S.CoA are transferred to the thiol group of acyl carrier
protein by the enzymes *malonyl transacylase* and *acetyl transacylase*:

$$CH_3CO \sim S.CoA + ACP.SH \rightleftharpoons CH_3CO \sim S.ACP + CoA.SH$$

$$HOOC.CH_2CO \sim S.CoA + ACP.SH \rightleftharpoons HOOC.CH_2CO \sim S.ACP + CoA.SH$$

Both transacylases are heat-labile, contain thiol groups which are essential for
activity and, in consequence, are extremely sensitive to thiol inhibitors. They
can, however, be partly protected against such inhibition by prior incubation
with their respective acyl CoA compounds which suggests that reaction occurs
with the thiol group of the enzyme, presumably to yield acyl$\sim$S-enzyme as an
intermediate.

The acyl carrier protein is heat stable with a molecular weight of about
10,000 and has been highly purified and its amino acid sequence determined.
Its thiol group is in a single pantetheine residue attached by a phosphodiester
link to the protein. It is thus analogous to coenzyme A. By experiments with
isotopes it has been shown that reaction with $[1\text{-}^{14}C]$acetyl$\sim$S.CoA leads to
the stoicheiometric disappearance of the free thiol group. The carrier protein-
substrate compound formed is radioactive whereas no radioactivity is found
when acetyl$\sim[^{14}C]$S.CoA is used. These observations accord with the
mechanism proposed.

A third enzyme catalyses the condensation of acetyl S.ACP and malonyl. S.ACP to form acetoacetyl.S.ACP with the elimination of one molecule each of CO_2 and ACP:

$$CH_3CO \sim S.ACP + HOOC.CH_2CO \sim S.ACP \rightleftharpoons CH_3CO.CH_2CO \sim S.ACP + CO_2 + ACP.SH$$

The molecule of CO_2 so released contains the same carbon atom that was introduced in the acetyl CoA carboxylase reaction, which explains why when $^{14}CO_2$ is used the isotope never appears in the fatty acid finally synthesized; it plays a catalytic role and is regenerated as each two-carbon unit is added. The condensation-decarboxylating reaction has been shown to be the chain-elongating reaction of fatty acid synthesis.

The acetoacetyl.S.ACP now undergoes reduction with $NADPH_2$, catalysed by *β-ketoacyl-ACP-reductase*, to form the D-isomer of β-hydroxybutyryl.S. ACP:

$$CH_3CO.CH_2CO \sim S.ACP + NADPH_2 \rightleftharpoons CH_3CHOH.CH_2CO \sim S.ACP + NADP$$

This is another distinguishing feature from fatty acid oxidation in which the β-hydroxyacyl.S.CoA derivatives are all L-isomers.

Dehydration to crotonyl.S.ACP follows, catalysed by *enoyl-ACP dehydratase*:

$$CH_3CHOH.CH_2CO \sim S.ACP \rightleftharpoons CH_3CH=CH.CO \sim S.ACP + H_2O$$

and then further reduction to butyryl.S.ACP by *crotonyl-ACP reductase* which also employs $NADPH_2$:

$$CH_3CH=CH.CO \sim S.ACP + NADPH_2 \rightleftharpoons CH_3CH_2CH_2CO \sim S.ACP + NADP$$

This reductive step differs from the reverse oxidation which occurs in fatty acid breakdown by not employing a flavoprotein.

Butyryl.S.ACP can now condense with another mole of malonyl.S.ACP and, by undergoing a similar sequence of reactions, two further carbon atoms are added to the fatty acid. The cycle is repeated until a fatty acid of the required chain length is obtained and then the fatty acid dissociates from the acyl carrier protein by the action of a hydrolytic deacylase. It will be noticed that acetyl~ S.CoA furnishes the two carbon atoms at the methyl end of the fatty acid chain and malonyl~S.CoA contributes the rest of the carbon atoms. The overall reaction for the synthesis of palmitic acid (C_{16}) may be summarized as:

$$8 \text{ Acetyl} \sim S.CoA + 7 ATP + 14 NADPH_2 \longrightarrow$$

$$CH_3(CH_2)_{14}COOH + 7 ADP + 7 \textcircled{P} + 8 CoASH + 14 NADP$$

The reducing power required in the form of $NADPH_2$ is principally generated by the oxidation of glucose 6-phosphate to ribulose 5-phosphate and CO_2 in the pentose cycle (p. 174).

$CH_3CO \sim S.CoA + CO_2 \longrightarrow COOH.CH_2CO \sim S.CoA$

$CH_3CO \sim S.CoA + ACP.SH \longrightarrow CH_3CO \sim S.ACP + CoA.SH$
$COOH.CH_2CO \sim S.CoA + ACP.SH \longrightarrow COOH.CH_2CO \sim S.ACP + CoA.SH$

$CH_3CO \sim S.ACP + COOH.CH_2CO \sim S.ACP \longrightarrow CH_3CO.CH_2CO \sim S.ACP + CO_2 + ACP.SH$
$CH_3CO.CH_2CO \sim S.ACP + NADPH_2 \longrightarrow CH_3CHOH.CH_2CO \sim S.ACP \sim NADP$
$CH_3CHOH.CH_2CO \sim S.ACP \longrightarrow CH_3CH{=}CH.CO \sim S.ACP + H_2O$
$CH_3CH{=}CH.CO \sim S.ACP + NADPH_2 \longrightarrow CH_3CH_2CH_2CO \sim S.ACP + NADP$

$CH_3CH_2CH_2CO \sim S.ACP + COOH.CH_2CO \sim S.ACP \longrightarrow CH_3CH_2CH_2CO.CH_2CO \sim S.ACP + CO_2 + ACP.SH$

etc.

Figure 18. Biosynthesis of fatty acids. Simplified representation of the reactions leading from acetyl-coenzyme A to long chain fatty acids and involving the acyl carrier protein (ACP.SH).

In *Escherichia coli* extracts the major product of the fatty acid synthase system is *cis*-vaccenic acid, the Δ^{11} isomer of oleic acid:

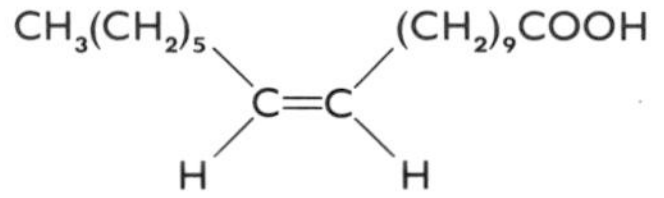

$$CH_3(CH_2)_5 \qquad\qquad (CH_2)_9COOH$$
$$C{=}C$$
$$H \qquad\qquad H$$

cis-Vaccenic acid.

Present evidence from experiments with isotopes using *Clostridium butyricum* indicates that while octanoate (C_8) and decanoate (C_{10}) are precursors of both saturated and unsaturated fatty acids, C_{12} and C_{14} saturated acids are incorporated only into saturated acids. These observations suggest that the pathways for synthesis of saturated and unsaturated fatty acids diverge at the C_{10} and possibly at the C_8 level:

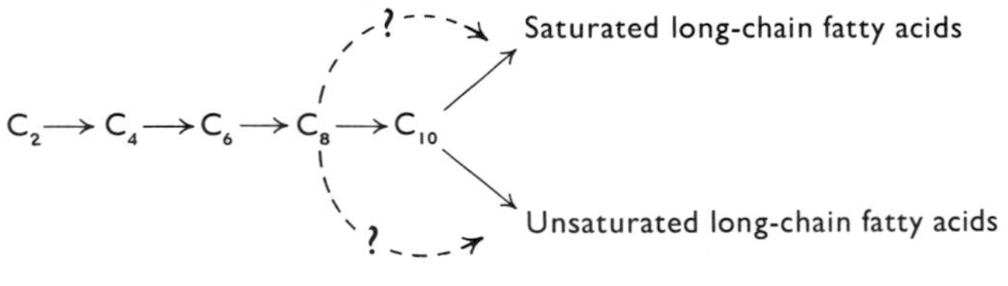

$$C_2 \rightarrow C_4 \rightarrow C_6 \rightarrow C_8 \rightarrow C_{10}$$

? Saturated long-chain fatty acids

? Unsaturated long-chain fatty acids

A feature of fatty acid synthesis is the absence of free intermediates in the process; once acetyl $\sim$ S.ACP and malonyl $\sim$ S.ACP have condensed to form acetoacetyl $\sim$ S.ACP the whole sequence of reduction to butyryl $\sim$ S.ACP and further chain elongation and reduction takes place on the same acyl carrier protein molecule.

Although in animal cells the first enzyme of fatty acid biosynthesis, acetyl CoA carboxylase, is controlled by an allosteric activation with citrate, this enzyme in *E. coli* does not appear to be subject to regulation. Instead it seems possible that acetyl CoA is diverted from the tricarboxylic acid cycle to fatty acid biosynthesis by the feedback inhibition of citrate synthase by α-oxo-glutarate. Thus, under conditions where there is production of α-oxoglutarate in excess of the biosynthetic requirements of the cell it will serve as the signal to switch acetyl CoA from entry to the tricarboxylic acid to fatty acid synthesis.

Unlike the bacterial systems which have been fractionated, the fatty acid synthesizing system of yeast functions as a single complex in which it is believed that seven enzymes are arranged round a central functional thiol group in such a manner that the intermediates, bound covalently to this central thiol group, are held in close proximity to the active centres of the enzymes.

We may now consider the pathways of biosynthesis of lipids and phospho-lipids which are shown in Figure 19. L-α-Glycerophosphate reacts with two molecules of the coenzyme A derivatives of fatty acids under the influence of appropriate *acyl~S.CoA glycerophosphate acyltransferases* to form L-α-phosphatidic acid. There appears to be some specificity exhibited in the esterification process since it is usually found that saturated fatty acids occur in the α- and unsaturated fatty acids in the β-position of the phosphatidic acid. In *E. coli* and some other bacteria the fatty acyl donor in the formation of phosphatidic acid is fatty acyl carrier protein rather than the coenzyme A derivative.

At this point the biosynthetic pathways for lipids and phospholipids diverge. For lipid synthesis a phosphatase removes the phosphate group to yield a D-α,β-diglyceride which can then react with a molecule of fatty acyl~S.CoA (or fatty acyl~S.ACP) to yield a triglyceride.

The reactions leading to phospholipid synthesis in both mammalian and bacterial systems have been shown to involve cytidine 5′-triphosphate (CTP) as coenzyme although the reaction pathways differ. CTP has the structure shown below and in its co-enzyme function the terminal phosphate is replaced with a group such as choline, ethanolamine or diglyceride. Thus, catalysed by an

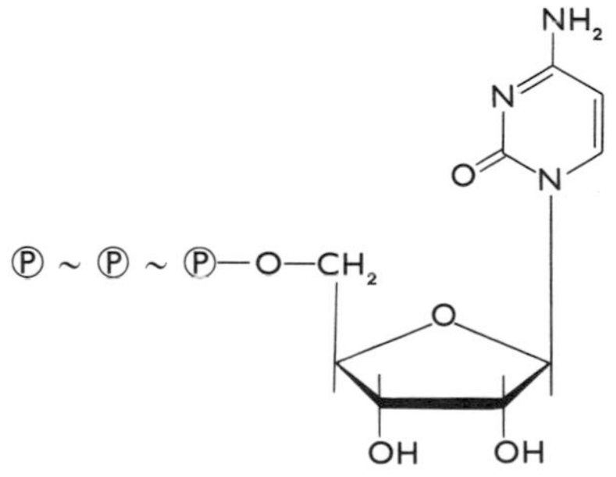

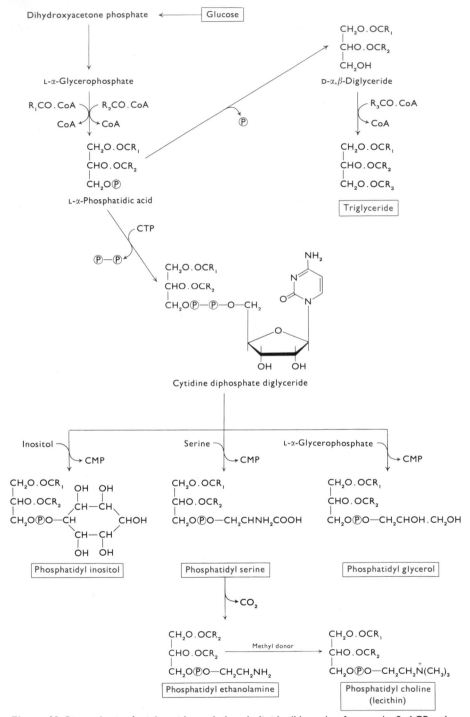

Figure 19. Biosynthesis of triglycerides and phospholipids. (Note that fatty acyl ∼S. ACP rather than fatty acyl ∼S. CoA functions as the fatty acyl donor in some micro-organisms.)

appropriate enzyme, L-α-phosphatidic acid reacts with CTP to yield cytidine diphosphate diglyceride and pyrophosphate:

$$\text{L-}\alpha\text{-phosphatidic acid } + \text{ CTP } \longrightarrow \text{ CDP-diglyceride } + \text{ ℗—℗}$$

In mammalian tissues cytidine diphosphate-ethanolamine reacts with D-α,β-diglyceride to yield phosphatidyl ethanolamine directly, according to the equation:

$$\text{CDP-ethanolamine } + \text{ D-}\alpha,\beta\text{-diglyceride } \longrightarrow \text{ phosphatidyl ethanolamine } + \text{ CMP}$$

In *Escherichia coli*, however, this reaction does not occur and instead phosphatidyl serine appears to be an essential intermediate, undergoing decarboxylation to yield phosphatidyl ethanolamine. The following reactions have been demonstrated in extracts of the organism:

$$\text{CDP-diglyceride } + \text{ L-serine } \longrightarrow \text{ phosphatidyl serine } + \text{ CMP}$$

$$\text{Phosphatidyl serine } \longrightarrow \text{ phosphatidyl ethanolamine } + \text{ CO}_2$$

The first is catalysed by L-*serine phosphatidyltransferase*; the second reaction is catalysed by a highly active enzyme, *phosphatidyl serine decarboxylase*, which does not decarboxylate free L-serine. Although much work remains to be done with other bacteria it seems highly likely that, in view of the type of phospholipids encountered, CDP-diglyceride-requiring pathways will be involved.

Escherichia coli has been found to contain two distinct types of phospholipid. One, lipid A, which is part of the endotoxin complex of the organism is rendered chloroform-soluble only by acid hydrolysis, whereas the other is readily extracted by chloroform-methanol. Lipids of this latter type consist mainly of phosphatidyl ethanolamine with much smaller amounts of phosphatidyl glycerol and phosphatidyl serine.

Experiments using cultures of *E. coli* pulse-labelled with ^{32}P revealed that the phosphatidyl ethanolamine fraction is relatively stable during growth whereas the phosphatidyl glycerol fraction loses ^{32}P rapidly and is presumably involved in turnover of biosynthetic reactions.

FURTHER READING

I Magasanik B. (1962) Biosynthesis of purine and pyrimidine nucleotides. In *The Bacteria*, Volume 3: Biosynthesis, (eds. Gunsalus I. C. and Stanier R. Y.), p. 295. Academic Press, New York.

2 Umbarger H. E. and Davis B. D. (1962) Pathways of amino acid biosynthesis. In *The Bacteria*, Volume 3: Biosynthesis (eds. Gunsalus I. C. and Stanier R. Y.), p. 167. Academic Press, New York.

3 Kates M. (1966) Biosynthesis of lipids in micro-organisms. *Ann. Rev. Microbiol.* **20**, 13.

4 Stadtman E. R. (1966) Allosteric regulation of enzyme activity. *Adv. Enzymol.* **28**, 41.

5 White A., Handler P. and Smith E. L. (1964) *Principles of Biochemistry.* McGraw-Hill, New York.

6 Mahler H. R. and Cordes E. H. (1966) *Biological Chemistry.* Harper and Row, New York, Evanston and London.

7 Cohen G. N. (1967). *Biosynthesis of Small Molecules.* Harper and Row, New York, Evanston and London.

8 Umbarger H. E. (1969) Regulation of Amino Acid Metabolism. *Ann. Rev. Biochem.* **38**, 323.

9 Lehninger A. L. (1970) *Biochemistry.* Worth, New York.

Chapter 5
Class III Reactions: The Structure and Synthesis of Nucleic Acids

BASE-PAIRING

The structure, properties and synthesis of DNA and RNA depend fundamentally on the ability of adenine to pair specifically with thymine (or uracil) and of guanine to pair with cytosine. The pairing occurs through the formation of hydrogen bonds as shown in Figure 1. The bases themselves are flat molecules and the hydrogen bonding occurs in the same plane as the bases, so that the base-pairs as a whole are planar. In the adenine-thymine and adenine-uracil pairs two hydrogen bonds are formed, whereas in the guanine-cytosine pair there are three. This difference accounts at least in part for the GC pairing being rather more stable than the AT and AU pairings, as will be seen later. Although each base can assume different tautomeric forms (Figure 5, p. 385) only the forms shown in Figure 1 are important in the pairing mechanisms. The specific recognition between A and T or U and between G and C is referred to as *base complementarity*. In fact a considerable number of other types of base-pairing are stereochemically feasible, and some are believed to occur in the structure of tRNA (p. 263) and in codon-anticodon recognition (p. 269), but apart from these special instances there is little reason to believe that they play any wider role in living systems.

Two further features of the base-pairings represented in Figure 1 are worth noting. Firstly, the distance between the sugar moieties in the paired nucleotides is independent of the type of base-pair, and the same is true of the spatial orientation of the sugar residues with respect to each other and with respect to the plane of the paired bases. Secondly, the relative position of the sugar residues remains the same if the positions of the bases in a given type of nucleotide-pair are reversed. One can sum up these facts by saying that if a nucleotide is part of a polynucleotide chain and its complementary nucleotide is associated with it by hydrogen-bonding, then the position of the attached nucleotide in relation to the chain is fixed independently of the bases involved. This relationship is fundamental to the role played by nucleic acids in living systems. It explains why DNA molecules from different sources can show infinite variety in their base sequences while preserving a very stable, characteristic macromolecular

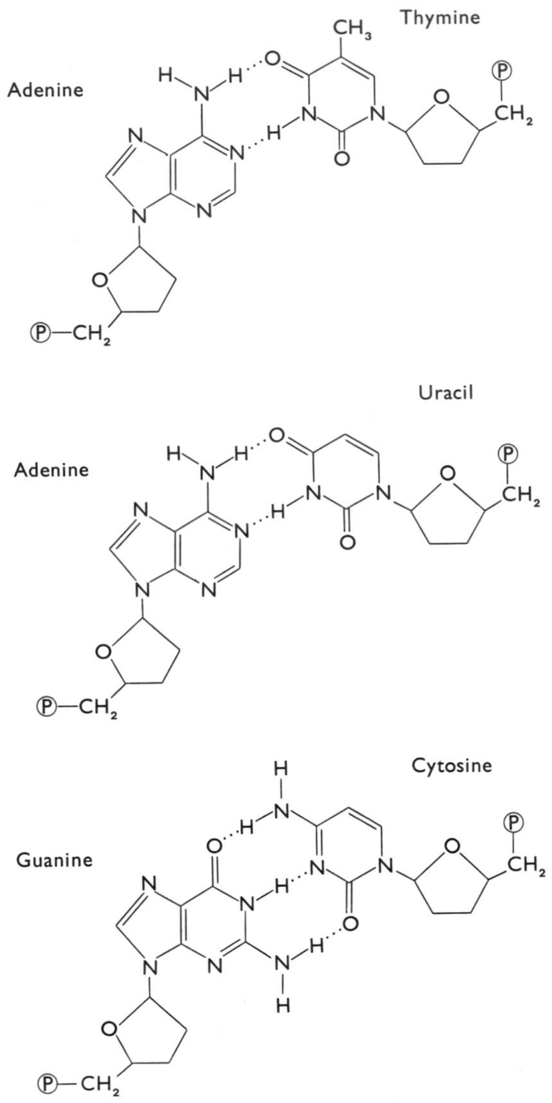

Figure I. Complementary base-pairing between nucleotides.

structure. It also provides a sound basis for the replication and transcription of nucleotide sequences since the enzymic formation of the phosphodiester linkages in the products can proceed by the same type of mechanism in all cases.

STRUCTURE OF DNA

The primary structure of DNA consists of a linear unbranched chain of deoxyribonucleotides joined by phosphodiester linkages between the 3' and 5' positions of the 2'-deoxyribose units (Figure 3, p. 17). It may be considered as a continuous deoxyribose phosphate backbone from which the nitrogenous bases project in linear array. It is the precise sequence of the bases which endows the molecule with its biological significance; this sequence represents the encoding of the genetic information which characterizes the organism in which the DNA is contained. The molecular mechanisms by which this coded sequence is duplicated and expressed will be considered later; at this stage it is appropriate to describe the form in which DNA exists in the cell, for the molecular organization of the DNA molecule is directly concerned with its unique biological functions.

The size of DNA molecules

In vivo, DNA (as well as RNA) does not exist as the 'naked' molecule but is involved in interactions with other substances, particularly proteins, many of which may be presumed to have regulatory functions. Much of the protein is basic, so that its positively charged groups help to neutralize the negatively charged phosphates of the nucleic acid, but the organization of the DNA-protein complexes in bacteria does not lead to the appearance of well-defined chromosomes such as are seen in the nuclei of higher organisms. In addition to its interaction with proteins the DNA also seems to be associated with the cell membrane; this is significant with respect to the mechanism and control of its replication (pp. 118 and 122).

Solutions of purified DNA are highly viscous, the viscosity of a particular preparation depending very much upon how carefully it was treated during the extraction and purification procedures. Although, as Cairns has shown, it is possible to liberate the total DNA of *Escherichia coli* as an apparently continuous molecule of molecular weight approximately $2 \cdot 5 \times 10^9$, most preparations of bacterial DNA have molecular weights in the region of 10^7 after they have been purified to remove protein and RNA. This is because the enormous size of DNA molecules renders them highly susceptible to hydrodynamic shear forces; the mechanical agitation with organic solvents necessary to remove proteins during DNA purification is quite sufficient to produce breakage into fragments. Even so, DNA samples isolated from bacteria are among the largest molecules which can be obtained in reasonably pure form from living materials. The enormous size of DNA can be used to advantage in purifying it because when cell extracts are treated with ethanol or *iso*-propanol the DNA precipitates in the form of fibrous threads which can be collected on a glass rod and thus selectively removed from smaller macromolecules.

Two techniques of visualizing DNA molecules have been developed so that their size and conformation may be revealed. In Cairns' method the DNA is first labelled *in vivo* with [³H] precursors such as [³H] thymidine; it is then very gently released from the cells and overlaid with a photographic emulsion. After exposure for several weeks the film is developed, revealing the positions of DNA molecules by near-continuous lines of silver grains produced by decay of the incorporated tritium atoms. Kleinschmidt's method does not require that the DNA be radioactively labelled; the DNA is simply allowed to spread over a film of basic protein molecules which act as a supporting medium, the film is caught on an electron microscope grid and shadowed with a heavy metal to outline the DNA molecules covered with basic protein, and the grid is then observed in the electron microscope. Not only have these experiments shown that the DNA of a bacterial 'nucleus' or of a virus can be extracted in one piece, but it has also been demonstrated that an unbroken molecule of *Escherichia coli* DNA appears to be circular, in agreement with the circular nature of the bacterial genome as deduced from genetic studies (see Chapter 7). It cannot yet be concluded, however, that the DNA of a bacterium has an uninterrupted phosphodiester backbone. There may exist breaks at certain points in the molecule, or even occasional non-nucleotide 'linkers', for instance amino acids or short peptide chains. Some viruses, however, have been found to yield truly circular DNA molecules, and here it can be stated with much greater certainty that the circles do not contain breaks or 'linkers'.

Secondary structure of DNA

The great viscosity of DNA solutions is due to the stiff rod-like character of the molecules, and this in turn is due to their highly organized secondary structure stabilized by base-pairing between strictly complementary nucleotide sequences.

Table I. Base-composition of bacterial DNA.

	Proportion of base, moles %				GC content (% G+C)	$\dfrac{A+T}{G+C}$
	Adenine	Thymine	Guanine	Cytosine		
Clostridium perfringens	35	35	15	15	30	2·33
Proteus vulgaris	30	29	20	21	41	1·47
Escherichia coli	25	25	25	25	50	1·00
Aerobacter aerogenes	21	22	29	28	57	0·75
Micrococcus lysodeikticus	14	14	36	36	72	0·39

Values are given to the nearest integer.

Although the gross base-composition of DNA from different bacteria varies between wide limits, there is always a close correspondence between the contents of adenine and thymine and of guanine and cytosine (Table 1). The significance of this relationship was not fully appreciated until the work of Watson and Crick on the secondary structure of DNA. These workers studied the X-ray diffraction patterns produced by oriented fibres of DNA and proposed the now-famous double-helical model for the structure of DNA. By suggesting that DNA consists of two polynucleotide chains wound into a helical form with hydrogen-bonding between complementary bases in the opposing strands,

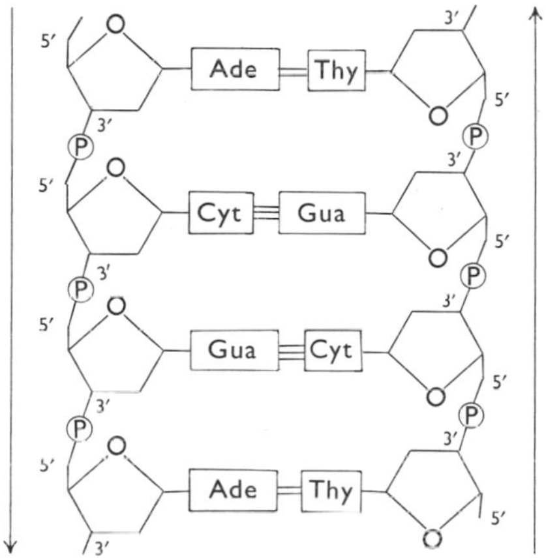

Figure 2. Sequence complementarity between opposite strands of DNA is determined by hydrogen bonding between bases.

   Note the opposite polarity of the strands, indicated by the direction of the 3′–5′ linkages and the position of the oxygen in the deoxyribose ring. Evidence for this antiparallel arrangement is described on p. 291.

Watson and Crick not only accounted for the observed equivalence of the amounts of A and T and of G and C, but they laid the foundation for a whole new understanding of the molecular events concerned in the storage, replication and expression of genetic information. The essential postulate in the Watson–Crick model is that the *sequence* of bases in one strand is exactly the complement of that in the other, so that every nucleotide participates in a sterically similar hydrogen-bonded association with a nucleotide in the opposite strand (Figure 2). Since, as pointed out earlier, the stereochemistry of the hydrogen-bonding is such that the relative orientation of the sugar-phosphate residues is not affected by the nature of the base-pairs, the duplex structure takes up the same conformation irrespective of the actual base-sequences involved although it obviously requires the A=T and G=C equalities. The conformation assumed is that of a right-handed double helix having the sugar-phosphate backbones on the periphery and the hydrogen-bonded base-pairs stacked in the core of the molecule with their planes perpendicular to the helix axis (Figure 3).

The two polynucleotide chains are antiparallel, i.e. they run in opposite directions (see Figures 2 and 3), so that rotation through 180° does not alter the appearance of the molecule. There are ten base-pairs for every complete turn of the helix. A better impression of the structure can be gained from Figure 4 which shows a double helix constructed from space-filling atomic models.

Figure 3. Diagrammatic representation of a DNA double helix.
   The direction of the arrows corresponds to the direction of the 3′–5′ phosphodiester linkages as represented in Figure 2. Note that the molecule appears to be traversed by two grooves of unequal widths. This feature arises because the polynucleotide chains are separated by a distance less than half that required for a complete turn of the helix (34 Å). The difference in size of the grooves has been exaggerated here; compare Figure 4.

Since each nucleotide is involved in hydrogen bonding the stability of the structure is very high. As well as the hydrogen bonding it is likely that other forces, arising from the ordered hydrophobic stacking of the base-pairs, also contribute to the stability of the DNA double helix. Be that as it may, hydrogen bonding is surely the principal element which determines strict accuracy of base-pairing and is thus of paramount importance for the biological role of DNA.

Denaturation

Disruption of the helix is a dramatic event which may be triggered by raising the temperature above a critical value or by exposing the DNA to extremes of pH. The change which takes place, called *denaturation*, is more or less irreversible and shows the characteristics of a cooperative process, i.e. the rupture of a critical number of hydrogen bonds facilitates the breakage of neighbouring bonds leading to the collapse of the entire double helix. Denaturation may be observed by measuring the ultraviolet absorbancy of the DNA; when the helix breaks up there is an increase of some 40% in the absorbancy at 260 nm (Figure 5). In addition to this *hyperchromic effect* there is a pronounced decrease in the

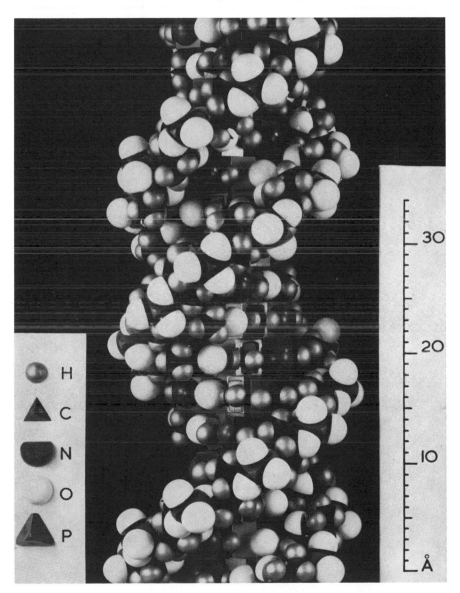

Figure 4. Scale model of the DNA molecule. Photograph kindly provided by Dr Watson Fuller.

viscosity of the solution as the strands separate and the buoyant density of the DNA in CsCl density gradients is increased. If the DNA carries genetic markers which can be studied in transformation experiments (Chapter 7) a considerable loss of transforming activity is observed.

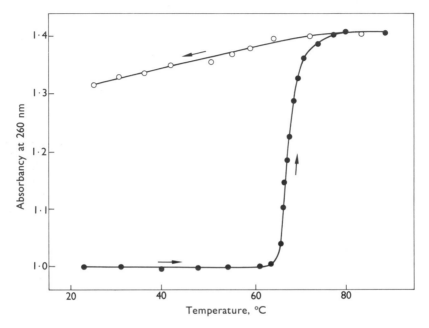

Figure 5. Thermal denaturation of double helical DNA.
A sample of DNA from bacteriophage T2 was heated to 90° and then cooled, as indicated by the arrows. The absorbancy at the various temperatures is expressed relative to the starting value. In this experiment the 'melting' temperature is 67·2°. The relatively small, gradual decrease in absorbancy on cooling represents the development of random-coil structure by the separated strands (Waring, unpublished data).

The temperature at which denaturation occurs (often called the 'melting' temperature) depends upon the ionic strength of the medium. While the dimensions of the double helix are not significantly affected by the nature and concentration of salts present, the stability of the structure depends very much upon the ability of salts to neutralize the repulsive forces between the negatively-charged phosphate groups of the two strands. Marmur and Doty have shown that there is an approximately linear dependence of melting temperature on the logarithm of the cation concentration. Furthermore, the melting temperature under given ionic conditions increases in proportion to the GC content of the DNA; this provides evidence that the GC pairing is more stable than the AT pairing, as would be expected if the GC pair had three hydrogen bonds while the AT pair had only two.

Once the strands of the helix have become separated the resulting chains of denatured DNA are free to assume whatever conformation is energetically most favourable. Generally they simply collapse into a tangled mass called a *random coil* but the nature of the random coil structure they take up is strongly dependent upon the salt concentration. At low ionic strengths the neutralization of the repulsive charges on the phosphate groups is poor and the coiled structure remains relatively open; at higher ionic strengths the coil becomes more compact. These differences in structure are evidenced by variations in the ultra-violet absorption of the DNA, which suggest that a certain amount of interaction occurs between the nucleotide bases in denatured DNA. Quite probably some hydrogen-bonding takes place between complementary bases which happen to come close together, but long helical regions like those of the native double helix are highly improbable. It should be emphasized that base interactions in denatured DNA must be virtually entirely *intra*-strand and that interactions between different strands probably play very little or no part in the formation of random-coil structures.

When the hydrogen bonds of a base-pair have been disrupted the amino groups of the bases become available for reaction with reagents such as formaldehyde or formamide. Once they have reacted in this way the modified bases are prevented from subsequently forming H-bond interactions with other bases, so that the denatured, single-stranded state of the polynucleotide chain is permanently 'frozen'. This fact has been exploited by a number of workers, especially Inman, to study denaturation. If DNA is treated with formaldehyde under conditions where denaturation is just beginning and then is prepared for electron microscopy by the Kleinschmidt technique, partially denatured states of the DNA molecules can be visualized (Figure 6). It can be seen that local separation of the strands occurs earlier in some regions of the molecules than in others. These 'early melting' regions are presumed to be regions relatively rich in AT pairs, i.e. base-pairs with the lowest stability, while the regions which resist separation are thought to be relatively GC-rich. Denaturation maps can be constructed showing characteristic locations of early melting regions along the length of the molecules.

Single-stranded DNA is found naturally-occurring in certain bacteriophages (such as ϕX174) which infect *Escherichia coli*. The physical properties of DNA extracted from this phage clearly resemble those of the denatured DNA produced by heat or alkali treatment of ordinary double-helical DNA, and moreover ϕX174 DNA does not contain equivalent amounts of A and T or of G and C—a feature which has proved useful for showing the complementary relationship between template and product in reactions catalysed by the nucleic acid polymerases (see p. 289). DNA from ϕX174 is circular as well as being single-stranded. It is resistant to digestion by exonucleases (enzymes which progressively remove nucleotides from the ends of nucleic acid molecules), suggesting that it has no free ends, and it appears circular when shadowed preparations are viewed in the electron microscope. Immediately after infection of *E. coli* by ϕX174, the entering (+) strand serves as a template for the synthesis of a complementary (−) strand, resulting in a double-helical, circular *replicative form* (RF). This is then replicated to produce a number of RF molecules which serve as templates for the synthesis of progeny (+) strands; these are packaged into newly-synthesized phage heads and finally the cell bursts releasing the progeny phages.

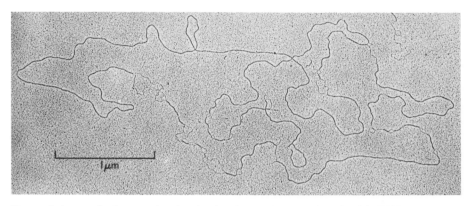

Figure 6. A partially denatured molecule of replicating bacteriophage lambda DNA.

The DNA was extracted from cells of *Escherichia coli* infected with the phage; it shows an apparently circular double helix which has been partly replicated (compare the autoradiograph of the *E. coli* chromosome in Figure 20). Alkali treatment was used to effect partial denaturation, and at several points it can be seen that the bold line of the intact double helix is interrupted by a region where the individual strands have become separated and are visible as two thinner lines. Photograph kindly provided by Dr R. B. Inman.

Renaturation

Although complete denaturation of double-stranded DNA is effectively irreversible under most conditions, re-formation of a true double helix from separated single strands (*renaturation*) can be observed in special circumstances. The problem is to re-align the strands with their complementary base-sequences in register and this can only be achieved by repeated trial-and-error collisions between polynucleotide chains. Conditions must be adjusted so that when a pair of complementary nucleotide sequences happen to meet a 'nucleation' occurs and the helical structure then 'zippers' up along the remaining length of the strands. In practice, the heat-denatured DNA has to be cooled very slowly or held for a long time at a temperature some 25° below the melting temperature of native DNA; under these conditions the reformation of helical structure is energetically favoured once nucleation has occurred, and collisions due to thermal agitation give a reasonable frequency of nucleation. The renaturation reaction proceeds with second-order kinetics, as would be expected for a reaction requiring collision between two complementary species. Of course, the problem of getting complementary nucleotide sequences into register becomes greater as the size and complexity of the strands increases. Because of this a high degree of renaturation has only been observed with DNA from relatively simple organisms such as viruses and bacteria.

STRUCTURE OF RNA

The primary structure of RNA is closely similar to that of DNA; it contains a linear arrangement of ribonucleotides linked through the 3′ and 5′ positions of

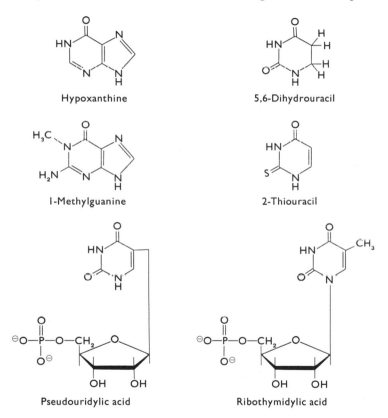

Hypoxanthine

5,6-Dihydrouracil

1-Methylguanine

2-Thiouracil

Pseudouridylic acid

Ribothymidylic acid

Figure 7. Some unusual bases and nucleotides found in RNA.

    A more comprehensive list of 'minor' bases and their derivatives is given by Zachau (*Angewandte Chemie Internat. Ed.* **8**, 711 (1969)). The ribonucleoside and ribonucleotide containing hypoxanthine are known as *inosine* and *inosinic acid* respectively. Note that ribothymidylic acid is the analogue of thymidylic acid (as found in DNA) containing ribose in place of 2′-deoxyribose. The base in pseudouridylic acid is normal uracil joined to the ribose by C_5, instead of by N_1 as in uridylic acid.

the ribose residues (Figure 4, p. 18). As well as having a 2′ hydroxyl group in the sugar, RNA differs from DNA by having uracil in place of thymine. Certain types of RNA, particularly tRNA, also contain a significant proportion of unusual nucleotides such as pseudouridylic acid and ribothymidylic acid (Figure 7). As with DNA, no evidence has been found for the existence of branching in the molecule.

In bacteria, as in other types of organism, three kinds of RNA can be distinguished; ribosomal (rRNA), amino acid transfer (tRNA) and messenger (mRNA). Their secondary structure is not nearly so well defined as that of DNA but rather resembles the structure assumed by denatured, single-stranded DNA, which is consistent with the belief that RNA molecules are single polynucleotide chains. Indeed, the simple fact that most forms of RNA do not have the $A=T$, $G=C$ equality of base-composition shows that they could not be entirely double-helical molecules. However, it is now clear that rRNA and tRNA do take up a partially ordered configuration in solution and that this molecular organization consists of short and possibly imperfect double-helical regions formed by looping of the chain, connected by non-helical stretches of the polynucleotide.

When solutions of RNA are heated the secondary structure of the molecules is disrupted and, as seen with DNA, there is a hyperchromic effect in the ultraviolet absorption. Unlike DNA, however, the increase in UV absorption occurs much more gradually and extends over a wide temperature range showing that the interactions between the bases vary widely in stability to heat. This would be expected if there existed helical regions which had different lengths or imperfections or, very possibly, a mixture of both. The RNA 'melting' curve is more or less fully reversible, for the recovery of native-like secondary structure requires only the reformation of *intra*-strand interactions rather than collision between separate complementary strands in register—cf. renaturation of DNA. Only in special circumstances, e.g. when some tRNA's are heated in the absence of Mg^{2+}, has a sort of irreversible denaturation and loss of biological activity been observed (see below).

Nucleotide sequences

The pioneering work of Sanger, Holley and their colleagues has led to the development of methods for determining the nucleotide sequences of RNA molecules. In the main the approach is analogous to that which proved so successful in determining the amino acid sequences of proteins, though recent work on sequences of bacteriophage RNA molecules synthesized synchronously by purified enzymes *in vitro* has yielded remarkable results. Sanger's methods involve enzymic cleavage of ^{32}P-labelled RNA into small oligonucleotides which are separated, identified, and sequenced by end-analysis, and reconstruction of the starting sequence by study of large fragments produced from the original molecule by restricted cleavage with nucleases. In principle, similar methods could be applied to the determination of sequences in DNA, but the enormously greater size of DNA molecules renders this a daunting prospect.

Transfer RNA's have provided the focus of most attention; they are relatively small (about 80 nucleotides long) and their rare modified bases greatly facilitate identification of overlapping sequences. To date, the sequences of over 30 species have been determined, from higher organisms as well as bacteria. Some examples are shown in Figure 8. A striking feature is the substantial homology between different molecules: not only do all have the 3′ terminal sequence —pCpCpA,* but they also have, with one known exception, a G residue (bearing a 5′ phosphate group) at the 5′ end, and they all possess the same sequence —pGpTpψpCpPu— at about the same position along the length of the molecule. Even more striking homologies become apparent when the sequences are written in the form of a *clover leaf* arrangement (Figure 8) which is widely believed to represent the secondary structure of all tRNA's. Written in this form, the tRNA molecule contains four helical segments: one terminates in the —pCpCpA sequence to which the amino acid becomes esterified (the amino acid arm), the next sustains a fairly large loop in which dihydrouridylic acid residues often occur (the dihydroU loop), the next sustains a loop of seven nucleotides of which the middle three are believed to be the anticodon (the anticodon loop), and the fourth sustains a seven-nucleotide loop starting with the universal —TpψpC— sequence (the TψC loop). The region between the anticodon arm and the TψC arm is very variable between different tRNA's and may characterize each molecule for recognition, e.g. by the amino acid activating enzymes. Each helical region contains a characteristic number of base-pairs, though in some molecules the pairing seems to be imperfect and apparent pairing between G and U (which is theoretically possible if small perturbations of a regular structure are allowed) seems to be necessary. Structure-function relationships of tRNA in the context of its role in protein synthesis will be found in Chapter 6; evidence for its conformation is discussed below.

Determination of the nucleotide sequences of the larger 23 S and 16 S rRNA's presents formidable difficulties, though the sequences of a few fairly large tracts are known. However, the sequence of the smallest ribosomal RNA, 5 S RNA, from *E. coli* has been elucidated (Figure 9). As yet the function of this molecule is obscure, but its structure reveals a curious feature in that there is considerable internal homology of nucleotide sequence which becomes evident if the chain is considered in two parts. Could this internal homology mean that 5 S rRNA is the product of a gene which, during the course of evolution, became duplicated and then the sequences of the two halves diverged by mutational drift?

* See p. 338 for explanation of this notation.

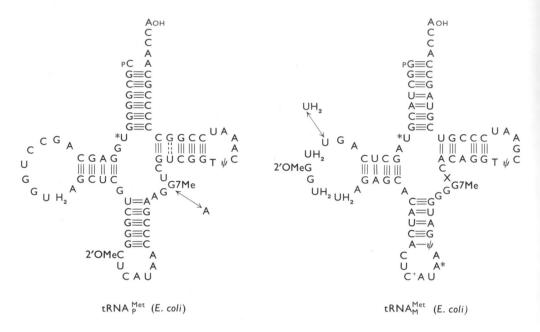

tRNA$_P^{Met}$ (*E. coli*)   tRNA$_M^{Met}$ (*E. coli*)

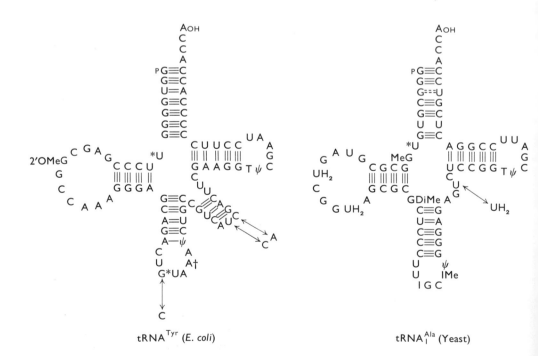

tRNA^{Tyr} (*E. coli*)   tRNA$_I^{Ala}$ (Yeast)

Secondary structure

The probable nature of helical regions in natural cellular RNA's has been deduced mainly from X-ray diffraction studies on two unusual RNA's: first, the completely double-helical RNA which constitutes the genetic material of reovirus, an animal virus, and second, the synthetic double-helical complexes which are formed by mixing equimolar proportions of synthetic homopolymers such as polyriboadenylic acid and polyribouridylic acid (see p. 276). These materials are better suited to X-ray diffraction work because of their long-range regularity of structure which makes them pack nicely into fibres. So far as it goes, the limited information from X-ray diffraction of fragments of natural rRNA yields similar results. It is found that RNA helices are *A-type*; they differ from the DNA helix (*B-type*) in that the base-pairs are tilted and further away from the helix axis, and that there are eleven or twelve base-pairs per turn (Figure 10). These differences apparently result from the presence of the 2′ hydroxyl group on the ribose sugar, which imposes a constraint on the puckering of the sugar ring preventing the RNA ribose-phosphate backbone from adopting the same conformation as the deoxyribose-phosphate backbone of DNA. Significantly, the normal B-type DNA helix can 'click' into an A-type conformation resembling the RNA helix—it does so in fibres of packed DNA molecules when the relative humidity is lowered—which may be important in the transcription of RNA from a DNA template (p. 306).

The clover leaf conformation of tRNA has already been mentioned. Evidence that tRNA's do indeed possess the helical regions defined by the clover leaf comes primarily from work on nucleolytic or chemical attack on the molecule. It is found that nucleotides which would appear to be involved in the helical regions of the clover leaf are relatively resistant to the action of nucleases

---

Figure 8. Nucleotide sequences of some transfer RNA's.

Abbreviations for nucleotides:

A:	adenylic acid
G:	guanylic acid
C:	cytidylic acid
U:	uridylic acid
T:	ribothymidylic acid
I:	inosinic acid
UH_2:	5,6-dihydrouridylic acid
ψ:	pseudouridylic acid

X represents an unknown nucleotide. Special modifications of nucleotides are represented by the substituent or by * or † where the nature of the modification is not certain. The 5′ ends of the sequences bear a phosphate group represented by p, and the 3′ ends are shown A_{OH}. In some cases variant or mutant sequences have been found; here the nucleotide replacements are indicated by a double-headed arrow. The G* → C replacement in tRNA^{Tyr} alters its coding properties such that it acts as an *amber* (UAG) suppressor (see p. 339). From Waring, *Ann. Reports Chem. Soc.* **65B**, 551 (1968).

and various chemical reagents. This is in accord with the known specificity of these reagents: in general they work poorly if at all on nucleotides which are involved in base-pairing or are otherwise 'buried' in the secondary structure. However, the whole story is not quite so simple because it is also found that certain regions, such as the TψC loop, which one would expect to be readily attacked are not. For this and other reasons there is speculation that tRNA

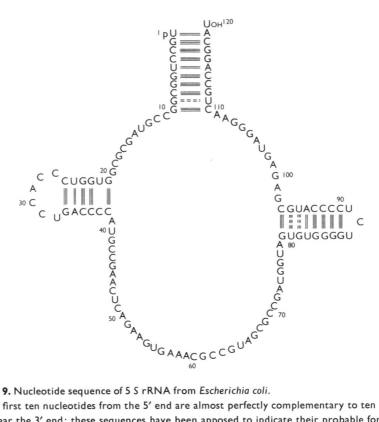

Figure 9. Nucleotide sequence of 5 S rRNA from *Escherichia coli*.

The first ten nucleotides from the 5' end are almost perfectly complementary to ten nucleo-tides near the 3' end: these sequences have been apposed to indicate their probable formation of a helical region, but the actual conformation of the molecule is not known. From Brownlee *et. al., Nature* **215**, 735 (1967).

may possess some additional higher-order tertiary structure formed by inter-actions between the arms of the clover leaf. The existence of further structural forces, perhaps involving interaction with Mg^{++} ions, might account for the 'irreversible' denaturation mentioned earlier. Various theoretical models for the tertiary structure of tRNA can be formulated, but it is likely that definitive evidence will have to wait for X-ray analysis of crystalline tRNA's which have recently been prepared.

One model, for the anticodon loop of tRNA, is of considerable interest

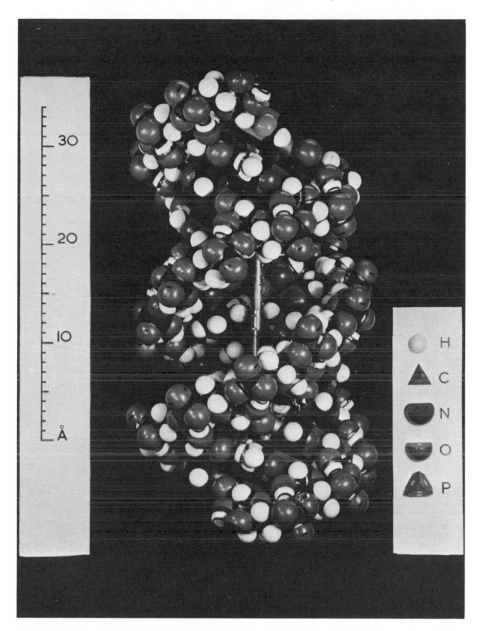

Figure 10. Space-filling model of a helical RNA molecule. Photograph kindly provided by Dr Watson Fuller.

(Figure 11). The five nucleotides on the 3′ side of the loop are stacked on top of the helical stem so as to continue one side of the helix, and the remaining two pyrimidines are loosely arranged to connect the top nucleotide of the stack back to the remaining end of the helical stem. This places the three nucleotides of the

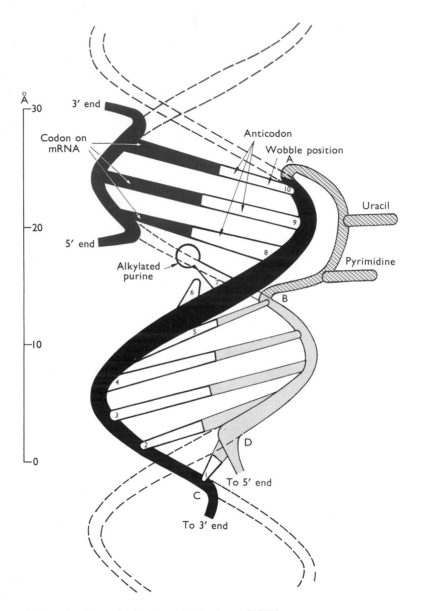

Figure 11. A molecular model for the anticodon loop of tRNA.

The base-pairs of the helical stem of the anticodon arm are arranged in the 11-fold A-type helical conformation and numbered 1–5. Bases 6–10 are stacked on top as if to continue this helix in a regular fashion. Dashed lines indicate the regular geometry of the extrapolated helix. The three bases of the anticodon, at the top of the stack, are able to form regular base-pairs with a mRNA strand whose three-nucleotide codon is positioned as if it formed part of the complementary strand of the continued helix. From Fuller & Hodgson, *Nature* **215**, 817 (1967).

presumed anticodon at the top of the stack, where they can interact *via* hydrogen bonding with a triplet of nucleotides in a mRNA strand as if that strand formed the opposing strand of a regular helix. The nucleotide at the top of the tRNA stack bears the first base of the anticodon, which is known to engage in unusual 'wobble' pairings (see Chapter 6, p. 352) with the third base of the mRNA codon yielding the characteristic pattern of degeneracy in the genetic code (p. 339). Being at the top of the stack, this first anticodon base enjoys special freedom of movement which can account very nicely for its ability to pair with any of two or even three different bases in the third place of the mRNA codon. Although this model for the tRNA anticodon loop is far from proved, its ability to account so well for the degeneracy of the genetic code gives it the ring of truth.

NUCLEIC ACID HYBRIDIZATION

The ability of two polynucleotide chains to interact and form a base-paired helical structure represents a sensitive test for complementarity in their base-sequences, and since the base-sequence of a polynucleotide is the chemical expression of its genetic information content the test becomes a measure of the 'genetic relatedness' of the two strands. The phenomenon of renaturation of DNA, which has already been described (p. 260), is an example of sequence matching between strands which presumably exhibit perfect complementarity; however, the test can also be applied in less extreme cases where sequence complementarity is only partial, the one criterion for successful reaction being that such helical regions as can form must be stable under the experimental conditions. Thus the possibility exists, at least in principle, of comparing the genetic information content of DNA from different sources. Moreover, sequence-matching is not restricted to DNA–DNA interactions but can also be observed between single-stranded DNA and RNA. In this case a DNA–RNA 'hybrid' results the structure of which resembles that of double-stranded DNA except that one strand is a polyribonucleotide (with uracil) and the other is a polydeoxyribonucleotide (with thymine). As discussed earlier, the substitution of uracil for thymine would not be expected to modify the stereochemistry of hydrogen-bonding, so the formation *in vitro* of DNA–RNA hybrids is not surprising on theoretical grounds. Interestingly, the form of a DNA–RNA hybrid helix has been found by X-ray diffraction to be the 11 or 12-fold A-type, which shows that the inability of the ribose sugar ring to assume the puckered conformation needed to give a 10-fold B-type helix dominates and determines the conformation of the deoxyribose-phosphate backbone in the DNA strand.

Analysis of relationships between nucleic acids using these tests of base-sequence complementarity has been immensely valuable in probing the problems concerned in the genetic control of growth and metabolism. Experimental approaches fall into two categories, depending on whether the process of sequence-matching occurs in free solution or with one of the reacting poly-nucleotides (always the DNA strand) immobilized on a support. The first category is exemplified by the density gradient method of Schildkraut, Marmur and Doty, where one of the reacting species of polynucleotide is obtained from organisms grown in the presence of a heavy isotope (often ^{15}N) so that its density is slightly higher than that of the normal polynucleotide. Normal ('light') and labelled ('heavy') nucleic acids can then be distinguished by cen-trifuging at high speed for a long time in concentrated solutions of caesium salts (about 7·7 molal). The centrifugal force tends to sediment the heavy Cs$^+$ ions, but is opposed by the tendency of the salt to diffuse back and maintain uniformity of concentration throughout the system. The net result is that at equilibrium a concentration gradient exists, with the density of the solution in-creasing in proportion to the distance from the centre of rotation. The range of densities covered can be chosen so that each type of nucleic acid (containing ^{14}N or ^{15}N) comes to an equilibrium position in the gradient determined by its characteristic buoyant density. Thus to investigate possible sequence comple-mentarity between two samples of DNA, one of the samples is obtained in a 'heavy' form. It is mixed with an equal quantity of the other 'light' sample, the mixture is heated to denature the native helical structures, the denatured material is incubated under conditions suitable for sequence-matching and 'renaturation', and finally the nature of the products is analysed in a CsCl density gradient. Pairing between complementary strands derived from differ-ent samples is demonstrated if a native-like hybrid fraction containing mole-cules with one ^{15}N strand and one ^{14}N strand is found (Figure 12). In the case of *complete* nucleotide sequence homology between the samples, three bands are observed corresponding to renatured ^{14}N-DNA, the hybrid, and renatured ^{15}N-DNA. They are found in the proportion 1:2:1, which is the proportion expected for random pairing of strands in a mixture containing equal quantities of complementary ^{14}N and ^{15}N strands. Less than 100% homology between the strands of the two samples results in the appearance of a much lower pro-portion of hybrid molecules.

Density-labelled DNA can also be obtained by growing organisms in a medium containing 5-bromodeoxyuridine (BUDR) which efficiently substitutes for thymidine in DNA synthesis, leading to the formation of 'heavy' molecules containing bromouracil (BU) in place of thymine. The density gradient tech-nique is also applicable to the study of base-sequence complementarity between DNA and RNA. In this case density-labelling is not necessary because the buoyant density of RNA in CsCl is considerably higher than that of DNA, so the DNA–RNA hybrid is readily resolved. The samples of DNA and RNA for analysis are mixed, heated and 'annealed' to form the hybrid under much the same conditions as are used for renaturation of DNA, and then RNAase is added to digest away unhybridized RNA. RNAase does not attack RNA if it is hydrogen-bonded to another polynucleotide, whether DNA or RNA.

DNA–DNA or DNA–RNA hybrids formed in free solution may also be separated by chromatography on hydroxyapatite columns. This method has the advantage that it is readily applicable on the preparative scale, so that relatively large quantities of hybrids may be obtained. In this way, hybrids between

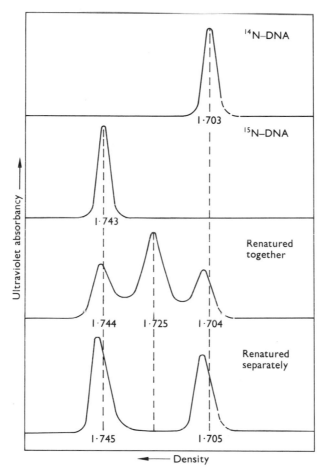

Figure 12. Renaturation between homologous samples of DNA, one density-labelled and the other unlabelled.

The top two curves show the characteristic banding patterns of double-helical DNA from *Bacillus subtilis*, unlabelled and density-labelled. The lowest curve shows the result obtained when samples of each type of DNA are renatured separately and mixed just prior to centrifugation. The curve next to the bottom shows the 1:2:1 banding pattern which results when the two DNA's are renatured together. Renatured samples must be treated before centrifugation with a phosphodiesterase which specifically attacks single-stranded DNA and thus removes unreacted denatured DNA. After Schildkraut *et al.*, *J. Mol. Biol.* **3**, 595 (1961).

rRNA and DNA, or tRNA and DNA, have been fractionated and purified by several successive cycles of hybridization between the appropriate purified RNA and denatured DNA followed by hydroxyapatite chromatography. The products constitute essentially pure preparations of the bacterial DNA cistrons which code for rRNA's or for tRNA.

A third technique, applicable to the detection and estimation of DNA–RNA hybrids, consists simply of filtering the solution containing the hybrids through nitrocellulose membrane filters. These filters have the curious property that they absorb denatured DNA and DNA–RNA hybrids, but not RNA or native (or renatured) DNA. Nowadays nitrocellulose filters are mainly used in a different way for hybridization experiments (see below), but recently they have proved invaluable for yet another remarkable property, namely that they also absorb the *lac* repressor (a protein) which binds to DNA containing the *lac* operon, so that labelled DNA which has formed a complex with the repressor is retained by the filter while free, uncomplexed native DNA passes through. This provides a rapid and simple means of measuring repressor–operator gene interaction (see p.451).

A natural disadvantage of all free-solution hybridization reactions is that renaturation of the separated DNA strands competes with the hybridization reaction. For this reason, most hybridization work is performed using our second category of methods where the denatured DNA is trapped on some supporting material. The support can be an agar gel or, better, nitrocellulose filters. Denatured DNA can be baked on such filters which are then incubated with labelled RNA, treated with RNAase, and washed. It only remains to count the filters to assess the extent of hybrid formation. Only slight modification is needed to enable the same procedure to be used to measure hybridization with labelled denatured DNA. Generally, comparison between the nucleotide sequences of two species of polynucleotides is effected by a process of competition, e.g. by using reaction mixtures containing fixed amounts of denatured DNA on filters and of one species of RNA radioactively labelled, and including varying amounts of the other RNA in unlabelled form. To the extent that the nucleotide sequences of the 'cold' competitor RNA resemble those of the 'hot' RNA they will successfully compete with the 'hot' RNA for binding to the DNA and thus reduce the amount of radioactivity which becomes associated with the filters.

Nucleic acid hybridization has proved a powerful tool for investigating genetic relatedness among microbial species, relationships between bacteriophages and their hosts, and the mechanisms by which the bacterial chromosome is replicated and directs the activities of the cell through the synthesis of RNA. More will be said about these experiments in later sections of this chapter and in Chapter 7 and Appendix A.

It has also made possible two notable advances in the study of gene structure at the level of the DNA molecule: (1) direct mapping of deletion mutations along the length of the molecule, and (2) isolation of the *lac* operon DNA of *E. coli*.

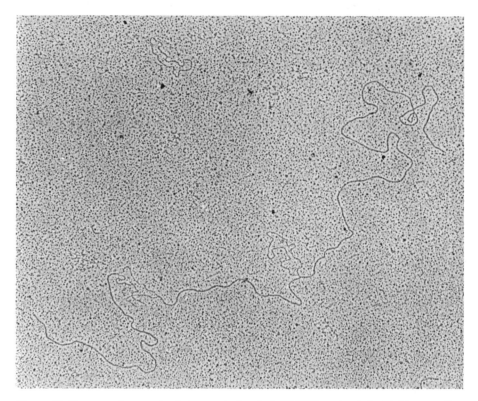

Figure 13. Electron micrograph of a heteroduplex (hybrid) DNA molecule formed by annealing denatured wild-type λ DNA with denatured DNA from a double mutant, λb2b5. The b2 mutation is a deletion of slightly more than one-tenth of the total nucleotide sequence, revealed here as a loop located approximately half way along. The b5 mutation is a substitution in which a sequence present in the wild-type has been lost and replaced by a shorter, unrelated sequence: it is visible here as a non-helical region near the bottom right-hand end of the photograph. Note the difference in lengths of the mutant and wild-type single strands. At the top of the picture is a small circular single-stranded DNA molecule (φX174 DNA) included to act as a length calibration; its circumference is 1·9 μm. Photograph by Dr M. Wu kindly provided by Prof. N. Davidson.

(1) If DNA's from a deletion mutant and the corresponding wild-type organism are denatured and annealed together, the resulting hybrid molecules are visibly shorter than the wild-type DNA in the electron microscope and a 'loop' is visible at the position of the deletion (Figure 13). The 'loop' is formed by the sequence in the wild-type strand which is missing in the strand from the mutant.

(2) Beckwith and his colleagues isolated two specialized transducing phages which had incorporated the *E. coli lac* operon into their DNA's in opposite

orientations. DNA from each phage was denatured and the complementary H and L strands separated by complexing with synthetic poly U,G (for the basis of this separation see p. 304). The H strands from each phage, which were complementary for the genes of the *lac* operon but not for the phage genes, were annealed to allow the *lac* sequences to interact and form a double helix, and the four non-complementary single-stranded tails of the helical section were finally digested away with a single-stranded DNA-specific nuclease (Figure 14). The

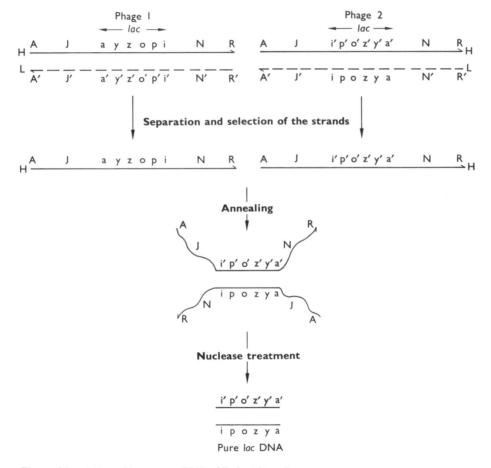

Figure 14. Isolation of *lac* operon DNA of *Escherichia coli*.

  DNA molecules which constitute the genomes of two bacteriophages, A and B, are shown. Each contains the *lac* operon, but the orientation of the *lac* region is opposite in the two genomes as shown by the order of the genetic markers. A, J, N and R are markers on the phage chromosome. The *lac* markers are the repressor structural gene (*i*), the promoter (*p*), the operator (*o*), the β-galactosidase structural gene (*z*), the *lac* permease structural gene (*y*), and the galactoside transacetylase structural gene (*a*). Complementary sequences are indicated by primes. The 5′ end of each DNA strand is indicated by an arrowhead, and H and L designate the complementary strands. From Shapiro *et al.*, *Nature* **224**, 768 (1969).

remaining material was a duplex DNA containing most of the genes of the *E. coli lac* operon.

Synthetic polynucleotides

From what has already been said it will be appreciated that synthetic poly-nucleotides have contributed significantly to our understanding of the structure and function of natural nucleic acids. Outstanding, however, is the role they have played in the elucidation of the genetic code (p. 338). The first synthetic polynucleotides to be made were homopolymers—essentially RNA molecules containing only a single type of nucleotide. These are readily synthesized *in*

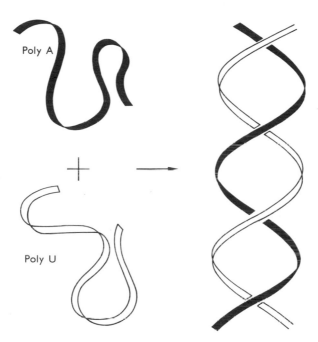

Figure 15. Schematic drawing of the combination of poly A and poly U to form a double helix.

vitro by polynucleotide phosphorylase, an enzyme which is abundant in many species of bacteria but whose role *in vivo* is obscure. It catalyses the following reaction:

$$nXDP \rightleftharpoons (XMP)_n + n \, \text{\textcircled{P}}$$

X can be a nucleoside containing adenine, guanine, uracil or cytosine—or indeed hypoxanthine or other bases. The products are polyadenylic acid, etc. (poly A, poly G, poly U and poly C) with IDP, the nucleoside diphosphate containing the base hypoxanthine, giving polyinosinic acid (poly I). Poly I is structurally analogous to poly G since hypoxanthine has a 6-hydroxy group although it lacks the 2-amino group of guanine. Because poly G is difficult to prepare, poly I is often used instead in model systems. When more than one nucleoside diphosphate is present the product of polynucleotide phosphorylase action is a *copolymer* with a random sequence of nucleotides.

Double-stranded helical structures form when 'equimolar' solutions of poly A and poly U, or of poly I and poly C, are mixed. 'Equimolar' in this case means solutions which contain equal quantities of the corresponding nucleotides, rather than equal numbers of polymeric molecules which may be of different sizes. In effect, the complementary polynucleotides simply wrap round each other in a helical configuration as illustrated in Figure 15. Here there is no question of aligning the molecules so that uniquely complementary sequences come into register, and the intermolecular reaction proceeds very rapidly and efficiently at room temperature. The structures formed have been characterized as hydrogen-bonded double helices by the usual criteria, including X-ray diffraction patterns and thermal denaturation behaviour showing a sharp melting temperature at which a large hyperchromic effect occurs as the strands separate. Heat-denaturation of these helical structures is, of course, readily and completely reversible. The evidence is consistent with the supposition that the strands are associated by Watson–Crick base-pairing. It should be noted, however, that the cytosine-hypoxanthine base-pair can only involve two hydrogen bonds, like the A=U pairing, whereas the G≡C pairing can form three. This correlates with the observed thermal stabilities of the homopolymer mixtures. The $(A+U)$ and $(I+C)$ helices break down at about the same temperature, while the $(G+C)$ ribohomopolymer helix only dissociates at a much higher temperature. These observations suggest that the known increase in melting temperature of DNA (and, so far as can be judged, of RNA) as the GC content rises can be at least partially explained by the capacity of the GC base-pairs to form three hydrogen bonds.

More recently, small oligodeoxyribonucleotides have been prepared by chemical synthesis, and it turns out that they act as excellent templates for reiterative copying by DNA and RNA polymerases (see below). By this means an impressive array of synthetic polynucleotides has been prepared, including some which contain repeating di, tri- or tetra-nucleotide sequences, both deoxyribo- and ribo-. These, too, have proved invaluable in checking the genetic code (p. 340). A prodigious extension of this approach has resulted in the total synthesis, in Khorana's laboratory, of the gene for a tRNA. This is a DNA double helix 77 base-pairs long whose nucleotide sequence corresponds precisely to that determined by Holley for the alanine tRNA of yeast. It was assembled (Figure 16) by synthesizing pairs of complementary oligonucleotides which formed small helical sections with protruding ends; each end was made complementary to that of the next section of the molecule such that the sections could associate by hydrogen bonding, and then the gaps in the chains were sealed by the polynucleotide ligase enzyme (see below). The availability of this synthetic gene, as well as the isolated *lac* operon referred to above, opens up exciting possibilities for the study of gene transcription and translation *in vitro*.

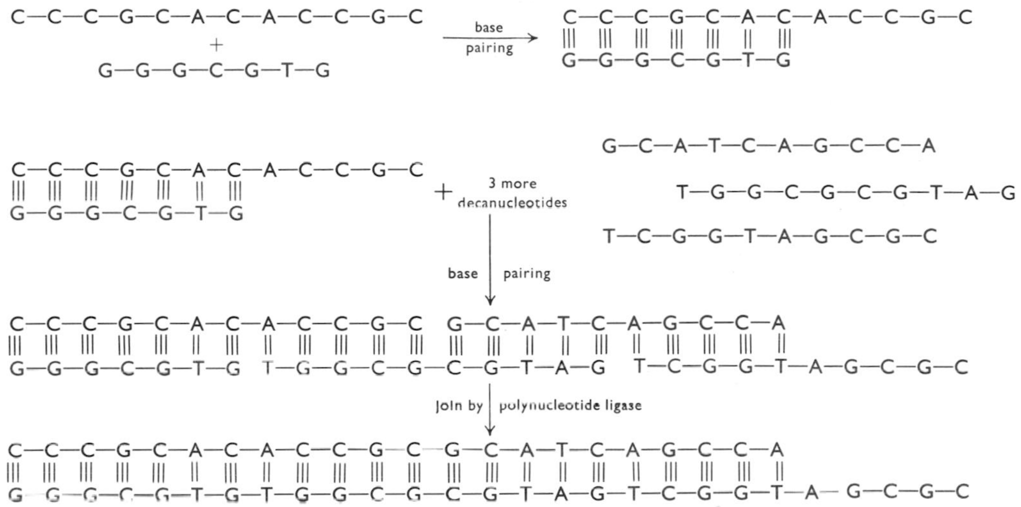

C—C—C—G—C—A—C—A—C—C—G—C
        +
    G—G—G—C—G—T—G

$\xrightarrow{\text{base pairing}}$

C—C—C—G—C—A—C—A—C—C—G—C
||| ||| ||| ||| ||| || |||
G—G—G—C—G—T—G

G—C—A—T—C—A—G—C—C—A

C—C—C—G—C—A—C—A—C—C—G—C
||| ||| ||| ||| ||| || |||
G—G—G—C—G—T—G

+ 3 more decanucleotides

T—G—G—C—G—C—G—T—A—G

T—C—G—G—T—A—G—C—G—C

base | pairing ↓

C—C—C—G—C—A—C—A—C—C—G—C G—C—A—T—C—A—G—C—C—A
||| ||| ||| ||| ||| || ||| || ||| ||| ||| ||| ||| ||| || || ||| || ||| ||| ||| ||
G—G—G—C—G—T—G T—G—G—C—G—C—G—T—A—G T—C—G—G—T—A—G—C—G—C

Join by | polynucleotide ligase ↓

C—C—C—G—C—A—C—A—C—C—G—C—G—C—A—T—C—A—G—C—C—A
||| ||| ||| ||| ||| || ||| || ||| ||| ||| ||| ||| ||| || || ||| || ||| ||| ||| ||
G—G—G—C—G—T—G—T—G—G—C—G—C—G—T—A—G—T—C—G—G—T—A—G—C—G—C

Figure 16. Assembly of a portion of the synthetic gene for yeast alanine tRNA. After Agarwal *et al., Nature* **227**, 27 (1970).

SYNTHESIS OF DNA

The genetic material of a cell has two primary functions; (i) it must be able to undergo accurate self-replication, and (ii) it must be able to provide information for the synthesis of the determinants of cell structure and function, i.e. the proteins. The physical basis for both these properties is now believed to depend upon the capacity of uniquely ordered sequences of bases in DNA to act as templates for polynucleotide synthesis. We have already seen how the phenomenon of complementary base-pairing enables DNA molecules to assume a rigidly defined, highly stable secondary structure and we can now consider how the same base-pairing mechanisms underlie the processes of replication and transcription of the coded nucleotide sequence.

The scheme of DNA replication outlined on pp. 37–9 is now almost universally accepted. Because the two strands of the original double helix become separated and are subsequently incorporated into different daughter double helices the mechanism is called *semiconservative* to distinguish it from a fully *conservative* mechanism where both strands of one duplex would be freshly synthesized and the other duplex would consist of the old strands associated together in their original state. The principal evidence which demonstrates the semiconservative mode of DNA synthesis in growing bacteria comes from the classic experiment of Meselson and Stahl in 1958. These workers grew *Escherichia coli* in a medium containing $^{15}NH_4Cl$ as the sole source of nitrogen; after 14 generations of growth in this medium the nitrogen of the bases in the DNA of the organisms was virtually all ^{15}N so that this DNA was 'heavy' and could be separated from normal ^{14}N-containing ('light') DNA in a CsCl density gradient. The organisms were then abruptly changed to a medium containing ^{14}N and samples containing equal numbers of bacteria were removed at intervals

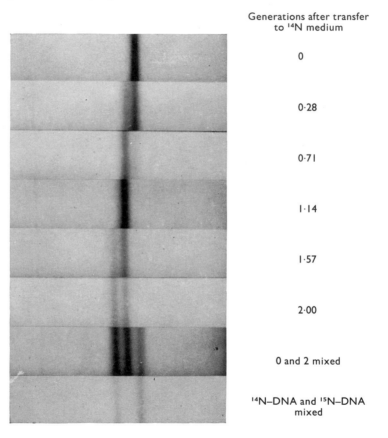

Generations after transfer
to ^{14}N medium

0

0·28

0·71

1·14

1·57

2·00

0 and 2 mixed

$^{14}N–DNA$ and $^{15}N–DNA$
mixed

Figure 17. CsCl density gradient analysis of *Escherichia coli* DNA after transfer of ^{15}N-labelled organisms to ^{14}N-medium.

The photographs, taken with ultraviolet light, reveal the positions of the DNA molecules in the ultracentrifuge cell as dark bands. The density increases from left to right. Photographs kindly provided by Dr M. Meselson.

over the next few generations. The bacteria in each sample were lysed by addition of detergent and the crude lysate was centrifuged in concentrated CsCl until the DNA had banded at its characteristic density (Figure 17). After a short period of growth in the new medium the band formed by the $^{15}N–DNA$ of the

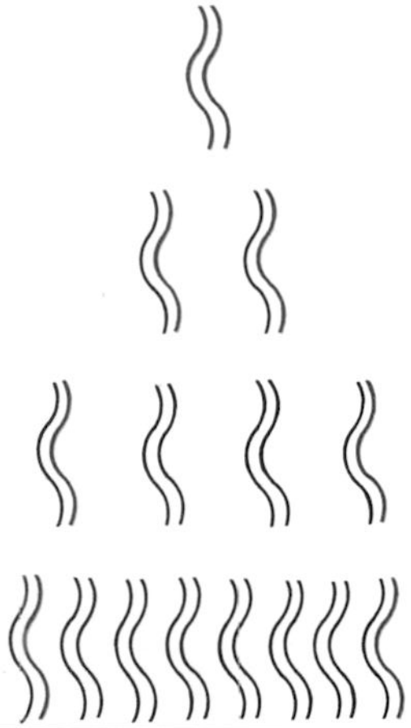

Semiconservative

Generation	^{15}N	$^{15}N/^{14}N$ hybrid	^{14}N
0	100	0	0
1	0	100	0
2	0	50	50
3	0	25	75

Observed

^{15}N	$^{15}N/^{14}N$ hybrid	^{14}N
100	0	0
0	100	0
0	50	50
0	(25)	(75)

Conservative

^{15}N	$^{15}N/^{14}N$ hybrid	^{14}N	Generation
100	0	0	0
50	0	50	1
25	0	75	2
$12\frac{1}{2}$	0	$87\frac{1}{2}$	3

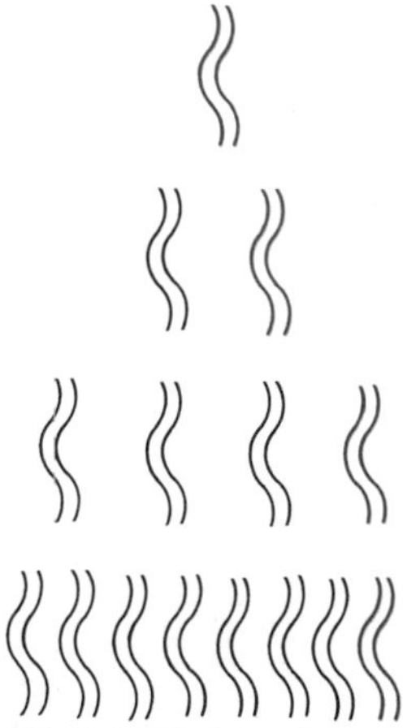

Figure 18. Semiconservative and conservative mechanisms of DNA replication. The table on the left lists the predicted results of the Meselson–Stahl experiment according to the semiconservative mechanism, while that on the right shows the predictions of a hypothetical conservative process. The centre table contains the experimental results (*cf.* Figure 17).

organisms began to diminish and a new band appeared at a density exactly half-way between that of ^{15}N–DNA and that of ^{14}N–DNA. This new band, formed by DNA molecules containing equal proportions of ^{15}N and ^{14}N, became stronger until after one complete generation it was the only form of DNA present. During the next generation a band appeared corresponding to DNA containing ^{14}N only; this band increased and the hybrid band decreased until at the end of the second generation equal quantities of the two types were present. During subsequent generations the ^{14}N–DNA band continued to increase in relation to the hybrid band.

The results are in exact accord with the predictions of a semiconservative replication mechanism, as illustrated in Figure 18. The appearance of a band corresponding to hybrid molecules is itself strong evidence against a fully con-servative mechanism, and the quantitative agreement between predicted and observed values for the proportions of the various bands establishes the semi-conservative mechanism in a most convincing fashion. The constitution of the hybrid band was investigated by heating this DNA beyond the 'melting tem-perature' so that the individual strands were separated, and analysing the result-ing denatured DNA again in a CsCl density gradient. Two bands were observed, having the densities of denatured ^{15}N–DNA and denatured ^{14}N–DNA, and they were found in equal proportions. Thus the identification of the original hybrid band as DNA containing ^{15}N strands and ^{14}N strands was confirmed, but to prove that the hybrid material consisted of molecules with one ^{15}N strand hydrogen bonded to a complementary ^{14}N strand it was necessary to show that the subunits produced by denaturation had half the molecular weight of the intact molecule.

The mechanism of semiconservative replication

For the two strands of a parent double helix to appear as components of the daughter double helices it is necessary that the original strands become separa-ted either before or during the replication process. Since strand separation requires an expenditure of energy (shown by the need for a high temperature to provide enough thermal energy to dissociate the strands of DNA *in vitro*), it is difficult to imagine how the DNA strands could unwind *in vivo* independently of some energy-producing reaction. Moreover, there has never been any evi-dence to suggest that much of the DNA in a bacterium does exist in a single-stranded form. A much more likely mechanism is that depicted in Figure 19 where strand separation and synthesis of new DNA proceed concurrently; in this case the energy requirement for separating the strands is compensated by the formation of new base-paired structures and the overall energetics of the process can be considered in terms of the synthesis of relatively 'low-energy' phosphodiester linkages at the expense of 'high-energy' pyrophosphate bonds in the nucleoside triphosphates. From the standpoint of energy relationships this seems a much more feasible scheme and it also explains why only an imperceptibly small fraction of the total DNA need exist in a single-stranded form during active DNA replication.

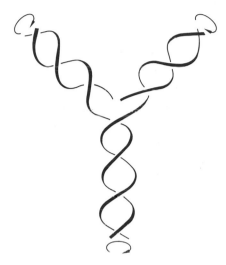

Figure 19. Replication of double-stranded DNA. From C. Levinthal and R. K. Crane, *Proc. Nat. Acad. Sci. U.S.* **42**, 436 (1956).

The synthesis of DNA at a growing-point which moves along the chromosome forming a Y-shaped structure still presents certain formidable problems, however. Because the double helix is a *coiled* structure it means that there must be physical rotation of the unreplicated portion relative to the new daughter helices (Figure 19), and while the uncoiling of the parent helix can be coupled to the coiling of the daughter helices there must still be a considerable viscous drag opposing the rotation of such enormous molecules. This problem seems all the more complex when one considers that the DNA of *E. coli* appears to be a single molecule 1·1–1·4 mm long, that it is circular, and that it is packed into a cell about 0·5 μm wide × 2 μm long. If very large DNA molecules contained occasional single-strand breaks or non-nucleotide 'linkers' providing free rotation at well-spaced points in the molecule this difficulty would be overcome but as yet there is no compelling evidence for the existence of points of free rotation in bacterial DNA. As for the problems raised by the circular structure of DNA, autoradiographic studies by Cairns on the DNA of *E. coli* have shown how a circular chromosome can be replicated. Cairns' experiment also shows that the entire chromosome can be replicated from a single growing point. A thymine-requiring strain of *E. coli* was grown in ^{3}H-thymidine for about two generations; the labelled DNA was then extracted extremely gently, embedded, and allowed to expose a film as described earlier (p. 254). After the first generation the DNA of the organisms had one strand radioactive and the other not; during the next cycle of growth these 'hybrid' duplexes were replicated in a semiconservative fashion to yield one daughter helix with both strands labelled and the other with only one strand labelled. The half-labelled and doubly-labelled duplexes were distinguishable by autoradiography because the density of silver grains produced by a given length of the former was only half that produced by an equal length of the latter. An example of the structure resulting from this treatment, and a representation of the suggested replication mechanism, are illustrated in Figure 20.

The first point to note is that this experiment confirms the semiconservative replication mechanism (since strands with two-fold differences in density of labelling are evident) without fragmentation into much smaller pieces as was necessary in the Meselson–Stahl experiment. Secondly, it appears that the daughter duplexes remain attached at the point where duplication commenced until the replication cycle is complete. This junction point is believed to act as a swivel about which the unreplicated stretch of the parent helix can rotate. A further possibility is that the swivel might itself divide when replication is

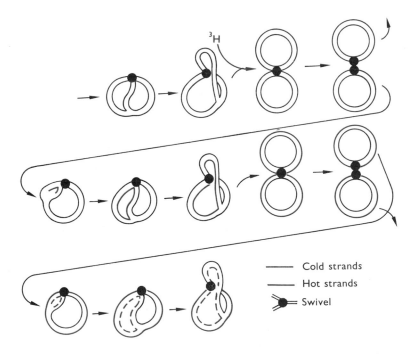

Figure 20. (a) Replication of the circular chromosome of *Escherichia coli* studied by autoradiography.

A representation of the replication mechanism, based on the assumption that each round of replication begins at the same place (the swivel point, represented by ●) and proceeds in the same direction.

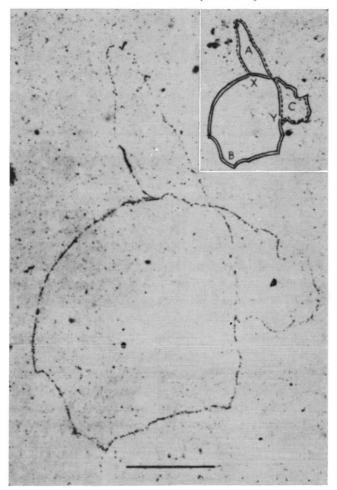

Figure 20. (b) Autoradiograph of the chromosome of *Escherichia coli* K 12 Hfr labelled with
³H-thymidine for two generations. The scale shows 100 μm. *Inset*, the same structure represented
diagrammatically (labelled strands ——, unlabelled strands ————) and divided into three sections
(A, B and C) which arise at the two forks X and Y. About two-thirds of the chromosome (XBY)
has been replicated. Part of the still unreplicated section (Y to C) has one strand labelled but the
rest (C to X) has both labelled. This happened because the labelled thymidine was not added
exactly at the beginning of a round of replication. One can deduce therefore that X is the starting
and finishing point of duplication (the swivel) and that Y is the growing point.

   Photographs kindly provided by Dr J. Cairns.

complete, and perhaps facilitate removal of the daughter chromosomes to
regions of the cell destined to form each of the progeny.

Growth of DNA chains at the replicating fork

The circular *E. coli* chromosome shown in Figure 20 has only a single growing-
point but it is not *a priori* necessary that there be only one such replicating fork

in any cell at a given time. Since DNA synthesis in growing *E. coli* appears to proceed continuously throughout the cell cycle it is pertinent to ask how the rate of DNA synthesis is geared to different rates of growth and cell division. One could visualize that some mechanism might adapt the rate of DNA replication at a single growing-point to match the rate of cell division (or *vice versa*) or, alternatively, that the rate of replication at a growing-point remain constant and the *number* of growing-points be varied. Experiments described in Chapter 2 show that the latter explanation holds true in *E. coli*. Once initiated, a replicating fork traverses the length of the chromosome in 40 min and cell doubling times shorter than this are made possible by the timed appearance of fresh growing points. With this information we can calculate the rate of growth *in vivo* of daughter DNA chains at the replicating fork. The molecular weight of a non-replicating chromosome has been found to be $2 \cdot 5 \times 10^9$, in agreement with Cairns' measured length of $1 \cdot 1 – 1 \cdot 4$ mm (Figure 20), which corresponds to 8×10^6 nucleotides. If a single fork traverses this in 40 min, the rate of incorporation of new nucleotides is 2×10^5 per min, or 3300 per sec. Since two new chains are growing at each fork, the rate of growth of each chain is about 1700 nucleotides per second. Incidentally, by assuming that the DNA helix has ten base-pairs per turn we can also estimate the rate of unwinding of the helix required by the mechanism depicted in Figure 19, p. 281. It is close to 10,000 rev/min.

The complementary strands of the DNA helix are antiparallel, i.e. their $3'$ to $5'$ sugar-phosphate linkages run in opposite directions (Figures 2 and 3). Consequently, simultaneous replication of both strands at a single growing-point moving along the molecule in one *sidereal* direction must mean that the two daughter strands are growing in different *chemical* directions. One is being extended at its $3'$ end while the other is growing at its $5'$ end (Figure 21 (a)). As will be seen, there is abundant evidence for growth of polynucleotide chains in the $5' \rightarrow 3'$ direction, both *in vivo* and with enzymic systems *in vitro*, but sequential addition of nucleotides to the $5'$ end of a pre-existing chain (giving growth in a $3' \rightarrow 5'$ direction) has never been observed. Attempts to account for $3' \rightarrow 5'$ synthesis by invoking a role for nucleoside $3'$-triphosphates have proved groundless. This appeared to be a central paradox of DNA synthesis: one of the newly synthesized strands seemed to be growing at its $5'$ end, yet no mechanism for adding nucleotides to this end could be detected.

The key to this problem was provided by the discovery of Okazaki and his colleagues that a substantial proportion of newly synthesized DNA in T4-phage-infected cells occurs as small pieces about 1000 to 2000 nucleotides in length. Moreover, it could be shown that these small pieces, which have come to be known as Okazaki fragments, are synthesized in the well-established $5' \rightarrow 3'$ direction. The direction of synthesis was established by pulse-labelling cells with ^{3}H-thymidine for very short periods (as little as 2 sec), usually at low temperatures to slow down the rate of incorporation of nucleotides, then isolating the nascent DNA fragments as single strands by centrifugation in an alkaline sucrose gradient, and treating them with exonucleases. Two enzymes were used; *E. coli* exonuclease I which progressively removes nucleotides from the 3′

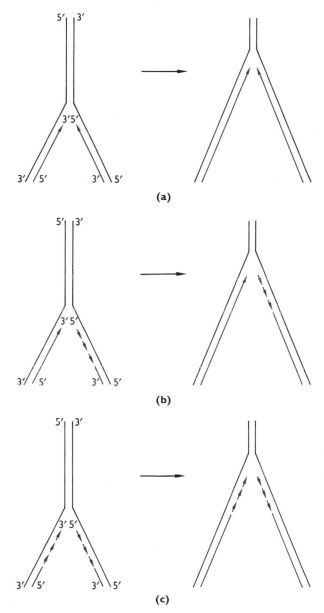

(a)

(b)

(c)

Figure 21. Possible mechanisms of elongation of DNA chains at the replicating fork.
  (a) By direct addition of nucleotides to the growing ends of both daughter strands. (b) By discontinuous synthesis of one strand by joining of fragments. (c) By discontinuous synthesis of both daughter strands.

end of single-stranded DNA, and a nuclease from *Bacillus subtilis* which attacks
in a similar fashion from the 5' end (Figure 22). It was found that release of the
newly incorporated ³H-labelled nucleotides preceded release of the bulk of the
nucleotides when *E. coli* exonuclease was used, and followed it when *B. subtilis*
nuclease was used. Thus most, if not all of the DNA chains were synthesized in
the 5' → 3' direction. Similar results were obtained with pulse-labelled DNA
fragments isolated from uninfected bacteria.

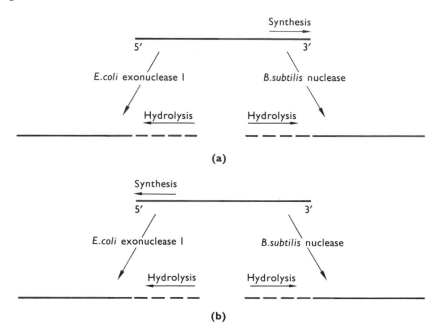

Figure 22. Analysis of 'Okazaki fragments' to determine the direction of synthesis.
    An 'Okazaki fragment' is shown half-labelled by a short pulse of a radioactive precursor. The
black stretch is the portion of the fragment present before addition of the label; the red stretch
is the radioactive portion synthesized after the labelled precursor was added. In (a) the fragment
was synthesized in the 5' → 3' direction; in (b) it was produced by (hypothetical) 3' → 5'
synthesis. The actual direction of synthesis may be established by comparing the rates of release of
radioactive nucleotides by digestion with exonucleases of known end-specificity, as shown.

Okazaki postulated that these fragments could explain the apparently
anomalous synthesis of a daughter strand in the 3' → 5' direction if this
strand were synthesized by a *discontinuous* mechanism (Figure 21 (b)), involving 5' → 3' synthesis of fragments which could subsequently be joined together
by the polynucleotide ligase enzyme described below. In support of this notion
he showed that radioactive nucleotides originally incorporated into the fragments became incorporated into much longer DNA strands if replication was
allowed to continue.
    Other workers have confirmed the finding of newly-synthesized DNA in

small fragments, though in many cases it appears that substantially more than 50% of the newly incorporated nucleotides appear in this form after short pulses, which would not be expected from the mechanism depicted in Figure 21 (b) where one strand grows continuously in the 5′ → 3′ direction. It has been suggested therefore that *both* strands may be synthesized by a discontinuous mechanism as illustrated in Figure 21 (c). Discontinuous synthesis of one or both daughter strands can be distinguished by determining whether the Okazaki fragments are complementary to only one or both of the parental strands. With bacteriophage T4 DNA the complementary strands can be separated and isolated by the synthetic polynucleotide binding technique (p. 305), and it has been found that pulse-labelled Okazaki fragments from T4-infected cells hybridize equally well with either strand of T4 DNA, indicating that the fragments contain sequences complementary to both strands and supporting mechanism (c). With uninfected bacteria the position is less clear-cut; relatively long strands as well as Okazaki fragments have been reported after short pulses of labelled precursors, which would tend to favour the model shown in Figure 21 (b) involving discontinuous synthesis of only one daughter strand.

Polynucleotide ligase

A mechanism for joining DNA strands together is not only required for replication, but also for processes involved in genetic recombination (p. 390), the repair of DNA molecules (p. 297), and a number of other phenomena including

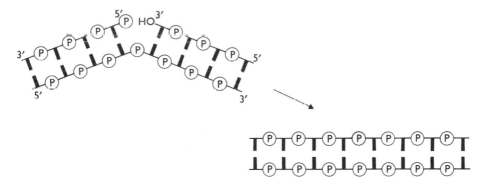

Figure 23. Joining of polydeoxyribonucleotide strands by the action of polynucleotide ligase.

interaction of the bacterial chromosome with prophages, sex factors, etc. (Chapter 7). It may be premature to attribute all these joining reactions to a single enzymic activity but such an enzyme has been discovered in *E. coli* and in T4 phage-infected *E. coli*. Both enzymes catalyse the same reaction: formation of a regular 3′–5′ phosphodiester linkage between two DNA strands, one bearing a 5′-phosphate and the other a free 3′-hydroxyl group. The two strands must be exactly adjacent and hydrogen-bonded to a third, continuous DNA strand (Figure 23).

The reaction is highly specific—it will not work for instance with a 3'-phosphate and a free 5'-hydroxyl—and a cofactor is needed: ATP for the T4-induced enzyme, and NAD for the *E. coli* enzyme. In both cases an enzyme-AMP intermediate is involved. The substrate in Figure 23 is in effect a 'nicked' double-helical DNA, for a break or 'nick' like that illustrated is produced by a single cleavage with an endonuclease; such nicks are readily repaired by the ligase.

The involvement of polynucleotide ligase in DNA replication has been neatly demonstrated using a temperature-sensitive mutant of T4 which produces a ligase active at 20° but inactive at 43–44°. Cells were infected with the mutant, or with wild-type T4, and incubated at 20° until phage DNA synthesis was well under way. The temperature was then raised to the non-permissive level of 43–44°, where the mutant-infected cells were found to accumulate large amounts of Okazaki fragments while, in the wild-type infected cells at the elevated temperature, the fragments were only seen with short pulses and rapidly became transferred into longer strands. Only on cooling to the lower temperature, where the temperature-sensitive ligase recovered its activity, did the Okazaki fragments accumulated by the mutant disappear.

DNA polymerase

Enzymes which catalyse DNA synthesis *in vitro* have been isolated from a variety of sources including bacteria. Attention will be restricted here to the DNA polymerase from *E. coli*, which has been most thoroughly investigated. Its substrates are deoxyribonucleoside 5'-triphosphates and the reaction is conveniently followed by measuring the incorporation of radioactivity from a labelled triphosphate into acid-insoluble material; e.g.

$$
\begin{array}{l} n_1\ dATP \\ n_2\ dGTP \\ n_3\ d*CTP \\ n_4\ dTTP \end{array} \xrightarrow[Mg^{2+}]{DNA} \left[\begin{array}{l} n_1\ dAMP \\ n_2\ dGMP \\ n_3\ d*CMP \\ n_4\ dTMP \end{array} \right] + (n_1 + n_2 + n_3 + n_4)\ \textcircled{P}\!-\!\textcircled{P}
$$

$$
\begin{array}{cc} \text{(acid-} & \text{DNA} \\ \text{soluble)} & \text{(acid-insoluble)} \end{array}
$$

* Indicates the presence of radioactive isotope in some part of the dCTP molecule other than the two terminal phosphoryl groupings.

The reaction is completely dependent upon the presence of Mg^{++} and preformed DNA as well as the substrates. However, under artificial conditions in the absence of DNA two curious reactions can occur after a long lag period: with dATP and dTTP alone the polymerase catalyses *de novo* synthesis of a high-molecular weight dAT copolymer containing dAMP and dTMP residues in perfectly alternating sequence, while with dGTP and dCTP alone it synthesizes homopolymeric poly dG and poly dC. These synthetic products are quite useful, but their origin seems to have little to do with the normal DNA-dependent reaction.

Table 2. DNA polymerase: relationship between the base-compositions of template and product.

	Proportion of base, moles %			
	Adenine	Thymine	Guanine	Cytosine
Escherichia coli DNA				
template	25	25	25	25
product	26	25	24	25
Bacteriophage T2 DNA				
template	33	33	17	18
product	33	32	17	18
Synthetic dAT copolymer				
product	50	48	<1	<1
Bacteriophage ϕX174 DNA				
template	25	33	24	19
product (limited synthesis)	31	24	20	25
product (extensive synthesis)	27	29	21	22

Two sets of results are shown for reactions with a ϕX174 DNA template. In the limited synthesis reaction the amount of product formed was only one-fifth the amount of ϕX174 DNA added. In the extensive synthesis reaction the amount of product was six-fold greater than the amount of template.

The product of the normal polymerase reaction is high-molecular weight DNA whose properties reveal that the pre-existing DNA acts as a template for its synthesis. DNA from any source, natural or synthetic, will act as a template, and the complementarity of the product can be seen in its base-composition (Table 2). With single-stranded ϕX174 DNA as template, which does not show the usual A=T, G=C equality, the product of limited synthesis has a clearly complementary base-composition; but if synthesis is allowed to proceed it can be seen that the total base-composition of the product tends towards new values with A=T and G=C showing that the enzymically synthesized complement of the original strand is also able to act as a template. The fidelity of the polymerase to its template can be scrutinized by providing it with a chemically pure homopolymer as template and looking for incorporation of non-complementary nucleotides into the product: errors occur at a frequency less than once in 10^6 nucleotides incorporated.

The correspondence between template and product is further evidenced by the equivalence of their *nearest-neighbour nucleotide frequencies*. Measurement of these frequencies represents an ingenious approach to characterizing nucleic acids in terms of the arrangement of their nucleotides: it consists of measuring the relative frequencies of occurrence of the 16 possible dinucleotide sequences in a polynucleotide. The frequencies are measured by synthesizing DNA in a polymerase reaction using nucleoside triphosphates of which one is labelled

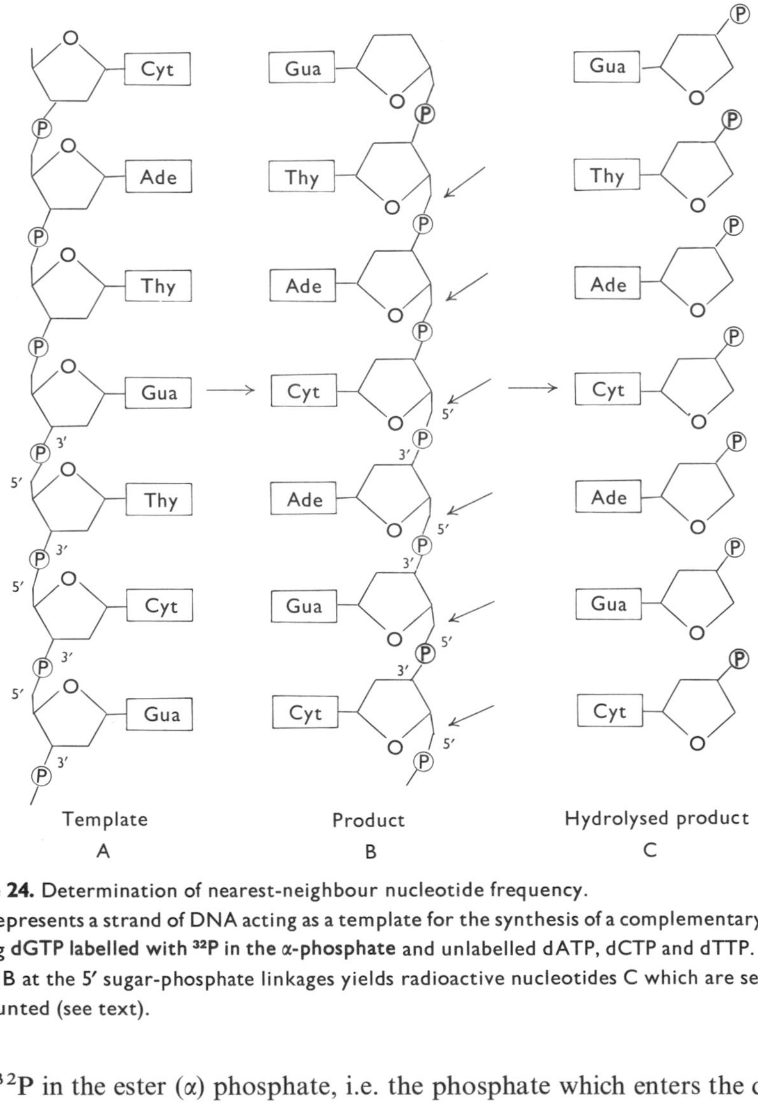

Template   Product   Hydrolysed product

A   B   C

Figure 24. Determination of nearest-neighbour nucleotide frequency.

A represents a strand of DNA acting as a template for the synthesis of a complementary strand B, using dGTP labelled with ^{32}P in the α-phosphate and unlabelled dATP, dCTP and dTTP. Hydrolysis of B at the 5′ sugar-phosphate linkages yields radioactive nucleotides C which are separated and counted (see text).

with ^{32}P in the ester (α) phosphate, i.e. the phosphate which enters the deoxyribose-phosphate backbone of the new DNA. The product is then hydrolysed with enzymes which split specifically at the 5′ sugar-phosphate link and the liberated 3′-mononucleotides are separated and their specific radioactivities determined (Figure 24). Thus if the label were given in the form of [^{32}P] dGTP, the radioactivity would appear in the various nucleotides in proportion to the frequency with which they occurred on the 5′ side of a dGMP residue. The measurements are performed in a replicate series using a different labelled deoxyribonucleoside triphosphate in each case, and in this way the frequencies of occurrence of the 16 dinucleotide sequences are obtained. The DNA thus characterized can then itself be used as a template for DNA polymerase and the nearest-neighbour nucleotide frequencies of the product determined and compared with those of the template. The correspondence between the values obtained can be seen in Table 3.

Table 3. Nearest-neighbour nucleotide frequencies in the template and product of a DNA polymerase reaction with calf thymus DNA as template.

Nearest-neighbour sequence	Frequency per 1000 dinucleotides	
	Template DNA	Product DNA
d(ApA), d(TpT)	89, 87	88, 83
d(CpA), d(TpG)	80, 76	78, 76
d(GpA), d(TpC)	64, 67	63, 64
d(CpT), d(ApG)	67, 72	68, 74
d(GpT), d(ApC)	56, 52	56, 51
d(GpG), d(CpC)	50, 54	57, 55
d(TpA)	53	59
d(ApT)	73	75
d(CpG)	16	11
d(GpC)	44	42

The template DNA was itself produced in a DNA polymerase reaction and only contained 5% of the original calf thymus DNA used to prime its synthesis.

Not only does this type of experiment reveal the influence of the template over the sequences of dinucleotides in the product, but it also provides proof that the two strands in the native DNA double helix are of opposite polarity, or 'antiparallel'. Consider the model structures represented in Figure 25. Opposite polarity (case A) requires that the sequence ApG appear with the same frequency as CpT, or that GpA appear as often as TpC. These relationships must hold irrespective of the length of the double helix. If the strands were of similar polarity (case B) the sequences ApG and TpC would have to appear with equal frequencies, as would GpA and CpT. Moreover, case (B) requires that the frequencies of TpA and ApT be equal, whereas this is not a *necessary* prediction of case (A). The experimental values given in Figure 18 clearly support (A), the Watson–Crick model with strands of opposite polarity.

The direction of synthesis of DNA strands by the polymerase is exclusively

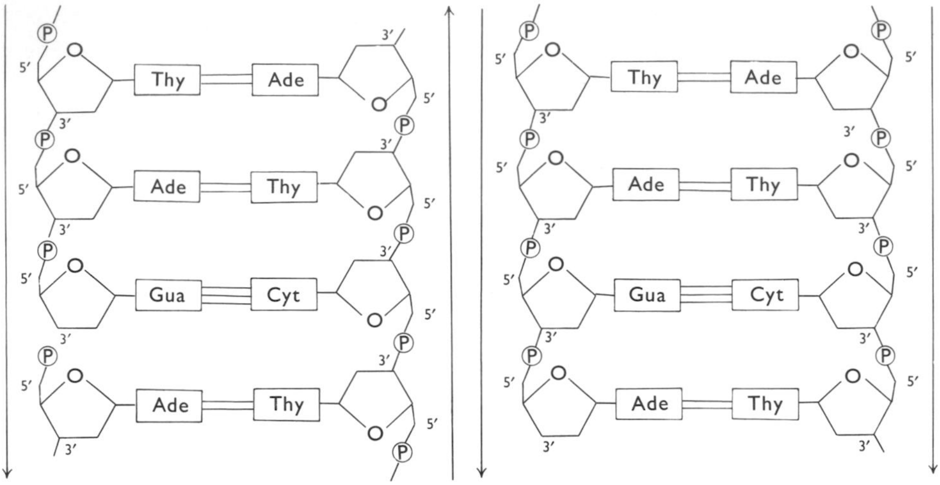

Opposite polarity

TpA(0·012) = TpA(0·012)
ApG(0·045) = CpT(0·045)
GpA(0·065) = TpC(0·061)

Similar polarity

TpA(0·012) = ApT(0·031)
ApG(0·045) = TpC(0·061)
GpA(0·065) = CpT(0·045)

Figure 25. Contrast of a Watson–Crick DNA model (having strands of opposite polarity) with a model having strands of similar polarity.

　　Predicted matching nearest-neighbour nucleotide frequencies for the two models are given. Values in parentheses are frequencies obtained in a DNA polymerase reaction with *Mycobacterium phlei* DNA as template. From A. Kornberg, *Enzymatic Synthesis of DNA* (Wiley, New York, 1961).

$5' \rightarrow 3'$. In chemical terms, the mechanism of chain growth is believed to occur via nucleophilic attack by the 3'-hydroxyl of the terminal nucleotide of the growing chain on the 5'-pyrophosphate of the incoming nucleotide (see Figure 26).

　　Certain features of the polymerase reaction show that its dependence upon pre-formed DNA is not simply due to its requirement for a template to copy.

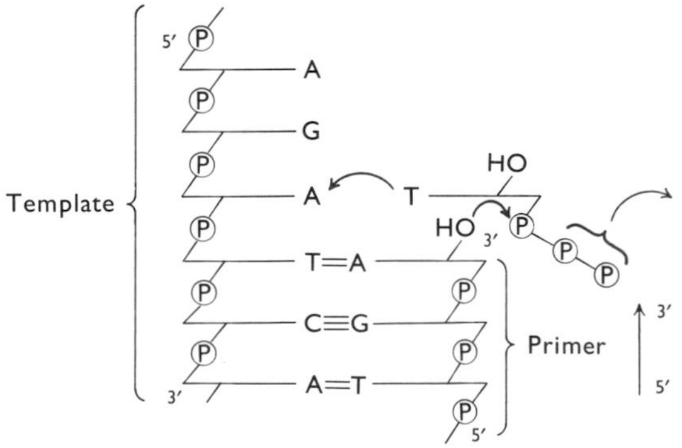

Figure 26. Mechanism of chain growth catalysed by DNA polymerase. After A. Kornberg, *Science* **163**, 1410 (1969).

It also requires a *primer*, i.e. a pre-existing DNA strand to which the new material is covalently linked. Thus the polymerase cannot initiate synthesis of a DNA strand *de novo*: the product is always found to be attached to a pre-existing strand, although this requirement may apparently be met by any small oligonucleotide. Nevertheless, the primer requirement is absolute and is quite specific inasmuch as the primer strand must be hydrogen-bonded to the template strand, and its 3′-terminal nucleotide must bear a free 3′-hydroxyl group (Figure 26). This explains why some DNA's are only poorly active or even inactive as templates for synthesis by the polymerase; the enzyme binds well to single-stranded regions in DNA molecules, which it recognizes as suitable template strands, but only catalyses synthesis if a free 3′-hydroxyl-ended primer strand is available as well. Thus closed circular duplex DNA molecules like the

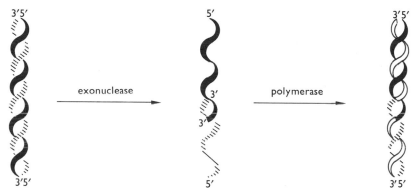

Figure 27. Repair by DNA polymerase of DNA partly degraded by exonuclease III. After C. C. Richardson *et al., J. Mol. Biol.* **9**, 46 (1964).

replicative form of φX174 DNA (p. 260), which are completely double-helical without ends or nicks (p. 288), do not bind the polymerase at all and are inactive as templates. They can be activated for polymerase action by the introduction of a nick in one strand by an endonuclease. One enzyme molecule can then bind to each nick, but synthesis can only take place if the nick is produced by cleavage of the 3′-sugar-phosphate bond leaving a 3′-hydroxyl group and a 5′-phosphate group. The role of the primer is well illustrated by the action of the polymerase on DNA which has been degraded with exonuclease III of *E. coli* (Figure 27). This enzyme degrades the strands progressively from their 3′ ends but stops when it reaches the middle of the molecule, leaving two single strands held together only by a small region of intact double helix. The resulting molecule is an excellent substrate for the polymerase, which efficiently 'repairs' it (Figure 27).

It was noted above that a double-helical molecule containing a nick like that illustrated in Figure 23 can act as a substrate for synthesis, but how does this occur? The answer resides in the finding that the purified polymerase has, in addition to its polymerizing activity, two exonuclease activities: it can degrade DNA strands in both the $5' \rightarrow 3'$ and $3' \rightarrow 5'$ directions. The latter activity may serve an 'editorial' function, enabling the enzyme to excise a mis-matched nucleotide at or near the primer terminus before synthesis proceeds. However, it is the $5' \rightarrow 3'$ nuclease activity which accounts for the ability of the polymerase to commence polymerization at a nick. In the first stages of the reaction there is a burst of hydrolysis as nucleotides are released from the template-primer, exposing the bases of the template strand which are then available to direct the polymerization process (Figure 28). The nuclease activity soon stops and the 5'-ended strand seems to become displaced from the template strand, allowing polymerization to proceed while the 5'-ended strand is peeled back. At

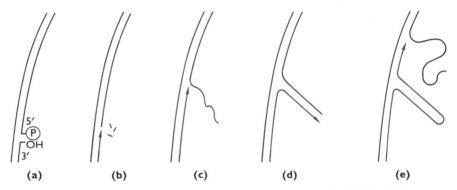

Figure 28. Scheme for synthesis by DNA polymerase on a nicked duplex DNA.

some stage this strand may compete successfully for the template function by attracting the growing strand to switch templates, producing a fork in the molecule. If, when it reached the end of the loose strand, the growing strand were itself to compete for template function, the sequence of events in Figure 28 could occur. The product would be a branched molecule with self-complementary regions present in the newly synthesized strand. This mechanism accounts convincingly for the branched appearance and readily renaturable character of DNA synthesized by the polymerase on a helical template *in vitro*.

From these and other observations, Kornberg has built up a picture of the active centre of the enzyme containing five sites (Figure 29). The molecule consists of a single polypeptide chain with a molecular weight of 109,000. It can be cleaved by limited proteolytic action into two active fragments of unequal size: the large one retains the polymerizing and $3' \rightarrow 5'$ nuclease activities, while the small one has the $5' \rightarrow 3'$ nuclease (excision) activity. Thus the latter activity is clearly identified with an independent site. It is likely that the $3' \rightarrow 5'$ nuclease activity, and related pyrophosphorolytic and pyrophosphate-exchanging activities, are catalysed in the vicinity of the triphosphate and primer terminus sites, as presumably is also the polymerizing reaction. There is only a single site for binding of nucleoside triphosphates, for which all four triphosphates compete; this finding agrees with the supposition that the polymerase itself plays no direct part in the selection of nucleotides for incorporation but merely acts to form a phosphodiester bond from the primer terminus to a triphosphate which is correctly positioned by virtue of strict Watson–Crick base-pairing with the template strand.

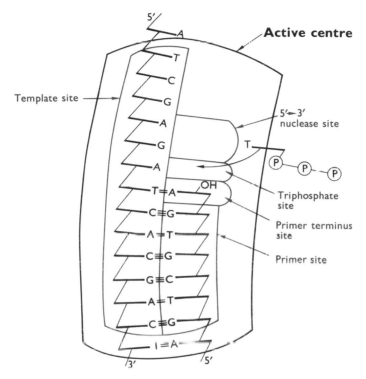

Figure 29. Kornberg's model to account for multiple catalytic sites within the active centre of DNA polymerase. After A. Kornberg, *Science* **163**, 1410 (1969).

A landmark in the DNA polymerase story was the demonstration by Kornberg and his collaborators in 1967 of the synthesis *in vitro* of biologically active, infective φX174 DNA. The procedure is outlined in Figure 30. Single-stranded circular viral (+) strands were incubated with DNA polymerase and ligase to produce circular (−) strands density-labelled with bromouracil (see p. 270). Because the template (+) strand was circular, having no free ends, a small oligonucleotide had to be added to act as a primer. The partly synthetic RF was nicked with an endonuclease to permit separation of the strands, and the synthetic (−) strands were purified on a CsCl gradient. Half of them were circular and the other half linear, because the endonuclease would nick either strand of the RF with equal probability. The density-labelled (−) strands then

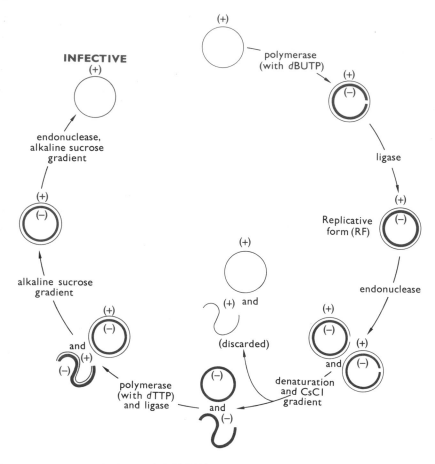

Figure 30. Synthesis of infective φX174 DNA *in vitro*.

dBUTP = bromodeoxyuridine triphosphate. Density-labelled strands are drawn with a thick line. Synthetic strands are shown in red. After M. Goulian *et al.*, *Proc. Nat. Acad. Sci. U.S.* **58**, 2321 (1967).

served as templates for a second round of the same procedure, leading to the production of totally synthetic non-density-labelled (+) strands, which were shown to be infective for *E. coli*. This remarkable experiment provides impressive evidence of the fidelity of complementary copying by the polymerase, for each strand contains 5000 nucleotides and the biologically active sequence was preserved through two cycles of synthesis.

It is tempting to conclude from this last experiment that DNA polymerase must be the enzyme which replicates DNA *in vivo*. However, there is serious reason to doubt whether this really is the case. A major objection is that the rate of polymerization of a DNA strand catalysed by the polymerase is at most 10–20 nucleotides per second, i.e. only 1% of the rate required *in vivo* (p. 284). A second difficulty is its obligatory demand for a primer: this would effectively rule out a mechanism as depicted in Figure 21(c), though it could be made consistent with mechanism 21(b) by postulating repeated endonuclease cleavage at the fork of the newly-synthesized strand in Figure 28(d).

On the other hand, the properties of the Kornberg polymerase would suit it admirably for a role in the repair of damaged DNA molecules rather than in replication of the chromosome.

REPAIR OF DNA

Work on the resistance of certain bacteria to the lethal effect of ultraviolet irradiation has shown that a mechanism exists whereby radiation-induced damage to DNA can be repaired *in vivo*. The defects introduced by irradiation

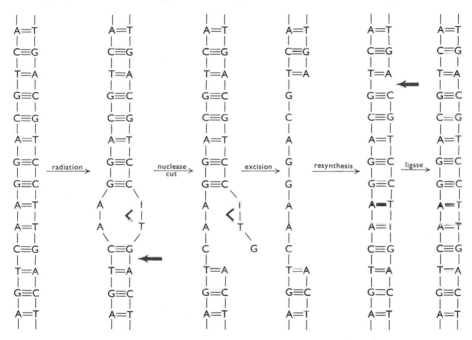

Figure 31. Postulated mechanism for repair *in vivo* of DNA damaged by irradiation.

Radiation causes formation of a thymine dimer. The region containing this is excised and repaired. The figure is diagrammatic. The number of nucleotides excised is estimated to be 20–500, but their relation to the damaged region is uncertain.

are principally covalent linkages between adjacent thymine residues. These thymine dimers can be excised, together with a considerable number of neighbouring nucleotides, and the gap in the double-stranded structure closed up by resynthesis of the missing region, using the base-sequence in the complementary undamaged strand as template (Figure 31).

An endonuclease which specifically introduces single-strand cuts into irradiated helical DNA has been isolated. The last stage of the process could easily be effected by polynucleotide ligase. It is the two intervening operations which could employ the properties of DNA polymerase. Provided that the initial endonuclease treatment left a 3′-hydroxyl end resynthesis would be easy. More importantly, the excision operation would provide a function for the $5′ → 3′$ nuclease activity of the polymerase, and indeed Kornberg's group have shown that the polymerase will efficiently catalyse excision of thymine dimers (and other mis-matched sequences), in the form of oligonucleotides, from irradiated DNA or deliberately mis-matched synthetic DNA's.

DNA-replicating enzymes

Late in 1969 de Lucia and Cairns reported the isolation of a DNA polymerase-deficient mutant of *E. coli*. Extracts of this mutant contain less than one per cent of the normal level of polymerase activity, yet the mutant multiplies and replicates its DNA at normal rates. It has, however, an increased sensitivity to killing by ultraviolet light, which is not due to failure to excise thymine dimers but can be traced to enhanced degradation of its DNA after irradiation—evidently the defect is in the resynthesis phase of its repair mechanism. These facts again implicate the polymerase in repair rather than replicative synthesis of DNA.

The discovery of Cairns' mutant has sparked off an intensive search for an alternative enzyme which could catalyse DNA replication. The mutant has played an important part in the search by providing a system in which the activity of other DNA-synthesizing enzymes would not be masked by the overwhelming DNA polymerase activity of normal cells. Several groups have reported finding novel DNA-synthesizing activity associated with the bacterial cell membrane. The new enzyme is membrane-bound and it is not dependent upon added DNA because the membrane preparations already contain a substantial fraction of the cell's DNA. The rate of synthesis is rapid—about 1500 nucleotides per chain per second, comparable to the rate *in vivo* (p. 284)—but falls off within a few minutes for unknown reasons. The substrates are the four deoxyribonucleoside triphosphates, and (ribo-)ATP stimulates the reaction. The product is high-molecular weight double-stranded DNA formed by a semiconservative mechanism. In many respects the properties of this enzymic system *in vitro* resemble those of a replicating activity which can be detected in toluene-treated whole cells of the DNA polymerase-deficient mutant.

The question remains, however, whether this novel polymerization reaction is completely independent of Kornberg's DNA polymerase or not. The new activity is inhibited by reagents which attack —SH groups and is unaffected by the presence of antibody specific to DNA polymerase, in contrast to what is found with the purified Kornberg enzyme, but all workers are careful to note that their results do not exclude the possibility that the synthesizing enzyme is DNA polymerase associated with other components in a special complex such that it acquires altered characteristics. Whether it will turn out that Kornberg's enzyme is concerned with repair, replication or both remains to be seen.

SYNTHESIS OF RNA

The metabolism of RNA is intimately tied up with the process of protein synthesis, for the sole role of RNA *in vivo* seems to be that of acting as genetically coded molecules which participate directly in the protein-forming reactions. To be sure, the ultimate repository of the cell's hereditary information is its DNA but the DNA itself does not apparently take part directly in the chemical combination of amino acids to form polypeptide chains. The immediate control over the all-important amino acid sequence in nascent polypeptides is exerted by the different classes of RNA molecules, and it is only by directing the synthesis of the RNA that the DNA determines the structure of the proteins whose properties are responsible for the phenotype of the organism. For clarity of presentation, therefore, the synthesis of RNA insofar as it relates to the synthesis of proteins is dealt with in the chapter on protein synthesis (pp. 333 *et seq.*). We may concern ourselves here with the mechanisms by which the coded nucleotide sequence of DNA determines the nature of the RNA and in particular with the enzymic processes involved.

It is now well established that in bacteria the synthesis of all three classes of RNA is directed by the DNA of the cell. More explicitly, there are genes or regions of the DNA whose nucleotide sequences specifically determine the nucleotide sequences of the different kinds of RNA. This by itself does not mean that all RNA molecules must be synthesized in association with a DNA template for it is quite possible to imagine that once a few prototype molecules had been made they could themselves 'prime' for further synthesis of similar molecules by RNA-dependent RNA-synthesizing reactions. However, there has never been any conclusive evidence for the existence of RNA-directed RNA synthesis in bacteria, except in cells infected with RNA viruses, so we must look to DNA-directed RNA synthesis as the sole source of RNA in normal growing organisms.

SYNTHESIS OF rRNA AND tRNA

Genes which specify the nucleotide sequences of tRNA and the three classes of rRNA (23 *S*, 16 *S* and 5 *S*) were originally discovered by DNA–RNA hybridization studies using purified preparations of each class of RNA. We have already seen how the formation of these hybrids has been exploited to enable isolation of DNA cistrons coding for the various RNA's (p. 271). rRNA's and tRNA are relatively rich in guanine (Table 4). From values like those shown in the Table it may be calculated that the cistrons in double-helical DNA coding for the larger rRNA's must have a G+C content of 53–54%; thus in bacteria whose overall DNA base-composition differs substantially from this value it can be predicted that fragments of their DNA containing the rRNA cistrons

Table 4. Size and base-composition of *Escherichia coli* nucleic acids.

	Molecular weight	Proportion of base, moles %			
		Adenine	Thymine or uracil	Guanine	Cytosine
DNA		25	25	25	25
23 S rRNA	$1·1 \times 10^6$	26	20	33	21
16 S rRNA	$0·55 \times 10^6$	24	22	31	23
5 S rRNA	40,000	19	17	35	29
tRNA	25,000	20	19	33	28

Values for base-composition include modified bases (Figure 5) counted under the parent base, and percentages have been rounded to the nearest integer.

will be significantly more or less dense than the average in CsCl gradients and will have a correspondingly different 'melting' temperature. These facts have proved useful in investigating the disposition of the 23 *S* and 16 *S* rRNA cistrons within the chromosome. In DNA of exponentially growing *E. coli* just under 0·4% of the sequences form hybrids with these rRNA's, from which it may be calculated that there are 6 or 7 copies of the cistrons for each of the larger rRNA's per chromosome. In other bacteria the number seems to be about the same. The cistrons are located in two discrete clusters on the genetic map. By studying hybridization between 23 *S* or 16 *S* rRNA and fragments of single-stranded DNA sheared to molecular weights as low as $0·4 \times 10^6$, it has been found that the 23 *S* and 16 *S* cistrons seem to be arranged in tandem, i.e. in contiguous pairs containing one cistron for each rRNA. However, hybridization to larger DNA fragments shows that the tandem pairs within a given cluster on the chromosome appear to be scattered, i.e. interspersed with sequences not present in the mature rRNA's.

The cistrons which code for tRNA are also clustered on the genetic map of the chromosome. tRNA forms hybrids with 0·04–0·05% of the DNA of *E. coli*, implying the existence of some 60 cistrons, which, compared with other estimates that there are about the same number of different species of tRNA in *E. coli*, suggests that most tRNA molecules are probably coded for by a single cistron. Again, there is evidence that tRNA cistrons in DNA fragments may be arranged contiguously.

It is now well established that in higher organisms rRNA molecules are formed from larger precursor molecules by a process of *post-transcriptional modification*. Recent evidence indicates that something similar occurs in bacteria. After short pulses of labelled precursors such as ^{32}P-phosphate, radioactivity is found in precursor RNA molecules (p23 and p16) which are slightly larger, by about 150 nucleotides in each case, than the mature 23 *S* and 16 *S* rRNA's. The radioactive label in p23 and p16 RNA's can be 'chased' into the respective mature forms and the kinetics are typically those of a precursor-product relationship. An intriguing possibility is that an even earlier precursor might exist, consisting of a transcript of a tandem pair of 23 *S* and 16 *S* rRNA cistrons, which could undergo successive cleavage yielding first p23 and p16 RNA molecules and eventually mature 23 *S* and 16 *S* rRNA's. Another suggestion is that one of the products of cleavage might represent a short-lived precursor of 5 *S* rRNA. To complete the picture, a species of rapidly-labelled RNA which behaves like a precursor of tRNA has recently been described.

Post-transcriptional modification also involves formation of the methylated and otherwise altered nucleotides (Figure 7, p. 261) which are characteristic of rRNA and tRNA. It is believed that the immediate product of transcription of the rRNA and tRNA cistrons contains only the four normal nucleotides and that, during the process of maturation, sequence-specific enzymes convert particular nucleotides at defined points in the molecule into the modified, mature forms. The majority of the modifications, e.g. methylation, are simple, while others appear more complicated, e.g. conversion of uridine to pseudouridine or dihydrouridine. Enzymes which catalyse some of these processes have been characterized.

Messenger RNA

rRNA and tRNA cistrons together account for no more than one per cent of the bacterial chromosome. Of the remainder, a large fraction will hybridize efficiently with pulse-labelled RNA synthesized in cells during a short exposure to a radioactive precursor *in vivo*. Much of this RNA must surely be mRNA transcribed from structural genes, but it is most unlikely that all of the remaining DNA, i.e. 99% of the total, is active or even potentially active in mRNA synthesis. Part of this DNA is presumably in the form of regulator genes rather than structural genes, and there may be regions with other genetic functions which need not necessarily involve the production of mRNA. Several approaches to the investigation of mRNA synthesis are possible, but all are hampered by the heterogeneity and metabolic instability of this class of RNA, in contrast to the rRNA's and tRNA which are stable and easy to isolate because of their defined molecular size. This topic is considered further in Chapter 6, where the role of mRNA synthesis in relation to protein synthesis is described, and in later sections of the present chapter.

Inhibition by drugs which bind to DNA

Compelling evidence that all forms of RNA are synthesized by DNA-dependent mechanisms in normal bacteria has come from studies on the action of the antibiotic actinomycin D. This substance completely blocks the synthesis of all forms of RNA in susceptible organisms without apparently affecting any other metabolic process to anything like the same degree. Furthermore, it forms a stable complex with native, double-helical DNA (but not with RNA) as a result of which the capacity of the DNA to act as a template for RNA polymerase is drastically reduced. Perhaps surprisingly, DNA can still act as a template for DNA synthesis by DNA polymerase in the presence of actinomycin, though this may be correlated with the fact that DNA polymerase does not act on a truly double-helical template (p. 293).

However, since binding to DNA appears to be the principal if not sole interaction between the antibiotic and a cell receptor, the fact that actinomycin blocks all *de novo* RNA synthesis in bacteria implies strongly that all RNA synthesis is directed by DNA. This interpretation of the action of actinomycin has found wide application: for instance, the antibiotic has been widely used as a tool for demonstrating the involvement (or non-involvement) of DNA-dependent RNA synthesis in a great variety of living processes. It has also been employed to investigate the rate of breakdown of messenger RNA and to reveal the occurrence of RNA-dependent RNA synthesis in a number of virus-infected systems.

Some bacteria are not sensitive to actinomycin, most probably because they have a permeability barrier which prevents the antibiotic from entering the cell. *Escherichia coli* is insensitive for this reason. Nevertheless, the production of RNA by these organisms can be inhibited by another class of drugs, the acridines. One of the most widely used acridines is proflavine (2:8 diaminoacridine). Like actinomycin, this drug forms complexes with DNA and, when sufficient drug molecules are bound, the synthesis of RNA is completely stopped. Inhibition by proflavine has been used to measure the rate of decay of mRNA activity in *E. coli*. Compared with actinomycin, however, proflavine is a much less specific tool for interfering with DNA-dependent RNA synthesis; its interaction with DNA results in at least as powerful inhibition of DNA synthesis as RNA synthesis, and it also forms complexes with RNA. There are also differences between the complexes formed by the two kinds of drug with DNA: actinomycin binds specifically to sites containing deoxyguanosine residues and blocks the template activity of the DNA when relatively few sites have become occupied, whereas complex formation with proflavine shows no marked base-specificity and a large amount of drug must be bound before template activity is lost. The proflavine-DNA complex is of special interest for other reasons, however. The acridines are powerfully mutagenic drugs and it seems that their mutagenicity can be correlated with the nature of their interaction with DNA: being planar molecules they slot in or *intercalate* between the stacked base-pairs of the double helix in such a fashion that the affected base-pairs become

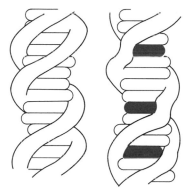

Figure 32. Sketches representing the secondary structures of normal DNA (left) and DNA containing intercalated proflavine molecules (right).

   The helix is drawn as viewed from a remote point, so that the base-pairs and the intercalated proflavine appear only in edgewise projection, and the phosphate-deoxyribose backbones appear as smooth coils. From Waring, *Symp. Soc. Gen. Microbiol.* **16**, 235 (1966).

separated by twice the normal distance (Figure 32). This mode of interaction can account for the striking capacity of acridines to generate frame shift mutations, i.e. mutations resulting from the insertion or deletion of a base-pair in the DNA of the mutant. Acridine-induced frame shift mutants have played an important part in establishing the triplet nature of the genetic code (p. 411).

Transcription from a unique strand

Only one of the paired DNA strands acts as a template for RNA synthesis. The reason for this seems fairly clear; if both strands coded for the synthesis of RNA there would be two kinds of RNA produced by each cistron in the DNA and this would lead to a curious degeneracy in the genetic code if two different forms of mRNA were both to be translated at the level of the ribosome and to give rise to the same type of polypeptide. There would also be a definite possibility of the two (complementary) RNA's from a single cistron finding each other and forming a perfectly base-paired double helix. This has never been observed in normal bacteria although such structures are known to occur in some RNA viruses. Of course, it is possible that both DNA strands could be 'copied' and then one of the RNA products immediately degraded but there is no evidence for a rather wasteful mechanism of this kind. Finally, the cell-free RNA polymerase system (see p. 308) can be made to transcribe only one of the DNA strands *in vitro*, and since this enzyme contains no detectable nuclease activity it can be said with reasonable certainty that the 'other' strand was never synthesized.

Three lines of evidence show that only one of the DNA strands codes for synthesis of RNA *in vivo*.

(1) In *E. coli* infected with bacteriophage ϕX174, phage-specific messenger RNA is only synthesized after the infecting (+) strand has been converted to the circular double-helical replicative form (RF). This RNA does not form hybrids with pure (+) strands of DNA isolated from phage particles, but it hybridizes readily with denatured RF–DNA. Evidently the template for its synthesis was the (−) strand of the RF.

(2) In certain instances it is possible to resolve the complementary DNA strands into separate peaks by centrifuging to equilibrium in CsCl gradients. This can be done with the DNA of several bacteriophages, notably α, which infects *Bacillus megaterium*, and SP8, which infects *Bacillus subtilis*. The basis for the separation resides in the fact that the A = T, G = C relationship required of the double helix does not necessarily fix the base-compositions of the individual strands—one may be relatively purine-rich and the other correspondingly pyrimidine-rich. This is the case for the two phages mentioned, and it is found that the phage-specific mRNA synthesized in infected cells is complementary only to the pyrimidine-rich strand.

(3) A similar, though less complete, fractionation of the complementary strands can be effected with denatured pneumococcal transforming DNA in a CsCl gradient. When the two fractions, enriched for one or the other of the strands, were used to transform recipient cells it was found that one fraction caused the appearance of phenotypic changes one generation time before the other, suggesting that the 'slower' fraction had had to participate in a cycle of replication before its genetic information could be expressed.

The above examples, and others given earlier (pp. 273 and 287), show that the ability to isolate the complementary DNA strands has proved a valuable tool in the study of macromolecular synthesis. The most effective means of achieving complete resolution of the strands derives from an observation of Szybalski, who found that one of the strands frequently interacts preferentially with a synthetic polyribonucleotide, so that it bands at a substantially more dense position in a CsCl gradient than the other, less reactive strand. Guanine-rich ribopolymers have proved particularly effective in separating the strands of several phage DNA's, apparently by complexing with sequences rich in dC residues which are asymmetrically distributed between the complementary strands. In many cases it could be shown that the strand which preferentially binds poly G is the strand which is predominantly transcribed into RNA *in vivo*. The relation between these findings may not be fortuitous; it has been suggested that pyrimidine-rich clusters may serve to identify the transcribing strand *in vivo* by acting as initiation sites to which the transcribing enzyme(s) might bind and commence synthesis of an RNA chain.

The strands of phage T4 DNA are efficiently separated by complexing with the synthetic random copolymer poly U,G. During the lytic infection cycle, at least three distinct classes of T4-specific RNA are synthesized: up to 2 min after infection only *immediate early* RNA is made; this is followed by *delayed early* RNA; and after 10–12 min *late* RNA species appear. Both species of early RNA are found to be complementary to the *C* strand of T4 DNA (the one which interacts preferentially with poly U,G), whereas late RNA is complementary to the other strand (*W*). Thus, while the rule is still obeyed that only one strand of a given DNA sequence is transcribed into RNA, the transcribed sequences need not all be located on the same physical strand of the DNA. Evidently in T4-infected cells the synthesis of RNA directed by late cistrons involves a switch of the transcribing enzyme(s) from the c strand to the w strand. The mechanism by

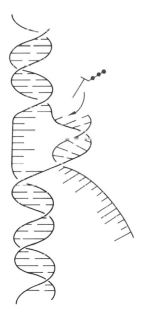

Figure 33. Synthesis of RNA by Watson–Crick base-pairing with a DNA template.

DNA strands are shown in black, with the double helix in the B-form having ten base-pairs per turn. The growing RNA strand is shown in red, with a ribonucleoside triphosphate about to be incorporated at the growing 3′-hydroxyl end. The RNA forms a transient hybrid helix with the template DNA strand, which is drawn with tilted base-pairs to indicate its likely A-form conformation having eleven base-pairs per turn. When the RNA is displaced from the hybrid the ten-fold DNA helix re-forms as the region of local unwinding moves away.

which this switch occurs presents an interesting problem which will be considered later (p. 313).

We can picture the events during synthesis of RNA on a double-helical
DNA template as follows (Figure 33). (1) The DNA helix opens up to allow the
bases of one strand to interact by Watson–Crick base-pairing with incoming
ribonucleoside triphosphates. (2) The growing RNA strand forms a transient,
perhaps helical, DNA–RNA hybrid with the transcribing DNA strand. (3) An
exchange of strands occurs, releasing the growing RNA strand and re-forming
the DNA helix as the growing-point moves away. It is worth noting that this
scheme involves two changes in helical involvement for the transcribing DNA
strand. If the transient hybrid were to assume the 11-fold A-type helical conformation found with DNA–RNA hybrids *in vitro* (p. 269), the necessary conformational changes in the sugar-phosphate backbone of the DNA strand as it
switched from B → A → B-type helical states could play a significant part in
the overall process of synthesis. In particular, these conformational changes
might play a major role in assuring the strand-exchange required to release the
RNA free from its template (process (3) above).

The scheme in Figure 33 is consistent with all the available evidence and it
attributes the determination of specificity in nucleotide incorporation to
Watson–Crick base-pairing with the template strand. It may be mentioned, however, that another scheme has been proposed in which the DNA helix remains
intact throughout and an RNA strand grows in one of the grooves of the helix,
incoming nucleotides being selected by hydrogen-bonding to the intact base-
pairs of the helix. There is little evidence to support this suggestion.

VISUALIZATION OF BACTERIAL GENES IN ACTION

Electron microscopy on cell contents, released by osmotic lysis of fragile cells of
E. coli, has provided striking confirmation of widely accepted ideas about gene
action. Some photographs are shown in Figure 34. Long DNA fibres can be
seen, to which are attached granular strings of varying lengths; the strings are
believed to be polyribosomes, i.e. RNA molecules in process of synthesis
covered with ribosomes which have already commenced translation of the
nascent RNA strands into protein (Figure 34(a) and (b)). Important conclusions are: (1) Long stretches of the DNA appear to be genetically inactive, at
least so far as RNA synthesis is concerned. (2) In the active regions, the lengths
of the attached polyribosomes increase in a regular fashion along a portion of
the chromosome, often with an irregularly shaped granule which could be an
active RNA polymerase molecule at one end. (3) Free polyribosomes, not
attached to DNA, are rarely seen. (4) Presumptive tandem 23 *S* and 16 *S* rRNA
cistrons may be recognized by the much larger number of fibrils growing from

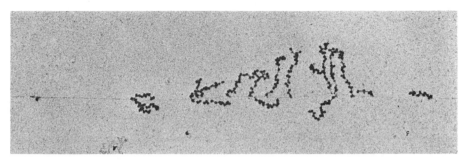

Figure 34. Visualisation of bacterial genes in action.

(a) Polyribosomes attached to an active portion of the *E. coli* chromosome, an unidentified but undoubtedly large operon. The fainter granule at the extreme left is a putative RNA polymerase molecule presumably at or very near the initiation site for this operon. Polyribosomes exhibit imperfect gradients of increasing lengths as they become more distal to the initiation site. The shorter, most distal polyribosomes may have resulted from mRNA degradation. 83,000 ×.

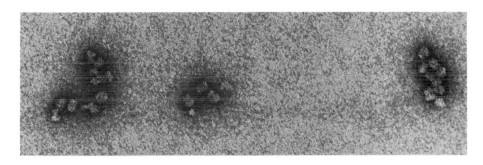

(b) An active segment of the *E. coli* genome at higher magnification showing polyribosomes which are attached to the chromosome by RNA polymerase molecules. 270,000 ×.

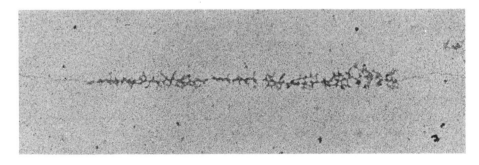

(c) A portion of an *E. coli* chromosome showing presumptive rRNA genes in action, with gradients of ribonucleoprotein fibrils on the 16 S and 23 S cistrons. 119,000 ×.

Photographs kindly provided by Drs O. L. Miller, Jr., and B. A. Hamkalo [see *Science*, **169**, 392 (1970)].

them (Figure 34(c)), and these fibrils seem to be picking up protein rather than ribosomes, which agrees with evidence that ribosomal proteins become associated with the rRNA's as they are synthesized. (5) The presumptive tandem rRNA cistrons appear to be about the expected length, approx. 1·5 μm, and are not clustered but quite widely spaced along the chromosome.

RNA polymerase

Bacteria, unlike eukaryotic cells, are believed to have a single enzyme for the synthesis of all classes of RNA. This RNA polymerase has been isolated from a number of organisms. Its properties are in many ways analogous to those of DNA polymerase: the substrates are ribonucleoside 5'-triphosphates, a metal ion is required (Mn^{++} is preferred to Mg^{++}), and the reaction is completely dependent upon the presence of a DNA template. The incorporation of a labelled nucleotide (indicated by *) into acid-insoluble material provides a convenient assay:

$$
\begin{array}{l}
n_1 \text{ *ATP} \\
n_2 \text{ GTP} \\
n_3 \text{ CTP} \\
n_4 \text{ UTP}
\end{array}
\xrightarrow[\text{Mn}^{2+}]{\text{DNA}}
\left[
\begin{array}{l}
n_1 \text{ *AMP} \\
n_2 \text{ GMP} \\
n_3 \text{ CMP} \\
n_4 \text{ UMP}
\end{array}
\right]
+ (n_1 + n_2 + n_3 + n_4)\; \textcircled{P}\!\!-\!\!\textcircled{P}
$$

$$
\begin{array}{cc}
\text{(acid-} & \text{RNA} \\
\text{soluble)} & \text{(acid-insoluble)}
\end{array}
$$

The RNA polymerase from *Escherichia coli* has been purified to virtual homogeneity, and in what follows attention will be concentrated on this enzyme. It is composed of five subunits: two α-chains, each of molecular weight 41,000, one β-chain of molecular weight 155,000, one β'-chain of molecular weight 165,000, and a σ-subunit of molecular weight 86,000. This constitutes the *holoenzyme*, designated $\alpha_2\beta\beta'\sigma$. An additional minor subunit (perhaps two) of molecular weight approx. 12,000, designated ω, is present in most preparations but it is not yet known whether ω is a true functional component of the enzyme molecule. The σ component is readily dissociated from the holoenzyme, leaving a second form of the enzyme, $\alpha_2\beta\beta'$, termed *core polymerase*. Core polymerase retains the polymerizing activity and requirements for reaction noted above. In fact, most methods of RNA polymerase purification yield a mixture of holoenzyme and core polymerase, and it is now clear that many (if not all) of the general properties of the enzyme described below are primarily those of the core polymerase rather than the holoenzyme. After these properties have been described we will return to σ and discuss its role in the activity of the holoenzyme.

The DNA requirement of RNA polymerase *in vitro* can be satisfied by either double-stranded or single-stranded DNA, though the enzyme exhibits some preference for the former and its synthesizing activity in presence of double-helical DNA is more likely to approximate to its action *in vivo*. In either case the product is high-molecular weight RNA which is released free from the DNA template, though the initial product formed with a single-stranded template is a base-paired DNA–RNA hybrid. The product RNA is not covalently linked to any pre-existing polynucleotide material of any kind, showing that there is no need for a primer in the true sense. The role of the DNA as a template

Table 5. RNA polymerase. Relationship between the base-composition of DNA from various sources as template and the RNA product synthesized by polymerase from *Escherichia coli*.

	Proportion of base, moles %			
	Adenine	Thymine or uracil	Guanine	Cytosine
Mycobacterium phlei				
DNA template	16	16	34	34
RNA product	14	16	36	34
Escherichia coli				
DNA template	25	25	25	25
RNA product	23	26	24	27
dAT copolymer	50	48	<1	<1
RNA product	52	48	0	0
Bacteriophage ϕX174				
DNA template	25	33	23	19
RNA product	32	25	20	23

for RNA synthesis is shown by the usual criteria: (1) The base-composition of the product is strictly determined by that of the DNA (Table 5)—it is in fact complementary, as can be seen from the data with the single-stranded ϕX174 DNA template. (2) The nearest-neighbour nucleotide frequencies of the template DNA are faithfully reproduced in the product. (3) When denatured and annealed together, the template and product form DNA–RNA hybrids with high efficiency.

However, it is clear from the above observations that *both* strands of the DNA can act as templates for RNA synthesis, and that the enzyme is capable of synthesizing RNA molecules complementary to *all* nucleotide sequences in the DNA presented to it. This discrepancy with the strand-selectivity known to occur *in vivo* points to the loss of controlling factors which must be present *in vivo*, and is characteristic of the core polymerase. The role of σ and other factors in curbing the indiscriminate copying of all nucleotide sequences will be described later.

RNA polymerase synthesizes chains exclusively in the $5' \to 3'$ direction, the mechanism presumably involving nucleophilic attack by the 3'-hydroxyl of the terminal nucleotide of the growing chain on the 5'-pyrophosphate of the incoming nucleotide (cf. DNA polymerase). The direction of synthesis was established by changing the specific radioactivity of the nucleoside triphosphates during the course of a polymerase reaction, hydrolysing the product, and

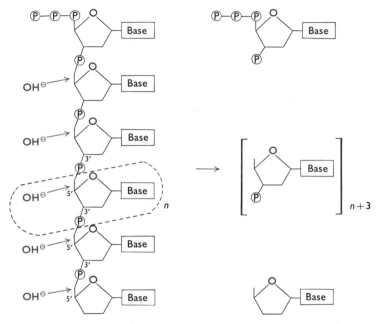

Figure 35. Alkaline hydrolysis of RNA formed in the DNA-dependent RNA polymerase reaction. The 5' end of the chain yields a nucleoside tetraphosphate, the 3' end yields a nucleoside, and the residues in between give nucleoside monophosphates (an equilibrium mixture of 2' and 3').

determining the specific activity of nucleotides derived from the 5' and 3' ends of the RNA chains. The end nucleotides are readily identifiable in alkaline hydrolysates of the product because the 5'-end nucleotide is released as a tetraphosphate, the 3'-end nucleotide as a nucleoside, and all intervening nucleotides as 2' or 3' nucleoside monophosphates (Figure 35). The specific activity of nucleotides derived from the 5'-ends was found to reflect the specific activity of the substrate nucleotides during the first stage of synthesis, while that of nucleotides from the 3'-ends reflected the specific activity of the substrates after it had been changed. It is interesting that the 5'-triphosphate grouping of the first nucleotide incorporated remains intact during subsequent growth of the chains: this provides a parallel with the genetic material of RNA bacteriophages, many of which have been shown to start with the sequence pppGp---

at their 5′ ends. The parallel goes further, for RNA molecules synthesized by the polymerase *in vitro* are also predominantly initiated with a purine nucleotide. Rates of polymerization *in vitro* are rapid, from 40 to 100 nucleotides per second, and comparable with estimates of the rate of RNA chain growth *in vivo*.

Various workers have obtained evidence for the transient existence of a hybrid between the growing end of the RNA chain and one of the DNA strands during polymerase-catalysed synthesis on a double-helical template *in vitro*. This is consistent with a model for RNA polymerase action analogous to that represented in Figure 33, where synthesis occurs via base-pairing with the transcribed strand in a locally unwound region of the helix which progresses along the DNA as synthesis proceeds. According to this model the enzyme would contain at least three distinct sites: a DNA strand-separation site, a polymerization site, and a strand-exchange site where the growing RNA strand would be displaced from the hybrid with concomitant re-formation of the DNA helix. The inhibition of RNA synthesis by drugs such as actinomycin and proflavine, which is reproduced with the polymerase reaction *in vitro*, is explained by their binding to the DNA helix and forming a more or less long-lived block preventing movement of the polymerase along the template, perhaps by making it more difficult to separate the DNA strands.

We must now explain the basis of asymmetric transcription *in vivo*, i.e. the mechanism which restricts RNA polymerase to copying only the correct strand of a DNA duplex. A hint has already been given that the σ factor present in the *E. coli* holoenzyme plays a part in this process. The first evidence for a role of σ in transcription was the finding of Burgess, Travers and their collaborators that the rather inefficient action of *E. coli* core polymerase with a T4 DNA template was markedly stimulated by addition of σ. It was subsequently found that, whereas core polymerase transcribed sequences more or less at random from both strands of T4 DNA, addition of σ factor restricted transcription to one strand only and the RNA produced behaved in hybridization tests like the 'early' RNA synthesized in T4-infected cells (p. 305). A similar result has been found with the circular double-helical RF–DNA of phage fd (analogous to φX174 RF) as template: again, core polymerase copied both strands whereas addition of σ factor restricted the enzyme to transcribing only the (−) strand which is the one transcribed *in vivo*. In general, it appears that with virtually any phage DNA as template the core enzyme synthesizes RNA randomly from the two strands, while in the presence of σ factor only those RNA species synthesized in infected cells immediately after infection are made.

Clearly the presence of σ in the *E. coli* holoenzyme imparts to it specificity enabling recognition of the correct DNA strand for transcription, and though this evidence comes from work with phage DNA's it must be presumed that σ exercises the same function in uninfected bacteria. The effect of σ is mediated at the level of initiation of RNA chains, most probably by assisting the holoenzyme to recognize correctly nucleotide sequences in the DNA which act as signals for initiation—the *promoter* sites defined in genetic analysis. Initiation is the rate-limiting step in the synthesis of RNA molecules; it can be dissected into a number of sequential stages: (1) A rapid, readily reversible interaction between the polymerase and the DNA template. (2) Formation of a comparatively stable initiation complex, probably involving σ factor-dependent recognition of a promoter sequence in the DNA. This step requires physiological temperatures and may involve local 'melting' of the helix as a prelude to binding of the first nucleotide. The antibiotic rifamycin, which is known to inhibit initiation (but not propagation) of RNA chains catalysed by RNA polymerase, prevents formation of this initiation complex provided it is added before the temperature is raised. Rifamycin binds very tightly to bacterial RNA polymerase, but not to the enzyme from resistant mutants which is found to have an altered β subunit. Thus the β subunit, as well as the σ factor, may play a direct part in the formation of the polymerase-promoter site complex. (3) Binding of the initiating nucleoside triphosphate, usually (perhaps necessarily) ATP or GTP. (4) Commencement of polymerization and movement of the whole transcription complex along the template. At some stage, possibly during steps (2) or (3), the σ factor seems to dissociate away from the complex because it becomes available to participate in renewed initiation with another core polymerase molecule before the newly growing RNA molecule is completed.

To account for the synthesis of RNA molecules of defined length it is necessary that the polymerase also be able to recognize defined termination signals in its DNA template. Release of free RNA chains and polymerase molecules from a phage DNA template is observed *in vitro*, but the RNA chains are often longer than those synthesized *in vivo*. A soluble protein factor designated ρ has been purified which leads to the synthesis of more life-sized chains by RNA polymerase *in vitro*; it may assist the enzyme to recognize the correct signals for termination.

Recent work makes it clear that the asymmetric transcription of DNA observed *in vivo* is merely a gross manifestation of the existence of an intricate system of control elements which continuously monitor the transcription of different cistrons by RNA polymerase. Control over the recognition of different promoter sequences by RNA polymerase molecules provides an obvious potential site for regulation of the pattern of gene transcription, as does the Jacob–Monod model of repressor-operator gene interaction (p. 440), and other mechanisms may well exist. Here we will refer briefly to three examples of control mediated by alterations in RNA polymerase behaviour.

rRNA cistron transcription

These cistrons constitute less than 1% of the bacterial chromosome, yet rRNA makes up the bulk of the RNA of the cell. Consequently the cistrons must be transcribed extremely rapidly. However, when *E. coli* DNA is used as a template for RNA polymerase (core or holoenzyme) *in vitro*, synthesis of rRNA is not detectable. Travers and his colleagues have isolated a protein factor ψ_r from *E. coli* which stimulates the holoenzyme (but not core polymerase) to synthesize substantial amounts of rRNA *in vitro*. This factor, which is also a component of the Qβ replicase enzyme (p. 314), is not a new species of σ factor because it requires the presence of σ for its activity. It might be an example of a new class of positive control factors which specifically programme RNA polymerase to recognize a whole class of initiation promoter sites. The action of the ψ_r factor is inhibited by guanosine tetraphosphate, ppGpp, a nucleotide whose appearance in amino acid-starved 'stringent' strains of bacteria is correlated with cessation of rRNA and tRNA synthesis.*

Sporulation

The process of sporulation in *Bacillus subtilis* involves a major change in the pattern of gene transcription. Some genes concerned with vegetative cell growth, including the rRNA cistrons, cease to be expressed while other genes which specify functions required for spore formation become activated for transcription. At the same time the cell loses its capacity to support growth of the bacteriophage ϕe, and the lesion is due to failure to transcribe RNA from the phage DNA after it has entered the cell. These changes can be partially accounted for by a structural alteration of the RNA polymerase in sporulating cells: one of the β subunits of the enzyme is substantially smaller. If, as seems likely, this alteration results in a modified pattern of promoter recognition by the enzyme we may conclude that the β subunit of the core polymerase plays an active role in recognizing correct promoter sequences in the DNA, perhaps in conjunction with a modified σ factor or factors.

Lytic bacteriophage development

On page 305 the changing pattern of RNA transcription in T4-infected cells of *E. coli* is described. With other bacteriophages a similar picture is seen, i.e. sequential, sometimes overlapping, synthesis of quite different classes of RNA molecules. Infection of *E. coli* with T7 provides a fairly simple example; two distinct classes of phage-specific RNA are made, 'early' and 'late'. In all cases it seems that the first class of RNA to appear is transcribed from the phage DNA by the host cell holoenzyme. The appearance of subsequent classes of RNA is dependent upon protein synthesis directed by the first class of phage-specific RNA. This suggests a plausible explanation for the mechanism by which transcription is switched from one class of genes to another, i.e. that the switches involve production of protein factors which modify the promoter-recognition properties of the pre-existing polymerase. Alternatively the new protein could be an entirely new polymerase with a different built-in specificity

---

* However, the role of ψ_r has more recently been questioned.

of promoter recognition. The latter is the case in T7 infection; synthesis of 'late' RNA is completely dependent upon the appearance of a new phage-coded RNA polymerase which contains only a single large polypeptide chain and specifically transcribes 'late' RNA from a T7 DNA template *in vitro*.

In T4 infection the situation is more complex. Apparently a totally new polymerase is not synthesized, and rifamycin is found to inhibit phage production when added at any stage during the growth cycle—this would suggest that the original host core polymerase, or at least its β subunit which is needed for rifamycin to act, is required throughout the whole infective cycle. Travers has isolated a factor from T4-infected cells which directs the *E. coli* host polymerase to transcribe delayed early RNA species from a T4 DNA template *in vitro*. Under comparable conditions the σ factor of the host enzyme restricts its activity to synthesis of the immediate early species, so the T4 factor could provide the first transcriptional switch, inhibiting initiation at immediate early promoters and facilitating initiation at delayed early promoters. The mechanism of subsequent transcriptional switching, enabling synthesis of late RNA species, is not clear; it may be correlated with a number of successive modifications to the α, β' (and ω?) subunits of the core polymerase which have been observed to occur fairly early in infection.

VIRAL REPLICASES AND POLYMERASES

Although not strictly relevant to the processes of nucleic acid synthesis in bacteria, some mention must be made of the existence of two novel types of nucleic acid-synthesizing enzymes associated with viruses. The first type, known as RNA replicase, is essential for the growth and replication of RNA viruses and bacteriophages such as Qβ. The genetic material of RNA phages is single-stranded RNA. After entering the host cell, the ($+$) strand acts as a template for the synthesis of a complementary ($-$) strand, and the resulting double-stranded molecule then acts as a template for asymmetric synthesis of many more ($+$) strands to yield the progeny phages. Both these enzymic processes are mediated by a single replicase enzyme which is synthesized by the host cell ribosomes using the original infecting ($+$) strand as messenger RNA. At no stage is DNA involved in the polymerization reactions, which therefore proceed in a novel fashion inasmuch as the sequence of nucleotides in the growing RNA strand is directly determined by the sequence in an RNA template strand. Biochemically the reactions are perfectly analogous to the DNA-dependent polymerase reactions: the substrates are the four ribonucleoside triphosphates and the polymerization of growing RNA chains proceeds in the $5' \to 3'$ direction. Qβ replicase contains four polypeptide subunits, but of these only one is coded for by the RNA genome of the phage: the other three derive from the activity of host genes, and one of them seems to be identical with the ψ_r factor which stimulates rRNA synthesis (p. 313).

The second novel type of enzyme is an RNA-dependent DNA polymerase activity associated with the virions of certain RNA tumour viruses which cause malignant transformation of mammalian cells, such as the Rous sarcoma virus and a number of leukaemia viruses. The transformed cells are genetically stable cancer cells, yet for generations they continue to yield small amounts of the transforming virus. The recent discovery of the novel enzyme was heralded by the work of Temin, who argued that the properties of the transformed cells could best be accounted for by postulating that the viral RNA served as a template for synthesis of a piece of DNA which then became stably integrated into the genetic material of the cells. His view is now vindicated. The novel polymerase employs the viral RNA strand as a template and converts it into a DNA RNA hybrid, the DNA strand of which may then serve as a template for the synthesis of a truly double-helical DNA.

There is little reason to doubt that the reactions catalysed by RNA replicase and RNA-dependent DNA polymerase involve normal Watson–Crick base-pairing as do the better known DNA-dependent polymerase reactions. Lacking evidence to the contrary, it seems reasonable to conclude that the same base-pairing forces are the primary determinants for selection of nucleotides in *all* the nucleic acid polymerizing reactions. Thus the discovery of the two last mentioned enzymes serves to underline the striking versatility of base-pairing as a force dictating the structure and function of nucleic acids. Base-pair complementarity accounts well for the structures of DNA and RNA and for all four directions of flow of sequence information between them.

FURTHER READING

1   Watson J. D. (1970) *Molecular Biology of the Gene*, 2nd edition. Benjamin, New York.
2   Lewin B. M. (1970) *The Molecular Basis of Gene Expression*. Wiley, London.
3   Symposium (1968) Replication of DNA in microorganisms. *Cold Spr. Harb. Symp. Quant. Biol.* **33**.
4   Kornberg A. (1969) Active Centre of DNA polymerase. *Science*, **163**, 1410.
5   Bonhoeffer F. and Messer W. (1969) Replication of the bacterial chromosome. *Ann. Rev. Genetics*, **3**, 233.
6   Lark K. G. (1969) Initiation and control of DNA synthesis. *Ann. Rev. Biochem.* **38**, 569.
7   Symposium (1970) Transcription of genetic material. *Cold. Spr. Harb. Symp. Quant. Biol.* **35**.
8   Attardi G. and Amaldi F. (1970) Structure and synthesis of ribosomal RNA. *Ann. Rev. Biochem.* **39**, 183.
9   Geiduschek E. P. and Haselkorn R. (1969) Messenger RNA. *Ann. Rev. Biochem.* **38**, 647.
10 Travers A. A. (1971) Control of transcription in bacteria. *Nature New Biology*, **229**, 69.
11 Richardson J. P. (1969) RNA polymerase and the control of RNA synthesis. *Progr. Nucleic Acid Res. and Mol. Biol.* **9**, 75.
12 Waring M. J. (1968) Drugs which affect the structure and function of DNA. *Nature*, **219**, 1320.

Chapter 6
Class III Reactions: Synthesis of Proteins

The outlines of the mechanism of protein synthesis have been sketched in Sections 5 and 6 (pp. 34–43) and it will be recalled that after activation the amino acids combine with specific transfer-RNA molecules and that these interact on a messenger-RNA template which is associated with ribonucleo-protein particles called ribosomes. There are hundreds of different kinds of proteins in each cell and each is made of polypeptides which are linear polymers of unique composition and sequence. In this chapter we will consider the components and reactions in more detail.

It is useful to have some idea of the distribution of the components in a bacterial cell and the Tables 1 and 2 give this information in various ways—percentages, numbers of molecules, numbers of subunits, total weights per cell (in daltons, where 6×10^{23} daltons are equivalent to 1 g). The numbers in these tables are of necessity approximations but they do give an indication of the make-up of a bacterium and their significance will be increasingly apparent.

FRACTIONATION OF BACTERIAL CELLS

In order to analyse the components of an organism it is first necessary to take it to pieces. Bacterial cells can be disrupted by mechanical, chemical or enzymic means and the resulting *lysate* can then be fractionated in various ways. Low speed centrifugation gives a pellet consisting of unbroken cells, fragments of walls and membranes, perhaps some highly polymerized DNA, some ribosomes, and any large granules. The supernatant contains the rest of the ribosomes, all the soluble proteins, soluble RNA, and small molecular weight compounds such as salts, amino acids and nucleotides. If this supernatant is centrifuged at high speed (10,000 g) the ribosomes spin down but the supernatant still contains enzymes and RNA.

Much has been learned about cellular components and the mechanism of protein synthesis generally by the use of zone-centrifugation in sucrose density gradients and this technique is so useful that it will be described here. A mixture of particles of different sizes can be separated by centrifugation through a solution of graded density. The sample in a small volume is loaded on the surface of the solution in a swinging-bucket centrifuge tube and this is then spun at an appropriate speed. Particles of like sedimentation coefficient tend to form bands which move towards the bottom of the tube as centrifugation continues. After a suitable time the contents are removed as a series of fractions. Often this is done by puncturing the bottom of the plastic centrifuge tube and collecting drops

Table I. Nucleic acid distribution in a bacterium

Volume of cell	10^{-12} ml.	
Dry matter per cell	$2 \cdot 5 \times 10^{-13}$ g.	
	$(1 \cdot 5 \times 10^{11}$ daltons)	

DNA per cell: % of dry weight	$1 \cdot 5 – 2 \cdot 0\%$ (say $1 \cdot 67\%$)	
weight	$2 \cdot 5 \times 10^9$ daltons	
no. of nucleotides	8×10^6	
length	1100–1400 μm	
Size of average structural gene	1250 nucleotides	
	4×10^5 daltons	
Total no. of genes per cell	6400	
Genes for tRNA	50 (0·02% of total DNA)	
Genes for 16 S rRNA	10 (0·1% of total DNA)	
Genes for 23 S rRNA	10 (0·2% of total DNA)	
Genes for mRNA	thousands	

	Sedimentation coefficient	Molecular weight daltons $\times 10^{-3}$	Nucleotides per molecule	Total no. molecules per cell	% total RNA	Total nucleotides $\times 10^{-6}$	Total weight daltons $\times 10^{-8}$
rRNA	5 S	40	125	12,000	80	64	200
	16 S	550	1700	12,000			
	23 S	1100	3400	12,000			
tRNA	4–5 S	25	70–80	150,000	15	12	36
mRNA	8–30 S	100–1500	300–4500	500–1000*	1–2	0·8–1·6	2·5–5
				Total RNA		80	250

Total RNA 10–20% of dry weight of cell, say 17%.

* Mean life of mRNA about 4% of the mean generation time of the bacteria. Hence number of molecules of mRNA synthesized per cell per generation = 12,500–25,000. The figures given are based on a variety of sources but may be approximately correct for a bacterium of the size of *Escherichia coli.*

Table 2. Protein distribution in a bacterium

Total protein per cell	% of dry weight	60%
	weight	9×10^{10} daltons
	amino acid residues	8×10^8
Size of average polypeptide	amino acid residues	200
	weight	22,500 daltons
No. of polypeptide chains per cell		4×10^6
No. of species of proteins		say 1000
No. of molecules of each species		$1–10^5$
70 S ribosomes　no. per cell		12,000
protein/ribosome		40% (w/w)
		$1\cdot1 \times 10^6$ daltons
		10^5 amino acid residues
protein units/ribosome		*c.* 55
Total ribosomal protein per cell		$1\cdot32 \times 10^{10}$ daltons
		$1\cdot2 \times 10^8$ amino acid residues
		$6\cdot6 \times 10^5$ molecules
		14·5% of total cell protein
'Soluble' protein per cell	% of total	75%
	weight	$6\cdot7 \times 10^{10}$ daltons
	amino acid residues	6×10^8
Size of average soluble		
polypeptide chain	amino acid residues	230
	weight	25,000 daltons
No. of chains per cell		$2\cdot7 \times 10^6$

'Structural' protein (wall and membrane) say 10% of total cell protein

The figures given are based on a variety of sources but are approximately correct for a bacterium of the size of *Escherichia coli*.

into a series of tubes (Figure 1). The fractions can be analysed, for instance, for absorbance at 260 nm (indicative of nucleic acids) or for radioactivity, etc. If 0·1 ml of a whole cell lysate is layered on 5 ml of a 5% to 20% linear gradient of buffered sucrose and spun for 45 minutes at 37,000 r.p.m. in a centrifuge, and the contents are then divided into 40 fractions, a graph like Figure 2a might result. A gradient of 15% to 40% sucrose might give a curve like Figure 2b. Had the lysate first been treated with ribonuclease in media of various Mg^{2+} concentrations, results like those in Figure 3 might be obtained. The polyribosomes would have been degraded by ribonuclease to 70 *S* ribosomes and low Mg^{2+} would have caused these to dissociate (see below).

Thus it is possible by separating appropriate fractions and then dialysing or centrifuging them, to get various components in more or less pure form. In the study of protein synthesis, experiments can be carried out *in vivo* followed by

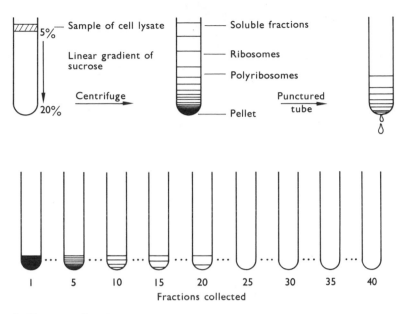

Figure I. Zone-centrifugation.

This technique is useful for fractionation of components such as ribosomes and polyribosomes or of RNA's of different molecular weight. Note that centrifugation is *not* continued until equilibrium. (See R. J. Britten and R. B. Roberts (1960) *Science*, **131**, 32–33.)

fractionation to determine what has happened and where; alternatively, cell-free systems can be reconstructed from the bits and pieces and then tested for activity. These *in vitro* experiments usually consist of incubating the components with a radioactive amino acid, adding trichloroacetic acid to precipitate protein, and measuring the radioactivity in the precipitate. The conversion of non-precipitable to precipitable counts is not, however, adequate proof that an amino acid has been incorporated by peptide linkage into polypeptide. There is an excellent account in a review by Loftfield (see reference on p.366) of the criteria which have to be satisfied.

RIBOSOMES AND POLYRIBOSOMES

In Chapter 1 it was shown that the cytoplasm of bacteria appears to contain thousands of electron-dense particles about 10–20 nm in diameter. When cells are disrupted by mechanical, chemical or enzymic means these are released and can be isolated by differential centrifugation. Preparations can be purified by various techniques and the particles can be spun down at 100,000 g in one to two hours. They are found to consist of protein and ribonucleic acid and in 1958 they were given the name *ribosomes*. Frequently they are characterized by their behaviour in an ultra-centrifuge and are, therefore, referred to by their sedimentation coefficient, e.g. 70 *S* ribosomes. Electron microscopy of isolated ribosomes and of sections of bacterial cells indicate that the predominant form is 70 *S* but that this is composed of two subunits—a 30 *S* and a 50 *S* piece. These can be separated (Figures 4a and 4b). The intact 70 *S* particle can be made to dissociate reversibly by altering the ionic composition of the medium in which

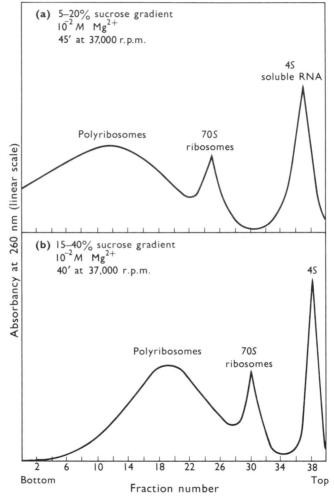

Figure 2. Separation of polyribosomes, ribosomes and soluble fractions by zone-centrifugation in a sucrose gradient.

The sample to be separated is loaded on the surface of a sucrose gradient. The steepness of the gradient and the duration of centrifugation determine the degree of separation of the components.

(a) 5%–20% gradient; 45 minutes at 37,000 r.p.m. Some of the larger polyribosomes have reached the bottom of the tube.

(b) 15%–40% gradient; 40 minutes at 37,000 r.p.m. The tail of the polyribosome peak can be seen but there is less separation of components.

it is suspended. In particular, reducing the Mg^{2+} concentration from 10^{-2} M to 10^{-4} M tends to cause dissociation into the subunits:

$$70\,S \;\rightleftharpoons\; 30\,S + 50\,S$$

Bacterial ribosomes all have approximately the same chemical composition—about 40% protein and 60% RNA—and a total molecular weight of about 2.9×10^6 daltons. Each $30\,S$ subunit has a weight of 1.0×10^6 daltons and contains one molecule of $16\,S$ rRNA of molecular weight about 0.55×10^6 daltons.

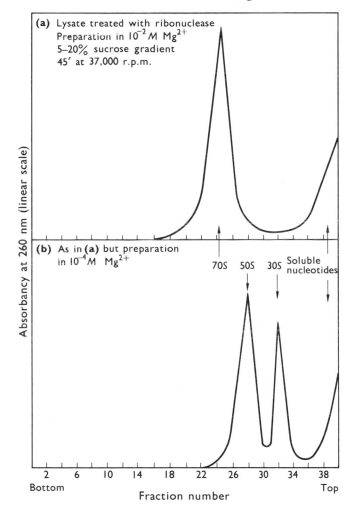

Figure 3. Effects of ribonuclease and reduced concentration of magnesium on ribosome pattern.

(a) Ribonuclease converts polyribosomes to $70\,S$ ribosomes and soluble nucleotides. The soluble RNA ($4\,S$) is also degraded. Preparation in $10^{-2}M\,Mg^{2+}$.

(b) Same preparation in $10^{-4}M\,Mg^{2+}$. The $70\,S$ peak of ribosomes is replaced by two peaks consisting of $50\,S$ and $30\,S$ subunits respectively. This dissociation is reversible: $70\,S \rightleftharpoons 50\,S + 30\,S$. All of the 260 nm-absorbing material is precipitable by 5% TCA except the nucleotides.

Each 50 S particle is about $1\cdot8 \times 10^6$ daltons, having rRNA of 23 S, molecular weight $1\cdot1 \times 10^6$ daltons, but also a molecule of 5 S rRNA (see Figure 9, p. 266 for structure). 16 S and 23 S rRNA differ slightly in nucleotide composition and the latter is not, as was once suggested, a dimer of the former although many of its sequences occur twice.

The proteins of ribosomes have all been separated and the 30 S subunit is found to contain about twenty different species and the 50 S subunit about thirty-six. Not more than one molecule of each protein occurs in each particle and none is common to both 30 S and 50 S subunits. Molecular weights of these proteins range from 10,700 to 65,000 and nearly all have as N-terminal amino acid methionine (45–49%) or alanine (36–40%) (see below p. 343). Many enzymic activities have been found associated with ribosome preparations but it is not always certain which are an integral part of the structure and which are only transiently attached. Since ribosomes are involved in the formation of all kinds of proteins, one would expect to find traces of nascent enzymes on them.

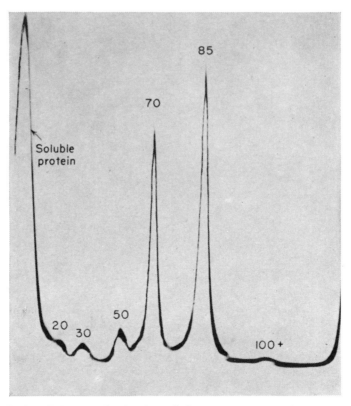

Figure 4a. Analytical ultracentrifugal analysis of ribosomes.
    Pattern of ribosomes in a lysate from *Escherichia coli* in the exponential phase of growth. Preparation in medium containing 10^{-2}M Mg^{2+}. Schlieren plate taken 5 minutes after 50,740 r.p.m. was reached. The peaks are labelled with nominal sedimentation constants.

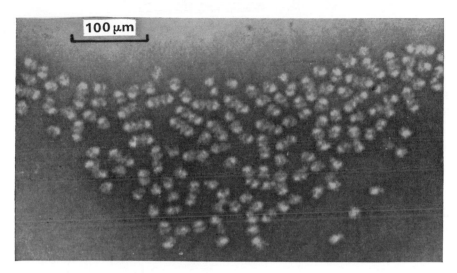

Figure 4b. Ribosomes from *Escherichia coli*.

Electron micrograph of preparation fixed in appropriate medium and 'negatively stained' by drying down in a thin film of sodium phosphotungstate. The preparation contains 70 S and 100 S particles. 70 S ribosomes appear to contain two unequal subunits (30 S and 50 S) and the 100 S particles consist of two 70 S particles attached with their smaller subunits apposed.

The combination of proteins and RNA to form ribonucleoprotein particles requires cations, and Mg^{2+} and K^+ play this role *in vivo* with, perhaps, polyamines such as putrescine, spermidine and spermine also contributing to the neutralization of the many phosphate groups in the RNA. Suspension of ribosomes in media lacking Mg^{2+} or containing chelating agents such as ethylenediamine tetra-acetate (EDTA) can lead to disintegration of the particles.

It is now possible to reconstitute biologically active 30 *S* and 50 *S* subunits from the component rRNA molecules and ribosomal proteins. The required sequence of addition is known and studies *in vitro* have shed light on the process of assembly *in vivo*.

It is not possible by microscopy of living cells to see the form of ribosomes and all micrographs are of fixed and usually sectioned preparations. However, it is generally thought that these particles do exist *in vivo*, although in a hydrated form. When very gentle methods of cell disruption are used (e.g. lysis of protoplasts or spheroplasts—see pp. 87–9 in Chapter 1) most of the ribosomes are released as clusters joined together by a tenuous thread. It has been estimated that 80% or more of the ribonucleoprotein particles occur as these *polyribosomes*, also called polysomes or ergosomes. There is some evidence that the material linking the particles is RNA—probably messenger-RNA. If a growing culture of bacteria is treated for a few seconds with a radioactive precursor of RNA (e.g. uridine) radioactivity is found in the polyribosomes but not in the ribosomes

Chapter 6

which can be derived from them by treatment with ribonuclease (Figure 5). In other words, there is RNA in polyribosomes which is not the ribosomal rRNA itself. Moreover, much of this labelled RNA forms hybrids with homologous DNA in a manner consistent with its being mRNA (see pp. 269–72, Chapter 5). Finally, when further synthesis of RNA is inhibited by adding the antibiotic, actinomycin D (see p. 302, Chapter 5) much of the rapidly-labelled RNA is degraded during the next few minutes. Concomitantly the polyribosomes break

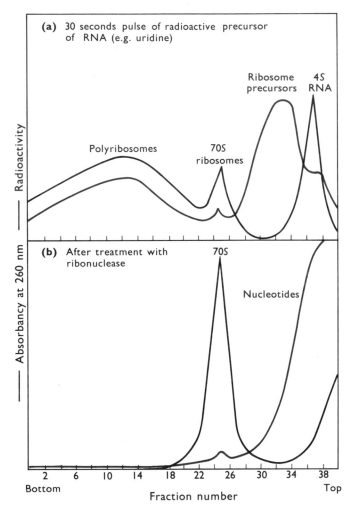

Figure 5. Presence in polyribosomes of an RNA component which is absent from 70 S ribosomes.

(a) A short pulse of a radioactive precursor to RNA (e.g. uridine) labels polyribosomes much more than 70S ribosomes. There is also a slower sedimenting peak of radioactivity which is known to contain ribosome precursors.

(b) Ribonuclease converts the polyribosomes to 70S ribosomes but their radioactivity is not in this peak; it is found in the TCA-soluble nucleotides.

down to yield single ribosomes. These findings all support the belief that polyribosomes consist of ribosomes strung together on a strand of mRNA.

Whereas prokaryotic organisms (bacteria and blue-green algae) contain 70 *S* ribosomes, eukaryotes (such as yeasts, protozoa, plants and animals) have particles of about 80 *S*, composed of subunits of 40 *S* and 60 *S* with correspondingly larger rRNA components. The ratio of protein to RNA is nearer 50/50 than 40/60. Eukaryotes, however, may have ribosomes resembling those of prokaryotes in such organelles as mitochondria and chloroplasts.

The site of protein synthesis *in vivo*

Just as radioactive precursors to RNA can be used, so can amino acids be added to label proteins. If they are added to growing cultures, radioactive protein will be found a few seconds later. The experiment can be done by squirting a solution of radioactive amino acid(s) into a vigorously stirred culture and then, after a few seconds, pouring the whole culture on crushed ice to reduce the temperature very rapidly and halt further metabolism. The cells are then broken and fractionated. Some species (e.g. *Bacillus megaterium*) can be lysed by adding the enzyme lysozyme a few seconds before cooling so that by the time the organisms have been centrifuged out of suspension (in the cold) their walls will have been digested and only fragile protoplasts will remain. These are highly sensitive to detergents such as Triton X100 which disrupts the cytoplasmic membrane releasing the contents.

The kinetics of transfer of radioactivity from amino acids through intermediates to the end-product protein can be studied by taking samples at intervals and examining them by this kind of technique. If this is done it is found that radioactive peptides occur first in association with ribosomes and polyribosomes. The activity here increases to a maximum in a few seconds and subsequent 'chasing' by adding an excess of non-radioactive amino acids causes replacement of labelled material on the ribosomes and appearance of radioactivity in the soluble protein fraction of the cell. This is illustrated in Figure 6. It seems, therefore, that nascent protein has a transient existence on ribosomes and determination of the flux through this stage suggests that probably all proteins are assembled in this manner before being released and going to their functional sites.

AMINO ACID ACTIVATION

Although particulate material is involved in the making of proteins, soluble components are also essential as can be demonstrated by reconstruction experiments. Amino acids become incorporated into polypeptide only if both

Chapter 6

particulate and soluble fractions are present. As has been mentioned (p. 316) the soluble fraction contains both enzymes and soluble RNA. The latter is seen as a 4 *S* peak of material absorbing at 260 nm near the top of the sucrose gradient (see Figure 2, p. 320). Both it and enzymes are implicated in the early stages of protein synthesis.

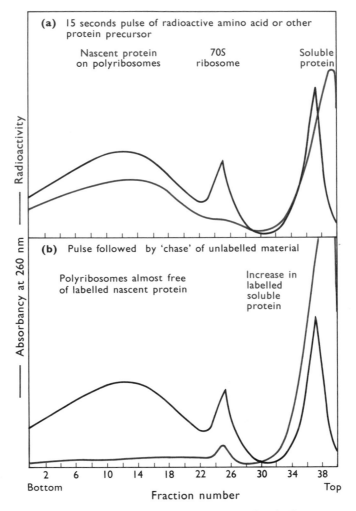

Figure 6. Appearance of nascent protein first in association with polyribosomes.

(a) Addition of radioactive amino acid to a culture growing exponentially. In a few seconds some soluble protein has become labelled but the bulk of the radioactivity which has been converted to peptide form is associated with the ribosomes and polyribosomes.

(b) If a short pulse of radioactive amino acid is followed by a large unlabelled 'chase' of the same amino acid, the radioactivity is swept on out of the polyribosome region and appears as soluble protein near the top of the gradient.

The nascent protein is transiently attached to the ribosomes; the soluble protein is the end product.

Soluble enzymes are able to catalyse a reaction of amino acids with adenosine triphosphate (ATP) as follows:

$$NH_2CH.COOH + ATP \longrightarrow NH_2CH.CO \sim AMP + \textcircled{P}—\textcircled{P}$$
$$\overset{|}{R} \qquad\qquad\qquad \overset{|}{R}$$

The products are inorganic pyrophosphate and an amino acyl-AMP in which there is a high energy bond between the carboxyl of the amino acid and the phosphate of the AMP:

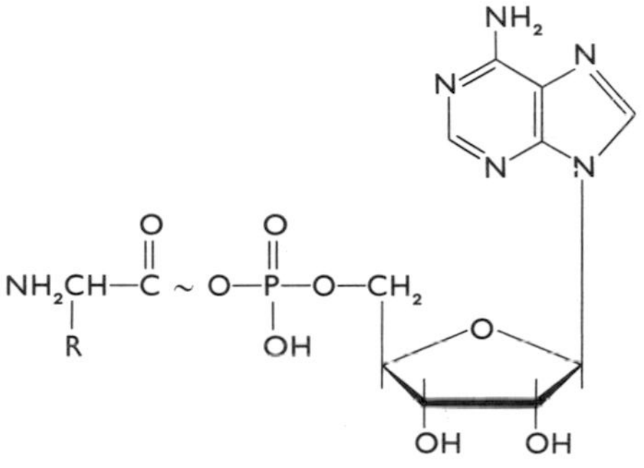

Amino acyl $\sim$ AMP

The formation of amino acyl-AMP is not easy to demonstrate as it remains bound to the enzyme. However, this complex will react with hydroxylamine to give a coloured amino acid hydroxamate which can be estimated:

$$Enz\,(NH_2CH.CO \sim AMP) + NH_2OH \longrightarrow NH_2CH.CO.NHOH + AMP + Enz$$
$$\qquad\quad \overset{|}{R} \qquad\qquad\qquad\qquad\qquad \overset{|}{R}$$

Another way of demonstrating the occurrence of the enzyme activity depends on the fact that the reaction is reversible. Labelled inorganic pyrophosphate is added to the reaction mixture and an exchange reaction occurs in which ATP becomes labelled. This labelling is dependent on the presence of the amino acid suggesting the following mechanism:

$$NH_2CH.COOH + ATP + Enz \longrightarrow Enz\,(NH_2CH.CO \sim AMP) + \textcircled{P}—\textcircled{P}$$
$$\overset{|}{R} \qquad\qquad\qquad\qquad\qquad\qquad \overset{|}{R}$$

$$Enz\,(NH_2CH.CO \sim AMP) + \textcircled{P}^*—\textcircled{P}^* \longrightarrow Enz + NH_2CH.COOH + ATP^*$$
$$\qquad\quad \overset{|}{R} \qquad\qquad\qquad\qquad\qquad\qquad\qquad \overset{|}{R}$$

There is at least one specific activating enzyme for each of the twenty amino acids and most have been obtained pure. The molecular weight is usually *c*. 100,000. These enzymes are present in the soluble fraction of cells and can be precipitated at pH 5 so they are sometimes referred to as the 'pH 5 precipitable fraction' or as 'pH 5 enzymes'.

AMINO ACID TRANSFER RNA (tRNA)

When unpurified soluble fractions are used as sources of amino acid activation, the reaction goes further and the amino acids are transferred to specific soluble RNA acceptors:

$$\alpha\alpha \;+\; \text{ATP} \;+\; \text{Enz} \longrightarrow \text{Enz}\,(\alpha\alpha \sim \text{AMP}) \;+\; \textcircled{P}\!\!-\!\!\textcircled{P} \xrightarrow{\;\text{tRNA}\;} \alpha\alpha\text{-tRNA} \;+\; \text{AMP} \;+\; \text{Enz}$$

These RNA's are variously called soluble-RNA, sRNA, acceptor-RNA, amino acid transfer-RNA, tRNA. We shall use the latter terms. Each tRNA is specific for a single amino acid. The transfer RNA's can be purified from the soluble fraction but are *not* precipitated at pH 5. It is thus possible very easily to separate the activating enzymes from them. It turns out that the same enzyme activates an amino acid by forming the amino acyl-AMP and then transfers the amino acid residue to the tRNA, i.e. catalyses the whole of the reaction:

$$\alpha\alpha \longrightarrow \alpha\alpha \sim \text{AMP} \longrightarrow \alpha\alpha\text{-tRNA}$$

Sometimes tRNA must be present for the enzyme to catalyse a reaction between the amino acid and ATP—perhaps the sequence of binding of substrates to the enzyme is important.

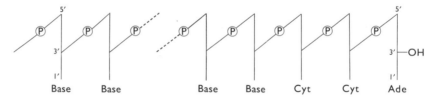

Figure 7. Diagrammatic representation of transfer-RNA.

    The amino acid transfer-RNA's are about 80 nucleotides in length and have the sequence -cytidylyl-cytidylyl-adenosine at one end. There is at least one specific tRNA for each of the 20 amino acids. The vertical lines represent ribose residues which are joined by phosphate through position 3′ and 5′. They have a purine or pyrimidine base attached to position 1′.

    In most bacterial cells the bulk of the ribonucleic acid is rRNA in ribosomes which can be sedimented. The soluble RNA is largely tRNA but there may also be ribosome precursors and traces of mRNA. The term transfer-RNA (tRNA) is a functional description whereas soluble-RNA (sRNA) is an operational definition. All transfer-RNA's are similar in many properties. They have sedimentation coefficients of about $4\,S$; their molecular weights are about 25,000; they are about 80 nucleotides in length and contain pseudo-uridylic acid residues and methylated purines and pyrimidines (see Chapter 5, p. 264); and they have considerable secondary structure. Finally, they all appear to have a sequence of -cytidylic acid, cytidylic acid, and adenosine at one end of the molecule (Figure 7). The amino acid is linked to the ribose of the terminal adenosine, being esterified to the 3'-OH. Some antibiotics which inhibit protein synthesis are analogues of amino acyl-adenosine, e.g. puromycin (Figure 8). The ester bond in $\alpha\alpha$-tRNA has not the high energy of the mixed acid anhydride bond in $\alpha\alpha \sim$ AMP but is of higher energy than most simple esters because of the properties of the 2' and 3' hydroxyls of the furanose ring.

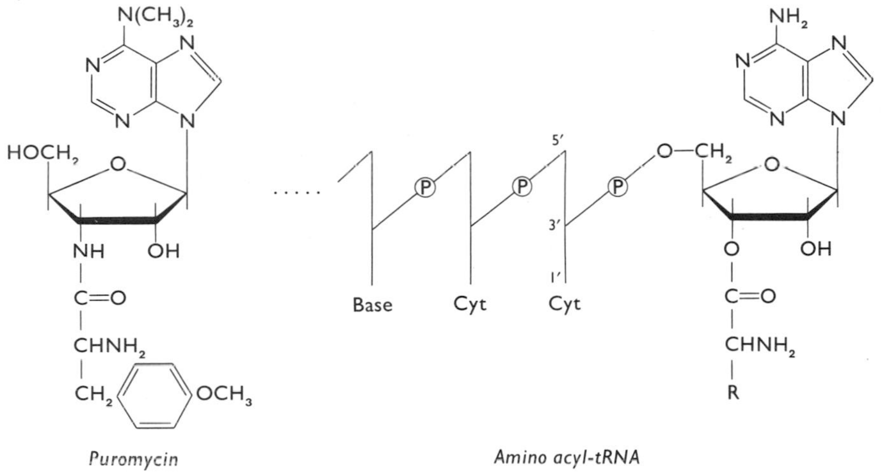

Puromycin   Amino acyl-tRNA

Figure 8. Comparison of structures of puromycin and the terminal sequence of amino acyl-tRNA. Puromycin is an antibiotic which interferes with the synthesis of proteins by acting as an analogue of amino acyl-tRNA.

There exists at least one tRNA for each amino acid and for some several have been demonstrated. However, only one activating enzyme for each amino acid may occur. These enzymes are highly specific for each of the two stages but more so for the second. Thus the isoleucine enzyme activates isoleucine and valine and the valine enzyme activates valine and threonine but neither of the 'wrong' amino acids gets transferred to tRNA. Sometimes 'alien' amino acid does become incorporated into protein—7-azatryptophan and 2-azatryptophan (tryptazan) are incorporated in place of tryptophan having been activated by its enzyme. However, 5-methyl-tryptophan is neither activated nor incorporated but it does inhibit synthesis of protein. In general the fidelity of synthesis is very high indeed. The elucidation of primary, secondary and tertiary structures of tRNA's (see Chapter 5, pp. 262–9) helps us to understand why. All end with the same -cytidylate–cytidylate–adenosine sequence but significant differences reside elsewhere apart from the anti-codon triplet (see p. 352). The terminal sequence can be split off fairly readily and there are cytoplasmic enzymes which successively add back the three residues using nucleoside triphosphates as precursors (Figure 9).

Which species of tRNA occur in a bacterial cell depends on the stage and

Chapter 6

conditions of growth, and infection by bacteriophage may result in synthesis of phage-specific tRNA's and inactivation of pre-existing host species.

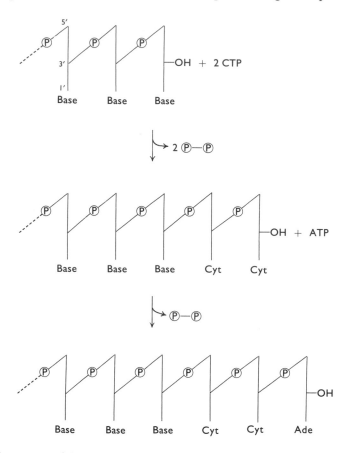

Figure 9. Restoration of the terminal sequence of transfer-RNA.

All sixty or so tRNA's have the terminal sequence -cytidylyl-cytidylyl-adenosine and this can be split off by enzymes which inactivate the tRNA. Two enzymes add back the residues successively using the nucleoside triphosphates and eliminating inorganic pyrophosphate.

The anti-codon triplet is *not* the site recognized by the activating enzyme and, for instance, the pure *Escherichia coli* serine enzyme charges tRNA^{Ser} whether it has U C A or A G U as the anti-codon. Enzymes and tRNA's from heterologous species may also interact—an *Escherichia coli* enzyme will charge a yeast tRNA.

The tRNA's are precipitated by cold 5% trichloracetic acid (TCA) and the amino acyl residue is not removed by such treatment. It follows, therefore, that a radioactive amino acid will become TCA-precipitable if combined with tRNA and that precipitability is not an adequate indication of incorporation into polypeptide. However, there are procedures which leave proteins intact but which degrade amino acyl-tRNA complexes rendering the amino acid moiety soluble in TCA. Treatment with ribonuclease or with dilute alkali has this effect, as has hot TCA (95° for a few minutes). It is thus possible to distinguish between the earlier stage of complex formation between the amino acid and ribonucleic acid, and the later stage of incorporation of amino acid into protein. The first part requires only the amino acid, the activating enzyme, ATP and the appropriate tRNA. No particulate fraction is needed. Indeed, the supernatant from high speed centrifugation (100,000 g for 1–2 hr) is an adequate source of the enzymes and tRNA's. The latter can be discharged of their amino acids by raising the pH value to 8·8 and incubating for 45 min before re-neutralizing. They can be subsequently recharged with labelled amino acids by incubating them in the presence of ATP. The tRNA's complexed with radioactive amino acids can then be isolated by extraction with phenol for use in the next stages.

INCORPORATION OF AMINO ACIDS INTO POLYPEPTIDE

Fractionation of disrupted organisms showed that for amino acid to become incorporated into polypeptides, the particulate fraction and the soluble fraction and also a supply of ATP are all necessary. If the soluble fraction was dialysed to remove small molecules, then GTP was also needed and all the amino acids had to be added. When it became possible to prepare tRNA loaded with radioactive amino acids it could be shown that these complexes were obligatory intermediates. Even the presence of a large excess of unlabelled free amino acids did not reduce the conversion of labelled amino acyl-tRNA to polypeptide. The complex is not, therefore, just acting as a source of free amino acids and the sequence must be as follows:

$$\alpha\alpha \longrightarrow \alpha\alpha \sim AMP \longrightarrow \alpha\alpha\text{-tRNA} \longrightarrow \text{Polypeptide}$$

That this is so has been shown unequivocally in cell-free studies but has not been convincingly demonstrated *in vivo*. The complexes cannot enter whole cells and the rate at which amino acyl-tRNA is formed and broken down is such

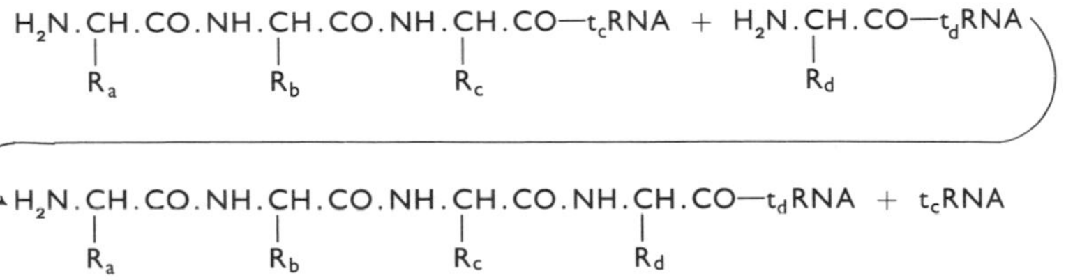

$$H_2N.CH.CO.NH.CH.CO.NH.CH.CO-t_cRNA + H_2N.CH.CO-t_dRNA$$
$$\quad\quad | \quad\quad\quad\quad | \quad\quad\quad\quad | \quad\quad\quad\quad\quad\quad\quad\quad | $$
$$\quad\quad R_a \quad\quad\quad\quad R_b \quad\quad\quad\quad R_c \quad\quad\quad\quad\quad\quad\quad R_d$$

$$H_2N.CH.CO.NH.CH.CO.NH.CH.CO.NH.CH.CO-t_dRNA + t_cRNA$$
$$\quad\quad | \quad\quad\quad\quad | \quad\quad\quad\quad | \quad\quad\quad\quad | $$
$$\quad\quad R_a \quad\quad\quad\quad R_b \quad\quad\quad\quad R_c \quad\quad\quad R_d$$

Figure 10. Elongation of a peptide chain.
The peptide combined via its carboxyl group to one tRNA reacts with an amino acyl derivative of another. The first is eliminated yielding an extended peptidyl-tRNA. The enzyme involved has been called *peptidyl transferase.*

as to have defied efforts to demonstrate its position as an obligatory intermediate. Molecules of amino acids have to be converted to protein at the rate of 8×10^8 per cell-generation, about $8 \times 10^8/50 \times 60$ molecules per cell per second. If there are $1 \cdot 5 \times 10^5$ molecules or tRNA per cell, this implies a turnover rate

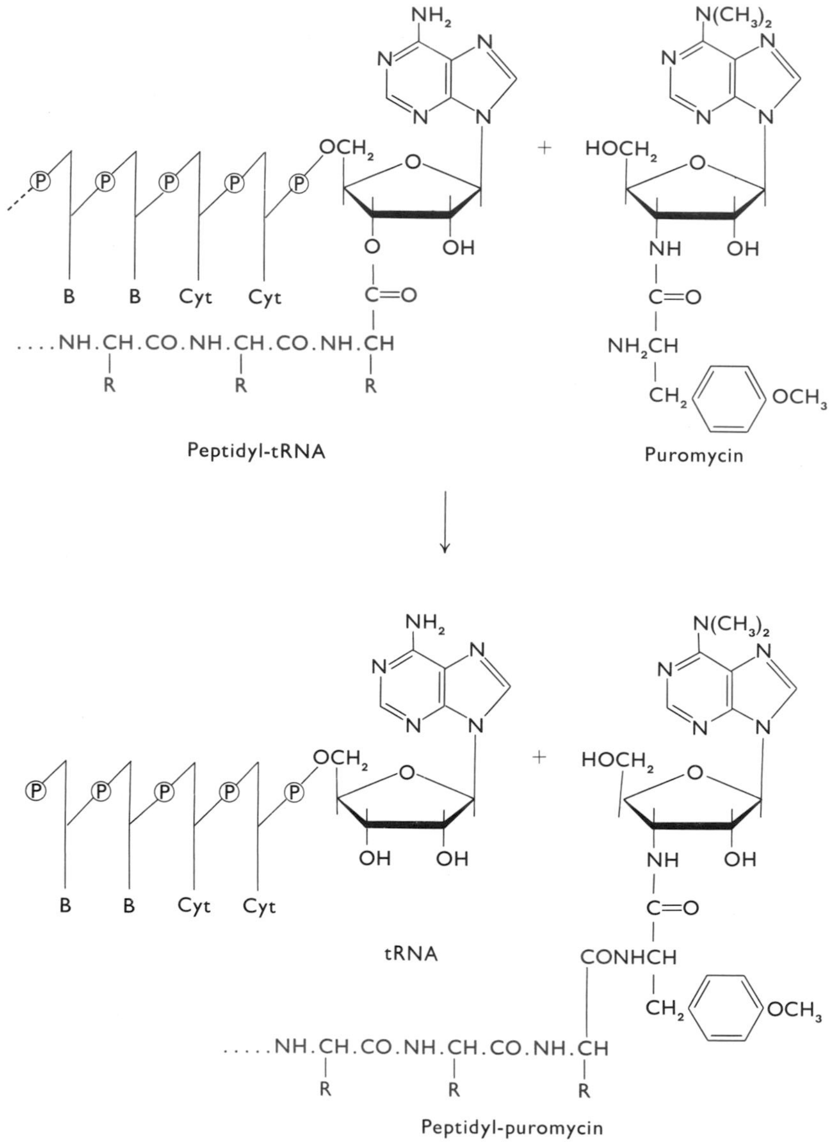

Figure 11. Premature termination of peptide chains by puromycin.

The antibiotic is an analogue of the amino acyl-adenosine end of an amino acyl tRNA and can react with a peptidyl-tRNA to yield peptidyl-puromycin and thus prematurely terminate the lengthening of the polypeptide chain.

of 8/4·5 per second. That is, each molecule of tRNA would become charged with an amino acid and discharged again about twice per second (see Tables 1 and 2, pp. 317 and 318). It is believed that this does happen *in vivo* but the difficulty of demonstrating it will be evident.

It has been established that polypeptides are built up from the *N*-terminal end by successive addition of amino acids carried by their tRNA's. Thus at any intermediate stage a peptide will carry a tRNA attached to the carboxyl of the most recently added amino acid and the next amino acyl-tRNA will react as in Figure 10. This process, catalysed by the ribosomal enzyme *peptidyl transferase*, would normally continue until the polypeptide was complete but the antibiotic, puromycin, acts as an analogue of amino acyl-tRNA and causes premature release of an incomplete polypeptide carrying the drug as its terminal residue (Figure 11). Such puromycin-terminated peptides have been found after addition of radioactively-labelled antibiotic to reconstructed cell-free systems. A useful application of puromycin in experimental systems is as a means of discharging nascent polypeptides from ribosomes. For it is in association with ribosomes that proteins are made and the newly-formed peptides can be separated still attached to these particles whether incorporation of amino acids has occurred *in vivo* or in a cell-free system.

The ingredients of cell-free systems for making polypeptides which we have established so far are the amino acyl-tRNA's, GTP, inorganic ions (particularly Mg^{2+} and K^+), 70 S ribosomes and soluble enzymes. ATP is necessary for activating amino acids but is probably not needed if amino acyl-tRNA's are provided. Soluble enzymes are not fully characterized but include at least two elongation factors which have been called T and G factors. The former is involved in binding an incoming $\alpha\alpha$-tRNA to the ribosome while the latter has GTP-ase activity and is involved in the translocation step (see p. 356).

MESSENGER-RNA (mRNA)

For some years it was thought that the ingredients mentioned in the last paragraph were adequate and, indeed, such reconstructed systems do convert amino acids to polypeptides. But it is difficult to see how a specific sequence of subunits can be assembled into a specific protein unless the 'programme' for this is present in the ribosome. For a time it was tacitly assumed that this was so and that the ribosomal-RNA contained the 'Message'. There are theoretical objections to this assumption and experimentally it has been shown to be wrong.

Cell-free incorporation systems prepared from bacteria (usually from *Escherichia coli*) had trivially low activity compared with intact cells which might be many orders of magnitude more active in polymerizing amino acids. Attempts were made by many people to increase *in vitro* performance by adding additional components to the usual amino acids, tRNA's, ribosomes, soluble enzymes, ATP, GTP, etc. Nirenberg and Matthaei in 1961 tried adding preparations of RNA of various kinds and they included polyuridylic acid (poly U) which had been synthesized enzymically from uridine diphosphate by polynucleotide phosphorylase (Figure 12). The addition of some natural RNA

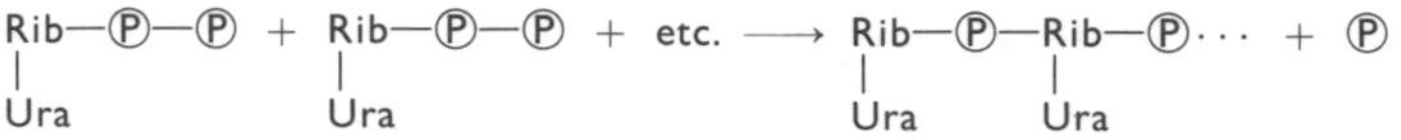

$$Rib—\textcircled{P}—\textcircled{P} \;+\; Rib—\textcircled{P}—\textcircled{P} \;+\; etc. \;\longrightarrow\; Rib—\textcircled{P}—Rib—\textcircled{P}\cdots \;+\; \textcircled{P}$$
$$\underset{Ura}{|} \qquad\qquad \underset{Ura}{|} \qquad\qquad\qquad \underset{Ura}{|} \quad \underset{Ura}{|}$$

Figure I2. Formation of poly U from UDP using polynucleotide phosphorylase.

preparations increased incorporation but the most dramatic effects were found with poly U. It caused enormous stimulation of incorporation *but only of one amino acid*, namely, phenylalanine. The product was shown to be poly-phenylalanine. Later many other polynucleotides of known base sequence have been found to enhance conversion of particular amino acids to polypeptides and the relationship of the nucleic acid bases (purines and pyrimidines) to the amino acids is fairly well established (see below, p. 339). It has also been shown that a natural RNA may direct the formation of a specific polypeptide in a reconstructed system. However, RNA extracted from purified ribosomes does not have this property and it is now believed that rRNA does not act as messenger-RNA. It is possible that nascent rRNA has activity which is lost subsequently on 'maturation'—precursor rRNA's are of greater molecular weight than those found in mature ribosomes and 10–20% of the bases are modified by, for example, methylation *after* the polynucleotide is formed. However, the 16 S rRNA precursor could be messenger for only three to five of the 30 S ribosomal proteins and the 23 S precursor for only six to ten of the 50 S ribosomal proteins. Thus at most about one-quarter of the 50–60 proteins could be specified by rRNA. The fact that ribosomes without addition of mRNA are active in promoting amino acid incorporation in cell-free systems is probably due to pieces of natural mRNA adhering to the particles.

We have seen above (p. 323) that ribosomes can exist as clusters held together by what is probably a strand of RNA and that it is on these polyribosomes that protein synthesis occurs *in vivo*. This can also be demonstrated *in vitro* and polyribosomes are often more active in promoting amino acid incorporation than are single ribosomes. Moreover, it is possible to aggregate single 70 S ribosomes into polyribosomes by adding suitable preparations of RNA, including polynucleotides such as poly U. An *in vitro* system appears, therefore, to require mRNA in addition to ribosomes and tRNA's.

THE GENETIC CODE

The information for determining the sequence of amino acids in a polypeptide chain resides ultimately in the DNA of a structural gene. The nucleotide sequence of this is *transcribed* into a complementary sequence in messenger-RNA by the mechanism described in the previous chapter (p. 299). This mRNA nucleotide sequence has to be *translated* into an amino acid sequence. There are several ways in which this 'code' has been deciphered. We shall first make some general statements which will be amplified later.

1. There are four species of nucleotide (dA, dT, dG, dC) in DNA and four (A, U, G, C) in mRNA. Both kinds of nucleic acid are linear polymers and the transcription process is as follows:

$$dA \longrightarrow U$$
$$dT \longrightarrow A$$
$$dG \longrightarrow C$$
$$dC \longrightarrow G$$

2. There are twenty different amino acids (see Figure 2, p. 14). Other amino acids found in proteins may be modifications occurring after polypeptide formation, e.g. hydroxylation of proline, methylation of lysine to ε-*N*-methyl lysine.

3. The code is triplet, non-overlapping, and is read sequentially.

$$\underline{dA \; dT \; dG} \quad \underline{dC \; dA \; dC} \quad \underline{dT \; dT \; dA} \quad \underline{dG \; dC \; dA} \quad \underline{dT \; dG \; dA} \quad \ldots$$

might correspond to

$$\alpha\alpha_1 \qquad \alpha\alpha_2 \qquad \alpha\alpha_3 \qquad \alpha\alpha_4 \qquad \alpha\alpha_5$$

4. The triplets do not recognize amino acids as such but, rather, the tRNA to which they are attached.

5. The code is universal, i.e. is the same for all organisms.

6. Of the 64 possible triplets (4^3) about 60 are meaningful for specific amino acids, i.e. the code is 'degenerate' in the sense that more than one triplet may represent the same amino acid. Some of the remaining triplets may have other functions such as to indicate 'chain termination'. Points 1 and 2 require no elaboration but we must consider the others.

Sections 5 and 6 (p. 41) mentioned the likelihood that the code is triplet in character since a doublet code only provides 4^2 different pairs of nucleotides whereas there are 20 amino acids. Conclusive proof that the code is triplet, non-overlapping, and is read sequentially has only been obtained very recently but earlier observations supported the postulate. Let us first consider the implications of point 3. Assuming that a DNA sequence of nucleotide is:

$$\underline{dA \; dT \; dG} \quad \underline{dC \; dA \; dC} \quad \underline{dT \; dT \; dA} \quad \underline{dG \; dC \; dA} \quad \underline{dT \; dG \; dA} \quad \ldots$$

and that the left-hand end is the beginning of the message in which triplets of nucleotides code for amino acids, we can predict the consequences of certain changes. For instance, removal or addition of a single nucleotide would radically alter the message from that point onwards:

<u>dA dT d$\bar{\text{G}}$</u> <u>dC dA dC</u> <u>dT$^+$ dT dA</u> <u>dG dC dA</u> <u>dT dG dA</u> ...

A deletion of the first dG would change the whole sequence:

<u>dA dT dC</u> <u>dA dC dT</u> <u>dT dA dG</u> <u>dC dA dT</u> <u>dG dA</u> ...

An addition of dX between the pair of dT's would cause alteration from that triplet onwards:

<u>dA dT dG</u> <u>dC dA dC</u> <u>dT dX dT</u> <u>dA dG dC</u> <u>dA dT dG</u> <u>dA</u> ...

However, the deletion together with the addition would only alter a small part of the sequence (the first three triplets in this instance):

<u>dA dT dC</u> <u>dA dC dT</u> <u>dX dT dA</u> <u>dG dC dA</u> <u>dT dG dA</u> ...

It is thought that some mutagenic chemicals such as acridine orange (see pp. 303 and 411) are effective because they cause the addition or deletion of a nucleotide pair from the DNA. If this is so and if the DNA is coding for protein, then three deletions at the places indicated:

<u>dA dT dG</u> <u>d$\bar{\text{C}}$ dA dC</u> <u>dT d$\bar{\text{T}}$ dA</u> <u>d$\bar{\text{G}}$ dC dA</u> <u>dT dG dA</u> ...

would give:

<u>dA dT dG</u> <u>dA dC dT</u> <u>dA dC dA</u> <u>dT dG dA</u> ...

Three additions at the same sites would give:

<u>dA dT dG</u> <u>dX dC dA</u> <u>dC dT dY</u> <u>dT dA dZ</u> <u>dG dC dA</u> <u>dT dG dA</u> ...

These triple mutants would not only have a small number of altered amino acids but would also have one fewer or one more amino acid than the wild-type. It has been found experimentally with acridine mutants of the bacteriophage T4 that single mutants were unable to grow, some double mutants could and some could not, and some triple mutants were semi-normal. Although there is no certainty as to which are additions and which are deletions, these findings were interpreted to mean that an addition plus a deletion could restore the bulk of the message to normal and that three additions or three deletions could do the same (see also Figure 16, p. 412). It was suggested also that the observations made it likely that the code was triplet or a multiple of that. As we shall see there are even firmer bases for the belief in the validity of point 3.

The genetic information we have been discussing must be transcribed as follows:

dA dT dG   dC dA dC   dT dT dA   dG dC dA   dT dG dA   ...   DNA sequence

U A C   G U G   A A U   C G U   A C U   ...   mRNA sequence

There is no evidence that a particular ribonucleotide triplet has any affinity for any particular amino acid. Rather, the recognition is of the transfer-RNA to which the amino acid is esterified. This has been demonstrated very beautifully in the cell-free system from reticulocytes (immature red blood cells). This preparation can synthesize the polypeptide chains of haemoglobin and can be supplied with the amino acids attached to their tRNA's. What was done was to modify the amino acid chemically *after* it had been attached to tRNA and to show that it was still incorporated into the peptide as if it had not been so altered. Cysteine attached to tRNA^{Cys} was treated with Raney nickel which converted the amino acid to an alanyl residue (Figure 13). When the product was added to the *in vitro* reticulocyte system it was found that the polypeptide produced contained alanine at some sites where cysteine was expected. Similar

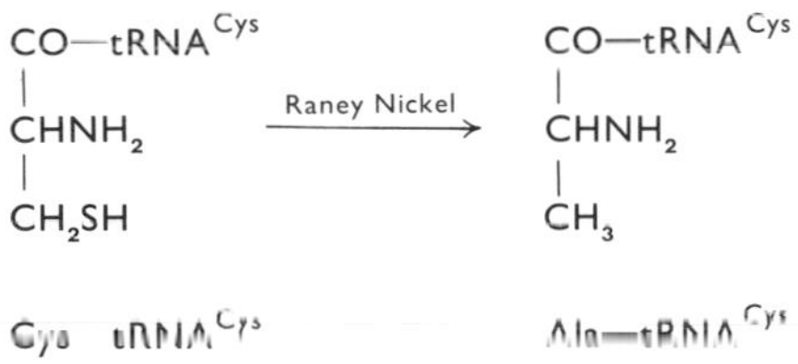

$$CO{-}tRNA^{Cys} \qquad\qquad CO{-}tRNA^{Cys}$$
$$| \qquad\qquad\qquad\qquad |$$
$$CHNH_2 \quad \xrightarrow{\text{Raney Nickel}} \quad CHNH_2$$
$$| \qquad\qquad\qquad\qquad |$$
$$CH_2SH \qquad\qquad\qquad CH_3$$

$$Cys\ tRNA^{Cys} \qquad\qquad Ala{-}tRNA^{Cys}$$

Figure 13. Conversion of the cysteinyl residue to the alanyl residue while still attached to the transfer-RNA specific for cysteine.

When the altered amino acyl-tRNA was used in a cell-free system making haemoglobin poly-peptide chains, some alanine was found in positions where cysteine was expected. This confirms that the tRNA (normally) recognizes the mRNA and so delivers the appropriate amino acid to the correct site.

experiments have been carried out using synthetic polynucleotides as mRNA and demonstrating that the product was determined by the tRNA rather than by the amino acid attached to it. This is excellent justification of point 4 above.

The universality of the code (point 5) is suggested by the results of *in vitro* experiments in which amino acids linked to tRNA from one species of organism are functional with ribosomes and mRNA of another, e.g. *Escherichia coli* and mammalian reticulocytes, respectively. There is an obvious evolutionary significance in this.

Deciphering the genetic code

The allocation to the 64 ribonucleotide triplets of specific functions is now complete. Table 3 lists these and it should be noted that the sequence of the three bases is important and that they are recorded in a conventional order. Thus A U G has phosphate diester links from the 3′ of the adenine nucleoside to the 5′ of the uracil nucleoside and from the 3′ of this to the 5′ of the guanine nucleoside whereas G U A would be the converse (Figure 14).

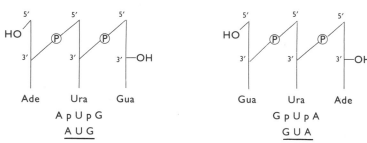

Figure 14. Conventional abbreviations for oligonucleotides.

   The form A U G implies a phospho-diester link between the 3′ and 5′ positions of the adenine and uracil nucleosides respectively. Similarly there is a link between the 3′ position of the uridine and the 5′ position of guanosine. A p U p G likewise means that the 5′ hydroxyl of the ribose attached to adenine and the 3′ hydroxyl of that attached to guanine are unsubstituted.

   We shall now discuss methods of deciphering the code. The most rigorous would be to establish the complete nucleotide sequence of a DNA structural gene or the corresponding mRNA and then to relate this to the amino acid sequence of the appropriate polypeptide. Techniques for determining polypeptide sequences have been available for some time and those for nucleic acids have now been developed. Many sequences of proteins and nucleic acids are known and the first total synthesis of a DNA structural gene (that for a yeast tRNA^{Ala}) was announced in June 1970 (see Chapter 5, p. 277). However, tRNA's do not act as messengers so there is no corresponding polypeptide. But the nucleic acids of RNA viruses do specify proteins and the *Escherichia coli* phages R 17 and Q β have been much studied. The complete sequence of over 3000 nucleotides in the RNA and the 1000 amino acids they specify in three separate proteins should be known before long (see below, p. 361).

   An early method used in decipherment was to relate amino acid incorporation by cell-free systems to the base sequence of synthetic polyribonucleotides acting as artificial messengers. Thus poly dA acts as a template for RNA polymerase to synthesize poly U and this acts as a code *in vitro* for the synthesis of poly-phenylalanine. It was concluded that the triplet U U U stands for phenylalanine. Similarly, poly A and poly C yield polymers of lysine and proline, respectively. Mixed polymers can also be synthesized by adding the enzyme polynucleotide phosphorylase to mixtures of ribonucleoside dis-phosphates. From the relative amounts of nucleotides in the product and on the assumption that the sequence is random, the frequency of all possible triplets can be calculated. If such a polynucleotide is used as an artificial messenger the amino acids incorporated will depend in kind and amount on the nature of the code and the frequency of occurrence of the various triplets. Thus if poly AU is a random sequence of A's and U's, all the following triplets will be present:

U U U, U U A, U A U, A U U, U A A, A A U, A U A and A A A.

The relative abundances of these can be predicted from the composition and it can be determined experimentally which amino acids become incorporated when this polynucleotide is used as messenger. It should be noted, however, that the three triplets U_2A will have the same frequencies as each other as will the three UA_2. In other words the sequence of the three nucleotides will not be apparent. From Table 3 it can be seen that the eight triplets will code for the amino acids Phe, Leu, Tyr, Ile, Asn and Lys. Another preparation of poly AU

Table 3. The Triplet Code

5'-OH Terminal base	Middle base				3'-OH Terminal base
	U	C	A	G	
U	UUU } Phe UUC UUA } Leu UUG	UCU UCC } Ser UCA UCG	UAU } Tyr UAC UAA Ochre UAG Amber	UGU } Cys UGC UGA Umber UGG Trp	U C A G
C	CUU CUC } Leu CUA CUG	CCU CCC } Pro CCA CCG	CAU } His CAC CAA } Gln CAG	CGU CGC } Arg CGA CGG	U C A G
A	AUU AUC } Ile AUA AUG Met	ACU ACC } Thr ACA ACG	AAU } Asn AAC AAA } Lys AAG	AGU } Ser AGC AGA } Arg AGG	U C A G
G	GUU GUC } Val GUA GUG	GCU GCC } Ala GCA GCG	GAU } Asp GAC GAA } Glu GAG	GGU GGC } Gly GGA GGG	U C A G

Most of the 64 triplets of nucleotides are allocated to specific amino acids. A few have special functions. Thus A U G represents the chain-initiating N-formyl-methionine. U A A, U A G and U G A indicate chain termination.

of different composition would have different abundances and the altered relative amino acid incorporations can help to indicate which triplet corresponds to which amino acid.

Polynucleotides of the form X U U U U U U ... have been prepared and used to investigate from which end the message is read and also to find out what X U U stands for. The resulting polypeptide should be either:

$$NH_2CH.CO—Phe—Phe—Phe...Phe.COOH$$
$$|$$
$$R$$

or

$$NH_2Phe—Phe—Phe—Phe...PheCO.NH.CH.COOH$$
$$|$$
$$R$$

The evidence indicates that the translation begins from the 5′ end of the polynucleotide and that this corresponds with the *N*-terminal end of the polypeptide (Figure 15).

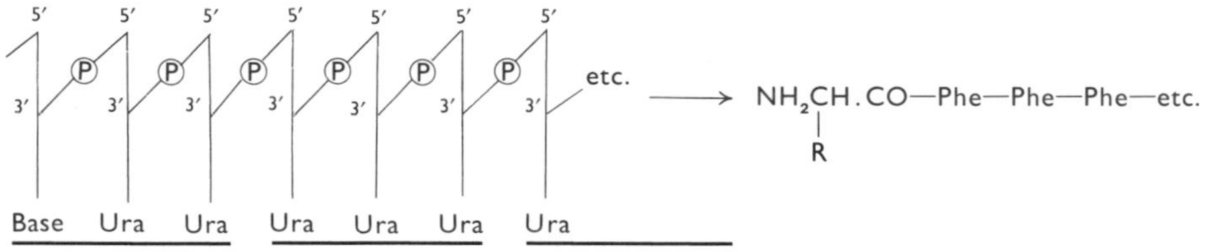

Figure 15. Direction of translation of mRNA.

A synthetic polyribonucleotide of the form X U U U U U U ... is used as artificial mRNA in a cell-free amino acid incorporating system. The product has an amino acid other than phenylalanine only at the *N*-terminal end. Since synthesis begins at the *N*-terminus, this implies that translation begins at the 5′ end of the mRNA.

Recently, polynucleotides of regular alternating sequence have been prepared, e.g. poly (UC), poly (AG), poly (UG) and (AC). Each of these, if added to an *in vitro* system (prepared from *Escherichia coli*) programmed the formation of a polypeptide consisting of only two alternating amino acids. Thus poly (UC) yielded poly (seryl-leucyl). This very strongly supports the non-overlapping triplet theory besides indicating the triplets for serine and leucine. Poly (UC) can be represented:

U C U   C U C   U C U   C U C   U C   ...

and of the only two triplets occurring, U C U probably codes for serine and C U C for leucine.

Poly (AG) gave alternating arginine (A G A) and glutamic acid (G A G); poly (UG) gave alternating cysteine (U G U) and valine (G U G); and poly (AC) gave alternating threonine (A C A) and histidine (C A C).

Even more striking was that a polynucleotide composed of repeating units of A A G acted as a messenger for three homopolymers—poly-lysine, poly-arginine and poly-glutamic acid. This is consistent with the following allocations of triplets:

A A G	A A G	A A G	A A G	...	A A G codes for lysine
Lys	Lys	Lys	Lys		

A G A	A G A	A G A	A G A	...	A G A codes for arginine
Arg	Arg	Arg	Arg		

G A A	G A A	G A A	G A A	...	G A A codes for glutamic acid
Glu	Glu	Glu	Glu		

Since for the series A A G A A G A A G A A G A A ... only the three triplets mentioned do occur and since no other amino acids are incorporated and since only homopolymers are formed, this is further evidence that the code is a non-overlapping, triplet one in which contiguous triplets are read sequentially without bypassing a single nucleotide.

A few polymers having repeating tetra-nucleotides have been prepared and are found to yield polymers of four amino acids in sequence; for instance poly (UAUC) codes for tyrosine, leucine, serine and isoleucine:

U A U	C U A	U C U	A U C	U A U	C U A	U C U	A U C	...
Try	Leu	Ser	Ile	Tyr	Leu	Ser	Ile	

A second method of approach for deciphering the code is one which is related more directly to natural systems. It is the study of amino acid replacements in proteins. A mutational change involving the *alteration* (not deletion or addition) of only a single nucleotide may result in the replacement of a single amino acid by another at a particular place in a polypeptide. Thus the DNA triplet dA dA dA is transcribed to the mRNA triplet U U U and this is translated as phenylalanine. A genetic alteration to dA dA dT would result in the RNA triplet U U A which represents leucine. Hence leucine would replace phenylalanine at *one* specific place in the appropriate protein. There are, of course, many possibilities:

DNA Triplet	mRNA Triplet	Amino Acid
dA dA dA	U U U	Phenylalanine
dA dA dT	U U A	Leucine
dA dT dA	U A U	Tyrosine
dT dA dA	A U U	Isoleucine

etc.

But with hindsight we can see that a *single change* should only produce a limited number of alternative replacements. This has been found to be so in naturally occurring mutants of this kind and the translation of the code must be consistent with these observations. Artificial changes of a similar kind can be induced by chemical modifications, e.g. nitrous acid may convert purine and pyrimidine amino groups to hydroxyls and thus alter the bases in DNA (see Chapter 7, pp. 380–4).

The final method to be described depends on a somewhat different principle. We have seen that mRNA can be attached to ribosomes and that amino acyl-tRNA appears to recognize an appropriate nucleotide region in the mRNA. The whole mRNA is not necessary for this association—a triplet of nucleotides is enough. Thus ribosomes plus the triplet U U U form a complex with phenylalanyl-tRNA but not with any other amino acyl-tRNA nor with any free amino acid. All of the 64 trinucleoside diphosphates (XpYpZ) have now been tested individually with all the amino acyl-tRNA's. In very many instances the results are clear-cut and indicate a specific relationship between a particular amino acid and one or more triplets. An advantage of this method is that it gives unequivocal information about the sequence of the three bases as well as the overall composition of the triplet.

It is gratifying that the results of *in vitro* incorporation studies, of amino acid replacements and the investigations on binding confirm each other in the main so that the translations given in Table 3 are probably correct. It should be noted that only three of the 64 triplets do not seem to code for an amino acid (see below) and also that certain generalizations can be made about those which do. In every instance XpYpU and XpYpC represent the same amino acid, i.e. either pyrimidine can be in the third position. Frequently this is also true for the purines, XpYpA and XpYpG being alternatives. Indeed for eight amino acids the code is essentially doublet rather than triplet in that the third position can be occupied by any one of the four bases. (This does not imply, however, that these alternatives may not have different functions *in vivo* and we shall return to this shortly.) Finally, the assortment of triplets does not seem to be random—those amino acids which are related structurally or metabolically often have related codes. A consequence of this which may be significant is that a mutation leading to replacement may cause substitution of a related amino acid rather than one which is quite dissimilar. This might well minimize the adverse effects.

The beginning and ending of polypeptide chains

Termination codons

When polypeptides are synthesized *in vitro* in a reconstructed, cell-free system it seems that the polynucleotide used as 'artificial' messenger can be read from various starting points. Thus the sequence A A G A A G A A G which we mentioned above can apparently be read as a sequence of A A G or A G A or G A A. Manifestly this would be undesirable *in vivo* with a natural messenger-RNA since it would involve the synthesis of a great deal of useless polypeptides. Messages must have a beginning and an ending. There is evidence that the three triplets which have not been assigned to amino acids (formerly called 'nonsense') represent termination codons and are normally present at the end of each message. They are U A A, A U G and U G A. Mutants have also been found in which a single base change converts an amino acid codon to one of these and this results in premature termination of polypeptide synthesis and release of incomplete peptide. The site of the mutation determines the amino acid at which the peptide ends and it has been established that the next triplet in the mRNA is U A A or U A G or U G A. Surprisingly these three termination codons do not have corresponding tRNA's but are recognized by specific termination proteins or release factors of which R_1 interacts with U A A or U A G and R_2 with U A A or U G A (see below, p. 357). However, an additional *suppressor* mutation (see p. 376) at another locus may over-ride the effect of a mutation such as we have been considering. The consequence of this is to cause the formation of a modified tRNA with an anti-codon complementary to the termination codon produced by the first mutation. Thus an amino acid is inserted and complete polypeptide is made (see Figure 16).

Some natural mRNA's carry information for more than one polypeptide chain, so-called *polycistronic messengers* (see also Chapter 8, p. 443) and have intra-cistronic as well as terminal stop codons (see p. 361).

Initiation codons

There are some curious features about the beginnings of polypeptide chains. We mentioned that the ribosomal proteins of *Escherichia coli* nearly all begin with methionine or alanine (p. 322). It is also found that the soluble proteins (the bulk of the total) of the same organism have *N*-terminal methionine (45%) or alanine (30%) or serine (15%). As there are probably of the order of a thousand different species of proteins it is certainly surprising to find so few of the twenty possible amino acids at the *N*-terminal ends of polypeptides.

There are two different transfer-RNA's for methionine and they have been called $tRNA_f^{Met}$ and $tRNA_m^{Met}$. When methionine is bound to the former it can be enzymically formylated both *in vitro* and *in vivo*, yielding *N*-formyl-methionyl-$tRNA_f$. The analogous methionyl-$tRNA_m$ cannot be so formylated. Moreover, the former complex is precursor to *N*-terminal methionine whereas the latter delivers methionine to other positions in the polypeptide chain. Both tRNA's have the same anti-codon triplet and interact with the codon A U G ($tRNA_f$ can also recognize the codon G U G which normally codes for valine) but there are differences in structure elsewhere. As far as is known all

(a) Wild-type Structural Gene   Gene for minor species of tRNA^{Tyr}

DNA   5′ ———— TAC ————▶   ———— GTA————▶ 3′

        3′ ◀——— ATG ————   ◀——— CAT ———— 5′

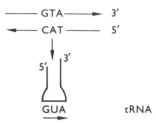

mRNA   5′ ————UAC————▶ 3′   GUA   tRNA

               Tyrosine codon U̲A̲C̲

                              A̲U̲G̲ Tyrosine anti-codon

Polypeptide   —αα$_p$—*Tyr*—αα$_r$—

(b) Mutation in Structural gene ('Amber')

DNA   5′ ———— TA**G** ————▶

        3′ ◀——— AT**C** ————

Altered
m RNA   5′ ————UA**G**————▶ 3′

               Termination codon UA̲G̲

                    - - - -   No corresponding anti-codon

Incomplete polypeptide   —αα$_p$ COOH

(c) Mutation in gene for minor species of tRNA^{Tyr}

DNA   5′ ———— TAC ————▶   ———— **C**TA ————▶ 3′

        3′ ◀——— ATG ————   ◀——— **G**AT ———— 5′

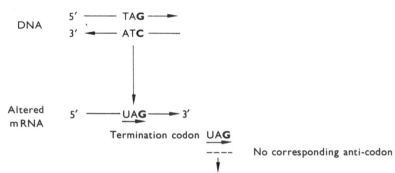

Unchanged   Altered tRNA
mRNA   5′ ———— UAC————▶ 3′   CUA   (unused)

               Tyrosine codon U̲A̲C̲ still recognised by

                  anti-codon A̲U̲G̲ of *major* species of tRNA^{Tyr}

Normal polypeptide   — αα$_p$— *Tyr*—αα$_r$—

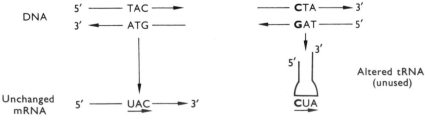

(d) Mutation in Structural gene *and* **Suppressor mutation** in gene for minor species of tRNA^{Tyr}

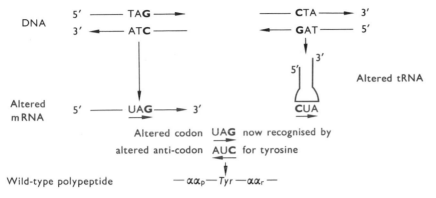

Altered codon UAG now recognised by

altered anti-codon AUC for tyrosine

Wild-type polypeptide — $\alpha\alpha_p$—*Tyr*—$\alpha\alpha_r$ —

Figure 16. Mutation in a structural gene and its suppression by mutation of a gene for a minor species of tRNA.

Mutation converting a tyrosine codon to UAG, a termination codon, yields an 'amber mutant' producing incomplete polypeptides terminating just before where a tyrosine residue occurs in the wild-type (b).

Mutation leading to an altered minor species of tRNA^{Tyr} has no effect since other species of tRNA^{Tyr} interact with the unchanged tyrosine codon (c).

The effects of the structural gene mutation giving UAG can be suppressed by the second mutation changing the tRNA. This results in the UAG being translated into tyrosine rather than acting as a terminator (d).

polypeptides are synthesized beginning with *N*-formyl-methionine and nearly half have the formyl group removed enzymically to yield proteins with *N*-terminal methionine. Others have the methionine removed also — by a specific aminopeptidase — yielding a protein with *N*-terminal alanine or serine or, less frequently, some other amino acid. Thus the RNA of the small phage f_2 acts, *in vitro*, as mRNA for the synthesis of a protein beginning

N-formyl-Met—Ala—Ser—Asn—Phe—Ser— · · ·

whereas the phage protein made *in vivo* begins

Met—Ala—Ser—Asn—Phe—Ser— · · ·

Further, in model systems with synthetic mRNA's the results shown in Figure 17 were obtained.

Synthetic mRNA						Product
AUG	AAA	AAA	AAA	...	AAA	(f)Met-(Lys)$_n$ on ribosomes
AUG	UUA	AAA	AAA	...	AAA	(f)Met-Leu-(Lys)$_n$ on ribosomes
AUG	UUU	AAA	AAA	...	AAA	(f)Met-Phe-(Lys)$_n$ on ribosomes
AUG	UUU	UAA	AAA	...	AAA	f-Met-Phe released from ribosomes

Figure 17. Initiation and termination codons (see next page).

Synthetic polyribonucleotides $AUG(A)_n$, $AUGU_2(A)_n$, $AUGU_3(A)_n$ and $AUGU_4(A)_n$ were used as artificial messengers in the presence of labelled amino acids, ribosomes and appropriate supplements.

With the first three, a peptide containing a labelled methionine (presumably formylated) remained attached to the ribosome. In addition it contained many lysine residues and, as indicated, no other residue, or one leucine, or one phenylalanine.

Only with $AUGU_4(A)_n$, which contains the termination codon U A A, was a peptide released from the ribosome—the dipeptide *N*-formyl-methionyl-phenylalanine.

This experiment proves that the code is triplet, non-overlapping, sequential, read from 5′ to 3′, and that A U G initiates and U A A terminates.

NATURAL MESSENGER-RNA

It should by now be abundantly clear that there is a component of bacterial cells to which the name messenger-RNA must be given. But it is an elusive substance which has been the subject of controversy. The existence of such a class of RNA was predicted before it was demonstrated in any convincing way. It was postulated that mRNA should have a composition like that of DNA since it was to transmit genetic information to the protein-forming systems; that it should be poly-disperse in size to account for the many different sizes of protein molecules; and that it should be capable of being used perhaps only a few times before being degraded—this because of the rapidity with which the production of particular proteins could be switched on and off in bacteria. These were all postulates and have had to be qualified.

The kinetics of enzyme synthesis after induction are discussed in Chapter 8 and it is shown that the maximum rate can be achieved in a minute or two and that synthesis can cease equally rapidly. Moreover, under certain conditions up to 5% of the total protein production may be turned over to the making of one species. These properties seem to demand great flexibility in formation, utilization and discarding of mRNA.

The concept of mRNA is of necessity a functional one—mRNA is that RNA which carries a message, i.e. the information for the formation of a poly-peptide. This has led to a number of misconceptions of which some are widely current. In fact mRNA does not have to have the composition of DNA although it has been referred to as DNA-like or D-RNA. It should, rather, be complementary in nucleotide composition to *one strand of DNA*—and therefore be like the other strand if the DNA is of the usual double-helical type. Much work has been carried out with *Escherichia coli* and it just happens that in this organism the composition of each strand is like that of the other as well as being complementary to it and, in addition, that for this DNA, $dA = dC = dG = dT$.

Secondly, one should not expect the composition of mRNA to resemble the total DNA of a cell, i.e. the thousands of genes of all kinds. Rather, a specific molecule of mRNA should be complementary in composition to one strand of the corresponding structural gene. If much of the genome (chromosome) is not composed of structural genes or if the relative abundances of the various mRNA's were different from those of the corresponding genes, one would not

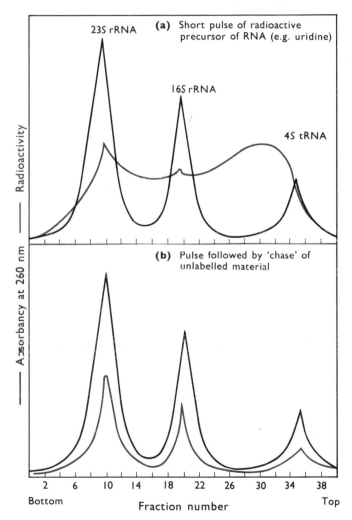

Figure 18. Sucrose gradient analysis of nucleic acids.

RNA can be extracted from bacteria with phenol and appears as three kinds—23 S and 16 S ribosomal RNA (rRNA) and 4 S soluble RNA (mainly tRNA). A short pulse of radioactive precursor shows a different distribution since the mRNA fraction, although heterogeneous in size and very small in amount, turns over very rapidly and accounts for about $\frac{1}{3}$ of the label in short times. After 'chasing' the pulse with unlabelled precursor the radioactivity is retained only in the tRNA and rRNA components.

necessarily expect overall relationships in composition between total DNA and total mRNA.

It is also misleading to suggest that all the RNA labelled during a short period of growth in the presence of a radioactive precursor (so-called pulse-labelled RNA or rapidly labelled RNA) is necessarily all mRNA or that all mRNA is obliged to have a short life. At any instant the fraction which can be identified by one or more criteria as being mRNA is only one or a few hundredths of the total cellular RNA. However, whereas ribosomal-RNA and amino acid

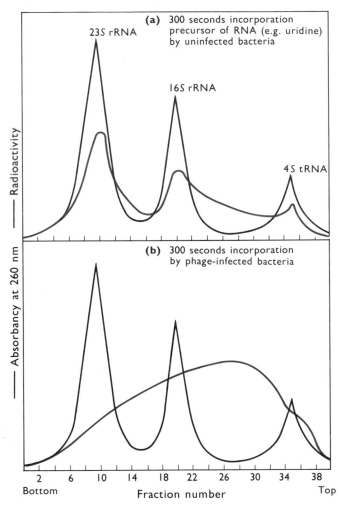

Figure 19. Labelling of RNA components in uninfected and phage-infected bacteria.

At early times in uninfected bacteria the label from radioactive precursors is found in all components, about $\frac{2}{3}$ of it being in rRNA (see Figure 18a). Later this proportion increases substantially (a). However, in cells infected with a virulent bacteriophage, little or no rRNA or tRNA is made and the label is found mainly as polydisperse mRNA even after quite long periods (b).

transfer-RNA are very long-lived, there is a fraction of RNA which does not survive very long—perhaps only a few minutes. This means that pulse-labelling will give a higher fraction of radioactivity in this labile RNA but even so short a pulse as 30 seconds may result in more than half of the radioactivity being in rRNA with a little in tRNA. Whether or not all the remainder is in mRNA is a matter for conjecture. If, after a labelling period of 30 seconds, excess of non-radioactive precursors are added to 'chase' the radioactivity out of the transient substances, it is found that rRNA's and tRNA's are stable end-products whereas the rest of the radioactivity is in a labile component. This is shown in Figure 18 where the total RNA extracted by phenol has been centrifuged through a sucrose gradient. The positions of the 23 S and 16 S rRNA's are seen as is the 4 S peak of tRNA's. The radioactivity after 30 seconds is distributed rather diffusely over a range of sizes. After 'chasing', only the rRNA and tRNA are labelled. It is highly probable that the transient component is mRNA as it can be shown to have a composition unlike that of rRNA or tRNA and to be hydridizable with DNA.

The differences between rRNA and tRNA on the one hand and what is likely to be mRNA on the other can also be seen by using a system in which little if any of the former are made and in which most RNA synthesized is probably mRNA. This occurs in bacteria infected with virulent bacteriophage. Figure 19 compares the radioactive distribution after 300 seconds in uninfected cells with that found after 300 seconds in phage-infected bacteria. In the former the label is in all components whereas in the latter it is mainly polydisperse—mRNA?

In general it is difficult to reduce the specific activity of the nucleotide pool by adding unlabelled substances and so the 'chase' has to be relatively long. There is, however, another way in which the disappearance of label from some types of RNA but not from others can be shown. This is by adding the antibiotic actinomycin D which prevents further synthesis of all kinds of RNA. Under these conditions breakdown of the labile RNA is roughly exponential with time and so it is possible to determine the mean half-life by studying the kinetics of loss of radioactivity from the unstable RNA fraction (Figure 20). The addition of the drug also brings about a progressive impairment in the rate of protein synthesis and this reduction is also exponential with time. Finally, there is an exponential decrease in the content of polyribosomes and a corresponding increase in the number of 70 S ribosomes. It would be remarkable if these phenomena were not interrelated and the simplest explanation is that they are consequent upon the breakdown of mRNA which holds 70 S ribosomes together as polyribosomes and which is necessary for synthesis of protein.

It is technically difficult to isolate natural mRNA from bacterial cells but the nucleic acids of some RNA viruses can readily be extracted and studied *in vitro*. These natural messengers have been shown to programme the formation

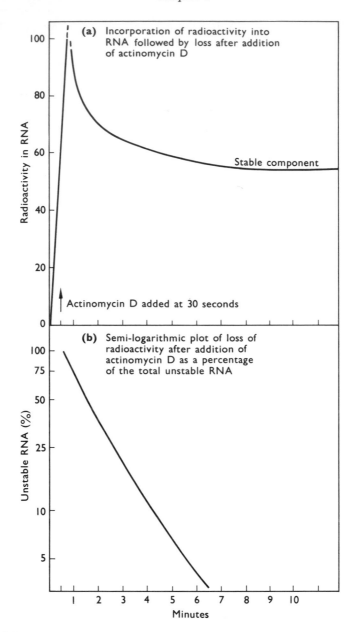

Figure 20. Breakdown of labile RNA.

(a) shows the rapid incorporation of label from radioactive precursors into RNA. This can be stopped almost instantly by addition of actinomycin D which prevents all synthesis. Subsequently about half of the label becomes soluble as RNA breaks down. This labile component is only a few percent of the total RNA and is probably largely mRNA.

(b) After subtracting the amount of the stable component (the asymptote in (a)) the decay of the labile component can be plotted. For a pulse of 30 seconds, the time for 50% decay is about 60 seconds and that for 80% is about 150 seconds.

of characteristic viral proteins. Such experiments are much more convincing than those using poly U as mRNA.

INTERACTION OF mRNA, tRNA AND RIBOSOMES

We have seen that ribosomes may be aggregated to form polyribosomes by addition of poly U and that natural polyribosomes are probably held together by mRNA. The binding *in vitro* can be shown to be to the 30 S subunit of the ribosome. Transfer-RNA, on the other hand, probably binds to the 50 S subunit and there is evidence that there are two such sites on the particle. These correspond to one for the tRNA at the end of the nascent polypeptide and the other for the next incoming amino acyl-tRNA (Figure 21).

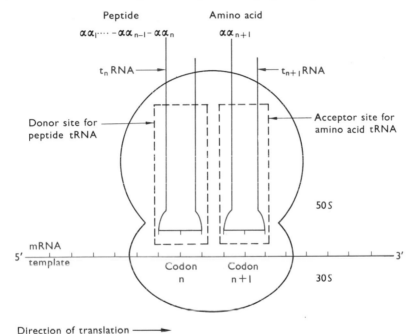

Figure 21. Representation of binding sites for transfer-RNA on 70 S ribosomes.

The mRNA is bound to the 30 S component. There are two sites for tRNA and these may involve both 30 S and 50 S subunits. One of them is an *acceptor site* for amino acyl tRNA (sometimes called the 'amino acid site'); the other is the *donor site* for the peptidyl tRNA (sometimes called the 'peptide site').

The antibiotic puromycin acts as if it could occupy part of the acceptor site and receive a peptide from the donor.

There must also be interaction between the mRNA and the tRNA if the latter is to place an amino acid correctly at the behest of the former. Since a triplet of nucleotides is now established as the determinant for an amino acid and since the only chemical explanation available for the recognition is that of base-pairing, it is not surprising that the tRNA has a triplet which is complementary to that of the mRNA sequence. The mRNA triplet is called a 'codon' and the complementary tRNA triplet is called an 'anti-codon'. The interactions are between anti-parallel strands just as they are between the two strands of DNA (see Chapter 5, p. 291). Relationship could be thus:

	dA dG dA	dT dT dG	dC dT dT
DNA	$\longrightarrow$	$\longrightarrow$	$\longrightarrow$
	dT dC dT	dA dA dC	dG dA dA
	$\longleftarrow$	$\longleftarrow$	$\longleftarrow$
mRNA codon	A　G　A	U　U　G	C　U　U
	$\longrightarrow$	$\longrightarrow$	$\longrightarrow$
tRNA anti-codon	U　C　U	A　A　C	G　A　A
	$\longleftarrow$	$\longleftarrow$	$\longleftarrow$
Amino acid	Arginine	Leucine	Leucine

where two of the six leucine codons are represented, the arrows indicating the $5' \rightarrow 3'$ direction.

When purified tRNA's with known anti-codons, became available it was possible to test their interactions with trinucleotides and it was found that some tRNA's could recognize several different codons. Moreover, many tRNA's contained the base inosine at the 5' position in their anti-codon. These observations led to the 'wobble' hypothesis which suggests that base pairings other than A—U and G—C can occur between the base at the 3'-end of the mRNA codon and that at the 5'-end of the tRNA anti-codon. The possible combinations are shown in Table 4.

Table 4. The 'wobble' hypothesis

Base in 5'-position of anti-codon	Base in 3'-position of codon
G	C or U
C	G
A	U
U	A or G
I	A, U or C

The base in the 5'-position of the anti-codon of tRNA may be able to form H-bonds with more than one base at the 3'-position of the codon of mRNA. Thus guanine in the 'wobble' position can pair with cytosine or uracil; uracil with adenine or guanine; inosine with adenine, uracil or cytosine. In the other two positions of the anti-codon only the usual C—G and A—U pairs are possible.

From the hypothesis it can be predicted that 31 is the minimum number of tRNA's which could react with all 61 amino acid codons and, more specifically, that three tRNA's should exist for the six serine codons (U C U, U C C, U C A, U C G, A G U and A G C). The latter prediction has been confirmed and it seems that organisms may indeed have more than twenty species of tRNA and may be able to translate any of the 61 codons.

As well as interacting with mRNA, the transfer-RNA has to have a site specific for the amino acid activating enzyme and a region which has affinity for some part of the 50 S ribosomal subunit. These properties may account for the size of these molecules (about 80 nucleotides in length) and for their having regions of similar structure (see Chapter 5, p. 264). There is much speculation as to how these interactions occur. A chain of 200 amino acid residues (molecular weight *c.* 22,500) requires an mRNA of 600 nucleotides, i.e. about 200 nm in length if there is one nucleotide every 3·4 Å, or double that if the

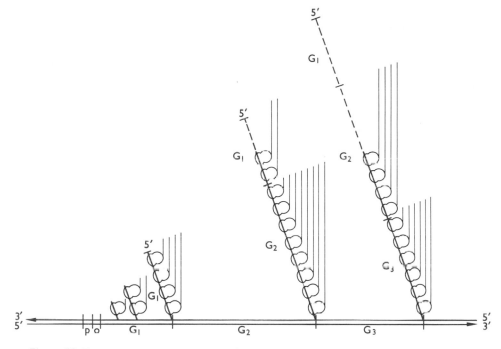

Figure 22. Diagrammatic representation of transcription and translation and degradation of mRNA.

Three structural genes, G_1, G_2 and G_3 form an operon with promoter, p, and operator, o. Transcription of one strand of DNA yields a polycistronic mRNA, the 5′-end of which is made first. Ribosomes become attached here and sequentially translate the three cistrons. Degradation from the 5′-end begins before transcription is complete and no further ribosomes can be added. Those already attached complete translation.

mRNA is in extended form. Since a 70 S ribosome is roughly 200×170 Å, it is clear that there must be relative movement between it and the mRNA so that the message can be read sequentially from beginning to end. A stretch of 15–20 nucleotides is protected by association with a ribosome against degradation by ribonuclease and it seems likely that ribosomes are closely packed along the length of the mRNA, perhaps one to every 100–200 Å. There is cumulative evidence that ribosomes begin translating mRNA while it is still being transcribed from DNA and recently it has been suggested that degradation from the 5′-end may also begin before transcription is complete but only after some dozens of ribosomes have sequentially begun translation. Such a situation might be represented diagrammatically as in Figure 22.

INITIATION, ELONGATION AND TERMINATION PROTEIN FACTORS

Many protein factors have been described which appear to be involved in the formation of polypeptides from amino acyl-tRNA's by complexes of ribosomes and mRNA. The peptidyl transferase itself is an integral part of the 50 S subunit but at least eight other proteins have been found in the high speed supernatant fraction of cytoplasm or can be washed off ribosomes by solutions of high ionic strength (e.g. 2 M NH$_4$Cl). Some of them are concerned in chain initiation, some in elongation and some in termination.

Initiation factors

Three factors have been separated from various organisms and shown to be involved in the formation of a complex of mRNA, 30 S subunit, N-formyl-methionyl-tRNA$_f$ (fMet-tRNA$_f$) and 50 S subunit with the concomitant breakdown of a mole of GTP to GDP:

$$30\,S + mRNA + fMet\text{-}tRNA_f + 50\,S + GTP \xrightarrow{\;F_1, F_2, F_3\;} [fMet\text{-}tRNA_f, 70\,S, mRNA] + GDP + ⓅＰ$$

The factors, F_1, F_2, F_3, have been given alternative names by different workers:

	Bacterial		Mammalian
F_1	A	F_i	M_i
F_2	C	F_{iii}	M_{iii}
F_3	B	F_{ii}	M_{ii}

Factor F_3 (molecular weight 21,000) also appears to cause dissociation of 70 S ribosomes into subunits so it may be that it is associated with the 30 S particle when it begins initiation. There are at least three protein fractions separable from *Escherichia coli* which have F_3 activity and at least one more is found in cells infected with phage T$_4$. These factors seem to discriminate between one kind of ribosome and another and also one kind of messenger and another. In fact it appears that F_3 selects a specific ribosome binding site on the 5'-side of the first initiation codon of the mRNA. Factor F_1 (a very small polypeptide, molecular weight about 8000) stabilizes the complex of mRNA and 30 S subunit:

$$[30\ S, F_3] + mRNA + F_1 \longrightarrow [mRNA, 30\ S, F_1] + F_3$$

Now fMet-tRNA$_f$ is prepared for reaction with this complex by first combining with GTP and F_2 (molecular weight about 75,000):

$$F_2 + GTP \longrightarrow [F_2, GTP]$$

$$[F_2, GTP] + fMet\text{-}tRNA_f \longrightarrow [F_2, GTP, fMet\text{-}tRNA]$$

Then the anti-codon of the tRNA binds to the initiation codon $\overrightarrow{A\ U\ G}$ (or $\overrightarrow{G\ U\ G}$) of the mRNA:

$$[mRNA, 30\ S, F_1] + [F_2, GTP, fMet\text{-}tRNA_f] \longrightarrow [mRNA, 30\ S, F_1, F_2, GTP, fMet\text{-}tRNA_f]$$

There has as yet been no breakdown of GTP and it can be replaced by its analogue (GMP–PCP) the hydrolysis of which is *not* catalysed by GTP-ase:

GTP: Guanine—Ribose—O—P(=O)(OH)—O—P(=O)(OH)—O—P(=O)(OH)—OH

GMP–PCP: Guanine—Ribose—O—P(=O)(OH)—O—P(=O)(OH)—CH$_2$—P(=O)(OH)—OH

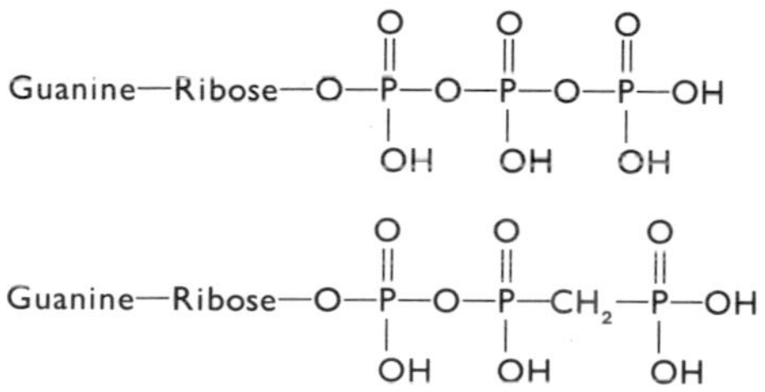

However, the F_2 factor (which may consist of two components) has GTP-ase activity when it is associated with the ribosome and the following reaction occurs:

$$[mRNA, 30\ S, F_1, F_2, GTP, fMet\text{-}tRNA_f] \longrightarrow$$

$$[mRNA, 30\ S, F_1, fMet\text{-}tRNA_f] + GDP + \text{\textcircled{P}} + F_2$$

The last step in these initial reactions is the attachment of a 50 S subunit to form the complete 70 S and the expulsion of F_1:

$$[mRNA, 30\ S, F_1, fMet\text{-}tRNA_f] + 50\ S \longrightarrow [mRNA, 70\ S, fMet\text{-}tRNA_f] + F_1$$

The result is that *N*-formyl-methionyl-tRNA$_f$ is in the donor site which is analogous to having a peptidyl-tRNA there. The subsequent set of reactions occurs whichever of these situations obtains and is repeated at each successive step in chain elongation.

Elongation factors (transfer factors)

Three factors have been separated and, again, have been given various names by different workers:

	Bacterial	Mammalian

$$T_s \} \atop T_u \} T \qquad \begin{matrix} S_1 & F_{is} \\ \\ S_3 & F_{iu} \end{matrix} \qquad TF_i$$

$$G \ S_2 \ F_{ii} \qquad TF_{ii}$$

The binding of amino acyl-tRNA first involves complex formation with GTP and factor T (cf. reaction of fMet-tRNA$_f$ with GTP and F$_2$):

$$[T_u, T_s] + GTP \longrightarrow [T_u, GTP] + T_s$$

$$[T_u, GTP] + \alpha\alpha\text{-}tRNA \longrightarrow [T_u, GTP, \alpha\alpha\text{-}tRNA]$$

The tRNA is selected by the next codon on the mRNA so that it is bound to the acceptor site on the ribosome and has its anti-codon associated with the mRNA codon. In the course of this binding reaction GTP is hydrolysed by the activity of T$_u$ associated with the ribosome:

$$[T_u, GTP, \alpha\alpha\text{-}tRNA] + mRNA, 70\,S \longrightarrow [mRNA, 70\,S, \alpha\alpha\text{-}tRNA] + [T_u, GDP] + Ⓟ$$

$$[T_u, GDP] + T_s \longrightarrow [T_u, T_s] + GDP$$

The GDP is first released as a complex with T$_u$ and then freed by T$_s$. The ribosome-mRNA complex has the amino acyl-tRNA in the acceptor site and already has peptidyl-tRNA (or fMet-tRNA) in the donor site. Transpeptidation now occurs, the peptide (or *N*-formyl-methionine) being joined *via* its carboxyl group to the amino group of the $\alpha\alpha$-tRNA, thus lengthening the peptide chain by one unit. A protein of the 50 S subunit, peptidyl transferase, catalyses this step. (The same enzyme will condense the peptide (or fMet) with puromycin if it is present—see Figure 11, p. 332):

$$[Pep\text{-}tRNA, mRNA, 70\,S, \alpha\alpha\text{-}tRNA] \longrightarrow [tRNA, mRNA, 70\,S, Pep\text{-}\alpha\alpha\text{-}tRNA]$$

The discharged tRNA is ejected from the donor site and the elongated peptidyl-tRNA is 'translocated' from the acceptor site to the donor site while the mRNA moves by one codon relative to the ribosome. These movements require the breakdown of another mole of GTP under the influence of factor G which interacts with the 50 S subunit at a site distinct from the peptidyl transferase:

$$[tRNA, mRNA, 70\,S, Pep\text{-}\alpha\alpha\text{-}tRNA] + GTP \xrightarrow{\ G\ }$$
$$[Pep\text{-}\alpha\alpha\text{-}tRNA, mRNA, 70\,S] + tRNA + GDP + Ⓟ$$

The result of these steps is an empty acceptor site, a donor site occupied by a peptidyl-tRNA, and release of GDP, phosphate and tRNA.

Binding of $\alpha\alpha$-tRNA, transpeptidation and translocation recur at each subsequent elongation and two molecules of GTP are cleaved (Figure 23).

At some stage before completion of the chain, the formyl group is removed by hydrolysis and the *N*-terminal methionine may also be eliminated.

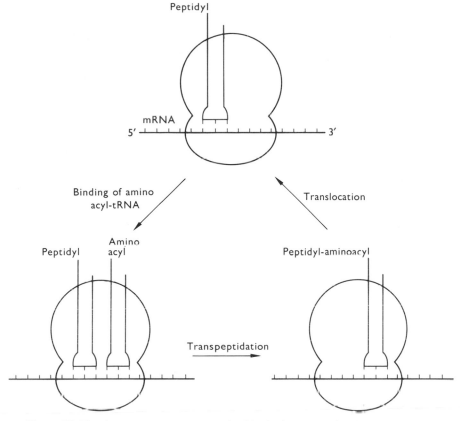

Figure 23. The three successive steps involved in the formation of each peptide bond.

Termination factors

Eventually the polypeptide is complete and still attached to the tRNA corresponding to its C-terminal amino acid. The subsequent codon in mRNA is a termination codon, U A A or U A G or U G A, for which there is no complementary tRNA but which is recognized by a protein termination factor or release factor. One of these is R_1, a single polypeptide chain of molecular weight 44,000 and it recognizes U A A or U A G; the other, R_2, is 47,000 and recognizes U A A or U G A. The activity of each is stimulated by a further protein factor (s) which has been called α or S protein. In the absence of a termination factor the polypeptidyl-tRNA remains attached to the ribosome; in the presence of the appropriate factor, hydrolysis occurs and the polypeptide and tRNA are released.

The ribosome may continue along a polycistronic messenger RNA and begin reading the next cistron at the next initiation codon but ultimately when the whole message has been translated the ribosome reaches the last termination codon. It then separates from the mRNA and becomes dissociated into its two subunits. There has been controversy as to whether this happens at, or subsequent to, separation and as to whether 70 S ribosomes or only mixtures of 30 S and 50 S particles exist *in vivo* unattached to mRNA. What seems certain is that elongation factor, F_3, can bring about dissociation stoichiometrically and possibly results in a complex of 30 S and F_3. In any event, there is rapid inter-

The mRNA under the influence of initiation factor F_3 binds a 30 S subunit at a specific site.

F_1 stabilizes this complex.

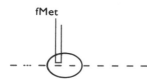

N-formyl-methionyl-tRNA$_f$ in complex with GTP and factor F_2 combines via its anti-codon to an initiation codon A U G (or G U G) of the mRNA.

GTP is split to GDP by F_2 in the presence of the 30 S subunit.

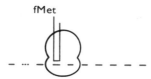

A 50 S subunit is now attached to give a complete 70 S ribosome with expulsion of F_1.

fMet-tRNA$_f$ is in the donor site.

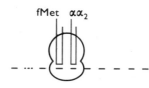

The second amino acid ($\alpha\alpha_2$) carried by its tRNA$_2$ and complexed with GTP and factor $T(T_u + T_s)$ is bound to the acceptor site by codon–anti-codon interaction.

GTP is split by ribosome-associated T_u.

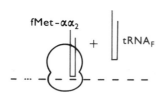

The carboxyl of N-formyl-methionine is joined to the NH$_2$ of $\alpha\alpha_2$ by peptidyl transferase, a protein of the 50 S subunit. The discharged tRNA$_f$ is ejected from the donor site and the dipeptidyl-tRNA is translocated from acceptor to donor site while mRNA moves one codon. This involves breakdown of GTP under the influence of factor G.

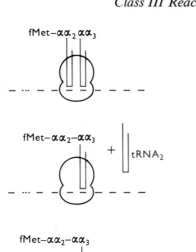

The third amino acid, $\alpha\alpha_3$, attached to tRNA$_3$ enters the acceptor site.

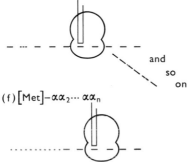

The peptide is elongated by $\alpha\alpha_3$ and tRNA$_2$ is eliminated.

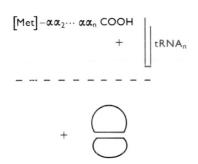

Translocation occurs

and so on

until the polypeptide is complete. At some stage the formyl group is removed as may be the initial methionine.

$[\text{Met}]-\alpha\alpha_2\cdots\alpha\alpha_n$ COOH

$+$ tRNA$_n$

Finally, a termination codon (U A A, U A G or U G A) is reached. There is no tRNA for these but a termination factor, R$_1$ or R$_2$, results in hydrolysis of peptidyl-tRNA$_n$ to yield completed polypeptide and tRNA$_n$. mRNA and the two subunits also separate, possibly under the influence of factor F$_3$.

$+$

Figure 24. Representation of steps in polypeptide formation.

change of subunits after completion of translation and before re-initiation for the next round:

$$[\text{Polypep-tRNA, mRNA, 70 S}] \xrightarrow[+S(\alpha)]{R_1 \text{ or } R_2} \text{Polypeptide} + \text{tRNA} + \text{mRNA} + 70 \text{ S}$$

$$70 \text{ S} + F_3 \longrightarrow 50 \text{ S} + [30 \text{ S, } F_3]$$

Figure 24 summarizes the steps in the biosynthesis of polypeptide chains so far as they are, at present understood.

Figure 25. *Message and Translation.* The probable sequence of the RNA of phage R17 and some suggested secondary structure together with the corresponding amino acid sequences of the three polypeptides into which it is translated. The initiation and termination codons are indicated. Note that several hundred nucleotides at beginning and end seem *not* to be translated.

Bacterial polycistronic messenger RNA's probably resemble this message in some but not all characteristics (see text).

TRANSLATION

It will have become apparent by now that synthesis of protein is not just a matter of linking together amino acids in the right order. In a living organism the process is vastly more complicated because of the operation of intricate systems of control (see Chapters 4 and 8). A message is not just a polyribonucleotide beginning and ending with initiation and termination codons. Many are polycistronic—sometimes they may be 'alien' as when phage DNA is transcribed or when an RNA phage infects a cell.

Protein factors ensure that an appropriate 30 *S* subunit becomes attached at the specific oligonucleotide region of the mRNA so that the initiation codon of the first cistron to be translated is positioned correctly. With polycistronic mRNA from bacteria this is probably the cistron nearest the 5'-end but even so a mechanism must exist to prevent random initiation of translation at A U G or G U G triplets other than those at the beginnings of messages. Studies of mRNA from the small RNA phages R 17 and Q β suggest that untranslated regions of the RNA and secondary structure may play important roles in these effects. The RNA of R 17 consists of about 3500 nucleotides and codes for three proteins, of which the 'coat protein' is the only one which occurs in the bacteriophage particle itself; another is the 'maturation' or 'A protein'; and the third is an RNA synthetase or replicase. They are produced in molar proportions of 20:1:5, the first two immediately after infection, but the synthetase only after a lag. The complete sequence of the 129 amino acids of the coat protein is known, as are parts of the other proteins and long stretches of the RNA sequence. Figure 25 illustrates some relationships between the polynucleotide and the polypeptides. Regions where base-pairing may give intra-chain helical regions are indicated although there is no direct evidence for their existence. There are many points of interest. It should be noted that there are several potential initiation codons (A U G or G U G) before the beginning of the A protein message at about the 300th nucleotide. Perhaps secondary structure as shown and the existence of specific ribosome binding sites ensure that translation begins appropriately. It is also interesting to note that there are two adjacent termination codons, U A A followed by U A G, at the end of the coat protein message. Further, there next comes what might be translated as a hexapeptide since it begins with an initiation codon, A U G, and ends with a termination codon, U G A:

A U G	C C G	G C C	A U U	C A A	A C A	U G A
fMet	Pro	Ala	Ile	Glu	Thr	

However, although looked for, this peptide has not been found. Finally, there are several hundred nucleotide residues on the 3'-side of the synthetase cistron which do not seem to be translated.

This tricistronic messenger of the bacteriophage may not be exactly typical of bacterial messengers and some of the untranslated portions may be sites for attachment of the RNA-primed RNA synthetase which replicates it. Such would not be necessary for bacterial RNA's which are only transcribed from DNA. However, it is probable that there are considerable similarities. Unfortunately there is not nearly so much known about bacterial genes as there is about the simple RNA phages. The *lac* operon of *Escherichia coli* and the histidine operon of *Salmonella typhimurium* have been studied in detail at the genetic level but only little is yet known of their molecular biology. The *his* operon has nine structural genes which are transcribed as a single polycistronic mRNA (see Chapter 4, p. 232) from which at least the second, third and sixth cistrons are translated with the same frequency. From a study of a frameshift mutant arising by a double deletion near the 3'-end of the second cistron, it is possible to speculate about the intra-cistronic divide between it and the third cistron. The wild type enzyme which terminates in —Glu—Glu—Ala is replaced by one ending —Ala—Ser—Leu—Thr (i.e. it has an additional as well as altered residues). Figure 26 indicates various possibilities for the intra-cistronic divide.

With the isolation of a pure operon DNA, the possibility of its transcription into RNA, and the techniques developed for sequence analysis, it should not be long before our understanding of translation becomes much greater.

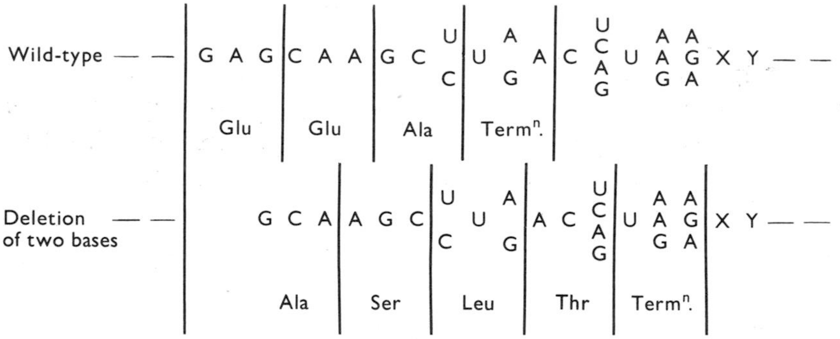

Figure 26. Possible sequences for the intra-cistronic divide between the second and third cistrons of the histidine operon.

Deletion of two bases alters the terminal sequence of an enzyme from —Glu—Glu—Ala to —Ala—Ser—Leu—Thr. The nucleotide sequence is shown (with alternatives where it is uncertain). The intra-cistronic divide could be, as a minimum, one or four nucleotides in length i.e.

only C or C $\begin{smallmatrix} U \\ C \\ A \\ G \end{smallmatrix}$ $\begin{smallmatrix} A \\ U \\ G \end{smallmatrix}$ A before the next initiation triplet AUG or GUG can begin the third cistron.

ANTIBIOTICS WHICH INTERFERE WITH SYNTHESIS OF PROTEIN

The complexity of the process of making a protein is indicated by the fact that more than 120 species of macromolecules (proteins and nucleic acids) are known to be involved. It is all the more surprising, therefore, that so seldom under physiological conditions does the 'wrong' amino acid become incorporate or does the process go awry in other ways. However, there are substances, including many antibiotics, which interfere specifically at one stage or another. Compounds which inhibit or modify nucleic acid replication and transcription are dealt with elsewhere (Chapter 5) but even with a correct message it is possible to get faulty translation. Aurintricarboxylic acid prevents the binding of mRNA to 30 S subunits *in vitro* and so prevents an early step. Streptomycin, kanamycin and neomycin can cause mis-reading of the mRNA. Thus whereas poly U normally promotes formation of poly-phenylalanine *in vitro*, the addition of streptomycin reduces this incorporation and stimulates that of isoleucine. A specific protein component of the 30 S ribosomal subunit is known to be involved in sensitivity or resistance to (or dependence on) this antibiotic and it is thought that the drug's effect might be to cause distortion of the mRNA so that codon/anti-codon interaction at the acceptor (or amino acid) site is abnormal. The formation of aberrant proteins *in vivo* has not been unequivocally established but would certainly be enough to cause death of the cells.

The tetracycline group of antibiotics, which includes aureomycin and terramycin, act by competing with the amino acyl-tRNA for attachment to the acceptor site on ribosomes. Possibly pactamycin and bottromycin also interfere with this binding reaction.

Chloramphenicol and, probably, sparsomycin, amicetin and gougerotin, inhibit the peptidyl transferase—they also prevent its catalysing the formation of peptidyl-puromycin (see p. 332).

Breakdown of GTP occurs at several stages in protein formation and these can be differentiated by the action of fusidic acid. It interferes with factor G by preventing the dissociation of GDP and phosphate from the ribosome after the action of G factor but it does not prevent the GTP-ase activity of initiation factor F_2 nor that of the ribosome-dependent activity of elongation factor T_u (see p. 356). Thus translocation is affected by fusidic acid and it is also prevented by erythromycin and spectinomycin—but not in these instances due to inactivation of G factor. The binding site on the 50 S subunit for G factor is inactivated by siomycin.

The probable sites of action of these substances are indicated diagrammatically in Figure 27.

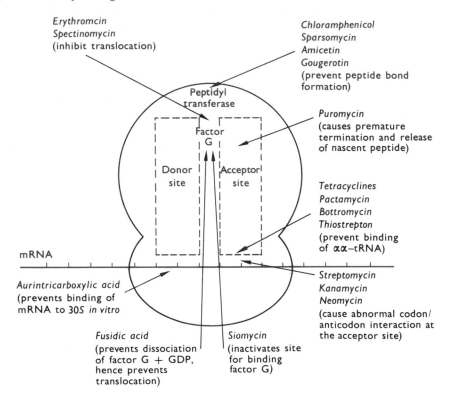

Figure 27. Probable sites of action of some antibiotics which interfere with synthesis of proteins.

ASSEMBLY OF NATIVE PROTEINS

Many proteins consist of more than one polypeptide chain and some enzymes are very large molecules containing a number of subunits which may be identical with each other so that the enzyme consists of an aggregate with several active centres. Thus β-galactosidase of *Escherichia coli* has a sedimentation coefficient of 16 S and a molecular weight of about 500,000. It can be dissociated by urea (which breaks H-bonds) into four subunits but is probably composed of 12 monomers of 40,000 molecular weight. The functional enzyme may be put together before release from the ribosomal site of synthesis and β-galactosidase activity can be found associated with 70 S ribosomes.

Tryptophan synthetase (see p. 410) is composed of two separable proteins components A and B and in *Escherichia coli* which is making the enzyme, some ribosomes have A protein and some B protein and some have both associated with them.

There is not a great deal which can be said about the mechanics of assembly of polypeptides into native functional proteins except that much may be in a sense 'spontaneous'. The ultimate secondary, tertiary and quaternary structures (see pp. 15–16) may be very largely determined by the amino acid sequence or primary structure of the constituent parts. Even elaborate cross-linking disulphide bridges may have an inevitability. This is said because some. proteins which have had their native structure destroyed have been shown to be capable of renaturation. For instance, the ribonuclease molecule is a single chain of 124 amino acid residues and is stabilized by the S—S bonds of four cystine bridges. These can be broken by reduction to cysteine residues in 8 M urea and all enzymic activity is lost. Re-oxidation restores activity almost completely and yields a product indistinguishable from the native protein. This is despite the fact that the reduced ribonuclease had eight sulphydryl groups which could have been combined in pairs in 105 different ways ($7 \times 5 \times 3$). Almost more surprising is the finding that heat-denatured β-galactosidase can be restored to activity.

It is now generally believed that there do not exist three-dimensional templates which force polypeptides to assume a particular tertiary structure but rather that chains once synthesized take up their configuration spontaneously since these are the only ones which satisfy the stereochemical requirements of the amino acid sequence.

ACKNOWLEDGMENTS

Figure 4a. From Carnegie Institution of Washington Yearbook **54**, p. 264.
Figure 4b. Micrograph supplied by H.E. Huxley and G. Zubay. Electron microscope observations on the structure of microsomal particles from *Escherichia coli*. *J. Mol. Biol.* **2** (1960) 10–18.

FURTHER READING

1 Loftfield R. B. (1957) The biosynthesis of proteins. *Prog. Biophysic. Biophys. Chem.* **8**, 347. (An excellent account of the position at that time and of the general implications.)

2 Perutz M. F. (1962) *Proteins and Nucleic Acids: Structure and Function.* Elsevier, London. (An important exposition.)

3 Maaløe O. and Kjeldgaard N. O. (1966) *Control of Macro-molecular Synthesis.* Benjamin, New York.

4 Ingram V. M. (1972) *Biosynthesis of Macromolecules.* Benjamin, New York.

5 Roberts R. B. (ed.) (1964) *Studies of Macromolecular Synthesis.* Carnegie Institution of Washington Publication No. 624, Washington, D.C.

6 Symposium (1963) Synthesis and structure of macromolecules. *Cold Spr. Harb. Symp. Quant. Biol.* **28**.

7 Symposium (1966) The genetic code. *Cold Spr. Harb. Symp. Quant. Biol.* **31**.

8 Symposium (1969) Mechanisms of protein synthesis. *Cold Spr. Harb. Symp. Quant. Biol.* **34**.

9 Symposium (1970) Transcription of genetic material. *Cold Spr. Harb. Symp. Quant. Biol.* **35**.

10 Attardi G. (1967) The mechanism of protein synthesis. *Ann. Rev. Microbiol.* **21**, 383 (A good review covering 1958–1967.)

11 Lucas-Lenard J. and Lipmann F. (1971) Protein biosynthesis. *Ann. Rev. Biochem.* **40**, 409.

12 Gauss D. H., von der Haar F., Maelicke A. and Cramer F. (1971) Recent results of tRNA research. *Ann. Rev. Biochem.* **40**, 1045.

13 Pestka S. (1971) Inhibitors of ribosome functions. *Ann. Rev. Microbiol.* **25**, 487.

14 Lengyel P. and Söll D. (1969) Mechanisms of protein biosynthesis. *Bacteriol. Rev.* **33**, 264.

15 Bretscher M. S. (1970) *Prog. in Biophys.* **19** Part 1, 175.

16 Nomura M. (1970) Bacterial ribosome. *Bacteriol. Rev.* **34**, 228.

17 Nirenberg M. W. and Matthae J. H. (1961) The dependence of cell-free protein synthesis in *E. coli* upon naturally occurring or synthetic polyribonucleotides. *Proc. Natl. Acad. Sci., U.S.* **47**, 1588.

Chapter 7
Genetics

Genetics is concerned with the *genotype*, or genetic constitution, and the way in which it determines the *phenotype*, or observable characteristics of an organism. At the population level, genetics deals with the changes which occur in the relative frequencies of different genotypes in populations as a result of mutation and selection. At the level of the individual cell, which is the aspect with which we are more particularly concerned in this book, it deals with the nature of the genetic material, its structure and mode of self-replication, and the way in which it carries out its directing function.

THE NATURE OF THE GENETIC MATERIAL

As we saw in Section 7, the most compelling evidence for the identification of the genetic material with DNA comes from experiments on *genetic transformation* in bacteria. In a number of genera, notably *Diplococcus* and *Haemophilus*, practically any genetic trait can be transferred from cells possessing it to cells lacking it by treating the latter with DNA isolated from the former. The actual procedure for carrying out a transformation experiment is usually complicated by the fact that only cells at a particular phase of growth are *competent* to assimilate the transforming DNA and the conditions for competence vary with the particular species being studied. When fully competent cells are used and proper care is exercised in the preparation of the DNA, transformation is extremely efficient. Some idea of the efficiency of the process can be obtained by using transforming DNA isolated from cells grown in medium containing radioactive phosphorus (^{32}P). Then the quantity of DNA taken up and retained by the recipient cells, as measured by their radioactivity, can be compared with the number of cells transformed. It turns out that the transformation of one cell in the population with respect to a single given genetic trait requires the uptake by the population of a quantity of DNA roughly equivalent to the amount contained within a single cell. Different experiments have shown that the yield of transformants is proportional to the concentration of transforming DNA within quite wide limits of DNA concentration (see Figure 1), and also to the amount of DNA actually taken up by the cells. The DNA supplied is certainly not exhausted during these experiments and the yield of transformants is

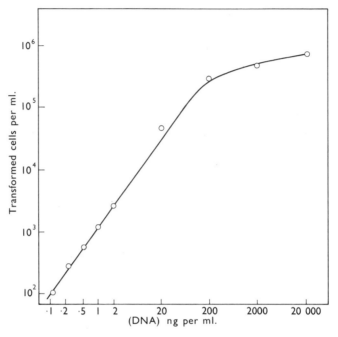

Figure I. Transformation of tryptophan-dependent *Bacillus subtilis* to tryptophan-independence by DNA from a tryptophan-independent strain as a function of DNA concentration in the medium. Data from J. Spizizen (*Proc. Nat. Acad. Sci., Wash.,* **44**, 1072).

limited, rather, by the chance of a given cell taking up the relevant molecule(s) during the limited time available. The first-order dependence of transformation on DNA concentration implies that a *single* piece of DNA suffices to transform a cell with respect to a single trait.

The DNA in transforming preparations usually has a mean molecular weight of 10^7 or more, but experiments using preparations in which the molecules have been broken down further by ultrasonic vibration have shown that pieces of DNA of no more than 10^6 molecular weight suffice for transformation. Such a small piece of DNA is, of course, only a very small fraction of the DNA of a bacterial cell and probably carries only a few items of genetic information from the complete donor genotype.

To isolate the small number of transformants (usually less than 1% for any one trait) from the mass of untransformed cells a selection method is necessary. For instance, in the example shown in Figure 1, only those tryptophan auxotrophs which have been transformed to wild-type are able to grow in the absence of the amino acid.

How can we be sure that the active material in transforming preparations really is DNA? Transforming activity is, indeed, still associated with DNA preparations after they have been purified as much as possible, but small traces of other kinds of macromolecules are very hard to exclude entirely. However, any possible effect of contaminating protein or RNA can be eliminated by treating transforming preparations with enzymes which will specifically degrade these classes of macromolecules. Pepsin, trypsin and other proteases are not found to have any effect on transforming activity, and neither has ribonuclease. Deoxyribonuclease, on the other hand, brings about a rapid destruction of the transforming power. Confirmatory evidence is obtained from the effects of heating transforming preparations. Transforming power is lost at the precise temperature at which the double-stranded structure of the DNA is 'melted' to single strands. It seems that single-stranded DNA is inactive or very poorly active in transformation, probably because it is not effectively taken up.

Many other lines of evidence confirm the genetic role of DNA. For instance, in infection of cells of *Escherichia coli* by bacteriophage T2 it has been shown in the classical experiment of Hershey and Chase that only the DNA of the virus (and very small traces of other substances) enters the bacterial cell. The proteins of the bacteriophage coat and tail are left outside. In fact bacterial protoplasts (i.e. bacteria with the cell wall digested away by lysozyme or lost as a result of growth in the presence of penicillin) can be infected with purified DNA of certain bacteriophages without the protein components being involved at all. In viruses which do not contain DNA, RNA takes over the genetic function. In a number of such viruses, including some bacteriophages and animal viruses, the isolated viral RNA has been shown to be infective. One of the most intensively studied examples of genetic RNA is that of tobacco mosaic virus, which infects tobacco and a variety of other plants. It is conceivable that RNA plays an auxiliary genetic role even in those organisms in which DNA is the principal genetic material. DNA is, of course, a universal major component of chromosomes, which are well established as the carriers of genetic factors in higher organisms of all kinds.

In bacteria and DNA-viruses the organization of the DNA is considerably simpler than it seems likely to be in the chromosomes of higher organisms. In the bacteriophage *lambda* it has been shown, most convincingly by electron microscopy, that the DNA is in the form of one continuous duplex strand capable of forming a closed loop through the pairing of complementary single-stranded pieces ('sticky ends') projecting from the ends of the double-stranded molecule. The total length of *lambda* DNA is 16 μm, corresponding to a molecular weight of 33×10^6. In *Escherichia coli*, the structure of DNA isolated from cells grown in the presence of the radioactive DNA precursor tritiated thymidine has been examined by autoradiography. Here too the chromosome* is a single enormous molecule of double-stranded DNA forming a closed loop of length 1100 μm (cf. p. 283 of this book). The simple conclusion that a complete set of genetic material (one *genome*) is contained in one very large DNA molecule probably holds for bacteria and DNA-viruses generally.

Experiments with transforming DNA preparations tell us that the genetic determinants of specific characters are contained within pieces of DNA which are very small compared with the total genome. There is no doubt that the DNA molecule which comprises the bacterial genome is made up of a great number of functionally differentiated segments (genes). The analysis of this differentiation within the genome presents a challenge which cannot be met by the methods of chemistry or physics. There is not enough to distinguish, from the point of view of chemistry, one segment of a DNA molecule from another. Only in rare and special cases has it so far been possible physico-chemically to separate specific genes from their neighbours. The base-pair sequence of the DNA even of simple organisms is far too long for complete chemical analysis by present methods.

The methods of genetics, however, do permit the mapping of the genome in very fine detail. In genetics, genes are distinguished from each other not by their chemical structures but by their effects on the phenotype and by their distinctive patterns of transmission from cell to cell. There are two essential stages in classical genetic analysis. First of all the gene to be studied has to be made distinguishable by a mutation. It is axiomatic that if a gene is present in identical form throughout the species one cannot tell, by the methods of genetics, that it is there at all. There has to be a *difference* residing in the gene before its inheritance can be followed. So the induction and isolation of mutants is commonly the first step in a genetic investigation. A piece of the genome carrying a mutation (i.e. a heritable change) with a readily observable phenotypic effect is thereby *marked*, and the mutation in question is often referred to as a *marker*. The second stage of genetic analysis is the determination of the relative positions of a series of markers by their modes of transmission. We will deal first with some practical and theoretical aspects of mutation.

MUTATION

The frequency of detectable spontaneous mutation in any given gene is very low, probably generally of the order of one per 10^5 to 10^7 cells per cell division.

---

* The term chromosome should perhaps be reserved for the more complex genetic structures of higher organisms but in practice it is now used for any structure carrying a linear array of genes.

But taking all genes into account, mutation is not such a rare event, and in a population of say 10^{11} cells, such as can easily be obtained in bacteria, mutants of practically any desired type will be present and can be isolated given a suitable selective technique. Frequencies of mutation can, furthermore, be greatly increased by the use of mutagenic treatments. We shall return to mechanisms of mutation as such later on in this chapter, but to begin with it seems more appropriate to consider the different types of mutant which can be obtained and the methods used for their detection and isolation.

On pp. 299 *et seq.* the enzymic mechanism for the transcription of DNA base sequence into RNA base sequence was described. So far as we know the primary effect of any mutation in the DNA will be to cause a complementary change in the base sequence of the RNA transcribed from it. This RNA is of three different kinds: ribosomal RNA, amino acid-transfer RNA and messenger RNA. The first two kinds of RNA, though they account for a high proportion of the total RNA, consist of relatively few kinds of molecule, and are transcribed from a correspondingly rather small fraction of the DNA. The mutations which we know most about are those which cause structural changes —generally simple amino acid substitutions—in proteins, and each of these is presumed to have its primary effect through a change in the base sequence of one or other of the hundreds or thousands of different kinds of messenger RNA molecules into which the greater part of the genetic DNA is transcribed. Mutations of this class are usually detected through their effects on specific enzymes.

Auxotrophic mutants: mutants with defective or altered Class II enzymes

The most useful and intensively studied kind of mutant in bacteria is the *auxotroph*. An auxotrophic mutant has an additional nutritional requirement, for example for an amino acid, a vitamin or a purine or pyrimidine base or nucleoside, over and above the *minimal* nutritional requirements of the wild-type organism. A medium containing only the minimum nutrients for the growth of the wild-type is called minimal medium, and in the case of *Escherichia coli* or *Salmonella* species, it contains only essential inorganic salts and a simple carbon and energy source such as glucose. A typical auxotrophic mutant fails to grow on minimal medium but grows as well as wild-type if the specific substance which it requires is added. More fundamentally, an auxotroph has usually lost the ability to make one of the Class II enzymes essential for the biosynthesis of the substance which it requires.

The selection of auxotrophic mutants might seem a difficult task in as much as it involves seeking out a few cells which cannot grow on minimal medium from among a large number of cells which can. The difficulty was overcome by making use of the fact that penicillin kills only growing bacterial cells. In the

penicillin enrichment method, cells of *E. coli*, usually after some sort of mutagenic treatment, are suspended in minimal medium plus penicillin. During a
subsequent period of incubation the nutritionally normal, or *prototrophic*, cells
start to grow and are killed by the penicillin. The auxotrophs, on the other hand,
are prevented from growing by their nutritional deficiencies and tend to survive.

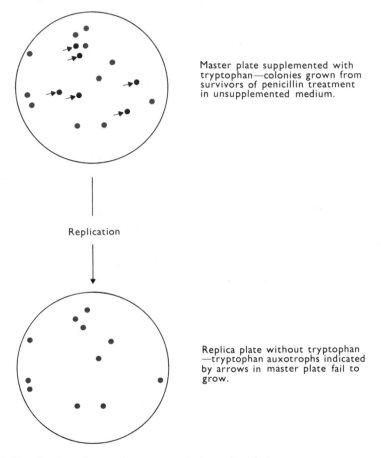

Master plate supplemented with
tryptophan—colonies grown from
survivors of penicillin treatment
in unsupplemented medium.

Replication

Replica plate without tryptophan
—tryptophan auxotrophs indicated
by arrows in master plate fail to
grow.

Figure 2. Identification of tryptophan auxotrophs by replica plating.

They are subsequently recovered by washing the cell suspension free of penicillin
and plating on supplemented medium. If only one class of auxotrophs is sought
a master plate is prepared containing minimal medium supplemented only
with the specific substance required by that class. Of the colonies which grow up
some are usually wild-types which have, for some reason, survived the penicillin,
but a good proportion of them are auxotrophs of the desired type. This method
is very rapid and efficient and has been used more than any other for the isolation of auxotrophs, not only in *E. coli* but in several other species as well. The
genuineness of the presumptively auxotrophic colonies which grow up on the

supplemented plates can be quickly checked by the technique of *replica plating*. In this method representative cells from the whole ensemble of colonies on the master plate are simultaneously transferred to a plate of minimal medium on the surface of a replicator which is usually a disc of sterile velveteen material supported on a round wooden block. The replicator is lightly pressed first on the surface of the master plate and then on the replica plate containing fresh medium. A pattern of colonies should grow on the replica plate identical to that on the master plate provided that all the colonies are capable of growth on the medium in the replica plate. Since the replica plate contains minimal medium, auxotrophic mutants from the master plate will not grow. In this way all the colonies on the master plate, numbering perhaps a hundred or more, can be tested in one simple operation (Figure 2).

On the other hand, if auxotrophs of a range of types are sought, the master plate should be supplemented with a wide variety of substances: amino acids, purines, pyrimidines, vitamins. Replicas from the master plate are then made on a series of replica plates each containing only one supplementary substance.

By special techniques it is possible to obtain mutants in which an enzyme is altered in its properties rather than merely absent. For instance, among temperature-sensitive auxotrophs—able to grow on minimal medium at a lower temperature (say 27°) but not at a higher temperature (say 37°)—one can find many which have unusually thermolabile Class II enzymes. Again, many growth-inhibitory metabolite analogues exert their effects through acting as 'false feedback inhibitors' of enzymes subject to allosteric regulation (see Chapter 8). A mutant selected as resistant to one of these drugs may have an altered enzyme no longer subject to end-product regulation of its activity. This mechanism of drug resistance should be clearly distinguished from resistance to false feedback *repression*; this latter type of resistance is due to a breakdown in the normal control of *rate of synthesis* of one or more enzymes and does not involve any change in enzyme structure. Further consideration of non-repressible, or constitutive, mutants is deferred until the following chapter.

Mutants defective in Class I enzymes

A second class of mutants includes those which are unable to catabolize some substance which can be utilized by the wild-type. For instance, whereas wild-type *E. coli* can utilize not only glucose but also lactose, maltose, arabinose or mannitol (among other substances) as sole carbon and energy source, mutants can be obtained which are unable to utilize any one of these less usual carbon sources. In principle it is possible to adapt the penicillin screening procedure to the isolation of such mutants, using a minimal medium with the substance in question as sole sources of carbon. However, a more commonly employed alternative procedure is to pick out the mutant colonies by eye after plating on a complex organic (broth) agar medium supplemented with the carbohydrate in question and with the dyes eosin and methylene blue. Such a medium containing, say, lactose is commonly referred to as EMB-lactose medium. Cells which can ferment the sugar give colonies of a deep purple colour on this medium. Cells not able to ferment the sugar grow on the other carbon sources in the broth, but they give white colonies which show up in striking contrast with the wild-type ones. Even though such mutant colonies occur only at the rate of one in several thousand they can easily be picked out. The EMB technique is also useful in that it permits the identification of sectored colonies in which only one-half or one-quarter of the colony is mutant (white) while the remainder remains wild-type (purple). Such sectoring indicates that the cell which gave rise to the colony was genetically mixed, containing both mutant and wild-type genes. Some circumstances in which this can occur will be mentioned later (cf. p. 387).

Mutants unable to utilize a given carbon source have most commonly lost the ability to make one of the Class I enzymes necessary for the catabolic pathway in question. For example lactose mutants may have lost β-galactosidase, while galactose mutants may have lost any one of the three enzymes specifically concerned in the conversion of galactose to glucose. In some cases, however, failure to utilize a carbon source may be due to an inability to concentrate it from the medium, such an inability being due to loss of a component of a specific transport system ('permease'). Since permease activity depends on specific proteins in the cell membrane it should be just as liable to mutational loss or damage as are enzymes. We may note here in passing that permease deficiency is responsible for some cases of drug resistance. For example the loss of an amino acid permease may confer resistance to a normally inhibitory amino acid analogue.

Sometimes mutational loss of an enzyme may be detected by a direct test of the enzyme activity itself. For example a large number of colonies can be tested simultaneously for the presence of alkaline phosphatase. This is done by spraying the surface of the plate with a phenol phosphate and then with a mixture of sulphanilamide and nitrous acid which couple with any phenol released by enzyme action to form a bright red azo-dye. Colonies deficient in the enzyme remain white and stand out strikingly among the red, wild-type colonies. Alternatively, colonies can be sprayed with dinitrophenyl phosphate which, on hydrolysis by the enzyme, colours the wild-type yellow with the ditrophenol released. Large numbers of mutants of *E. coli* deficient in alkaline phosphatase have been isolated by these two methods.

Mutants defective in Class III enzymes and other components of the protein-synthesizing system

Some of the most interesting and potentially valuable mutants are those defective in Class III enzymes—enzymes concerned with macromolecular synthesis.

Such mutants are, however, among the most difficult to obtain. If the Class III enzyme function is seriously depressed the mutant will usually not grow on any medium since the product of the enzyme activity will be a large complex molecule which is likely to be unstable and/or unable to enter the cell. The most generally fruitful approach to this problem is through temperature-sensitive mutants. Temperature-sensitive mutants usually have specific enzymes altered in such a way that, although functional at the lower temperature, they become inactivated as the temperature is raised to the upper part of the range compatible with wild-type growth.

There is, in principle, no restriction on the kinds of enzymes which can be affected in this way. It is easy enough to identify and collect many temperature-sensitive mutants by replicating large numbers of colonies to plates which are then incubated at the higher temperature. Among those temperature-sensitive mutants whose defects at the higher temperature cannot be repaired by supplying metabolites of low molecular weight are many affected in enzymes involved in nucleic acid or protein synthesis. It has been possible, for instance, to devise methods for picking out those which are deficient in DNA synthesis. Such mutants have been extremely useful in analysing the role which DNA synthesis plays in various cell processes. Other temperature-sensitive mutants have proved to have labile aminoacyl-tRNA synthetases, and the analysis of such mutants in *E. coli* has led to the identification of the genes coding for several of these enzymes.

Another approach to the isolation of mutants altered in macromolecular synthesis is through the isolation of drug-resistant mutants. We have already seen that resistance to metabolite analogues may be due to derepression of enzyme synthesis or to loss of a permease or to the alteration of a Class II enzyme to make it resistant to feedback inhibition. Another possible reason for resistance to an amino acid analogue is that the corresponding aminoacyl-tRNA synthetase has been altered so that it no longer activates the analogue, though continuing to deal adequately with the normal amino acid. To the extent that the inhibitory effect is due to the incorporation of the analogue into protein, such an alteration will confer resistance. For example a mutant of *E. coli* resistant to *p*-fluorophenylalanine was shown to have an altered phenylalanyl-tRNA synthetase.

Other drugs, such as streptomycin, spectinomycin and erythromycin, are inhibitory because of their high affinity for the ribosomes. Mutants resistant to these drugs are thought to have altered ribosomal proteins. For example, in *E. coli*, a specific ribosomal protein of the 50s subunit which is altered in an erythromycin-resistant mutant has been identified. The genes coding for this and other ribosomal proteins appear to be closely linked to each other and to the well-known gene governing streptomycin resistance.

There is no doubt that, through the twin approaches of temperature-sensitive and drug-resistant mutants, the genes for other ribosomal proteins and enzymes concerned in polypeptide chain initiation and elongation will be identified.

Turning now from proteins involved in nucleic acid and protein synthesis to genes for the nucleic acid components of the protein-synthesizing system, we might hope to be able to identify genes both for ribosomal RNA (rRNA) and for transfer RNA (tRNA). We have, in fact, direct evidence that genes for rRNA exist, since it can be shown that an appreciable fraction of bacterial DNA (0·3% in *E. coli*) will, after denaturation to single strands, combine specifically, through complementary base-pairing, with purified ribosomal RNA, both 16 *S* and 23 *S*. It may be assumed that the rRNA is transcribed from this part of the genome. The amount of this DNA is 10-fold greater than would be required for just one gene for 16 *S* and another for 23 *S* RNA, and it seems that there must be about 10 copies of each gene. Recent evidence suggests that these copies are scattered along the length of the bacterial chromosome rather than clustered together at one location. If, as seems to be the case, the bacterial chromosome is transcribed sequentially from one end to the other, this scattered arrangement of rRNA genes will ensure a more or less continuous production of rRNA throughout the transcription cycle. No mutations in these genes have as yet been identified. The multiplicity of copies would no doubt make a change in just one gene difficult to detect.

Genes for tRNA species, on the other hand, have been identified by mutation. The mutational effect which has been identified is that of genetic *suppression*. As will be explained in more detail later in this chapter, many mutants which lack enzyme activities do so because the mutation has created a chain-terminating codon—UAG, UAA or UGA—in place of what, in the wild-type, is a codon for an amino acid. Selection for revertants from such enzyme-deficient mutants leads to the isolation of some in which the original mutation has been truly reversed—or at least an acceptable amino acid codon substituted for the chain-terminating one—but also others in which the original mutation is still present together with a second (*suppressor*) mutation at a quite separate position which has the effect of overcoming, at least partially, the chain-terminating effect. The general way in which these suppressors of chain termination work is through the modification of the anticodon of a species of transfer RNA so as to make it recognize the chain-terminating triplet. In the first case of this kind to be fully analysed it was shown that a modified tyrosine-specific tRNA was responsible for the suppression (see Chapter 6, p. 343). The normal anticodon, GUA, which recognizes the tyrosine codons UAU and UAC, had been changed to CUA, which now recognizes the chain-terminating codon UAG. Thus this class of suppressor mutation identifies a gene for tyrosine-specific tRNA. Actually the tRNA involved here is only one of two or three tyrosine-specific varieties controlled by different genes, and it contributes only a minor fraction of the total tyrosine-tRNA activity. Were a mutation of this kind to occur in a unique tRNA the effect would normally be lethal, and it is only the fact that *E. coli* has a 'spare' tyrosine tRNA which permits the ready isolation of this class of suppressor. Nevertheless, mutant suppressor versions of normally unique tRNA's have been identified in special stocks in which the gene concerned is present in two copies.

Resistance to bacteriophage

Mutants resistant to virulent bacteriophages can be readily selected by exposing large populations to enough virus to kill all non-mutant cells. Such mutants have been of great importance in the development of bacterial genetics. Resistance to bacteriophage is probably most usually due to the loss or alteration of the specific component of the bacterial cell wall to which the virus normally attaches. Different bacteriophages evidently attach to different specific chemical groupings, since mutation to resistance to one phage type does not generally confer resistance to unrelated types.

Spontaneous mutation frequencies

Mutations arise constantly in cell populations and there is no known way of preventing them or of predicting whether a particular cell will suffer a particular kind of mutation at any specified time. This is what is meant by calling experimentally unprovoked mutations *spontaneous*, but the term should not be taken as implying that the process is necessarily uncontrollable or, still less, without definite chemical causes. It is difficult to give meaningful values for typical spontaneous mutation frequencies since everything depends on the kind of mutation being counted. Since a gene has of the order of a thousand nucleotide pairs, the mutation frequency per gene may be expected to be about a thousand times the frequency per nucleotide pair. But not all changes of single nucleotide pairs will have observable effects on the phenotype, and the proportion which do will be very different in different experimental systems. If one is looking for mutations leading to a loss of a given gene function it is likely that a change in any one of several hundred nucleotide pairs within the gene will have this effect. If, on the other hand, one is counting reversions from a non-functional mutant form of the gene to a functional one then only a further mutation in the originally mutated codon, and perhaps 'suppressor' mutations at relatively few other sites, will contribute to the score. Typical orders of magnitude might be 10^{-8} for spontaneous mutation frequencies per nucleotide pair per cell division, and 10^{-5} per gene, but the proportion of mutations in the DNA which will be actually observed will depend on the sensitivity to genetic change of the phenotypic character under observation.

Spontaneous nature of apparently adaptive mutations

Some mutations may enable the cell to grow better in the environment in which it happens to find itself. Such mutations tend, of course, to be automatically selected and to spread through the whole population as growth proceeds. For many years it was a matter of controversy whether such adaptive changes were really due to spontaneous mutations, unprovoked by the specific environmental conditions. This controversy is now settled to the satisfaction of most bacterial geneticists but since the answer is by no means self-evident and may, indeed, be contrary to intuitive expectation, it seems worth while to consider some of the relevant experimental evidence.

Two kinds of test have been decisive in favouring the hypothesis of selection of spontaneous mutations. The first is the *fluctuation test*, first devised by Luria and Delbrück in their experiments on resistance to bacteriophage T1 in *E. coli*. A number (say, eight) of identical small liquid cultures of the host bacterium are started from identical small inocula and, as a control, an equivalent number of bacteria in an equivalent volume of medium are grown in a larger *single* vessel. After all the cultures have grown measured samples are plated on nutrient medium in the presence of excess phage. In this way a measure of the incidence of phage-resistant mutants in each of the small cultures is obtained. In addition eight samples are taken from the single culture and plated to provide a measure of the fluctuation to be expected as a result of sampling error. If the mutations were provoked with a constant probability by contact with the phage, each sample should give the same number of mutant colonies, subject to the variation expected to result from ordinary statistical sampling error. If, on the other hand, mutations to phage resistance were arising all the time, irrespective of whether the phage was present or not, then eight plates seeded from eight separate small cultures would not start on an equal footing. A mutation might have occurred in some of these cultures but not in others, and among those where it did occur, the number of mutant progeny will be large if the mutation occurred early in the growth of the culture, and small, if it occurred late. The results usually demonstrate that mutations occurred spontaneously before exposure to the phage, the plates seeded from the series of small cultures showing great variation—far more than could possibly be explained by sampling error. The control series, inoculated with different samples from the same culture, shows only as much variation in number of mutants from plate to plate as would be expected to result from sampling from the same population. Some representative results are shown in Figure 3.

A more direct, and thus even more convincing, demonstration became possible with the invention of replica plating. An example illustrating the principle of the experiment is as follows. A master plate not containing the antibacterial agent (say streptomycin) is thickly seeded with wild-type bacteria and the resulting continuous lawn of cells is replicated to another plate containing a concentration of streptomycin sufficient to inhibit wild-type cells completely. Some orientation mark is necessary on each plate so that one can refer any resistant colonies growing on the replica to corresponding points on the master plate. Generally two replicas are made and when resistant colonies appear on both in exactly corresponding positions it is inferred that a spontaneously resistant mutant clone of cells exists at the corresponding point on the master plate. A sample of cells is then scraped from the surface of the master plate at this point and used as the inoculum for a second master plate, from which colonies

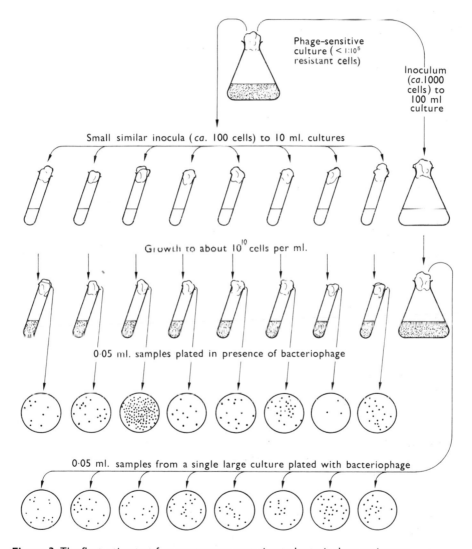

Figure 3. The fluctuation test for spontaneous mutation to bacteriophage resistance.

The incidence of phage-resistant mutants in a number of small cultures is compared with the incidence in replicate samples from a single large culture (see text).

are again replicated to streptomycin for identification of the positions of the mutant clones, which will now be numerous. The cycle of operations is repeated a few more times, the proportion of mutant cells being increased at each step until the stage is reached where it is possible to isolate a resistant clone in pure culture. The point to be emphasized is that the whole selection procedure is indirect. At each stage the cells selected for further growth are taken from the master plate which has never been exposed to streptomycin. The possibility that the mutation to resistance might have been induced by the drug does not arise. This kind of demonstration has been made for many different kinds of resistance mutant in several species of bacteria and there is no longer any doubt that spontaneous mutations occur in almost infinite variety, and that some of these do, fortuitously, confer a *preadaptation* to adverse conditions to which the bacterium may, in fact, never be exposed.

The chemical nature of mutations

Since the genetic material is DNA we would expect mutations to be either changes in single DNA nucleotide pairs or more complex changes involving sequences of nucleotide pairs. Various lines of evidence, some of which will be considered below, suggest that there are three principal kinds of mutation. Firstly, there are what Freese has termed *transitions*. These are changes in single nucleotide pairs such that a purine is substituted for a purine and a pyrimidine for a pyrimidine. Thus starting with an adenine-thymine (A-T) base pair a transition mutation would give a guanine-cytosine (G-C) base pair or *vice-versa*. Secondly, again using Freese's terminology, there are *transversions*, in which a purine is substituted for a pyrimidine and a pyrimidine for a purine. Thus A-T could mutate by transversion to either T-A or C-G, and G-C to either C-G or T-A. The third important class of mutations includes those involving changes in the total number of nucleotide pairs—that is either *additions* or *deletions* of nucleotide pairs or sequences of nucleotide pairs. When these additions or deletions involve single base pairs or short sequences of nucleotide pairs within genes they are likely to act as *frameshift* mutations, for reasons which are discussed below. On a larger scale one can occasionally find deletions of large parts of genes or of blocks of genes. Still rarer are transpositions of blocks of genes from one chromosomal position to another, or *inversions* of parts of the gene sequence. These grosser structural changes have been very important in the analysis of gene position in relation to function in *E. coli*, but it would take us too far afield to discuss them in detail here.

How may one distinguish experimentally between different kinds of mutation in the DNA? At present the only possible approach to this problem is through the use of mutagenic chemicals which have some degree of specificity in their action on the genetic material. Mutagenic treatments may be of two different kinds, radiation and chemical. Both generally result in some killing of the treated cells as well as in the induction of viable mutants, and in practice one always has to choose a dose of the mutagen which is high enough to give a useful yield of mutations among the survivors without being so high as to leave too few cells alive. More emphasis will be given here to chemical mutagens, since their action is more easily interpretable in chemical terms.

A great variety of chemicals were shown to be mutagenic even before very much was understood in detail about their probable mechanisms of action. With our present knowledge of the chemical properties of DNA we can now at least make good guesses about how each of the main classes of mutagens acts.

Two basically different kinds of chemical action on the DNA of living cells can be distinguished. Firstly, there are mutagens which act only at DNA replication. These include the base analogues, which can become incorporated into DNA in place of normal bases at replication, and drugs of the acridine class, which can become interpolated into the stack of paired bases in a DNA double helix and may throw out of phase the pairing of template and newly-synthesized strand during replication. The action of these compounds depends on growth of the cells, with replication of the DNA, during the treatment. The second class of mutagens act in a direct chemical fashion to modify the bases of non-replicating DNA. We will deal first with the base analogues.

Mutation induced by base analogues

The two base analogues which have been used more than any others are 5-bromouracil (or its nucleoside 5-bromodeoxyuridine) and 2-aminopurine. The former will become incorporated into DNA in place of thymine, especially when the organism's own synthesis of thymine is blocked by genetic mutation or inhibited by the drug aminopterin. Under favourable conditions 5-bromouracil (BU) can replace thymine in DNA almost completely. 2-Aminopurine (AP) is an adenine analogue which can apparently be incorporated into DNA in place of adenine, although its incorporation is small in comparison of that of BU. Once incorporated, these analogues act most of the time as if they were the corresponding normal bases, but it seems likely that they make 'mistakes' in pairing with an unusually high frequency. BU is thought sometimes to pair with guanine instead of with adenine at DNA replication, while AP is probably apt to pair with cytosine instead of with thymine (Figure 4). This ambiguity of hydrogen-bonded pairing properties can be plausibly explained on the basis of the structures and chemical properties of the analogues (see Figure 5), but the detailed explanations need not detain us here. It is enough to point out that the mistakes made probably do not violate the rule that a purine always pairs with a pyrimidine and *vice versa*, and thus can only lead to transition mutations and never to transversions. There are, in principle, two ways in which a base analogue can cause a mutation, and the direction of the induced transition will be different in the two cases. The mistake may be made in the incorporation of the analogue (*misincorporation*) or in the base-pairing one or more cell generations after the analogue has been 'correctly' incorporated (*misreplication*). For example, if BU is correctly incorporated instead of thymine and then misreplicates it will have the effect of bringing in guanine where adenine should be. After a further round of replication the transition A-T → G-C will have been completed, the intermediate stages being A-BU and G-BU. If, on the other hand, BU is misincorporated opposite guanine in place of cytosine, a further round of replication will establish the transition C-G → T-A, BU-G and BU-A being intermediate stages (Figure 4). By a similar argument, AP is

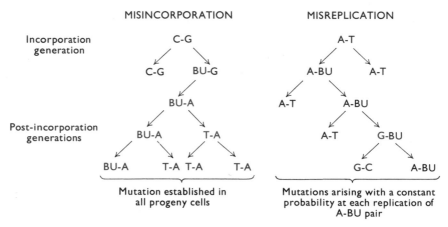

Figure 4. The different consequences of two different mechanisms of mutation induced by 5-bromouracil (BU) an analogue of thymine.

expected to induce G-C to A-T transitions by misincorporation and the reverse change by misreplication. Since mutants induced by either analogue are usually capable of being reverted to wild-type by both it seems that both analogues can act in both ways. It is, however, observed in experiments on T4 bacteriophage that mutants induced in bacteriophage by hydroxylamine (which is probably specific for G-C to A-T transitions—see below) are much less strongly revertible by BU than by AP and that most mutants induced by BU are not revertible at all by hydroxylamine. From this evidence it is concluded that BU acts preferentially by misincorporation to induce G-C to A-T transitions.

Acridine compounds as frameshift mutagens

The first acridine drug which was shown to be mutagenic was proflavine, which was used to obtain the first frameshift mutations in bacteriophage T4. Subsequently a variety of acridine compounds have been used for inducing frameshifts. One of the most widely used in recent years has been the acridine half-mustard, code-named ICR-191, the formula of which is as follows:

$$NH—CH_2—CH_2—CH_2—NH—CH_2—CH_2Cl$$

While the function of the 'mustard' part of the molecule in mutagenesis is not completely clear, it seems likely that the reactive chlorine enables the compound to combine with the DNA so as to increase the chance that the acridine part of the molecule will become inserted in the stack of base pairs.

It is not safe to assume that all mutations induced by a reactive compound like ICR-191 are of the frameshift type, but many are. In practice a mutant is thought likely to be a frameshift if it cannot be reverted to wild-type except by further treatment with the same or another acridine mutagen. The interpretation is strengthened if the mutation exerts a polar effect on translation within an operon (see p. 446).

Mutagens which act on non-replicating DNA

Among mutagens whose action does not depend on DNA replication, hydroxyl-amine, nitrous acid, and alkylating agents of various kinds are the most informative. Nitrous acid seems to react with DNA in various ways but its chief effect is probably to convert primary amino groups to hydroxyl groups and it thus tends to convert adenine to hypoxanthine and cytosine to uracil. Both these changes are expected to lead to transition mutations (in different directions) since uracil has the hydrogen-bonding properties of thymine rather than of cytosine, while hypoxanthine in its most stable tautomeric state would form hydrogen bonds more readily with cytosine than with thymine (Figure 5). These expectations about the action of nitrous acid are borne out by the finding that most mutants induced by nitrous acid are revertible by base analogues and *vice versa*. Hydroxylamine, which has a strong mutagenic effect on bacterio-phage, attacks only cytosine among the DNA bases so far as is known. It too induces transition mutations, as judged by the test of revertibility by base analogues, but probably only in the direction G-C to A-T.

Among the most carefully studied mutagenic alkylating agents are the ethylating agents diethylsulphate, ethylethane sulphonate (EES) and ethyl-methane sulphonate (EMS). It has been shown that both guanine and adenine can be ethylated but that the group most readily attacked is the 7 nitrogen of guanine, the product being 7-ethylguanine. An extremely potent alkylating agent of more recent introduction is *N*-methyl-*N'*-nitro-*N*-nitrosoguanidine which probably acts mainly by methylating guanine to form 7-methyl guanine. There are two possible ways in which this alkylation could be mutagenic. On the one hand, the 7-alkylguanine might have different pairing properties and thus bring about transition mutations. This is now thought to be the main mechanism. 7-Alkylated guanine is very readily hydrolysed from the polynucleotide chain to leave a gap where a base should be. If this happened one might expect that *any* base could be incorporated opposite the gap at the next replication and this could lead to either a transition or a transversion, but the extent to which mutations actually arise in this way is doubtful. It is indeed found in studies with bacteriophage that some mutations induced by ethylating agents are induced to revert by nitrous acid and base analogues, and are thus probably transitions, while others are not so revertible.

Radiations

Turning now to the mutagenic effects of radiations, ultraviolet light (UV) probably has a number of effects on DNA. The most studied effect is the linking of neighbouring pairs of pyrimidine residues (dimerization), and there is no doubt that this is one important mechanism of the killing of cells by UV. Bacterial cells have efficient enzyme systems for removing short stretches of DNA single strand containing pyrimidine dimers and then repairing the double helix by new synthesis (the so-called 'cut and patch' mechanism) or by the joining together of undamaged parts of different chromosomes (the recombination mechanism). There are strong indications that UV-induced mutations arise through errors in repair. The study of UV-sensitive mutants deficient in one or other aspect of repair is shedding a good deal of light on the connections between repair, recombination and mutagenesis.

X-rays and other ionizing radiations are both lethal and mutagenic, and can no doubt cause a variety of different chemical changes in DNA. To judge from work on fungi and higher organisms, both UV and X-rays bring about many structural rearrangements in the genetic material.

Mechanism of spontaneous mutation

We have said nothing so far about the probable chemical mechanism of spontaneous mutations. Watson and Crick, at the same time as they proposed the double-stranded structure and semi-conservative mechanism of replication of DNA, also suggested how spontaneous mutations might occur. They pointed out that the hydrogen atoms involved in the hydrogen-bonding of the paired bases of DNA might shift their positions by tautomerism. In fact, tautomerism is possible in each of the DNA bases, and to the extent that the tautomeric forms exist there is the possibility of some variation in pairing properties. For instance, the rare *imino* tautomeric variant of adenine, with a hydrogen atom in the N1 position and an imino group at C6, instead of no hydrogen at N1 and an amino group at C6, would pair with cytosine instead of with thymine (Figure 5). If this were the sole reason for spontaneous mutation spontaneous mutants should always be of the transition type and should thus be revertible

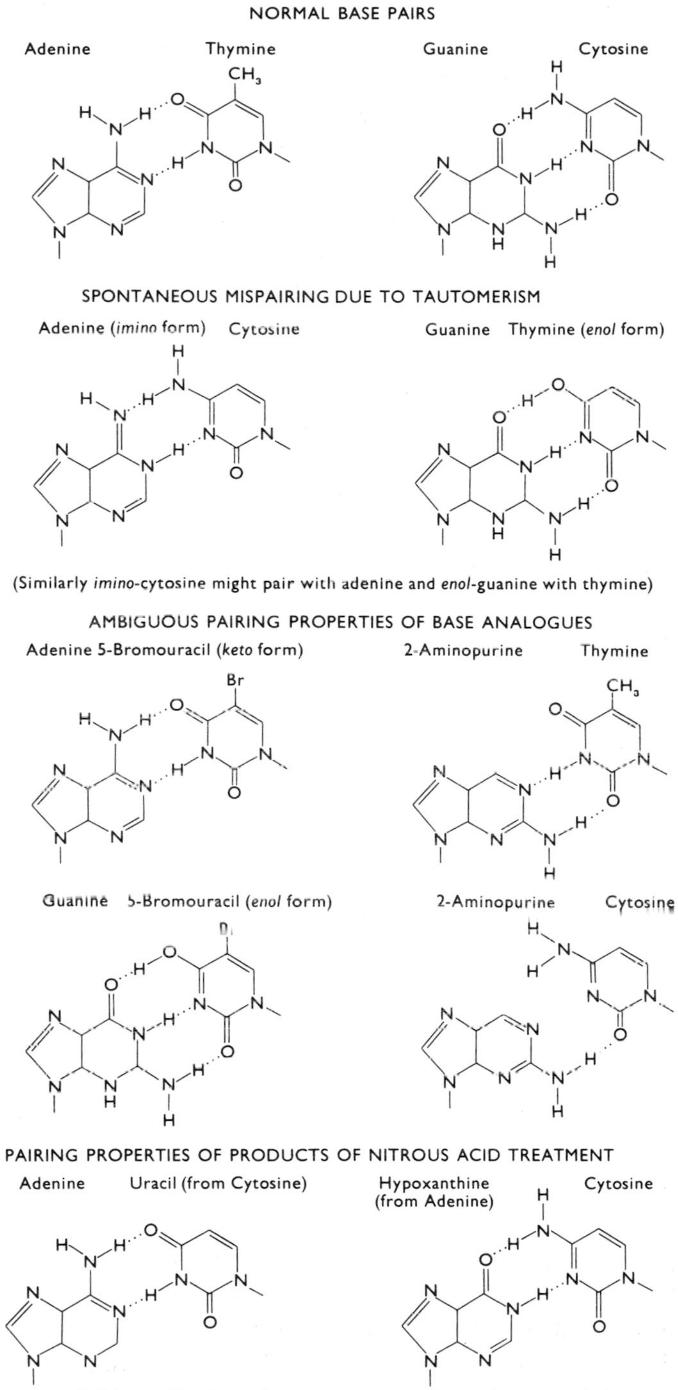

Figure 5. Some probable mechanisms of mutagenesis involving alterations in base-pairing properties (see text).

by base analogues or nitrous acid. That the tautomerism hypothesis is not the whole explanation is indicated by the finding that, at least in bacteriophage, most spontaneous mutations are *not* revertible by base analogues. Many spontaneous mutants may thus be transversions and some are certainly frameshifts. There is evidence that the latter occur through some kind of slippage of one DNA strand relative to the other in regions of fortuitously reiterated sequence.

Spontaneous mutation frequency, or at least the frequency of spontaneous transition mutations, is itself subject to genetic control. Several genes have been found in *E. coli* which increase mutability; one of these, that discovered by Treffers, promotes A-T to G-C transitions specifically. The mutation frequency in T4 bacteriophage has been shown to be controlled by the nature of the DNA polymerase produced by this virus. Mutant forms of the polymerase have been identified which determine both higher and lower frequencies than are found in the standard type. Presumably there is natural selection for a frequency of errors in DNA replication which strikes a balance between excessive inaccuracy and excessive evolutionary conservatism.

The expression of mutations

When a mutation arises in a bacterial cell it may have an immediate effect on the observable properties (i.e. the phenotype) of the cell or, more likely, it may not be expressed for a few cell divisions. There are a number of possible reasons for delay in expression. In the first place, many rod-shaped bacteria, including the most thoroughly studied species *Escherichia coli* and *Salmonella typhimurium*, tend to have two or more chromosomes per rod, the number depending on the stage of the cell division cycle (Figure 3, p. 144). Each rod is, in reality, equivalent to two or four cells, with one chromosome in each, though the convenient practice of referring to the whole rod as a 'cell' is very general. Products of chromosome replication are segregated into different quarter-cells, then, following cell division, into different half-cells, and, after a further division, into different rods altogether. If a mutation occurs in one strand of one chromosome it would be expected that it would take two to three cell divisions before a cell was formed with all its DNA descended from the mutant strand. It very generally happens that the mutation causes a *loss* of function of a protein—in other words it leads to a *negative* phenotype—and that the positive phenotype continues to be expressed so long as at least one non-mutant DNA molecule remains in the rod. Delay in the expression of a mutation due to this cause may be called *segregation lag*. A second reason for delay in the expression of a loss of function may be that the functional protein characteristic of the non-mutant cell may persist and function for some time after the genotype has become wholly mutant. This may be termed *phenotypic lag* and, if it occurs, it will be superimposed on the segregational lag. Finally, and perhaps

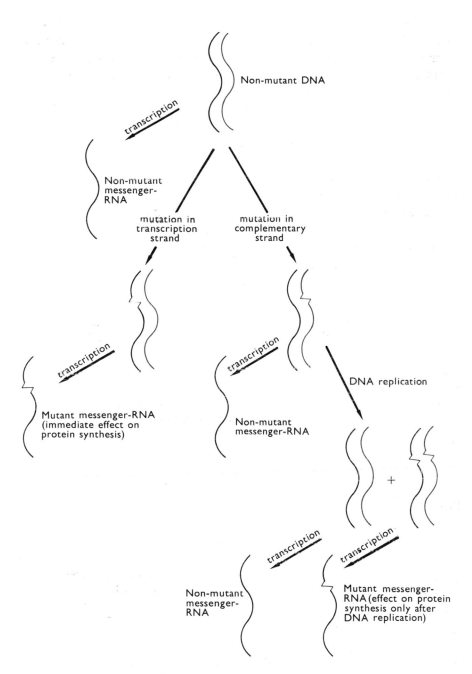

Figure 6. The steps intervening between a mutation (base substitution) and its expression in the phenotype.

most interestingly, the mode of replication and transcription of the DNA may itself result in a lag quite independent of the nature of the protein affected by the mutation. As was explained (Chapter 6), each messenger seems to be transcribed from only one of the two strands of the DNA, at least *in vivo*, and the strand used for transcription is always the same one. A mutation in the transcribed strand can have an *immediate* effect in the production of a new kind of messenger which, if it causes a *positive* change in the enzyme complement of the cell, can have an immediate effect on the phenotype. A mutation in the other strand, however, cannot produce any phenotypic effect until the DNA has replicated. Figure 6 illustrates the principle.

It should be mentioned that there are strong indications that a completely new base *pair* may sometimes be substituted by mutation in a DNA molecule, without the necessity for a further round of DNA replication. It seems quite possible that, following a change in one base of a hydrogen-bonded pair, its now mismatched partner may become vulnerable to enzymic removal and replacement even in the absence of extensive DNA synthesis.

We should not leave this discussion of the expression of mutations without mentioning the very important demonstration that the *expression* of certain mutations can be masked in the presence of 5-fluorouracil (FU), a uracil analogue which seems to interfere with transcription but which is not itself a mutagen. For instance, in *E. coli* it has been shown that FU will partially restore the ability to make alkaline phosphatase to certain mutants which are normally deficient in this enzyme. Suppression of a mutant phenotype by FU was first demonstrated among rII mutants of bacteriophage T4. The important property of such mutants is their inability to grow on the K12 (λ) strain of *E. coli*. A large number of rII mutants were identified as base-pair transitions on the basis of their revertibility with the base analogue 2-aminopurine (cf. p. 381). Within the transition mutant class a proportion were reverted strongly by 5-bromodeoxyuridine (BDU) and by hydroxylamine. For reasons which we discussed on

Table I. Properties of different classes of bacteriophage T4 transition mutants.*

Class	Revertibility by			Presumed base pair at mutant site	Phenotypic suppression by FU	Presumed base at mutant site on mRNA
	AP	BU	NH$_2$OH			
1	+	+ +	+ +	GC	0	G or C
2	+	±	0	AT	0	A
3	+	±	0	AT	+	U

* Based on data of S. P. Champe & S. Benzer (*Proc. Nat. Acad. Sci. Wash.*, **48**, 532).
Abbreviations: AP = 2-aminopurine
              BU = 5-bromouracil or 5-bromodeoxyuridine
              FU = 5-fluorouracil.

pp. 382–3 these mutants are likely to have had G-C at the mutant site. A second sub-class, roughly equal in number, were not reverted by hydroxylamine and were also insensitive, or relatively so, to BDU. They were presumed to have A-T at the mutant site. Within this second sub-class of mutants almost half were enabled to grow to a significant extent in K 12 (λ) when FU was added to the infected bacteria. With two doubtful exceptions none of the first sub-class responded to FU in this way (see Table 1). The interpretation given to these results is as follows. FU is probably incorporated into the messenger RNA in place of U, but once in the messenger it is occasionally translated as if it were C. It will tend to 'cure' any transition mutant with A-T at the mutant site when the A is on the transcribed strand. This is because, in such a mutant, the mutant base (A) will be transcribed in the messenger as U or FU, and FU will be some-times translated as if it were C, which is the base which would have been trans-cribed from the wild-type base G. Transition mutants with T at the mutant site on the transcribed strand will not be cured by FU since they would need to replicate before the FU would affect them, and the rII phenotype is such that they cannot replicate at all in K12 (λ) cells. It must be emphasized that the effect of FU is only on the *phenotype*; it modifies the effect of DNA mutations without altering the mutations themselves. Table 1 summarizes the interpretation.

GENETIC MAPPING

Chromosome pairing and exchange

In essence the mapping of the genome depends on much the same sort of pro-cedure as is used by biochemists for determining the amino acid sequence in a long polypeptide chain. As a result of a genetic exchange or breakage (Figure 7, p. 390), the genetic DNA can be divided into segments and one can, by special techniques, determine which genetic markers are present together in the same segment. Segmentation may occur at many different points (probably between any two adjacent nucleotide pairs), so originally linked markers are separated with frequencies dependent on their distances apart. Eventually all possible blocks of markers obtainable by segmentation can be arranged in sequence on the basis of the overlaps in their marker compositions and a map of the entire genome can thus be established.

How can one determine the marker composition of fragments of the genome? The information can only be gained through allowing the genetic fragment to be incorporated into another, distinctively marked, genome and observing its phenotypic effects after it has been replicated and transmitted to a numerous progeny. The incorporation occurs through genetic exchange in which the incoming chromosome segment replaces the equivalent resident

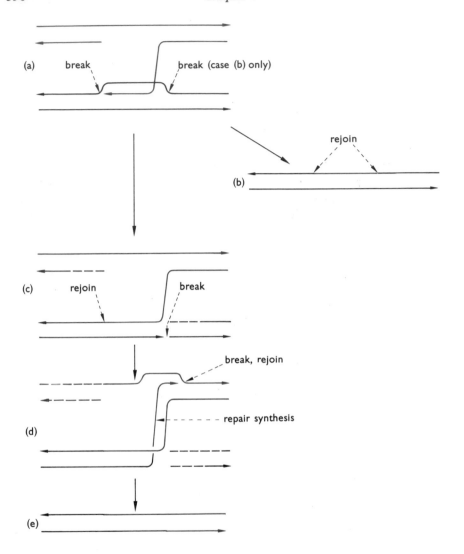

Figure 7. Possible modes of recombination. (a) Displacement of a length of single strand in one DNA duplex (chromosome) by a homologous single-stranded piece derived by breakage and unwinding from another chromosome (shown red). This leads either (b) to incorporation of a fragment of single strand into the otherwise intact recipient chromosome, or (c–e) to exchange (crossing-over) between chromosome segments. In (c) the complementary strand of the black chromosome also breaks and (d) by new synthesis, using the crossed-over red strand as a template, completes a double-stranded cross-over (e). Arrowheads indicate the opposed polarities of the paired strands, and different colours indicate material derived from, or copied from, different chromosomes. The necessary breakage and rejoining processes probably involve limited digestion and repair synthesis; the final rejoining may well be catalysed by the known enzyme DNA ligase. The diagram, which shows only one of several possible schemes, leaves open the question of whether a reciprocal cross-over product (i.e. left segment of red chromosome joined to right segment of black chromosome) is formed at the same time. Reciprocal crossing over must be

segment. It is thus on this process of exchange, or *recombination*, that all genetic mapping depends.

Recombination usually occurs only between *homologous* segments of chromosomes—that is, segments which have a very large proportion of their DNA base-sequences in common. It depends on getting two homologous, but differently marked, chromosomes or pieces of chromosome into the same cell; in bacteria a great diversity of methods can be used to achieve this end, as we shall see below.

Recombination between homologous chromosomes must depend on their becoming paired, with homologous sections in contact. The molecular basis of this pairing is not understood but it is evidently a very precise process since genetic exchange usually involves exactly corresponding segments. It seems very likely that hydrogen-bonding between stretches of complementary single-stranded DNA is part of the explanation.

However it occurs, homologous pairing always seems to lead to genetic re-combination. The mechanism of recombination is currently one of the most vexed questions of molecular genetics and it cannot be said that any of the hypotheses which have been put forward has yet been confirmed in detail by experiments.

Studies on the integration of transforming DNA into the recipient chromo-some have given valuable information about the recombination process. For example, experiments by Fox on transformation by DNA in *Diplococcus* have shown that stretches of single strands are removed from the chromosome of the recipient cell and replaced by corresponding stretches of DNA from the transforming preparation. How could such single-strand displacement occur? Virtually all theories agree that an initial step in recombination is a pairing of complementary *single* strands derived from different DNA molecules, forming a region of hybrid DNA (Figure 7a). After this a number of different possibilities can be envisaged and probably occur. One is that the incoming single-stranded fragment is effectively separated from its parent duplex and becomes incor-porated into the recipient duplex chromosome, displacing the corresponding 'resident' segment and becoming covalently bonded into the duplex through single-strand breaks and rejoining (Figure 7b). This is roughly what is thought to happen in transformation by DNA in pneumococcus and *Bacillus subtilis*. Alternatively, the single-stranded donor segment which enters the recipient chromosome may remain joined by one end to the DNA duplex from which it has become partially unwound (Figure 7c)—in this case the possibility exists that, through appropriate breaks, repair synthesis and reunions, the single-

---

assumed to occur in episome integration and excision (see Figures 11 and 14), but this may be a special case. Most recombination data from bacteria (in contrast to higher organisms) can be explained without the assumption of reciprocal crossing-over.

stranded cross-over will be converted to a double-stranded one. Figure 7(c–e) suggests one possible type of mechanism through which segments of two chromosomes of different origin may thus become joined end-to-end, probably with a short section of hybrid duplex in the joining region (Figure 7e). There are a number of plausible possibilities based on the properties of enzymes known to be present in bacteria but as yet there is no clear evidence in favour of any one detailed model.

Some of the enzymes involved in recombination appear also to function in the repair of damage to DNA caused by ultraviolet light, since some mutants selected for exceptional ultraviolet-sensitivity are also deficient in their ability to promote genetic recombination. For example, two classes of such mutants (*recB* and *recC*) have been shown to lack an ATP-dependent nuclease specific for double-stranded DNA. The search for enzymic deficiencies in other classes of *rec* mutants offers perhaps the best hope for further progress in the identification of the chemical steps involved in recombination.

General methods of mapping

We stated at the beginning of this section that genetic mapping depended on analysing the genome into overlapping segments and determining the composition of these segments with respect to genetic markers. One can distinguish two basically different kinds of mapping procedure depending on the way such segments are generated. First, in practically all systems of genetic recombination in bacteria there is a distinction between the donor and the recipient parent. Whereas a recipient cell contributes a complete genome to the immediate product of cross (the *zygote*), the donor contribution consists only of a fragment—sometimes a very small fragment—of its genome. In such systems genetic fragmentation has already occurred *before* the donor DNA has entered the recipient cell (or, in the case of conjugation—see below—as it is entering). Subsequent genetic exchange will then enter the analysis only as an essential step in integrating the donor fragment and thus in determining its marker constitution. This general kind of analysis is, as we shall see, an important element in mapping by transduction or transformation and it is the essence of the interrupted mating technique which is extremely important for mapping in *E. coli*. On the other hand, one can determine the order of markers within a donor chromosome, or chromosome fragment, by an analysis of the patterns of exchange which occur *after* it has entered the recipient cell. Markers which are close together on the donor segment are very likely to be integrated together into the recipient chromosome; markers which are farther apart are more likely to be separated—one integrated and the other lost. From the relative frequencies of different patterns of marker integration one can determine the marker order, the general principle being that a result which can be explained by a simple pattern of exchanges should be more frequent than one which requires a more complex pattern. This is the kind of reasoning which is used in mapping the very fine structure of the genome. It is analogous to the standard method of linkage mapping in higher organisms and fungi, in which prezygotic fragmentation of the chromosomes does not occur and where the parental contributions are usually equal.

A refinement of recombinational analysis is the method of *overlapping deletions* (Figure 8). A proportion of auxotrophic and other kinds of defective mutants have lost more or less extensive gene segments rather than merely having nucleotide-pair substitutions. Such deletions, if they are not to be unconditionally lethal, must not be too extensive; indeed, they are usually confined within the limits of single genes or operons. They may each, however,

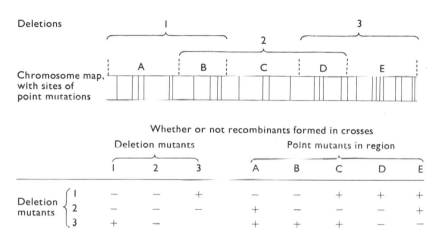

Figure 8. The principle of overlapping deletions, showing how point mutations can be assigned to sections of the chromosome defined by three deletions.

include the sites of many 'point' (i.e. transition and transversion) mutants. Two mutants whose deletions do not overlap will have a complete chromosome between them and will therefore be able to produce some non-mutant progeny by recombination when crossed together. If, on the other hand, their deletions overlap, the segment of the chromosome corresponding to the overlap can be contributed by neither mutant and a cross between them will be incapable of producing wild-type progeny. Merely by seeing whether *any* non-mutant recombinants are produced one can, without having to worry about the exact numbers of recombinants, see whether two deletion mutants overlap or not. Given a series of deletion mutants one can often arrange them in a unique linear order on the basis of the overlapping or non-overlapping relationships of the various different pairwise combinations. Furthermore, it is easy to determine whether a given point mutation falls within the segment missing in a given deletion mutant, simply by seeing whether any recombinants are formed from a cross between the two mutants (Figure 8). This is a powerful general method which is in principle applicable to any of the genetic systems which will be mentioned in the following pages. The classic example of deletion mapping is Benzer's analysis of the rII region in bacteriophage T4, but exactly the same principle can be applied to bacteria.

In the following pages we will review briefly the very diverse experimental procedures which are used for genetic mapping in different bacteria.

Conjugation

One-way transfer and time of entry experiments

The key to the understanding of mating in *E. coli* lay in the recognition that different strains of the organism are of different mating types. Cell conjugation, in which pairs of cells are temporarily joined depends on at least one of the partners possessing a *fertility factor*, called *F*. The ability to conjugate appears to be due to the presence of special filamentous cell wall appendages—sex-pili (see p. 106)—for the formation of which F is necessary.

When F^+ and F^- populations, differentiated by genetic markers, are mixed, one can demonstrate the formation of recombinants by plating the mixture, on a selective medium, after an hour or so of incubation. For example, if the F^- strain requires threonine and leucine (symbolized by *thr⁻ leu⁻*) and is streptomycin-resistant (symbolized by S^r), while the F^+ strain does not require the amino acids and is streptomycin-sensitive (*thr⁺ leu⁺ S^s*), plating of the mixture on minimal agar plus streptomycin will select for recombinants of type *thr⁺ leu⁺ S^r*, and colonies of this type are regularly found. If the mixed strains also differ with respect to markers which make no difference to growth on the selective medium (*unselective markers*), these markers have a strong tendency to be inherited by the recombinant colonies from the F^- parent. This is explained by the finding that the F^+ strain acts as the genetic donor, and normally donates only relatively small pieces of the genome; the selected recombinants therefore tend not to inherit unselective markers from the F^+ strain unless the markers happen to be quite closely linked to the selected ones.

The frequency of recombinants given by an ordinary $F^+ \times F^-$ mating is low—often of the order of 0·1%. It is now well established that this low fertility is the result of there being a great majority of F^+ cells which transfer no chromosomal markers at all to the F^- cells and a small minority which do so with a high efficiency. Cells of this latter type, known as *high-frequency-recombination* or *Hfr*, are to be found in all F^+ populations and arise from F^+ cells by a kind of mutational event the nature of which will be made clear in our further discussion. When Hfr cells are isolated by plating they give rise to colonies all the cells of which are of the Hfr type. In this way reasonably stable Hfr strains can be established and it is such strains, rather than ordinary F^+ stocks, which are always used nowadays for genetic analysis.

Isolation of individual cells resulting from the first few divisions after conjugation has shown that, in an $Hfr \times F^-$ mating, all the recombinants arise by subsequent division of the F^- cells and none from their Hfr partners. This shows that there is a one-way transfer of genetic markers from the Hfr ('male') to F^- ('female'). The Hfr markers are transferred in a definite time sequence as

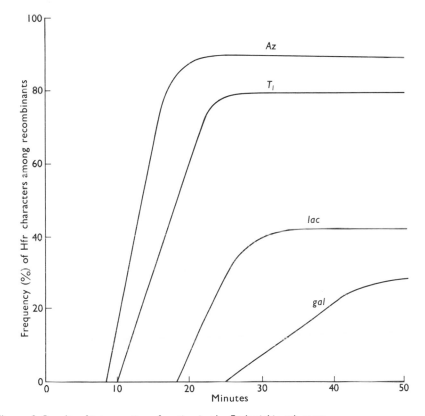

Figure 9. Results of interruption of mating in the *Escherichia coli* cross.

F   *thr   leu⁻ Azʳ T₁ʳ lac⁻ gal⁻ Sʳ × Hfr thr⁺ leu⁺ Azˢ T₁ˢ lac⁺ gal⁺ Sˢ*. Conjugation was interrupted at various times after mixing; the cells were plated on streptomycin medium devoid of amino acid to select *thr⁺ leu⁺ Sʳ* recombinants; the frequencies of the unselected markers among these recombinants were determined.

can be shown by the *interrupted mating* type of experiment. The principle of such an experiment is best illustrated by reference to one of the classical crosses. The F^- parent strain carried mutations resulting in resistance to streptomycin (S^r), azide (Az^r) and bacteriophage T1 ($T1^r$), nutritional requirements for threonine and leucine (thr^- leu^-) and inability to use galactose (gal^-) or lactose (lac^-). The Hfr strain had the contrasting wild-type characters (S^s Az^s

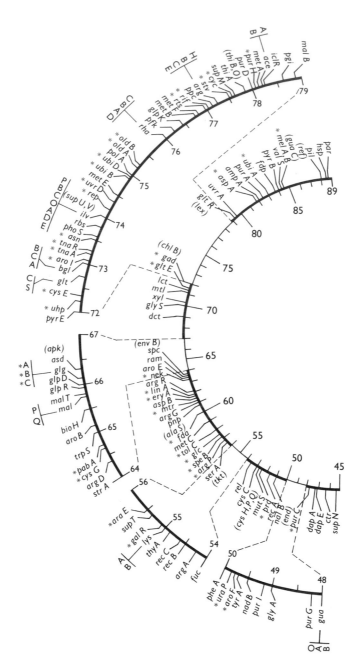

Figure 10. The circular linkage map of *Escherichia coli* K12.

The relative positions of markers and the approximate distances between them in time units are shown. From A. L. Taylor, *Bacterial Rev.* **34**, 155 (1970).

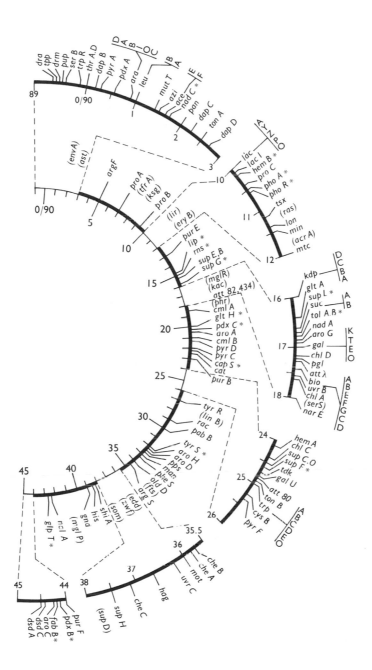

$T1^s$ *thr*$^+$ *leu*$^+$ *gal*$^+$ *lac*$^+$). Cells of the two strains were mixed in a suitable liquid nutrient medium and, at different times after mixing, samples of the mixture were violently agitated in a mechanical blendor to interrupt mating and plated on minimal (glucose) medium containing streptomycin. On this medium only *thr*$^+$ *leu*$^+$ S^r recombinants were able to grow to form colonies, the *Az*, *T*1, *lac* and *gal* markers being unselective. The mechanical agitation separated the conjugating pairs of cells so that only those F$^-$ cells which had received *thr*$^+$ and *leu*$^+$ from the Hfr conjugant *before* the time of blending were able to form recombinants of the selected type.

The markers *thr* and *leu* are, in fact, quite closely linked and could therefore be regarded to a first approximation as a single unit. It was found that when mating was interrupted after less than 8 min no *thr*$^+$ *leu*$^+$ S^r recombinants were recovered. Starting at about 8 min, however, such recombinants began to appear and increased in frequency with time. The recombinants formed after the earliest possible interruption tended to contain no Hfr markers other than *thr*$^+$ *leu*$^+$ but after about 9 min of uninterrupted mating Az^s began to make its appearance among the recombinants followed in sequence by $T1^s$, *lac*$^+$ and *gal*$^+$, starting at 11, 18 and 25 min respectively. The result is summarized in Figure 9, and it can only be reasonably interpreted as meaning that the Hfr markers enter the F$^-$ cell in a time sequence corresponding to their order on a linear chromosome.

The length of the sequence which is transferred depends on how long the cell pairs remain in conjugation. When the cells are free in liquid suspension, conjugation is usually broken off after only a fraction of the chromosome has entered the F$^-$ partner. The lower maximum frequencies among recombinants shown by the later-entering markers in Figure 9 is mainly due to the lower probability of entry the longer the time required. On the other hand, if the cells are held immobile on a filter the whole chromosome may be transferred.

The sequence of markers *thr-leu-Az-T*1*-lac-gal* is characteristic of one particular Hfr strain. Other Hfr strains have the same linear order of genes, but the transfer may start at different points in the order and may proceed in either direction. The comparison of the sequences of markers transferred by a number of different Hfr's shows that they often overlap. The overlaps can be used to build up a continuous map of the whole genome. The feature which caused most surprise when this map was first made was that when all the different sequences were put together they formed a closed loop. Circular linkage maps were at that time completely novel. It is now known that the *E. coli* genome consists of a closed loop of DNA (cf. p. 283). Figure 10 shows the linkage map of *E. coli* as determined by time of entry experiments. The distances between the markers are expressed in minutes. The entire map is about 90 minutes in circumference, this being the time necessary for the entry of the complete Hfr chromosome under standard conditions.

Conjugation in other bacteria

The other methods for mapping which we shall be mentioning, particularly that of transduction, are generally only useful for resolving the fine structure of very short segments. Unfortunately, mating systems of the Hfr − F⁻ type are not commonly found. A somewhat similar but less efficient system has, indeed, been described in *Pseudomonas aeruginosa* but the high fertility Hfr characteristic has been found to originate only in certain strains of *E. coli*. Hybridization between *E. coli* and *Salmonella typhimurium* has, however, resulted in the isolation of Hfr strains of the latter species, which can be used for time of entry experiments in the same way as *E. coli* Hfr's. Interestingly enough, the complete genetic maps of the two species turn out to be almost identical in gene content and marker sequence, though the time taken for entry of a given chromosome segment is somewhat greater in *Salmonella* than *E. coli*.

Properties of the F factor—episomes

The inheritance of the fertility factor F is of great interest. When an ordinary F^+ strain (i.e. not an Hfr) is mixed with an F^- strain, F^+ and F^- cells conjugate with high efficiency, but transfer of chromosomal markers occurs in only a small minority of cell pairs. However, there *is* a very efficient transfer of the F factor to F^- cells. The whole population tends to become F^+ within a few cell divisions. Thus F behaves as an infective factor transmitted independently of the chromosome. It is now known that the free F factor is a small closed loop of DNA of about 10^8 daltons in weight. In ordinary F^+ cells it is quite separate from the main chromosome and is an example of a class of extrachromosomal genetic elements known as *plasmids* (see p. 420).

The behaviour of Hfr cells with regard to the transmission of fertility is quite different. Almost all the recombinants from Hfr × F⁻ matings are F⁻. It is as if two simultaneous changes have occurred in the origin of the Hfr from the F^+ condition: the chromosome has become transferable and the fertility factor has become non-infective. The explanation is suggested by the finding that those few F⁻ recombinants which have inherited markers from the extreme trailing end of the Hfr chromosome—those in other words, which have received the entire chromosome from the Hfr parent—are often Hfr in type. The capacity for transfer is evidently determined by something at the trailing end of the chromosome and this is believed to be the F factor, integrated into the chromosome and therefore no longer independently infective. The integration of the F factor has occurred at different points in different Hfr strains—always at the point which is transmitted to the F⁻ cell last.

The integration of the F plasmid into the chromosome is thought to depend on its having relatively short base sequences in common with various segments of the bacterial chromosome. Reciprocal exchanges between these short homologous regions would then lead to integration of the small loops into the large one as shown in Figure 11. This kind of integration should be reversible, and it is indeed known that Hfr strains tend to revert to the F$^+$ condition as well as to arise from it. In its capacity to exist either as part of the chromosome or as an independently transmitted extra-chromosomal element the F plasmid displays the classical properties of an *episome*. Other examples of episomes are the genomes of certain temperate bacteriophages, such as *lambda*, which will be mentioned later.

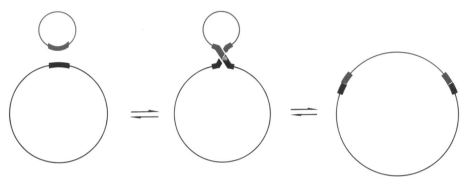

Figure 11. The suggested mechanism for the integration of an episome, for instance an F factor or prophage, into a bacterial chromosome. The thicker line indicates the region of homology between chromosome and episome.

The capacity of the integrated F factor to promote chromosome transfer, like its capacity to replicate autonomously, is probably connected with its capacity to initiate DNA replication. Though there has been some dispute on the question in the past, the evidence is now strong that the Hfr chromosome is replicated during its transfer to the F$^-$ cell. This *transfer replication* may be induced by conjugation and starts at the point where the F factor is integrated. Instead of the chromosome loop remaining closed, as in ordinary vegetative replication (cf. Figure 20, p. 282), separation of single DNA strands next to the F factor is thought to result in the formation of a free 5' end, which is then 'pushed' through the conjugation tube (which may be the sex pilus itself), by new strand synthesis at the fork in the duplex structure. Figure 12 illustrates the idea. There is some evidence that the donor DNA is transferred in single-stranded form and that only later, perhaps at the time of its integration into the recipient chromosome (cf. Figure 7, p. 390) is a complementary strand synthesized. An alternative view, which is still taken seriously, is that the new complementary strand is synthesized simultaneously with transfer.

Mapping by transduction

General transduction

While chromosome transfer of the F-mediated type is known only in a few bacteria, transfer of small pieces of the chromosome in bacteriophage particles (*transduction*) is known to occur in a wide range of species. The phenomenon of transduction was first described in *Salmonella typhimurium*, the bacteriophage agent in this case being PLT22 (P22 for short). The *Salmonella* phage P22 and

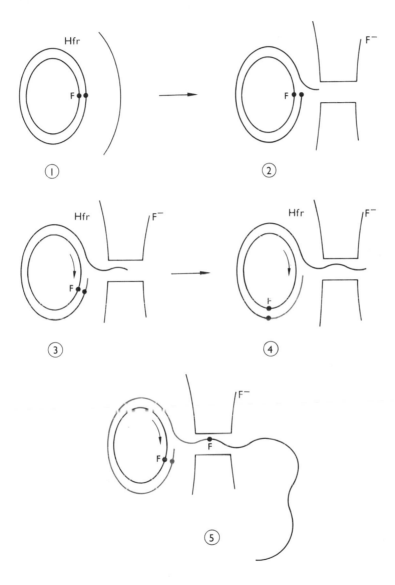

Figure 12. Hypothesis of DNA transfer during replication, in an Hfr × F⁻ mating.

The F factor is thought to provide the point where new synthesis is initiated, with generation of a free 5′ end. The F factor is replicated and transferred last. If conjugation continues long enough a second copy of the chromosome can begin to be transferred, as in 5. Newly synthesized strands are in red. The two DNA strands should really be helically wound, but are shown here as parallel lines the space between which is highly exaggerated.

the *Escherichia coli* phage P1 are examples of *general* transducing phages in that they are able to transduce any marker in the bacterial chromosome.

The procedure in a transduction experiment is to infect cells of one strain of the bacterium with bacteriophage, allow the cells to lyse, kill any cells which may have survived the infection with chloroform, and use the lysate to infect a second strain of bacteria carrying different genetic markers. Most of the infected cells of the second strain will probably be lysed, but among those which survive the infection some are found with markers acquired from the bacteria on which the infecting phage had been grown. Since the frequency of transduction with respect to any one marker is always low one has to use a selective method for the isolation of the transduced cells. In the most usual experimental design the second, or *recipient*, strain carries an auxotrophic marker which is not present in the strain (the *donor*) on which the bacteriophage was grown. By plating the surviving cells on medium lacking the supplement required by the original auxotrophic strain, one automatically selects those cells which have inherited the prototrophic character from the donor.

By choosing donor and recipient strains which are differentiated by several markers it is easy to show that, when transduction is mediated by P1 or P22, donor markers are generally transduced one at a time—it is quite uncommon to find that a recombinant cell has inherited more than one marker from the donor. This means that the donor chromosome is transferred to the recipient in small fragments, which is not surprising in view of strong evidence that the transduced material is carried within the very small particles of the transducing phage. A tendency of two markers to be simultaneously transduced (cotransduced) is evidence of their very close linkage. This point may be illustrated by reference to *E. coli* and the transducing phage P1. Using the phage grown on *thr*$^+$ *leu*$^+$ cells to infect *thr*$^-$ *leu*$^-$ cells, *thr*$^+$ transductants can be selected for by plating the recipient bacteria on medium containing leucine but devoid of threonine. In one experiment only about 4% of the selected transductant colonies were found to have received *leu*$^+$ as well as *thr*$^+$. When *leu*$^+$ transductants were selected by plating on medium containing threonine but lacking leucine it was found that 98% of them carried *thr*$^-$ and only 2% *thr*$^+$. Thus *thr*$^+$ and *leu*$^+$ show only about 2 to 4% cotransduction and the linkage of these two markers is very weak as judged by transduction tests. We saw, however, that *thr* and *leu* appear closely linked in interrupted mating experiments, differing in time of entry by less than a minute. This will give some idea of the resolving power of transduction analysis and, at the same time, its unsuitability for mapping more than very short genetic segments.

The situation is somewhat different with the *Bacillus subtilis* phage PBS1. The protein coat of this virus is capable of carrying considerably more bacterial DNA than either P1 or P22, and it has been calculated that as much as 8% of the *B. subtilis* genome can be transduced at one time. This permits linkage between much more widely spaced markers to be detected by transduction analysis in *B. subtilis* than in *E. coli* or *S. typhimurium*.

Transduction really comes into its own as a tool for the mapping of genetic fine structure, especially when one has a number of independently isolated and apparently similar auxotrophic mutants. The classical example of this type of

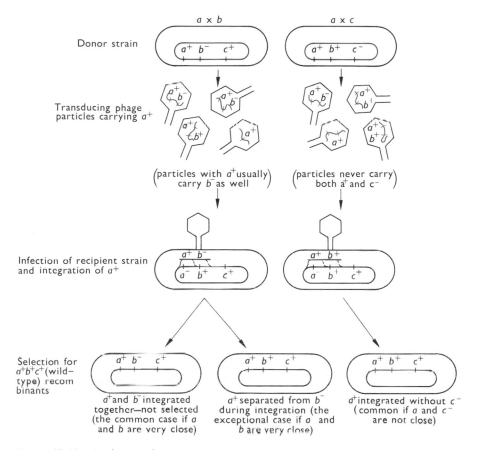

Figure 13. Mapping by transduction.

If *a* and *b* are closely placed and *a* and *c* relatively distant mutations, the cross *a* × *b* will yield few wild-type recombinants compared with *a* × *c*. The donor chromosome is shown in red and the recipient chromosome in black. Dashed lines indicate alternative possible positions for exchanges.

analysis is the very extensive work on histidine-requiring mutants of *S. typhimurium*. Almost all these mutants turned out to map at different sites, as shown by the fact that almost any two of them, when one was used as the donor and the other the recipient in a transduction 'cross' with phage P22, gave some histidine-independent recombinants. The appearance of wild-type recombinants means, of course, that the donor is able to contribute the wild-type equivalent of the recipient mutant site, and *vice versa*. The frequency of prototrophs was usually considerably less than would have been obtained had the same recipient been crossed with a wild-type donor. This indicated that the mutant sites in the two mutants must be closely linked, so that the donor mutant site was cotransduced with the donor wild-type site in a high proportion of transductants (Figure 13). The frequency of prototrophic recombinants evidently depended on the frequency with which the two donor sites became separated, either by fragmentation of the donor chromosome during the formation of the transducing phage or by recombination after transfer to the recipient cell. Whichever way the separation occurred it was expected to be more frequent the greater the physical separation of the sites and so recombination frequency could be taken as a measure of genetic map distance. By comparing the recombination frequencies given by crosses of numerous pairs of mutants it was possible to build up a linear map showing the approximate order of the sites of the *histidine* (*his*) mutations. All the sites fell within one comparatively short region of the chromosome. One source of uncertainty in this kind of mapping is that different transducing phage preparations may have different transducing efficiencies. This difficulty could be overcome to some extent by determining for each cross the frequency of transduction of a standard donor marker outside the *his* region and expressing recombination within *his* as a percentage of this frequency. Fortunately, also, a considerable number of overlapping deletion mutants turned up in the *his* series and these were used to confirm and correct the conclusions drawn from crosses between point mutants. The principles of deletion mapping were outlined on p. 393 (Figure 8) and apply just as much to transduction as to any other way of performing genetic crosses. Some of the results of the fine structure mapping of the *Salmonella his* region are shown in Figure 12, p. 233.

Another general method of establishing an unequivocal order of sites in fine structure mapping is through the use of *three point crosses*, in which the donor and recipient strains are distinguished by three closely linked markers. The principle used is that the simplest patterns of exchanges should give the commonest classes of recombinants; thus, if the order of the donor markers is *a–b–c*, *a* and *b* may relatively frequently be integrated independently of *c*, and *b* and *c* integrated independently of *a*, but the integration of *a* and *c*, leaving out *b*, should be much less common since this last pattern would require at least four exchanges in appropriate positions. Generally speaking they are not easy to set up since they depend on the availability of double mutants which, especially if they involve sites in the same gene, are likely to be difficult to distinguish and select, but they have been used to good effect in favourable circumstances.

Special transduction

Another group of phages of *Escherichia coli*, of which *lambda* is the best known example, carry out *special* transduction of genes from one small segment of the host genome. Transduction by *lambda*, which is normally confined to *bio*, *gal* and a few other nearby genes, is better understood than any other example and it is worth considering in some detail. Lambda, like all other transducing phages, including P1 and P22, is a *temperate* phage. By this is meant that it does not necessarily bring about lysis of the cell which it infects but sometimes establishes a balanced relationship with the host which enables it to be transmitted through an indefinite number of cell divisions in a latent condition. The condition in which a bacterial cell harbours a latent phage is known as *lysogeny*. Lysogenic cells have a certain chance of being lysed through the spontaneous uncontrolled multiplication of the phage but generally speaking such multiplication is inhibited and the phage genome multiplies only at the same rate as the host. Induction of uncontrolled phage multiplication, with resulting lytic release of infective phage particles by the entire lysogenic population can, however, be caused by various treatments of which the simplest is irradiation with low doses of ultraviolet light.

In *Escherichia coli* cells lysogenic with respect to *lambda*, it has been shown by genetic methods that the *lambda* genome is integrated into the bacterial chromosome. In its integrated state it is known as a *prophage*. The *lambda* genome is known, from electron microscopy, to consist of a single molecule of DNA which, although linear and two-ended in the infective phage particle, is capable of forming a closed loop through the interaction referred to on p. 369. During infection of the host cell these 'sticky ends' become stably joined by phosphodiester bonds to form a covalently closed loop. The integration of such a loop into the host chromosome can be pictured as occurring through a single exchange. The postulated mechanism is much the same as for the integration of the F factor in Hfr strains (Figure 11, p. 400) and may be a general one for episomes. After integration, the *lambda* prophage can be mapped by its pattern of segregation from crosses like any other genetic marker, with the proviso that it must enter the cross from the F$^-$ side; when a prophage on an Hfr chromosome is injected into a non-lysogenic F$^-$ cell the latter lyses, presumably because it has not developed the necessary system for repressing phage multiplication. Crosses between non-lysogenic Hfr and lysogenic F$^-$ cells show that *lambda* prophage maps between *gal* and *bio* (cf. Figure 10, p. 396) the same genes as *lambda* is able to transduce. It is, in fact, the general rule that phages of the *lambda* type are able to transduce only those chromosomal markers which are very close to their own sites of integration. For example, the rather similar phage $\phi80$ is integrated next to the group of genes concerned with enzymes of tryptophan synthesis (*try*) and transduces these and immediately neighbouring genes specifically.

Origin of transducing particles

How does the transducing phage particle originate? For *lambda* it has been well demonstrated that transducing particles are *defective* phages which lost a piece of the normal phage genome at the same time as they picked up the short segment of bacterial DNA. Such defective phages, though they are able to infect and transduce, are generally unable to complete another cycle of infective phage multiplication. A plausible suggestion is the following. The integration of the prophage always at the same site is due to its having a segment which interacts strongly with a segment of the bacterial chromosome adjacent to *gal*. Whether this interaction between specific bacterial and phage DNA segments is due to their sharing the same, or nearly the same, base sequence is in doubt.

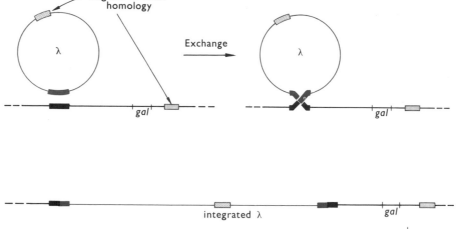

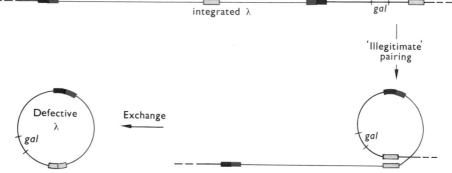

Figure 14. The suggested origin of the DNA of defective lambda phage capable of transducing *Escherichia coli gal* genes (Compare Figure 11, p. 400).

It has, in fact, been demonstrated that about 0·6% of the *E. coli* genome is able to hybridize with lambda DNA following denaturation and reannealing in a test tube; on the other hand recent experimental results suggest that the specificity of the interaction which leads to phage integration is due to the specificity of an 'integration' enzyme rather than to complementary base-pairing of the DNA regions. Be this as it may, integration of the prophage would follow a single exchange between the interacting segments. The release of prophage, which must occur prior to formation of infective particles, is usually due to further crossing-over between the same two segments. However, there are probably other nearby short segments of the bacterial chromosome which have a relatively weak interaction with other segments of the phage genome. 'Illegitimate' crossing-over involving these regions would result in the release of a free episome which had both acquired and lost genetic material in comparison with the prophage which was originally integrated. Although some aspects of this scheme are not yet completely clear—for instance the molecular basis of the postulated illegitimate exchanges has not been shown—extensive data are all consistent with the course of events being substantially as outlined in Figure 14.

Comparison of general and special transduction

It is apparent that general transduction is different in fundamental ways from the special transduction exemplified by *lambda*. Although recent work indicates that *Salmonella* lysogenic for P22 has the prophage integrated at a specific chromosomal site (close to *proA*) the apparent complete lack of specificity of P22-mediated transduction raises doubt as to whether recombination between phage and bacterial DNA plays any part at all in the formation of transducing particles. Indeed, it seems that P1 transducing particles contain little or no phage DNA but only random fragments of bacterial DNA encapsulated in the phage protein. Another difference between *lambda* and the general transducing phages is that whereas the former has to be stably integrated as a prophage before transducing particles can be formed (usually by induction with ultraviolet light), the latter can pick up chromosome fragments in promiscuous fashion during a primary lytic infection.

Finally, the *lambda* and P22 systems differ markedly in their effect on the transduced cells. A cell transduced by *lambda* can usually be shown to harbour the defective prophage which, with its associated bacterial DNA, constitutes an *addition* to the genome of the recipient cell. A colony of transduced cells is, in fact, diploid with respect to the chromosome segment carried by the prophage. With P22, on the other hand, the transduced material is not linked to any demonstrable prophage, and the transductants are haploid; the donor markers can only be replicated following integration into the recipient chromosome to the exclusion of the homologous markers of the recipient. In some recipient cells the integration of a donor fragment fails to occur. In such an *abortively* transduced cell the donor fragment can be transcribed but cannot replicate, being transmitted to one daughter cell at each cell division. Where the recipient strain is auxotrophic and the donor fragment permits growth, abortive transduction is shown by the formation of tiny colonies in which only the one cell containing the fragment is growing and dividing more or less normally.

Mapping using transforming DNA preparations

In principle, transformation with free DNA can be used for fine structure mapping just as transduction can. The two systems are formally similar in that relatively small fragments of the genome are transferred from donor to recipient. They generally share the disadvantage of being able to detect only close linkages, though the *Bacillus subtilis* transducing phage PBSI is an exception to this statement as already mentioned (p. 402). Fortunately other methods are now available. Two other methods for obtaining a more coherent map of the *B. subtilis* genome use transformation as a means of following the sequence in which genes are replicated.

The general approach was devised by Sueoka and Yoshikawa who based their reasoning on two plausible conjectures about the mode of replication of the *B. subtilis* genome. The first was that replication always began at the same point on the (presumably 'circular') chromosome and proceeded along the chromosome until completed. The second was that in a stationary culture, in which growth had ceased because of the exhaustion of the medium, all the cells would have completed their last round of DNA replication. There was, in fact, some evidence that both these assumptions might well be true. It was argued that in an exponentially growing culture there would be twice as many DNA molecules just starting replication as there were just finishing replication, since every molecule finishing gives two ready to start. Thus a genetic marker replicating near the starting point would have almost double the representation in the DNA of the whole population as a marker replicated just before the finishing point. In a stationary phase population, on the other hand, all markers should be equally represented. Thus, following extraction of the DNA, the ratio of transforming efficiency of exponential phase DNA to that of stationary phase DNA should be almost twice as great for markers replicated early as for markers replicated late.

Yoshikawa and Sueoka isolated DNA from samples of a *B. subtilis* culture at different stages of growth and used it to transform a multiple auxotrophic recipient strain. Some of their data are shown in Table 2. The ratios of transformation frequencies with respect to the different markers did indeed change with phase of growth in the way predicted. The strikingly consistent and orderly nature of the results permitted the placing of the markers in a linear order and vindicated the assumptions underlying the experiment. The markers *ileu* and *met* were very close together on the map obtained and they also show simultaneous transformation with significant frequency, confirming that they are contained within one relatively short piece of DNA.

A second ingenious method of mapping based on the idea of sequential replication is worth mentioning. It is technically more complicated than the last but gives rather more unequivocal results. The method depends on the fact that incorporation of 5-bromodeoxyuridine (BDU) instead of thymine markedly increases the buoyant density of DNA and renders it easily separable from DNA of normal density of gradient centrifugation in caesium chloride (see pp. 270–271). DNA which is the product of one round of replication in BDU is 'half-

Table 2. Mapping of the *Bacillus subtilis* chromosome by sequential replication of markers (Yoshikawa and Sueoka's method).

Wild-type (*ileu⁺ leu⁺ thr⁺ ade⁺ met⁺*) was used as the source of DNA for transforming, in different experiments, the triple mutants *leu⁻ met⁻ his⁻*, *leu⁻ met⁻ ileu⁻*, *leu⁻ met⁻ ade⁻* and *leu⁻ met⁻ thr⁻*. The ratios in each line were multiplied by a constant factor to give 1·00 in the last column.

Ratio of transformants	Phase of culture from which transforming DNA obtained			
	Exponential	End of exponential	Nearly stationary	Stationary
$met^+/ileu^+$	—	1·04	0·97	1·00
met^+/leu^+	1·37	1·24	1·05	1·00
met^+/his^+	1·55	1·28	—	1·00
met^+/thr^+	1·74	1·67	1·15	1·00
met^+/ade^+	1·96	1·92	1·22	1·00

Genetic map

Start of replication — met — ileu — leu — his — thr — ade — End of replication

heavy'—that is, it has one heavy and one light strand and forms a band in the ultracentrifuge intermediate in position between normal DNA and DNA with BDU in both strands—while after a second replication half of it becomes 'fully heavy'. In essence the procedure is to initiate a round of DNA replication by germinating spores of a thymine-requiring mutant in the presence of BDU, to extract and centrifuge the DNA from samples of the culture at different times thereafter, and to determine the distribution of different genes (assayed by transformation of multiple mutant recipient cells) between the different DNA bands. As each gene becomes replicated in turn it moves first from the normal to the half-heavy band, after a second replication, to the fully heavy band. The changes in the gene ratios in the different bands with time provide a sensitive indication of the replication sequence. This method has been used to show the order of the groups of genes cotransduced by phage PBSI.

GENE CONTROL OF AMINO ACID SEQUENCE

One gene—one polypeptide chain

So far as the distribution of mutant sites on the genetic map is concerned the bacterial chromosome seems to be a continuum. Functionally, however, it is obviously segmented. Mutations which are very closely linked tend to have similar physiological effects, while those which are loosely linked nearly always have obviously differing effects. The chromosome seems, in fact, to consist of discrete segments, each one distinguished by the characteristic effect of the mutants falling within it. It is very frequently found that all the mutations occurring within a given segment have effects on the formation of one specific enzyme and in some cases where the structure of the enzyme protein is well known it has become clear that one genetic segment determines the structure of one polypeptide chain. Such a segment is usually referred to by the term *gene.*

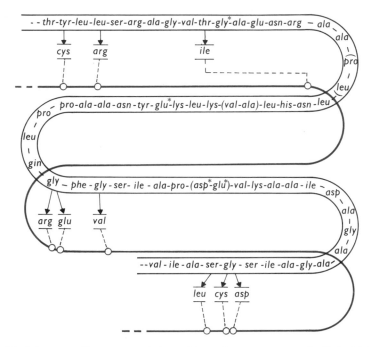

Figure 15. Colinearity. Comparison of the positions of mutant sites in the *Escherichia coli Trp A* gene and the corresponding amino acid substitutions in the polypeptide chain of the tryptophan synthetase A protein. The amino acid substitutions are connected to the corresponding sites on the genetic map by dotted lines. The mutant sites (open circles) on the genetic map are spaced according to the genetic map distance. Only a portion of the complete amino acid sequence is shown.

In several instances the effects of mutations within a gene on the corresponding polypeptide chain have been fully worked out. Perhaps the best example is Yanofsky's analysis of a large number of *Escherichia coli* mutants which require tryptophan because of the loss of tryptophan synthetase activity. This enzyme consists of two component proteins A and B which can be rather easily dissociated from one another and separately analysed. The A protein consists of a single polypeptide chain, and the B protein of two polypeptide chains of a different kind. The complete tryptophan synthetase molecule contains two A's complexed with one B, or two polypeptide chains of each kind. The tryptophan synthetase mutants fall into two classes, one defective in the A chain and one in the B chain and these two classes can be mapped by transduction in two non-overlapping though adjacent segments or genes. This is one of the most striking confirmations of the one gene—one polypeptide chain principle.

Yanofsky and his colleagues have concentrated their attention on mutations in the A gene, and have shown that many mutations within it result in alterations in the A protein. In every instance where the chemical analysis has been done the alteration takes the form of the substitution of one amino acid for another. Mutations mapping at different points within the gene cause amino acid substitutions at different points in the polypeptide chain. It is thus possible to compare the linear order and the spacing of the mutant sites in the genetic map with the order and spacing of the corresponding amino acid residues in the polypeptide. Figure 15 summarizes some of the data.

The two sequences correspond exactly as to order and very closely as to spacing. The gene obviously must constitute a linear code specifying the corresponding linear amino acid sequence. Similar, though less complete, data from a number of other gene-enzyme systems confirm that this is a general function of genes; indeed there is no reason to doubt that it is the function of all genes other than those which determine non-messenger RNA, such as tRNA and ribosomal RNA.

Genetic experiments on the nature of the code
Much can be deduced about the genetic code from the study of amino acid replacements in mutants. It has been found that very closely adjacent but separable mutations can cause different substitutions at the *same* amino acid position. Accepting that point mutations, in general, are changes in single nucleotide pairs, this shows that more than one nucleotide pair is involved in coding for each amino acid. This is, of course, what would be expected from the biochemical evidence that each codon consists of a *triplet* of nucleotides in messenger RNA.

It was a study of mutually compensating mutations within the bacteriophage T4 rII region which provided the first strong evidence from genetics for the triplet nature of the code. Crick, Brenner and their associates obtained a large number of proflavine-induced mutations which were originally selected as suppressing the effect of another mutation within the gene. All of them, by themselves, gave a defective phenotype. They fell into two mutually exclusive groups *plus* and *minus* on the basis of their interactions in double mutants. Most plus–minus combinations showed mutual compensation, giving a normal or near-normal phenotype, but plus–plus or minus–minus double mutants were always still mutant in phenotype. The plus mutants could be interpreted as equivalent to additions of single nucleotide pairs in the DNA and the minus mutants to single deletions, or *vice versa*. The crux of the experiment, however, came with the construction, through appropriate crosses, of triple mutants. It was found that most plus–plus–plus or minus–minus–minus triple mutants were again nearly normal in phenotype. This is exactly what would be predicted with a triplet code, since three successive additions (or deletions) would result in normal reading after the third mutant site had been passed if, but only if, reading was in threes or multiples of three (see also p. 336). The success of this experiment suggests that it is possible to make many different changes in the sequence of one part of the amino acid sequence of the presumed protein specified by the T4 rII gene without destroying its function, so long as a major part of the sequence is correct.

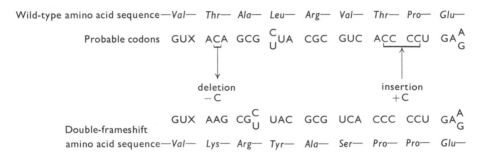

Figure 16. Result of an analysis by Yourno [*J. Mol. Biol*, **48**, 437 (1970)] of an altered histidinol dehydrogenase peptide resulting from two compensating frameshift mutations in *hisD* of *Salmonella typhimurium*. The first mutation, which not only eliminated histidinol dehydrogenase (controlled by *hisD*) but exerted a strong polar effect on the expression of the more distal genes of the *his* operon (cf. p. 233), was induced by ICR 191. The compensating mutation, which restored histidinol dehydrogenase activity and relieved the polar effect, occurred spontaneously. In the codon assignments X may be any one of the four RNA bases; base symbols written one above the other are indistinguishable alternatives.

Since the protein specified by the rII gene has still not been identified a direct verification of the predicted effects of plus and minus *frameshift* mutations on the amino acid sequence is not possible in this case. However frameshift mutations with similar genetic properties are known within another gene *e* of bacteriophage T4 coding for bacteriophage lysozyme, and in two bacterial genes, *his* of Salmonella and *trpA* of *E. coli*, coding for histidinol dehydrogenase and tryptophan synthetase respectively. These enzymes can all be isolated in pure form and their amino acid sequences are substantially known. Analyses of the altered sequences in double-frameshift strains have shown, as expected, that a sequence of several amino acids has been altered in each case. Knowing the amino acid code (cf. p. 339), it is possible to write down the sequence of bases in the wild-type messenger RNA which could generate the altered amino sequence by a double frameshift. Because of the degeneracy of the code there are numerous possibilities (cf. Table 3, p. 339) but choices are only possible in certain positions; many of the bases are fixed, given the amino acid sequences. Analyses of this type have identified many of the codons which are used in the cell: this is an interesting question since it is by no means certain that all the alternative codons which work in the cell-free systems (cf. p. 342) are actually used frequently in all organisms *in vivo*. Figure 16 shows an example of such a double-frameshift analysis. It is worth noting that the solution to the problem includes the direction of translation of the messenger (5' to 3') and the direction of synthesis of the polypeptide chain (N-terminal to C-terminal).

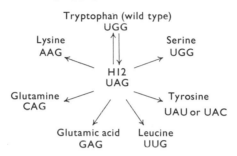

Figure 17. Identification of a UAG chain-terminating ('nonsense') codon in a mutant of *Escherichia coli* lacking alkaline phosphatase. Revertants from mutant H12, with alkaline phosphatase activity restored, had any one of seven different amino acids in one position in the polypeptide chain. The only codon from which codons for each of these amino acids could have been derived by a single base change is UAG. The wild-type, from which H12 was derived, has tryptophan in the position in question. After Weigert & Garen, *Nature*, **206**, 992 (1965).

The analysis of mutants has been of great help in the identification of codons which bring about termination of polypeptide synthesis as well as of codons for amino acids. A great many mutants which have enzymic functions missing fail to synthesize these enzymes because a mutation has created a chain-terminating codon somewhere in the RNA message, so that synthesis is discontinued prematurely. Starting with such a chain-terminating, or 'nonsense' mutant, one can select for regain of the enzymic function. Some of the revertants no doubt will be true reversals of the original mutation, restoring a normal wild-type enzyme. Others, however, are likely to have unusual, but acceptable, amino acids in the position corresponding to chain termination in the primary mutant. In some cases a whole family of amino acid replacements have been identified among revertants from a given chain-terminating mutant. Assuming that the codons responsible for all these amino acids are derived from the chain-terminating codon by a single base change (and, as we saw, mutations generally are of this nature) we may be able to deduce the nature of the original chain-terminating codon. Figure 17 shows an example of this type of analysis using reversion from absence to presence of alkaline phosphatase in *E. coli* as the experimental system. In this case the codon UAG was identified as the mutant chain-terminator, but in other mutants UAA has been shown to be the codon involved. It must be remembered, of course, that these identifications are of codons artificially introduced by mutation and are not necessarily used for normal chain-termination. Indeed, mutants with the UAG ('amber') codon can be suppressed so efficiently by alterations in tRNA that it seems extremely unlikely that it could be widely used as a natural chain-terminator—otherwise the suppressed strains could hardly be viable. The UAA ('ochre') codon seems a a much more likely candidate since 'ochre'-suppressors are never very efficient and even so cause considerable reduction in growth rate.

Identification of genes by complementation tests

Principle of the tests

From one point of view, then, a gene is a discrete segment of genetic DNA which constitutes a linear code for the amino acid sequence of one polypeptide chain. However, this is not of much help to the geneticist who wishes to be able to identify the functional genetic unit without committing himself to several years' work in protein chemistry. There is, fortunately, a short cut to the identification of genes, namely the complementation test.

The general question which one is constantly having to answer is whether two mutations are in the same or different genes. Defining the gene as the functional unit, this is equivalent to asking whether the mutations affect the same or different functions. If they affect different functions, then each mutant genome should possess the function which the other lacks, and, if put together in the same cell they should complement one another to produce a fully functional

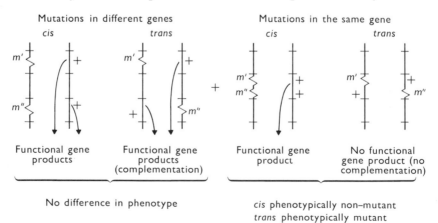

Figure 18. The principle of the complementation test.

phenotype even without any genetic recombination. If, on the other hand, they are defective in the same function, complementation under conditions precluding recombination is not expected. Thus, *prima facie*, complementation signifies mutation in different genes and non-complementation mutation in the same gene.

The most logically rigorous form of the complementation test is the *cis-trans* comparison, in which one compares the phenotype given by two genomes each carrying one of the two mutants (symbolized as $m' + / + m''$, and called the *trans* configuration) with that given by a doubly mutant plus a nonmutant genome ($m' m''/ + +$, the *cis* configuration). If the two phenotypes are similar the two mutations must be in different functional units, or *cistrons* (to use Benzer's term); if *cis* is phenotypically wild and *trans* is mutant, the mutations are in the same cistron (Figure 18). In practice the double mutant is usually troublesome to obtain and, in any case, the complete phenotypic dominance of a completely wild-type genome over a double mutant one is so much the rule that it can almost be assumed. Thus it is usual merely to examine the *trans* diploid, or partial diploid, to see whether the defects of the individual mutants are compensated for or not. The word *cistron* has been generally adopted to mean a genetic segment within which mutants do not complement one another. In its usual sense *gene* means the same thing.

A wide variety of different methods of complementation testing have been devised for different microbial systems.

Complementation tests in bacteriophage

Benzer's classical definition of the cistron arose from studies on bacteriophage T4. The rII series of mutants of this phage form large plaques in a lawn of cells of *Escherichia coli* strain B, but are unable to grow at all in cells of *E. coli* strain K12, lysogenic for *lambda*. Here the complementation test consisted in infecting K 12 (λ) cells with a mixture of two different rII mutants: growth of phage with lysis of the bacteria indicated complementation, while no lysis indicated non-complementation. The results showed a clear division of the rII mutants into two mutually complementing but internally non-complementing groups, representing two cistrons.

Complementation tests in bacteria—merozygotes and abortive transduction

In bacteria themselves complementation tests may be considered under two headings depending on whether or not the two chromosomes or chromosome segments whose complementation one is testing are able to replicate together to form a stable diploid (or partially diploid) clone. We will deal first with that kind of test in which a stable diploid is *not* formed.

F⁻ merozygotes from mating with Hfr are *transiently* diploid for a part of the genome and have been used for complementation tests in a few instances where complementation is shown by the formation of an enzyme for which there is a sufficiently sensitive assay method. Much more important is abortive transduction in *Salmonella*, which results in a persistent but non-multiplying merodiploid condition. As was explained on p. 407, a genetic fragment carried into recipient cells by phage P22 may be integrated into the recipient chromosome to give a stable haploid transductant but, more often, it fails to be integrated and persists as an extra non-replicating fragment transmitted to only one daughter cell at each cell division. Where the donor and recipient strains in the transduction 'cross' are both auxotrophic mutants, complementation is shown by the appearance of tiny prototrophic colonies after plating on minimal medium. These are much smaller than, and easily distinguished from, the large prototrophic colonies which result from complete transduction. Each tiny colony is fed by one cell which carries a complementary genetic fragment from the donor strain. While the presence of abortive transductants does not provide a quantitative measure of the efficiency of complementation, it is a very sensitive criterion of whether complementation is occurring at all. Large numbers of auxotrophic mutations of *Salmonella* have been classified on this basis.

Complementation tests in bacteria—merodiploids involving episomes
The second kind of complementation test, with diploid or merodiploid cells which replicate as such, depends on the use of episomes which have 'picked up' chromosomal material. Two kinds of episomes have been extensively used, *lambda* prophage and the F factor of *E. coli*. It was explained on p. 406 (Figure 14) that transduction by *lambda* was mediated by defective phage particles, with a piece of bacterial chromosome combined with a deficient prophage. In transduced recipient cells the incoming genetic material is not, in most cases, integrated into the chromosome *to the exclusion of* the homologous genes which were previously present but is preserved as an *additional* genetic segment. This probably occurs through the integration of the episome in the same way as normal *lambda* prophage is integrated. At all events it is certain that most gal^- recipient cells which are transduced to the gal^+ condition continue to carry the gal^- gene and are diploid with respect to the gal segment of the chromosome. This is shown by the fact that colonies arising from such transduced cells are not completely stable but give occasional gal^- sectors. The gal^- cells from these sectors still harbour the episome but now they carry the gal^- gene from the original recipient strain, both on the chromosome and on the episome, i.e. they have become homozygous for the mutation as a result of some kind of recombination. The defective episome cannot initiate a new cycle of free growth by itself, but it gets incorporated into large numbers of infective particles when lysis is induced by superinfection with normal *lambda* (not carrying gal at all)

followed by an inducing dose of ultraviolet light. Thus the particular *gal⁻* mutation now carried in the defective phage genome can be introduced by infection into any *E. coli* strain with a different *gal⁻* mutation, and the complementation or noncomplementation of the two *gal⁻* segments can be determined. In this way the *gal* region has been analysed into three different cistrons which turn out to be concerned with the synthesis of three different enzymes.

Lambda phage is only useful for the analysis of the *gal* and immediately neighbouring regions of the *E. coli* chromosome and the related phage $\phi80$ is specific for the region close to the *trp* genes. The F factor, on the other hand, is an episome with a considerable number of possible points of integration into the *E. coli* chromosome. Just as an integrated *lambda* prophage is sometimes released from the chromosome with an adjacent chromosomal segment attached to it, so augmented F factors (i.e. carrying an adjacent piece of chromosome) are formed with a low but significant frequency in Hfr strains. These are called *F-prime* (F′) factors, and can be selected relatively easily. The method is to interrupt an Hfr × F⁻ mating after a short time and to select for F⁻ cells which have received an Hfr marker which is close to the integrated F factor and would normally enter only with the trailing end of the chromosome. The rare colonies so selected, which have inherited the normally late-entering Hfr marker after an unusually short period of mating, are found to have the marker in question as part of the F′ factor. Following its release from the chromosome the F factor, with any attached chromosomal material, is able to 'jump the queue' and enter the F⁻ cell early. Like any ordinary F factor, an F′ episome can be introduced by conjugation into any F⁻ strain and is thereafter replicated as a stable extra piece of genetic material. One can thus easily test any mutation carried on the F′ for complementation with any other mutation in an F⁻ strain. Since a wide range of Hfr strains are available, with F integrated at different points, almost any chromosomal segment can be obtained in an F′ episome. A wild-type gene carried in such an episome can be replaced by any one of its mutant alleles by recombination. The F′ technique is certainly now the method of choice for making complementation tests in *E. coli*.

Adjacent genes distinguished by complementation tests

One of the best examples of the use of complementation analysis to identify genes is provided by the work on histidine-requiring mutants of *Salmonella*. The existence of well-defined genes with different functions within the *his* region of the *Salmonella* chromosome was not at first at all obvious. There were a large number of independently isolated mutants which all had a superficially similar phenotype—namely, a requirement for histidine—and which all mapped quite close together within one short chromosome segment, the *his* region. However, complementation tests by abortive transduction served to subdivide the

mutants into eight groups such that any mutant in each group would complement any mutant in any of the other groups, but none (or exceptionally a few)
of the mutants in the same groups. The exceptional complementation between
certain pairs of mutants assigned to the same group (cistron) spoiled the neat-
ness of the pattern to some extent but did not obscure the overall picture. Two
other lines of evidence confirmed the validity of the classification. First, mapping of the mutations by cotransduction frequencies and by overlapping deletion analysis showed that the cistrons established by complementation tests
corresponded to distinct, *non-overlapping* segments of the *his* region. Secondly,
biochemical analysis showed that the mutants assigned to the same cistron were
defective in the same enzyme of histidine synthesis, and that mutants in different
cistrons were defective in different enzymes. The biochemistry of the situation is
summarized in Figure 11, p. 230.

Allelic complementation

It was mentioned in the preceding paragraph that a group of mutants which the
main weight of the evidence would assign to the same cistron (that is, which
appear to carry different mutant forms or *alleles* of the same gene) may sometimes include some pairs which show complementation. This observation, at
first sight fatal to the whole cistron concept, has now been reasonably explained
in terms of the formation of hybrid proteins. There are two ways in which complementary interaction can occur at the polypeptide level.

The first, and probably more general, mechanism depends on the protein
concerned being an oligomer, that is composed of a number of subunits, with
the conformation of each subunit constrained by that of its neighbours. The
chains comprising an oligomeric protein molecule may be of more than one
kind but, perhaps more often, may be identical. In the latter case all the chains
in a oligomer will normally be the products of the same gene. But in a cell which
is diploid for the gene in question, and in which the two representatives of the
gene are different from each other because of mutation, two different variants of
the polypeptide chain may be produced in the same cell. In this case a proportion of the oligomers formed will be *hybrid*, containing chains of the two different kinds aggregated together. Thus there will arise the possibility of interaction
between two different mutant chains to give a conformation which is more
active than could be formed by either kind of chain by itself. That this kind of
interaction can really occur has been confirmed in several instances. Perhaps the
clearest example comes from work on mutants in the *pho* gene of *E. coli* deficient
in alkaline phosphatase activity. Many of the *pho* mutants produce abnormal
and inactive varieties of the enzyme and some of these are separable from each
other by electrophoresis. The enzyme is a dimer, containing two normally
identical polypeptide chains, and it is possible, by acidification, to cause the

purified protein to dissociate into monomers and then to re-form dimers by returning the pH to neutrality. One can thus form mixed dimers *in vitro* by dissociating and reassociating a mixture of two different mutant varieties of the protein. In a number of instances where a pair of mutants complemented each other in an F′ merodiploid strain with formation of active phosphatase, it was shown that the corresponding hybrid protein made *in vitro* did indeed have enhanced enzyme activity as compared with the two unhybridized mutant proteins. In one instance it was shown that the active enzyme formed by complementation in the merodiploid was indeed a hybrid; here the two mutants concerned produced mutant phosphatases which were quite distinct in electrophoretic properties and the active enzyme formed by their interaction was electrophoretically intermediate. The mechanisms by which different mutant polypeptide chains can, at least to some extent, compensate for each other's defects in a hybrid aggregate are not understood in detail. It seems very likely, however, that a mutual correction of tertiary structure is commonly involved, with the 'good' portion of each chain helping the corresponding 'bad' part of the other to fold to approximately the right shape. This type of explanation seems likely to be very generally applicable.

A second kind of explanation which applies in at least some instances is that active enzyme molecules can be reconstituted by the folding together of different parts of the amino acid sequence derived from different mutants. Here, although the enzyme may still be oligomeric, the complementation does not depend on interactions between monomers but rather on the formation of an inherently active monomer by piecemeal assembly. This rather surprising explanation has been shown to apply to the formation by complementation of *E. coli* β-galactosidase, where, for instance, a mutationally altered ('mis-sense') N-terminal section of the polypeptide chain, incompatible with enzyme activity, can be replaced in the folded structure by an N-terminal *fragment* of normal sequence produced by a chain-termination mutant. In this way a complete 'mis-sense' chain and a fragment can interact to produce an essentially normal and active structure—the displaced N-terminal mis-sense sequence apparently need not interfere with the correct folding of the rest.

Intracistronic (otherwise known as *interallelic* or *allelic*) complementation can probably be found within most cistrons if looked for hard enough. It occurs within several of the cistrons of the *Salmonella his* region, notably within *hisB* which determines the structure of imidazoleglycerol phosphate dehydrase. From its very nature, however, its incidence is sporadic and confined to certain pairs of mutants whose polypeptide products happen, by chance, to be complementing. The majority of pairwise combinations within a cistron remain noncomplementing. Thus cistrons can still be recognized by complementation tests even though their definition by these tests is not as neat and clear-cut as was once thought.

Functionally co-ordinated groups of genes (operons)

A more fundamental difficulty in defining genes by complementation tests is due to the fact that, in bacteria at least, neighbouring genes are not always functionally independent. Genes of linked function, for example specifying the different enzymes functioning in a single specialized metabolic pathway, frequently turn out to be adjacent on the chromosome and subject to a common control mechanism governing their activity. Certain mutations, occurring close to the end of such a co-ordinated group of genes (or *operon*), can render the whole group inactive or nearly so. Mutations of this type may thus fail to complement with mutations within any of the genes of the operon. An analysis of the effects on enzyme formation of the various mutations will usually serve to clarify such a situation. The point we wish to make here is that complementation tests, while convenient and generally reliable for defining genes (in the sense of genetic units specifying different enzymes), are occasionally misleading in the absence of detailed knowledge of the enzymes. Operons and the kinds of mutations which can effect their function are discussed much more fully in Chapter 8.

Plasmids

We have already seen, in the F factor of *E. coli*, an example of a DNA element, in the form of a covalently closed loop, additional to the regular chromosome and capable of replicating independently of it. Such extrachromosomal genetic elements are called *plasmids*, and during the last decade it has become recognized that bacterial plasmids are both widespread and important. The basic property that a plasmid must have is the capacity to initiate its own replication. At least some plasmids are able to become reversibly integrated into the chromosome and so conform to the classical definition of an *episome* (cf. p. 400) but this is not a universal property of plasmids and may be rather a special one. Many plasmids, especially in the enteric bacteria, also have the capacity, first seen in the F factor, to promote conjugation of the cells carrying them and thus to bring about their own transfer from cell to cell. Some of them also resemble F in being able to promote the transfer of the chromosome as well as of themselves but the frequency with which this occurs varies widely between different plasmids. The conjugation-promoting property is associated with the formation of sex-pili on the surface of the host cell, and two families of plasmids have been recognized producing two different kinds of pilus—one resembling the F-pilus in antigenic character and in specificity for bacteriophage attachment and one (type I) different in these respects.

In addition to a special site or region presumably necessary for the initiation of autonomous replication and, in many cases, a gene or genes determining sex-pilus formation, plasmids are found carrying a variety of other genetic determinants. These nearly always determine functions which are non-essential for the basic life of the cell but which are, or may be supposed to be, tactically advantageous for the bacterium under special circumstances. Thus the various plasmids which have been found inhabiting *E. coli* and *Salmonella* species have been found to carry an astonishing variety of determinants of resistance to anti-bacterial drugs such as penicillin, tetracycline, sulphonamides and strepto-mycin.

Another characteristic which is determined by some plasmids in *E. coli* is the capacity to produce *colicins*—that is, specific proteins which can kill other strains of bacteria but to which the cell producing them is immune. The property of colicinogeny is presumably an advantage to a bacterial strain in competition with other strains. The transferable colicinogenic factors were among the first plasmids to be identified; they turned out to resemble F in promoting cell con-jugation and, in some cases, chromosomal transfer.

These various plasmid properties are due to different genes which may be linked together in the same plasmid. The particular array of genes present is characteristic of the particular plasmid but it has been shown that different plasmids, when present together in the same cell, can sometimes undergo genetic recombination to give new plasmid types. Mapping of some plasmids has been carried out based on the pattern of deletions of blocks of genes which occur occasionally during plasmid transmission.

Physically, plasmids have been shown in a number of cases to be covalently closed DNA circles ranging in size from about 4×10^6 daltons (for the colicin E_1 factor) through around 5×10^7 for various drug-resistance transfer factors to about 10^8 for the F factor. One procedure for isolating plasmid DNA depends on the fact that it is often possible to transfer plasmids to bacterial species with DNA of quite a different base composition, and hence different buoyant density, from that of the plasmid. In this situation the plasmid DNA can be readily separated by density gradient centrifugation.

The degree of genetic relationship between plasmid DNA and the chromo-somal DNA varies from one case to another. In the case of the F factor in *E. coli* or *Salmonella typhimurium*, the base composition is similar and there is sufficient homology between the plasmid and various segments of the chromo-some to allow integration of the plasmid to take place at a number of different chromosome loci. In this case the plasmid is able to bring about transfer of the chromosome as well as of itself from cell to cell. For most plasmids, however, evidence for chromosomal integration is slight or lacking and promotion of chromosome transfer is relatively inefficient. Whether plasmid integration is essential for any promotion of chromosome transfer, or whether the promotion of conjugation is in itself sufficient to allow some chromosome transfer to take place is still undecided. The relationship of many plasmids to the cells in which they are currently found is probably quite remote. Not only, as mentioned above, do they often have distinct DNA base ratios but, in the case of the resistance transfer factors, the mechanisms of drug resistance which they determine may be quite different from those which can arise by mutation in the chromosome. Thus the plasmids of the enteric bacteria determining streptomycin resistance do so through enzymic modification of the drug, while chromosomal streptomycin resistance is associated with modification of the 30 S subunit of the ribosome so as to make it insensitive to streptomycin.

While plasmids have been most extensively investigated in the enteric bacteria they have been found in other groups as well, notably in Staphylococcus where they determine a wide array of drug resistances. The Staphylococcus plasmids do not bring about conjugation and have only been transferred from cell to cell by transduction, a mode of transfer which may well occur in nature.

Acknowledgements

Figure 3. After S. E. Luria and M. Delbrück. *Genetics* (1943) **28**, 491.

Figure 9. After F. Jacob and E. L. Wollman. *Sexuality and the Genetics of Bacteria* (1961) Academic Press, New York, Figure 21.

Figure 15. From Fincham. *Microbial and Molecular Genetics* (1965), from data of C. Yanofsky, B. C. Carlton, J. R. Guest, D. R. Helinsky and U. Henning. *Proc. Nat. Acad. Sci.*, *Wash.* (1964) **52**, 266.

FURTHER READING

 I   Adelberg E. A. (ed.) (1960) *Papers on Bacterial Genetics*. Methuen, London. (A selection of key papers covering much of the development of the subject).

 2   Campbell A. (1962) Episomes. *Advances in Genet.* **II**, 101.

 3   Curtiss Roy III (1969) Bacterial conjugation. *Ann. Rev. Microbiol.* **23**, 69–136.

 4   Gunsalus I. C. and Stanier R. Y. (eds.) (1964) *The Bacteria*, Volume 5: Heredity. Academic Press, New York. (Contains comprehensive reviews on conjugation, transduction, transformation, episomes and genetic fine structure).

 5   Hayes W. (1968) *The Genetics of Bacteria and Their Viruses*, 2nd edition. Blackwell Scientific Publications, Oxford. (An advanced textbook covering all the aspects dealt with in this chapter).

 6   Hayes W. (1965) The structure and function of the bacterial chromosome. In *Function and Structure in Micro-organisms* (eds. Pollock M. R. and Richmond M. H.), Fifteenth Symposium of the Society for General Microbiology, p. 294. Cambridge University Press.

 7   Ozeki H. and Ikeda H. (1968) Transduction mechanisms. *Ann. Rev. Genetics* **2**, 245–278.

 8   Spizizen J., Reilly B. E. and Evans A. H. (1966) Microbial transformation and transfection. *Ann. Rev. Microbiol.* **20**, 371–400.

Chapter 8
Co-ordination of Metabolism

INTRODUCTION

Considering the details of the various degradative and biosynthetic pathways, we may ask: how is the flow of metabolites through these pathways regulated, so that no compound is present in excessive or inadequate supply? More specifically, we might imagine that a bacterium must have acquired, through selection in the environments to which it is adapted, two main features of metabolic control: the ability to utilize as efficiently as possible (as regards both growth rate and yield) the nutrients normally available in those environments, and the capacity to respond rapidly to environmental changes. In the following discussion of the control mechanisms that are found in bacteria, their selective advantage in furthering one or other of these ends will in many instances be clear. However, we will not always be able to suggest plausible reasons for a particular observation, and in such cases we should remember that an adaptive character may persist in a changed environment where it is selectively neutral.

Control mechanisms may be divided into specific and non-specific; the former act by stimulating or depressing the flow through a specific pathway (although this may, indeed must, have repercussions throughout the cell's metabolism), while the latter may involve a wide variety of pathways or an entire class of macromolecules. We shall pay more attention to the specific controls, partly because there is more concrete information about them, and also because, having discussed them, we can consider the problem of what other controls might be looked for. Most of this chapter will therefore consist of a discussion of the mechanisms and patterns of specific controls operating in Class I and Class II pathways, with a final mention of Class III reactions and possible non-specific forms of regulation.

Specific control mechanisms are of two major types: control (activation or inactivation) of *enzyme activity*, and control (induction or repression) of *enzyme synthesis*. A minor type, controlled breakdown of a specific enzyme, not of cell proteins in general, will be mentioned, but is of much less importance in bacteria than in eukaryotic micro-organisms.

Control of enzyme activity

Here, a metabolite, usually unrelated in structure to the enzyme substrate, binds to an enzyme causing it to gain (*activation*) or lose (*inhibition*) in catalytic activity. This is most easily demonstrated *in vitro*, i.e. with cell extracts in which the relevant enzyme activity can be assayed, or with solutions of the purified enzyme; the addition of an *activator* metabolite causes the measured activity to increase, while an *inhibitor* causes this to decrease. Activators and inhibitors are together referred to as *effectors*, activation and inhibition sometimes as *modulation*. We shall mention some of the peculiarities of modulation of enzyme activity later when describing theories as to its mechanism, but one point may be made here, to clarify our later discussion on the kinds of control mechanism that are found in various pathways. The effect of modulation is not always to alter the maximum rate at which the enzyme acts when it is saturated with substrate, but is very often to increase or diminish its affinity for substrate, so that a higher substrate concentration is needed in the presence of an inhibitor, and a lower concentration in the presence of an activator, to attain the reaction rate found in the absence of these effectors.

The mechanism of control of enzyme activity

Enzymes the activity of which is subject to control by effectors have certain distinct biochemical characteristics which have led to a search for unifying principles to explain their behaviour. There are three distinguishing characteristics. Firstly, the effector binds to the enzyme at a site (called an *allosteric* site) distinct from the active site. This is suggested by the non-competitive kinetics found for inhibition by effectors which is only partially overcome by increase in substrate concentration, and it would in any case be expected from the fact that effectors are usually quite unrelated structurally to any of the substrates. It has sometimes been capable of experimental proof. For instance, in *Escherichia coli* aspartate carbamoyltransferase, which is *feedback inhibited* (i.e. inhibited by the end product of the reaction sequence involving it) by CTP (Chapter 4, pp. 240–1), is a multimeric* enzyme made up of two different types of polypeptide chain; after its disaggregation into monomers the polypeptide chains of one kind bind the substrates while those of the other kind bind the effector.

---

* A multimeric protein (or *multimer*) is one in which the molecule contains more than a single polypeptide chain (or *monomer*).

It may be asked how the effector operates if its binding site is distant from the active site. A possible answer is that the binding of the effector results in a change in overall conformation of the enzyme including the vicinity of the active site although, as we shall see below, one currently favoured model suggests otherwise. This is consistent with the fact that treatment with many agents that alter protein conformation (heat, high or low pH, or mercurials, for example) often results in desensitization to modulation while leaving the catalytic activity unimpaired, even though the effector may still bind at its allosteric site. For instance, the first enzyme of histidine synthesis in *Salmonella typhimurium*, if allowed to age, retains catalytic activity but loses its sensitivity to inhibition by histidine although histidine can still be shown to bind to it. However, we shall return to the question of mechanism below.

Secondly, plots of reaction velocity versus substrate concentration, for allosteric enzymes in the presence of effector (and usually in its absence as well), are *sigmoid*, instead of hyperbolic as expected for an enzyme showing classical Michaelis–Menten kinetics (see Figure 1a). This sigmoid effect indicates a *cooperative* effect, such that the affinity of the enzyme for substrate is not constant but, at low substrate concentrations, increases with substrate concentration (see below). Such a sigmoidal response may often be physiologically advantageous in conferring high sensitivity of reaction rate to small changes in substrate concentration; this gives almost a threshold effect, so that at low substrate concentrations, the reaction only proceeds very slowly with a concomitant sparing of substrate.

Sigmoid curves are also obtained if reaction rate is plotted against concentration of activator or inhibitor at constant substrate concentration (see Figure 1b).

Cooperative effects where the molecules capable of binding to an enzyme (*ligands*) are identical are said to be *homotropic*, while those produced by different ligands are termed *heterotropic*. It is clear that homotropic and heterotropic effects have much in common, and the term 'allosteric' is often applied to both types, although where substrates are concerned their effects may involve binding at the active site. In spite of this broadening of the use of the term, we shall call compounds capable of showing homo-or heterotropic effects *allosteric ligands*.

Thirdly, aspartate carbamoyltransferase exemplifies, as already mentioned, the general characteristic of multimeric structure because it appears that all allosteric proteins are built up of subunits. However, many are unlike aspartate carbamoyltransferase in that they contain only one kind of monomer. Furthermore, the binding between subunits seems to be affected by the presence of allosteric ligands. For instance, threonine dehydratase of *Escherichia coli* is reversibly dissociable in the presence of 1·5 M urea, but the extent of dissociation is affected by threonine, valine or isoleucine, the first two favouring dissociation and the last association.

Various schemes have been put forward to explain how the three characteristics outlined above are interrelated. A particularly influential model has been that of Monod, Wyman and Changeux, which we shall now describe.

An interesting point of this model is that it avoids taking as its starting point that the binding of any molecule to the enzyme causes a conformation change;

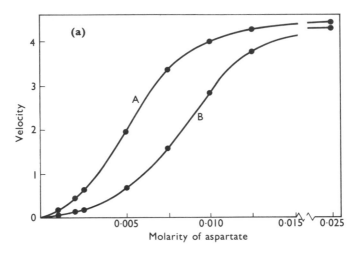

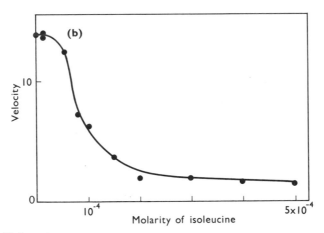

Figure 1. Multimeric enzymes.

(a) Sigmoid response of reaction velocity to substrate concentration for aspartate carbamoyltransferase of *Escherichia coli*. Curve A, no inhibitor present; curve B, the inhibitor CTP present at 2×10^{-4} M. From J. C. Gerhart and A. B. Pardee, *Cold Spring Harbor Symposia on Quantitative Biology* **28** (1963) 491.

(b) Sigmoid response of reaction velocity to inhibitor concentration at constant substrate concentration shown by threonine dehydratase of *Escherichia coli*. The concentration of substrate (L-threonine) was 2×10^{-2} M. From J. Monod, J. Wyman and J. P. Changeux, *Journal of Molecular Biology* **12** (1965) 88.

this is replaced by the concept of an equilibrium between molecules in different conformations. An allosteric enzyme is looked on as necessarily multimeric, with a symmetrical arrangement of identical *protomers*; these may themselves be monomers or may be some arrangement of identical or non-identical monomers. Four further assumptions are made: each protomer carries one binding site for each allosteric ligand (where this is a substrate showing homotropic interaction, the binding site is of course the active site); the conformation of a protomer is partly determined by its interaction with other protomers; the enzyme can exist in two or more interconvertible states which differ in the nature of the protomer-protomer interactions (thus the protomers adopt different conformations in the various states and the affinities of allosteric ligands for their binding sites may be dissimilar for the different states); and, finally, the symmetry of the enzyme molecule, including the symmetry of interactions between protomers, is conserved in transitions between the states. This last assumption is the most important of all, since it leads to the necessity for all protomers in one enzyme molecule to have the same conformation.

The deductions that can be made from this model may be illustrated in qualitative terms as follows. Suppose that we have an enzyme exhibiting a homotropic effect for a substrate S, and heterotropic effects for an activator A and an inhibitor I. Suppose also that the enzyme can exist in one of two states, such that in one of the states the protomers adopt a conformation with greater affinity for S than in the other (the states of the enzyme corresponding to greater and lesser affinity are conventionally termed 'R' and 'T', for *relaxed* and *tight*, respectively). In the absence of ligands, the R and T states will be in equilibrium. If a small amount of S is now added, there will be an increase in the proportion of enzyme molecules in the R state as opposed to the T state for purely thermodynamic reasons, because the affinity of S is greater for enzyme molecules in the former state than in the latter. This means that a low concentration of S causes the appearance of more available binding sites than were there in its absence (see Figure 2(a)); this in effect constitutes the observed cooperative effect with resultant sigmoid dependence of reaction rate on substrate concentration.

The activator A may be taken to bind to its own allosteric site more strongly when the enzyme is in the R state than when it is in the T state. The inhibitor I shows just the reverse effect: the enzyme has greater affinity for it in the T state than in the R state. Hence A increases the proportion of enzyme molecules in the R state, with consequent enhancement of binding of S and consequently of reaction rate, while I favours the T state, leading to diminution of binding of S and of reaction rate (see Figure 2(b), (c)). The sigmoid response of reaction rate to effector concentration can be explained in the same way as that to substrate concentration.

This elegant model can be used as a basis for quantitative calculations of

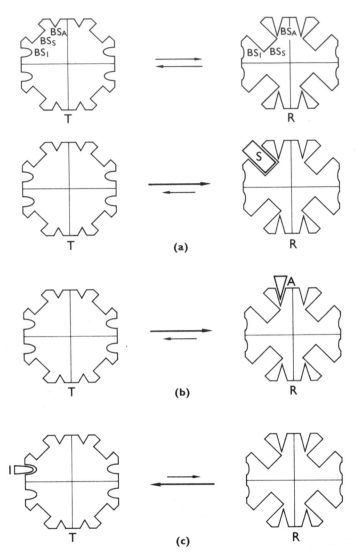

Figure 2. Monod–Wyman–Changeux model for control of enzyme activity.

A tetrameric enzyme is shown that can exist in 'tight' (T) and 'relaxed' (R) conformations. It has binding sites (BS_S) for substrate (S), and allosteric sites for an activator A (BS_A) and an inhibitor I (BS_I). The enzyme has greater affinity for S and A when in the R state, and for I when in the T state.

(a) In the absence of effectors, R and T are in equilibrium. If substrate is present this equilibrium will move towards a preponderance of R, creating more high-affinity binding sites for S so that a positive homotropic effect results.

(b) If activator is present, the R $\rightleftharpoons$ T equilibrium again shifts to favour R; more high-affinity binding sites for S appear and a positive heterotropic effect is observed.

(c) If inhibitor is present, the R $\rightleftharpoons$ T equilibrium alters, this time in the direction of T; the number of high-affinity binding sites for S diminishes, yielding a negative heterotropic effect.

various reaction parameters. Although the theoretical inferences are generally in agreement with observation, there seem to be certain enzymes the behaviour of which is not easily reconciled with the scheme. For instance, the model predicts that homotropic interactions should always be positive (i.e. binding of one substrate molecule facilitates binding of a second) and never negative (i.e. binding of one substrate molecule hinders binding of a second); and although this is frequently the case, there seem to be exceptions, for instance the threonine-inhibited homoserine dehydrogenase of *Rhodospirillum rubrum*.

Alternative models

Another influential approach is that of Koshland. This replaces the Monod–Wyman–Changeux postulate of alternative conformational states of the enzyme in a pre-existing thermodynamic equilibrium by an *induced-fit* model; enzyme molecules in the absence of ligand are supposed all to have the same conformation, but the binding of a ligand to a subunit induces a change in conformation of that subunit. When the ligand is a substrate the new conformation is one that facilitates the proper molecular alignment resulting in the enzyme-catalyzed reaction. Heterotropic activation arises when an allosteric ligand induces the same conformation change as substrate(s), heterotropic inhibition when an allosteric ligand holds the polypeptide chain in a conformation unfavourable for binding of substrate(s). Koshland's model also allows for sequential changes in subunit conformation, i.e. the binding of a ligand to a subunit does not necessitate a change in conformation of the remaining subunits. Cooperative effects arise if associations of subunits in the same conformation are energetically more favourable than associations of subunits in different conformations. This scheme allows for negative homotropic interactions.

Other models have been proposed, some of which involve broadening those outlined here or assimilating parts of each, others introducing in addition rather different possibilities of protein behaviour. It may be pointed out that since each enzyme has (presumably) evolved its pattern of modulation independently, there is no reason to suppose that any unified theory that does not allow for all potentialities of the protein molecule can explain in detail the properties of every enzyme.

A quite different means of controlling enzyme activity involves chemical modification of the enzyme. This is exemplified by the glutamine synthetase of *Escherichia coli* and other Gram-negative bacteria, which can be adenylated by the transfer of adenylyl groups from ATP. The *Escherichia coli* enzyme molecule accepts up to 12 such groups, attached to a tyrosine residue thus:

The adenylated enzyme shows altered kinetic parameters and affinities for divalent cations, and is more sensitive than the unadenylated enzyme to feedback inhibition by tryptophan, histidine and CTP, but less so by glycine and alanine. Removal of the adenylyl groups (which is not simply the reverse of the adenylation reaction, since the groups are liberated as AMP) is catalyzed by a separate enzyme inhibited by glutamine.

CONTROL OF ENZYME SYNTHESIS

Rapid and specific alterations in the rates of synthesis of particular enzymes attendant on changes in environment are a striking characteristic of most bacterial species. This type of regulation is of course entirely distinct from that exerted at the level of enzyme activity, and as we shall see, the description of its mechanism involves genetics as well as biochemistry. Before approaching the subject of mechanism, however, we shall discuss some physiological approaches.

Induction and repression

The usual finding is that an increased concentration of some compound in the medium is found to cause an increase (*induction*) or decrease (*repression*) in the cells' rate of synthesis of an enzyme, relative to total protein synthesis. If the rate of synthesis is little affected by the concentration of any compound the enzyme is said to be *constitutive*. The compound that causes induction is called an *inducer*; unfortunately the term 'repressor' has a special meaning that cannot be applied to a compound that causes repression (see below). A further ambiguity lies in the fact that the mechanism of induction or repression may not involve directly the inducer or repressing metabolite (see, for instance, the examples of *product induction*, p. 461 below).

In some instances, the growth response of bacteria to a carbon or nitrogen source provides presumptive evidence that one or more enzymes required for its metabolism are *inducible*. If the cells are inoculated into medium containing the compound as sole carbon or nitrogen source after many generations' growth in its absence, and it is then found that an appreciable lag ensues before growth occurs at the optimal rate for the medium, it is quite likely that enzymes required for utilization of the compound are induced by it. In some cases the inducible entity may be a *permease* required for transport of the compound into the cell, rather than an enzyme needed for its metabolism—see p. 475. Observations of lag periods were of considerable historical importance in the recognition of the phenomena of induction and repression. However, convincing evidence demands the determination of intracellular enzyme levels, preferably in cells grown under the various inducing or repressing conditions rather than in non-growing cells exposed to them. It is also better to use cell extracts, e.g. produced by ultrasonic disintegration, or cells rendered freely permeable, e.g. with toluene, rather than whole cells with intact permeability barriers.

Experimentally, we can show that a given substance induces (or represses) a given enzyme by demonstrating that the specific activity of the latter is higher (or lower) for cells grown in the presence of the substance than in its absence. Information about the kinetics of induction or repression, however, can best be obtained by the *differential plot* introduced by Monod. This is based on the fact that during balanced growth in a constant environment, for instance during early exponential growth or growth in continuous culture, cell proteins are synthesized in fixed proportions. The *differential rate of synthesis* of an enzyme is defined as the fraction which it constitutes of total newly-synthesized protein. To obtain a differential plot, samples are taken at intervals during growth of a batch culture, and enzyme activity and protein content are determined for each. Units of enzyme activity per unit volume of culture are then plotted against mass of protein per unit volume of culture. While culture conditions remain constant, a straight line is usually obtained with a slope that is proportional to the fraction which the enzyme constitutes of newly-synthesized protein and is therefore a measure of the differential rate of synthesis. Towards and following the end of exponential growth the plot may deviate from linearity as metabolites (which may themselves induce or repress) accumulate in the medium.

If a compound that induces or represses the enzyme is added during growth of the culture, the plot will show two linear regions with slopes that give different rates of synthesis in the absence and presence of the controlling compound. The extent of the curve joining these is also important for our understanding of the events that occur during the transition from one steady-state rate of synthesis to the other. Types of plot other than the differential plot, such as a plot of specific activity against time, give curves rather than lines following addition of the controlling substance; in spite of this, they are sometimes employed.

As an example of enzyme induction treated in this way, we shall consider the *Escherichia coli* enzyme β-galactosidase, which hydrolyzes lactose to glucose + galactose, and is *induced* by lactose or certain structurally related compounds. Some of these inducers are not metabolized (in which case they are said to be *gratuitous*); for instance, various alkyl thio-β-D-galactosides, such as the isopropyl derivative ('IPTG'), are excellent gratuitous inducers of β-galactosidase. Figure 3(a) shows a differential plot of β-galactosidase activity per ml. culture, IPTG having been added at a point during growth. The curved region between the two linear portions (the first corresponding to the very low basal level of synthesis in the absence of inducer, the second to the differential rate of synthesis following induction, some hundreds-fold greater) corresponds to only a few minutes; the significance of this will be discussed later.

Figure 3(b) shows a similar plot for a repressible enzyme, the ornithine carbamoyltransferase of *Escherichia coli*. The enzyme is involved in arginine biosynthesis (Chapter 4, p. 221), and its synthesis is repressed by the amino acid; hence following addition of arginine to a growing culture, the differential rate of synthesis, again after a transition lasting a few minutes, falls to a very low value.

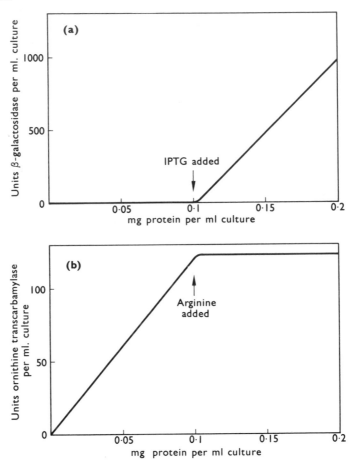

Figure 3. Kinetics of induction and repression.

(a) Differential plot of β-galactosidase activity against protein concentration for a culture of exponentially growing *Escherichia coli* ML 30 in minimal salts–maltose medium. The gratuitous inducer IPTG was added at the point indicated to give a final concentration of 10^{-4} M. The specific activity (given by the slope of the linear plot) of β-galactosidase in the presence of inducer is about 10,000 units/mg protein; in the absence of inducer it is about 10 units/mg, which is not discernible with the scale used.

(b) Differential plot of ornithine carbamoyltransferase activity against protein concentration for a culture of exponentially growing *Escherichia coli* K-12 in minimal salts–glucose medium. Arginine was added to repress enzyme synthesis at the point indicated. The specific activity of ornithine transcarbamoylase in the absence of arginine is about 1200 units/mg protein; in the presence of arginine it is about 1·5 units/mg, which is not discernible with the scale used.

The effect of inducible permeases on the kinetics of induction

In many inducible systems, the change from non-induced to induced rate of enzyme synthesis may in certain circumstances be much slower than the few minutes indicated in Figure 3. The explanation is often as follows. The move-ment of inducer into the cell occurs not only by diffusion but is also mediated by a specific protein (or proteins) named a *permease*. This protein is induced under the same conditions as the inducible enzyme under study; hence in a non-inducing situation, its level will be low, and inducer when first introduced into the medium will only enter the cell slowly, partly by diffusion and partly with the aid of the basal permease activity. Induction (both of the permease and of the enzyme) will thus build up slowly to give finally the optimum rates of synthesis of both under the conditions prevailing. An example is the β-galactosidase system of *Escherichia coli* (Figure 4).

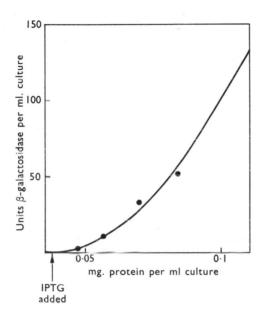

Figure 4. Kinetics of induction.

Differential plot of β-galactosidase activity against protein concentration for a culture of exponentially growing *Escherichia coli* ML 30 in minimal salts–maltose medium. IPTG was added at the point indicated to give a final concentration of 2×10^{-5} M. With this low inducer con-centration, the rate of β-galactosidase synthesis rises only gradually to its maximal value. From G. N. Cohen and J. Monod, *Bacteriological Reviews* **21** (1957) 169.

Coordinate control

Often a number of enzymes are induced or repressed in parallel. β-Galactosidase and β-galactoside permease have already been mentioned as being induced by the same inducers in *Escherichia coli*; a third protein, thiogalactoside trans-acetylase (whose physiological function remains obscure), is also induced under these conditions. Sometimes it is found that a strictly quantitative relationship holds between the specific activities of the proteins when the cells are grown under a variety of conditions of induction or repression, such that these activities are always in a constant ratio. This can be demonstrated for any pair of enzymes by plotting a *dispersion diagram*, in which each axis represents one of the enzyme specific activities; each point represents a pair of activities determined for a culture grown under a given set of conditions. If the two activities are always in a constant ratio this will result in a straight line which extrapolates through the origin, and in this case they are said to be *coordinately controlled*. The lactose proteins mentioned above, for instance, show coordinate control (Figure 5). Non-coordinate control results either in curves or in lines that do not pass through the origin.

Coordinate control is of interest in connection with the mechanism of

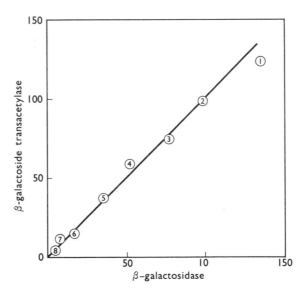

Figure 5. Dispersion diagram for β-galactosidase and β-galactoside transacetylase of *Escherichia coli* K-12 illustrating coordinate control. Specific activities of wild-type under conditions of complete induction are set arbitrarily at 100. Each point corresponds to a pair of specific activities obtained for a culture grown under particular inducing conditions, as follows:

1. Uninduced constitutive (i^-) mutant—see below, p. 436.
2. Wild-type, induced by IPTG at 10^{-4} M.
3. Wild-type, induced by methyl thio-β-D-galactoside at 10^{-4} M.
4. Wild-type, induced by methyl thio-β-D-galactoside at 10^{-4} M + phenyl thio-β-D-galactoside at 10^{-3} M.
5. Wild-type, induced by melibiose at 10^{-3} M.
6. Wild-type, induced by lactose at 10^{-3} M.
7. Wild-type, induced by methyl thio-β-D-galactoside at 10^{-5} M.
8. Wild-type, induced by phenylethyl thio-β-D-galactoside at 10^{-3} M.

From F. Jacob and J. Monod, *Cold Spring Harbor Symposia on Quantitative Biology* **26** (1961) 193.

regulation of enzyme synthesis, which will be discussed later. Its advantages to the organism are particularly clear when the enzymes come together to form aggregates. In an unbranched pathway, it would also seem desirable since an efficient balance of enzyme activities could be scaled up or down to deal with changing requirements for flow through the pathway. Inducible catabolic enzyme systems usually show coordinate control, especially where the genes concerned are clustered (see below p. 475), whereas repressible biosynthetic ones often do not; this may reflect the complicating factor of feedback inhibition in the latter case. An interesting example is the ornithine carbamoyltransferase in *Escherichia coli* K-12 (see Chapter 4, p. 221); this enzyme is *derepressed** by a factor of 100 when the cells are transferred from medium containing arginine to medium lacking it, while none of the other enzymes in this pathway is derepressed to the same degree (most of them only by a factor of about 10). This observation is sufficient to indicate that ornithine carbamoyltransferase is not controlled coordinately with the other enzymes. A possible explanation is that ornithine carbamoyltransferase competes for one of its substrates, carbamoyl phosphate, with the enzyme aspartate carbamoyltransferase, which catalyses the first step in pyrimidine synthesis (see Chapter 4, p. 240), and that under conditions of low intracellular arginine it is necessary to have a high level of ornithine carbamoyltransferase to ensure that sufficient carbamoyl phosphate is channelled into the arginine pathway. Of course, the reverse should also be true, and in fact aspartate carbamoyltransferase is derepressed in the absence of uracil or cytosine far more than the other pyrimidine biosynthetic enzymes.

THE MECHANISM OF SPECIFIC CONTROL OF PROTEIN SYNTHESIS

The elucidation of the mechanism whereby the synthesis of at least some proteins in some bacteria is regulated has provided a scheme that can be extended, on paper at least, to the analysis of problems as enormous and diverse as embryonic development and the formation of a tumour cell. Whether or not the entities postulated in the *operon model* of Jacob and Monod exist in higher organisms, this model has been of great value in demonstrating an actual mechanism for switching genes on or off, and in suggesting what controlling elements to look for and what experiments might be carried out to find them.

The operon model

The operon model was developed primarily with the proteins of lactose utilization in *Escherichia coli* K-12 as experimental system, and we shall first describe

* This term indicates a rise in rate of synthesis following the release of repression.

the results obtained with this system and then go on to discuss other systems that have been subjected to thorough study. As usual, most work has been done with the enterobacteria.

We shall begin by recapitulating information given earlier in this chapter but also introducing new material, some relating to genetic aspects:

(1) There are three proteins involved: β-galactosidase, β-galactoside permease and β-galactoside transacetylase. These proteins are induced co-ordinately by a variety of compounds structurally related to lactose, some of which (e.g. IPTG) are 'gratuitous' inducers.

(2) A very important point is that, on induction, protein is synthesized very rapidly *de novo* from amino acids, rather than being assembled from macro-molecules existing within the cell prior to induction. This was shown for β-galactosidase by adding inducer to a culture together with radioactively labelled sulphate. The methionine and cysteine of the enzyme carried the maxi-mum specific activity as soon as it is formed. The interpretation is clearly that control of rate of synthesis of an enzyme must act to regulate either the rate of transcription of the structural gene for the enzyme, or the rate of translation of these messenger RNA transcripts, or both (it should be noted that 'rate of transcription or translation' here and in ensuing discussions really means 'rate at which RNA polymerase molecules initiate transcription of the structural gene, or at which ribosomes initiate translation of the messenger RNA'—the rate of movement of RNA polymerase along the DNA, or of ribosomes along messenger RNA, being constant in a given culture, and a function of the growth conditions).

(3) The structural genes for the proteins (z for β-galactosidase, y for permease, and a for transacetylase) are *clustered*, i.e. they are arranged next to each other on the chromosome, the order being z–y–a running anticlockwise with the conventional orientation of the genetic map of *Escherichia coli* (see Chapter 7, pp. 396–7).

(4) Mutants can be obtained in which the wild-type pattern of inducibility is lost. In these mutants, synthesis of the lactose proteins is constitutive, pro-ceeding in the absence of inducer at the high rate of synthesis characteristic of the wild-type growing under conditions of maximum induction. The gene in which a mutation is presumed to have occurred is called i (for 'inducibility'). The wild-type inducible allele, following the normal convention, is called i^+, and the constitutive alleles are called i^-. The i gene is found to map close to the z–y–a gene cluster, the order being i–z–y–a. This group of genes is usually referred to as the *lac* cluster.

The first question asked was: what is the function of the product of the i gene? Jacob and Monod reasoned that this entity (whatever its chemical nature) must be acting in one of two ways, depending on the assumptions made.

Repressor model (negative control)

It might be assumed that gene expression occurs spontaneously unless some-thing inhibits it. The product of the *i* gene would then be a *repressor* acting to switch *off* expression of the structural genes in the *absence* of inducer (Figure 6(a)). No assumptions are made at this stage as to the step in protein synthesis at which the *i* gene product interferes. In this model the inducer is then assumed to bind to and thereby inactivate the repressor. Once this has happened, expression of the structural genes can proceed normally (Figure 6(b)). The constitutive mutant is assumed to have lost the ability to make functional repressor molecules and the enzyme is accordingly made whether inducer is present or not (Figure 6(c)).

Endogenous inducer model (positive control)

It might be assumed that gene expression does not occur unless something activates it. In this model there is a positive requirement for induction if the structural genes are to be expressed. In the constitutive strain the product of the *i* gene supplies this requirement directly or indirectly, i.e. it acts as, or gives rise to, endogenous inducer (Figure 6(f)). The inducible allele is assumed to lack the ability to make this inducer (Figure 6(d)). It will, therefore, produce no enzyme unless a galactoside inducer is supplied in the medium (Figure 6(e)).

Comparison of the two models

The models predict different dominance effects if i^+ and i^- are both present in the same cell, i.e. if a partial diploid (Chapter 7, p. 416) is constructed with two copies of the *i* gene, one in the wild-type i^+ allelic configuration, the other in the mutant i^- configuration.

In the negative control model the inducible strain produces a repressor, the constitutive strain produces an inactive repressor. Assuming, as seems likely, that the inactive species does not interfere with the functioning of the active species, it would be predicted that the partial diploid should be inducible, i.e. i^+ is dominant to i^- (see Figure 7(a)).

Conversely, in the positive control model the constitutive strain produces a molecule that activates or induces the structural gene whether or not inducer is present. The inducible strain carries the loss mutation and it would be expected that in the diploid the constitutive function would be dominant (see Figure 7(b)).

To ascertain which model is correct, all that is required is to determine whether the lactose proteins are inducible or constitutive in i^+/i^- partial diploids. The most readily available stable partial diploids formed by *Escherichia coli* are F′ strains (see Chapter 7, p. 416), and so this test can be carried out with such a strain in which the F′ factor includes the *lac* cluster (described as an F-*lac* factor). If the strain carries i^+ in chromosomal *lac* cluster and i^- in the F-*lac* it is denoted $i^+/F - i^-$; in the reverse situation it is $i^-/F - i^+$.

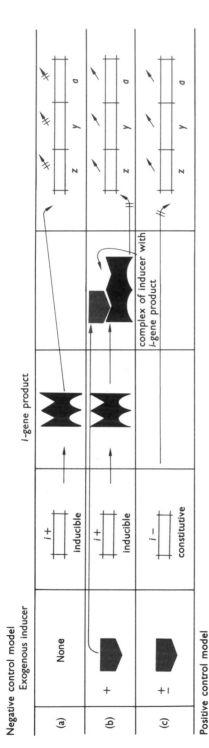

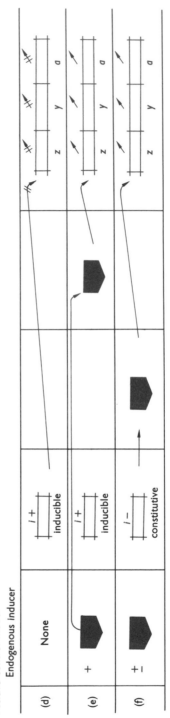

Figure 6. Models for the function of the *i*-gene product (Jacob and Monod).

Negative control model. (a) The *i*⁺ gene determines the structure of a product which passes into the cytoplasm and prevents the expression of the structural genes *z*, *y* and *a* which code for the lactose proteins β-galactosidase, permease and transacetzylase. (b) *i*⁺-gene product can combine with exogenously added inducer to give an inactive complex. This allows production of the lactose proteins. (c) In the constitutive strain (*i*⁻) the gene product is either non-functional or absent. Lactose proteins are produced whether inducer is present or not.

Positive control model. (d) The *i*⁺ gene does not determine the formation of an active endogenous inducer and no lactose proteins are produced. (e) Exogenously added inducer can act to cause induction. (f) In the constitutive strain (*i*⁻) an active inducer is produced endogenously.

It should be noted that in the negative control model it is the inducible strain which produces a regulatory product, whereas in the positive control model it is the constitutive strain *i*⁻ which produces a regulatory product.

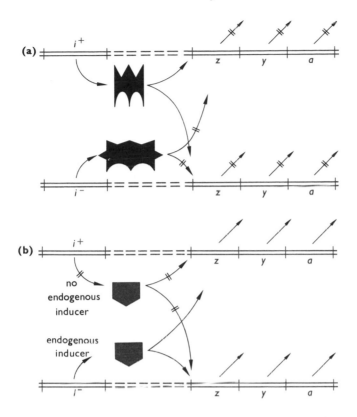

Figure 7. Consequences of negative and positive control models.

(a) *Negative control model.* The i^+/i^- diploid cell contains a functional i^+ product and a non-functional i^- product; in the absence of inducer, the former is capable of switching off expression of the structural genes z, y and a irrespective of the presence of the i^- product.

(b) *Positive control model.* In the i^+/i^- diploid in the absence of inducer, I^- determines the production of endogenous inducer even though i^+ does not. The lactose proteins are therefore synthesized even in the absence of exogenous inducer.

Both kinds of partial diploid are found to be inducible exactly as in the wild-type. In other words, the lactose system exemplifies negative control, and the i gene product can be taken to be a *repressor*. The chemical nature of the repressor is discussed later (p. 451).

Super-repressed mutants

The properties of another class of control mutants, termed *super-repressed* or i^s, are consistent with this picture. An i^s allele confers on cells carrying it almost complete inability to allow induction of the lactose proteins, though some induction of all three results with extremely high inducer concentrations. The latter fact, together with mapping data, shows that these mutations lie in the i gene. The simplest interpretation along the lines of the negative control model is that i^s directs the formation of an altered repressor that has largely lost its affinity for inducers which therefore lose their capacity to act as such. This leads to the prediction that the partial diploids carrying the pairs of alleles i^+/i^s should show the i^s phenotype, since in the presence of inducer, even through the i^+ product is inactivated, the i^s product remains active and represses expression of the structural genes (see Fig. 8). It is found that i^+/i^s partial diploids (constructed using an F-*lac* factor—see above) are in fact super-repressed, like i^s strains. Furthermore, i^-/i^s partial diploids would be expected to be, and are, super-repressed too.

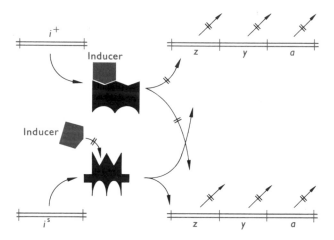

Figure 8. Effect of i^s mutation in i^+/i^s diploid. The i^s gene product does not interact with inducer, so that even in the presence of the latter, the i^s product represses expression of the structural genes z, y and a. The fact that the i^+ product combines with inducer, losing its capacity to repress thereby, is irrelevant.

The operator

The next point that concerns us is the mode of action of the repressor. Here Jacob and Monod used the striking coordinate control of the lactose proteins and clustering of their three structural genes to suggest that, since the system functions physiologically and genetically as a unit, it was to be expected that the repressor would act at a single specific site in switching off gene expression; this site they termed the *operator*. Clearly, the model as yet implied nothing as to the nature of the operator and of its interaction with the repressor. To gain some understanding of these things, it was necessary to obtain mutants with mutations in the *operator gene* (specifying the structure of the operator), in which the operator was altered so as to be unable to function normally. How might such mutants be discovered? The operator was (so far) a hypothetical entity with the sole necessary property of affinity for the repressor; therefore the obvious alteration to look for was a form to which the repressor could not bind. Mutants that had suffered such a change would evidently be constitutive, since the repressor could not function even in the absence of inducer and what is more would they be constitutive even in the presence of an i^s allele, which would be no more able to function than an i^+ allele. This presented a method for the selection specifically of such mutants: a culture of an i^s/i^s partial diploid which, having neither β-galactosidase nor β-galactoside permease, could not grow on medium containing lactose as sole carbon source, should when plated on a lactose medium give rise to rare colonies derived from such operator mutants—for if two i^s alleles were present, the back-mutation of one to i^+ or i^- would still be insufficient to permit induction and therefore growth; while the back-mutation of both i^s alleles simultaneously would be exceedingly unlikely.

When this selection procedure was carried out, the mutations in the strains isolated were found to have properties that strongly suggested a model for the mechanism of regulation of gene expression in this system. The mutations, presumed to lie in the operator gene, confer constitutivity, as expected, on the synthesis of all three lactose proteins, and are described as *operator constitutive*, or o^c (Figure 9). They map between i and z so that the gene order in the *lac* cluster is *iozya*. However, their most characteristic property is that of *cis-dominance*. This means that although o^c is dominant to i^+ and i^s (as it must be, from its manner of selection in partial diploids), this dominance is only exerted on the structural genes on the same chromosome (which is what is meant by the term 'in the *cis*-position'). Those structural genes on the other chromosome, in the *trans*-position to the o^c and *cis* to a wild-type operator gene (written o^+), are subject to the control of the i^+ or i^s allele. This can be shown by constructing partial diploids such as $i^+o^cz^+y^-/i^+o^+z^-y^+$—i.e. the o^c allele is *cis* to a wild-type β-galactosidase gene and to a mutant allele of the y gene directing formation of a non-functional permease, while the wild-type o^+ is *cis* to a mutant allele of the z gene, specifying a non-functional β-galactosidase, and to a wild-type permease gene. It is found that such a strain produces β-galactosidase constitutively but β-galactoside permease inducibly (see Figure 10).

These results imply that the function of the operator gene is somehow connected with its position at one end of the z–y–a gene cluster. Jacob and Monod's interpretation was as follows. They supposed that control is at the level of transcription; that is, the rate of enzyme synthesis is regulated by setting the rate at which messenger RNA transcripts of the enzyme's structural gene are produced. (It was argued that the alternative, control at the level of translation, would involve the wasteful production of messenger RNA molecules.) The operator 'gene' and the operator are one and the same, a stretch of DNA with two functions: it is the binding site for the repressor, and

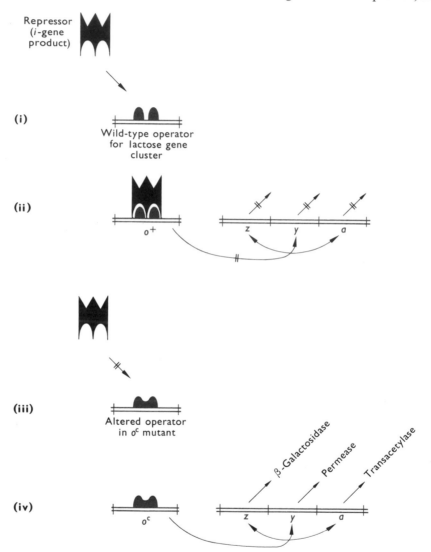

Figure 9. Model for operator function.

(i) The wild-type operator is the target for the repressor (*i* gene product); their combination (ii) in some way prevents the expression of the structural genes. The altered operator of an o^c mutant (iii) cannot combine with repressor. The uncombined state of the operator, which also results if the repressor is inactivated by inducer (not shown), permits expression of the structural genes and formation of the lactose proteins.

also the site at which RNA polymerase molecules attach before beginning transcription (but see below). Once an RNA polymerase molecule has bound to the operator, it moves along the z–y–a genes transcribing them in that order on to a single messenger RNA molecule which, since it carries the transcripts of more than one gene or cistron (Chapter 6, p. 353), is called a *polycistronic messenger RNA* molecule.

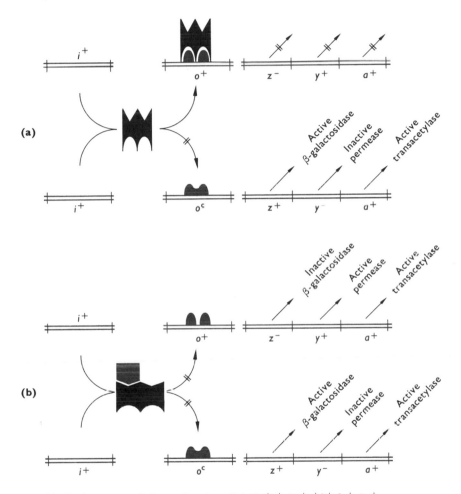

Figure 10. *Cis*-dominance of o^c mutations in a diploid $i^+o^+z^-y^+a^+/i^+o^cz^+y^-a^+$.

(a) In the absence of inducer, repressor acts at the o^+ operator but cannot bind to the o^c operator. The *cis*-dominance of o^c means that it affects only those genes on the same chromosome as itself, i.e. the z^+, y^- and a^+ genes *cis* to o^c are expressed but the z^-, y^+ and a^+ genes *cis* to o^+ are not. Hence in the absence of inducer, active β-galactosidase and transacetylase are made, but only the inactive permease.

(b) Combination of the repressor with inducer abolishes repression at both operators and so active β-galactosidase (encoded by the z^+ *cis* to o^c), permease (encoded by the y^+ *cis* to o^+), and transacetylase (encoded by the a^+'s *cis* to both operators) are produced.

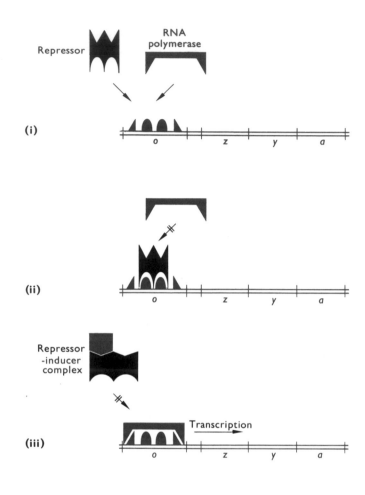

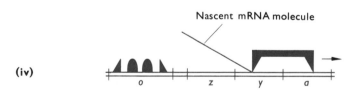

Figure 11. The operon model in its first complete form (*c.* 1961).

(i) Repressor and RNA polymerase compete for their common binding site, the operator *o*, on the DNA. Since the repressor has a much higher affinity for *o* than does the RNA polymerase, the latter is virtually excluded, and (ii) transcription of the structural genes *z*, *y* and *a* is thereby prevented. When the repressor is inactivated it does not bind to *o*, and (iii) the RNA polymerase binds to *o* and transcription of the structural genes is initiated. (iv) The RNA polymerase is shown with about half of the polycistronic mRNA transcribed.

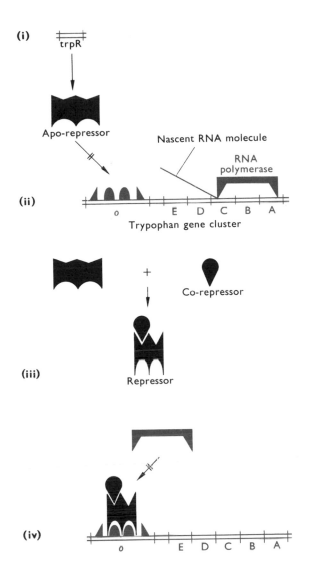

Figure 12. Application of the operon model to repressible systems.

In *Escherichia coli*, the terminal enzymes for tryptophan biosynthesis are coded for by five contiguous genes. In the wild-type, these enzymes are repressed by tryptophan. Mutants in which they are constitutive, i.e. formed at high levels whether tryptophan is present or not, may carry mutations in an unlinked regulator gene *trpR* or at an operator site *o* adjacent to the structural gene cluster (see below, p. 477). It is supposed that *trpR* gives rise (i) to an apo-repressor which by itself (ii) cannot combine with *o*; the RNA polymerase therefore binds at *o* with consequent transcription of the structural genes. However, if the apo-repressor combines with the co-repressor tryptophan (iii), it is converted into functional repressor with high affinity for *o* (iv), so that binding of RNA polymerase to *o* and consequent transcription are blocked.

When an i^+ strain is induced, the repressor binds inducer and is thereby converted into a form which no longer has affinity for the operator. The latter can therefore bind RNA polymerase molecules, and polycistronic messengers carrying transcripts of the z, y and a structural genes are then produced in rapid succession. In an i^- strain, the altered i gene product is presumably incapable of binding to the operator, while in an o^c strain the altered operator is unable to bind normal repressor.

In an i^+ strain in the absence of inducer, or in an i^s strain under all conditions, the repressor binds to the operator and thereby prevents RNA polymerase molecules from doing so. Transcription of the z, y and a genes is therefore very infrequent. That it occasionally happens (so that there are measurable basal, uninduced levels of the proteins) may result from the fact that the position of the equilibrium: repressor (free) + operator $\rightleftharpoons$ repressor bound to operator, does not lie totally over on the right-hand side; in other words, even in uninduced cells, the operator is for a small part of the time free for attachment of RNA polymerase molecules. The mechanism of control is summed up diagrammatically in Figure 11.

This scheme can be extended to repressible systems that also show gene clustering, regulator gene mutations leading to constitutivity that are recessive to the wild type allele, and o^c-type mutations. It is sufficient to suppose that the regulator gene product (or *apo-repressor*) is inactive on its own, but when combined with the repressing metabolite (e.g. end-product of a biosynthetic pathway) becomes a repressor and binds to the operator to block transcription (see Figure 12). A group of contiguous genes such as z, y and a, transcribed on to a single polycistronic messenger RNA molecule at a rate controlled by an operator situated at one end of the gene cluster, is termed an *operon*.

Polar mutations

Further evidence that the z, y and a genes are transcribed on to a common *polycistronic messenger* is provided by polar mutations. These mutations may occur in z or y; those in z lead to loss of β-galactosidase activity and diminished activities of permease and transacetylase, while those in y have no effect on β-galactosidase activity but lead to loss of permease activity and diminished transacetylase activity. Polar mutations (in this and other systems) thus lower the rates of synthesis of proteins coded for by genes in the cluster other than that in which the mutation occurs, but this property shows *polarity* in that it affects genes only on the *operator-distal* side (i.e. the side away from the operator) of the mutated gene, but not genes on its *operator-proximal* side (i.e. the side towards the operator). The activities of enzymes affected by a polar mutation are always a constant percentage of their activities under identical growth conditions in the normal strains and the polar effect is unrelated to metabolic control mediated by the induction or repression system.

This percentage depends on the polar mutations and it can vary from zero (no production of permease or transacetylase) to almost 100% with weakly polar mutations.

It is thought that the polar effect is caused by the appearance of a chain terminating triplet (see Chapter 6, p. 357) in the messenger RNA codon of the mutated gene. There is still some dispute as to how this results in the polar effect, but we shall give what seems the most likely hypothesis. It appears that as a messenger RNA molecule is transcribed, it peels away from the chromosome; ribosomes attach to its started end, one after another, and move along the messenger translating as they go, the leading ribosome following only a short distance behind the transcribing RNA polymerase. If only a single gene is being transcribed, the end of its transcription is followed by the arrival of the ribosomes, one by one, at whatever ribonucleotide sequence signals 'end of polypeptide chain'. At this point the ribosome, the completed polypeptide chain, and the messenger separate from each other. In the lactose system where a polycistronic messenger is involved, the transcript of the *z* gene is immediately followed by that of the *y* gene, and this by the *a* gene transcript. Here it seems (although various interpretations of the experiments are possible) that translation of the *y* transcript may be initiated either by ribosomes that have finished translating the *z* transcript or by new ribosomes, depending on the experimental conditions. At all events, the important points are that the messenger RNA is at all times quite closely covered with ribosomes, and that new ribosomes only attach to it at the beginning of a gene transcript.

Suppose now that there is a polar mutation in the *z* gene, so that the *z* transcript now carries a chain-terminating codon. The transcription-translation process will start normally, but as ribosomes reach the chain-terminating triplet they, and the truncated β-galactosidase polypeptide, will detach from the messenger. As transcription continues, a length of mRNA will be produced which does not carry any ribosomes, for these can only re-attach at the beginning of the *y* transcript.

We come now to the essential factor in the phenomenon of polarity, which seems to be the presence of an endoribonuclease that binds to any region of mRNA not protected by ribosomes and breaks the RNA chain; this is followed by rapid degradation in the 5' to 3' direction (which is also the direction of transcription). In polar mutants an exposed length of messenger will be formed. If the *y* gene transcript has already been started before the endoribonuclease attaches to the exposed section of messenger or before it catches up with the RNA polymerase, then ribosomes can attach and translation of the *y* and *a* transcripts can occur normally. If, however, the endoribonuclease reaches the RNA polymerase before the ribosomal attachment site at the beginning of the *y* transcript, then no sooner is more messenger transcribed than it is immediately broken down, and transcription (and therefore translation) of the *y*

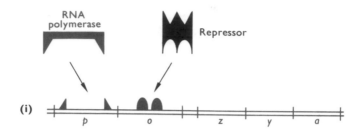

(i)

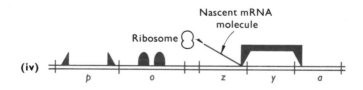

(ii)

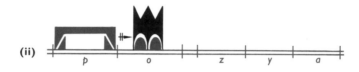

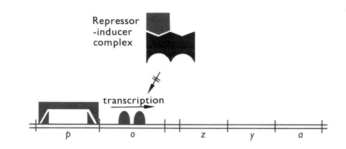

(iv)

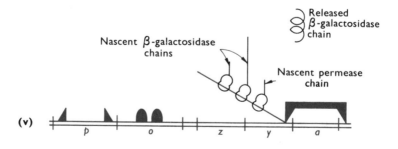

(v)

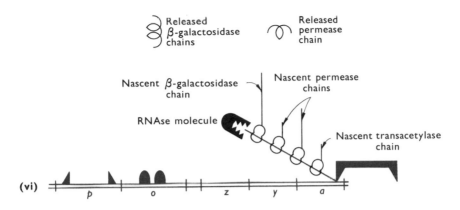

Figure 13. The current form of the operon model.

(i) The promoter *p* (RNA polymerase binding site) and the operator *o* (repressor binding site), are distinct regions of DNA, with *o* lying between *p* and the structural genes.

(ii) Binding of repressor to *o* seems not to prevent the attachment of RNA polymerase to *p* but rather to block its movement along the DNA.

(iii) When the operator is free, as when the repressor has been inactivated by inducer, movement of the RNA polymerase along the operon can occur and transcription commences.

(iv) As nascent messenger appears, ribosomes can attach to its 5' end.

(v) When the RNA polymerase has moved some way along the operon, several ribosomes may already have attached and be moving along the mRNA behind the RNA polymerase, each ribosome bearing a nascent polypeptide. When a ribosome reaches the end of a structural gene transcript, the completed polypeptide is detached.

(vi) Before the RNA polymerase has reached the 3' end of the operon it is thought that the messenger is attacked at the 5' end by the ribonuclease responsible for the characteristic lability of bacterial messenger RNA molecules (see also Figure 22, p. 353). Another RNA polymerase molecule may attach at *p* before the first has finished transcription of the operon; for the sake of clarity this is not shown.

and *a* genes never occurs. Clearly, the longer the length of unprotected mRNA, the more likely is the latter eventuality. This explanation is consistent with the fact that the polar effect of a mutation increases with its distance from the start of the next gene.

Whatever the precise mechanism of polarity may be, it seems certain that the occurrence of polar mutations in clustered genes indicates that the genes are transcribed on to a single polycistronic messenger.

The promoter

In the original form of the operon model, the operator was supposed to represent the binding sites for both the RNA polymerase and the repressor. However, certain observations have made it unlikely that, in the lactose system at any rate, these functions are shared by the same DNA sequence. In particular those o^c mutants that carry point mutations rather than deletions have reduced affinity for the repressor yet show the same rate of synthesis of the lactose proteins under conditions of induction as the wild-type. One might have expected that, if the same region of DNA were involved in binding the RNA polymerase, this binding would be weakened, transcription of the operon would be initiated less frequently and the maximum activities of the proteins, achieved in the presence of inducer when the repressor does not function, would be lower than in the wild type.

It was therefore suggested that there might be two separate DNA sequences, one for binding the repressor (which could retain the name 'operator'), and another for binding the RNA polymerase, to which the name *promoter* was given. In the lactose system, for instance, a mutation in the operator would lead to diminished affinity for repressor. In the absence of inducer the levels of the lactose proteins would be about the same as in the induced wild-type cells; in the absence of the inducer, repression would be partially or completely ineffective, so that the levels would remain higher than in the uninduced wild-type and might even approach those of induced cells—as found, in fact in o^c point mutations. A mutation in the promoter would have the opposite effect: in the presence of inducer the levels would be lower than in induced wild-type cells, but would drop in the absence of inducer by the same factor as in wild-type. This mutation should, like the o^c mutations, be *cis*-dominant. Such promoter mutations (p^-) have been found in the lactose system and shown to possess exactly the properties predicted. In addition they are *cis*-dominant like o^c mutations. Interestingly, they map between the i mutations and the o^c mutations. This suggests that when repressor is bound to the operator it prevents transcription by blocking movement of the RNA polymerase from the promoter to the structural genes. The current form of the operon model for the lactose system, with separate operator and promoter, is depicted in Figure 13.

Regulation at the level of transcription or translation?

Jacob and Monod originally suggested on *a priori* grounds, as pointed out above, that enzyme synthesis would probably be found to be controlled by regulation of the rate of transcription rather than of translation. This has since been shown to be so, at least in very large part, for the lactose system. The evidence is of two types: firstly, the *lac* polycistronic messenger has been assayed by DNA–RNA hybridization in cells under various conditions of induction and its amount is proportional to the rate of synthesis of the lactose proteins; and secondly, the repressor has been isolated and shown to bind specifically to DNA from the lactose region.

In experiments of the first kind, it is necessary to be able to prepare DNA in which the lactose operon sequence accounts for a much greater fraction than it does in chromosomal DNA. This is the case in the DNA of F-*lac* factors (see Chapter 7, p. 416), which can under certain circumstances be separated from chromosomal DNA. Another possibility is to use temperate phages of the λ type which have been integrated into the chromosome close to the *lac* cluster, and then detached carrying the *lac* cluster as part of their genomes (cf. the formation of λdg particles (Chapter 7, p. 417)). (For controls F-*lac* or phage was obtained from bacterial strains carrying deletions of the *lac* cluster.) Such '*lac*-enriched DNA' is then denatured (see Chapter 5, p. 256). RNA is purified from a culture usually harvested immediately after exposure to a pulse of radioactively-labelled uridine. The amounts of labelled RNA that can hybridize (see Chapter 5, p. 269) with the *lac*-enriched DNA and with similar DNA with the lactose cluster deleted are then estimated. The difference between these amounts is taken as the proportion of rapidly-labelled RNA that is *lac* mRNA.

When the RNA was obtained from i^+ cells with inducer present, or from i^- or o^c cells whether inducer was present or not, the *lac* mRNA made up an appreciable proportion of the labelled RNA. When, however, it was obtained from i^+ cells without inducer, or i^s cells whether inducer was present or not, no *lac* mRNA could be detected. These results are entirely consistent with control being at the level of transcription.

Experiments of the second kind involve the isolation of the *lac* repressor. This isolation was first carried out by Gilbert and Müller–Hill, who employed conventional techniques of protein separation together with an ingenious assay system making use of *equilibrium dialysis*. It was assumed that the repressor is the only protein inside the cell that binds the gratuitous inducer IPTG, so that if protein fractions were enclosed in dialysis bags and immersed in a solution of radioactively labelled IPTG, those bags containing fractions enriched for repressor would at equilibrium contain IPTG at a higher concentration than it is in the medium outside the bags. It is thus possible to show that the protein so isolated will bind to '*lac*-enriched' DNA. It does not do so at all if the *lac* region is deleted and only slightly if the *lac* region carries an o^c mutation. More striking still, it has been possible to constitute an *in vitro* transcription system consisting of '*lac*-enriched' DNA, cyclic 3',5'-AMP and its binding protein (see the discussion of the mechanism of catabolite repression, p. 452), the termination factor ρ (see p. 357), RNA polymerase, and the four ribonucleotides. This synthesizes *lac* mRNA which is detectable by hybridization. The addition of purified repressor greatly diminishes the production of *lac* mRNA and inducers such as IPTG overcome this repression.

In spite of this strong support for the major part of control of enzyme synthesis being at the level of transcription, there are several pieces of evidence suggesting at least some role for translational regulation. Some evidence is derived from the lactose system, both for induction and for catabolite repression, and some from other systems which will be discussed later.

Catabolite repression and its mechanism

Induction is not the only form of control over the rates of synthesis of the lactose proteins; the presence of carbon compounds other than inducers may have a striking effect on the production of these proteins. Organic compounds which can be metabolized more efficiently and support more rapid growth than lactose diminish the rates of formation of the lactose proteins. The first compound found to act in this way was glucose; for this reason, the phenomenon was originally known as *glucose repression* or the *glucose effect*. However, it has become apparent that most probably it is not the compound present in the medium but some intermediary metabolite(s) resulting from its breakdown that is responsible for the repression; for this reason, Magasanik's term *catabolite repression* is generally employed. As we shall see, catabolite repression is very common for inducible enzymes in Class I sequences.

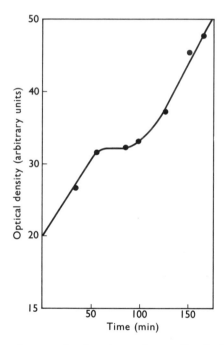

Figure 14. Biphasic growth curve showing glucose–lactose diauxie. Wild-type *Escherichia coli* K-12 was grown in minimal medium containing 10^{-3} M glucose and 10^{-2} M lactose. From W. F. Loomis, Jr. and B. Magasanik, *J. Bacteriol.* **93** (1967) 1397.

The term 'catabolite repression' covers three distinguishable phenomena, at any rate in the lactose system. There is (i) a severe but transient repression that occurs if glucose (or a related compound) is added to a growing fully induced culture; synthesis of the lactose proteins slows or even stops for a time, but is then resumed; (ii) the weaker permanent repression that follows release from transient repression, i.e. synthesis of the lactose proteins is resumed but at a lower rate than initially; and (iii) a severe permanent repression when the cells are exposed to low concentrations of inducer at the same time that, or after, they start to grow on glucose. In fact, if *Escherichia coli* uninduced for the lactose proteins is inoculated into medium containing a mixture of glucose and lactose, the former so severely represses synthesis of β-galactosidase and β-galactoside permease that no lactose is utilized until the glucose is exhausted; a biphasic or *diauxic* growth curve is thereby obtained (Figure 14).

We have implied that catabolite repression involves an antagonism of induction by intermediary metabolites accumulating through breakdown of a carbon source that can be rapidly metabolized. However, this may not apply to all three aspects of catabolite repression in the lactose system. In particular, it has been claimed that transient repression (type (i) above) can be caused by glucose analogues (such as 2-deoxyglucose) than can be phosphorylated but not further metabolized by the cell. As will be discussed below, this result is hard to interpret in the light of other evidence.

Type (iii) almost certainly results from catabolite repression combined with inhibition of the permease needed for inducer uptake so that the intracellular level of inducer remains low. This effect can be abolished in various ways: by increasing the inducer concentration, by exposing the cells to the inducer before introducing glucose, or by use of a constitutive mutant.

The important questions regarding those forms of catabolite repression that are mediated by intermediary metabolites are (a) which intermediary metabolite(s) actually effect the repression and (b) what genetic elements are needed for it to come about. It would appear that the same answer to these questions apply to both the less severe, permanent repression of type (ii) and the transient repression of type (i) (which is not easy to reconcile with the suggestion discussed above that the latter does not involve intermediary metabolites at all). Regarding the active metabolite, until 1968 there was only very ambiguous and conflicting circumstantial evidence that could be taken as implicating or ruling out various compounds; however, in that year it was shown that cyclic 3',5'-AMP relieves both types of repression, and even enhances the effect of inducers in the absence of catabolite repression. It has also been shown that growth on glucose produces lowering of the intracellular concentration of cyclic 3',5'-AMP and it is now proposed that this compound directly mediates catabolite repression.

Evidence as to the mechanism by which cyclic 3',5'-AMP achieves its effect has also been obtained only very recently. DNA–RNA hybridization studies indicate that, at all events in the lactose system, catabolite repression acts at the level of transcription. Both transient repression of type (i) and permanent repression of type (ii) are as marked in strains carrying i^- or o^c mutations as in the wild-type, so that neither the repressor nor the operator can be involved.

However, some promoter mutations lead to diminished sensitivity to both types of repression, suggesting that promoter plays a role in the mechanism. Further, it has been shown that *Escherichia coli* cells contain a protein which is capable of binding cyclic 3′,5′-AMP but which is neither an enzyme associated in its metabolism nor involved in its transport into the cell. Mutants lacking this protein are unable to be induced for the lactose proteins, nor for the proteins of the galactose, L-arabinose and maltose degradative systems which are also subject to glucose repression. On the basis of these results it is suggested that cyclic 3′,5′-AMP attaches to its binding protein, and the combination acts via the promoter to stimulate binding of the RNA polymerase. In support of this, it is found that the *in vitro* system for transcription of *lac* mRNA requires both cyclic 3′,5′-AMP and its binding protein. It seems that the complex of these binds to the promoter DNA, in some way facilitating attachment of the RNA polymerase.

PATTERNS OF CONTROL IN CLASS I PATHWAYS

We shall now describe the interplay of the kinds of control mechanism outlined above in the regulation of some Class I systems. Carbon sources, the regulation of whose breakdown has been studied in various organisms, include mono-, di- and polysaccharides, polyhydric alcohols, amino acids, peptides ('nutrient broth'), pyrimidine and purine bases, fatty acids, and a variety of cyclic compounds such as camphor, mandelate and *p*-hydroxybenzoate. Generally we can consider the degradative pathway as comprising a 'peripheral' sequence in which the carbon source is converted into an intermediate of one of the glucose degrading pathways or of the tricarboxylic acid cycle, followed by the remaining steps of these 'central' pathways (Figure 15).

It may be noted that not all of these sequences serve a purely catabolic function. The Embden–Meyerhoff pathway, for instance, works in reverse in cells grown on a TCA cycle intermediate to produce glucose to be incorporated into cell wall components. Pathways which can act either degradatively or biosynthetically are said to be *amphibolic*, and might be expected to show peculiarities as regards control. Another class of reactions with special function comprises those which replenish the TCA cycle when the carbon source is one that feeds into the glycolytic sequence at or prior to pyruvate; they are said to be *anaplerotic* (see Chapter 3, p. 190).

Control of the central pathways

The controls of activity of the enzymes in the Embden–Meyerhof pathway and tricarboxylic acid cycle, and of the anaplerotic enzymes pyruvate carboxylase and phosphoenolpyruvate carboxylase, are known in some detail. Since these pathways together provide, for most bacteria, the main route for the formation of carbon skeletons for biosynthesis by degradation of carbohydrates and related compounds and for the production of energy and reducing power needed in biosynthetic reactions, regulation of the flow through these pathways is clearly very important. As might be expected from these two functions of the

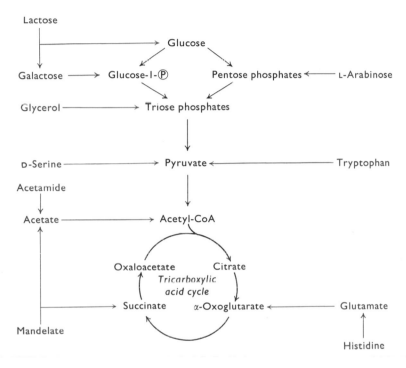

Figure 15. Patterns of catabolic pathways for carbon/energy sources. The 'central' pathways of glucose breakdown via the Embden–Meyerhof and pentose phosphate routes and the TCA cycle are shown in black, while the 'peripheral' sequences leading into these pathways from some less common carbon sources (we have chosen those which will be discussed in this chapter) are shown in red.

sequence, effectors of modulation tend to be either compounds that reflect the availability of metabolites at various points in the pathway, or compounds that reflect the energy state or level of reducing power of the cell. In a number of instances an enzyme is inhibited by a compound formed in a reaction many steps later in the pathway, or a compound immediately derived from such. Examples are the inhibition of phosphofructokinase by phosphoenolpyruvate in *Escherichia coli*; the inhibition of phosphoenolpyruvate carboxylase by aspartate (formed by transamination of α-oxoglutarate), malate, and fumarate in *Escherichia coli* and *Salmonella typhimurium*; and the inhibition of citrate synthase by α-oxoglutarate in the same two enterobacterial species. This is a form of end-product or feedback control, the function of which is to reduce the flow through the early part of a sequence in which products of the later steps are accumulating, so that wastefully excessive flux through the pathway is avoided. Another phenomenon sometimes noted is the activation of an enzyme late in the sequence by a metabolite coming some way before it, termed *precursor activation*. For instance, fructose 1,6-diphosphate activates one of the two pyruvate kinase isozymes of *Escherichia coli* and the phosphoenolpyruvate carboxylase of *Salmonella typhimurium*. Presumably this ensures that a key enzyme is mobilized to deal with a flow of substrate shortly to reach it from breakdown of the activating 'precursor'. The phosphoenolpyruvate carboxylase is also activated by acetyl coenzyme A, perhaps to facilitate a continued supply of oxaloacetate for condensation with the acetyl CoA to form citrate. Activation by the two effectors fructose 1,6-diphosphate and acetyl CoA is cooperative (Chapter 4, p. 203). It is worth pointing out, in view of our comments below on the un-representative nature of the enterobacteria, that in *Pseudomonas* phosphoenol-pyruvate carboxylase does not occur; its anaplerotic function is taken over by an ATP-dependent pyruvate carboxylase, which does not seem to be subject to control by any effectors.

The energy state of the cell is principally indicated by the concentration of ATP and other nucleoside triphosphates. The nucleoside mono- and diphos-phates may, however, be equally useful in a regulatory role, because of the ready interconversion of mono-, di- and triphosphates within the cell: a drop in ATP concentration, for example, is equivalent to a rise in that of the AMP or a somewhat smaller rise in that of ADP (see the discussion of *energy charge* in Chapter 3, p. 168). We find, for instance, that in *Escherichia coli* phospho-fructokinase is activated by ADP and GDP, and the pyruvate kinase *not* activated by fructose 1,6-diphosphate (see above) *is* activated by AMP. Both of these activations promote the flow through energy-yielding sequences in response to the accumulation of compounds signalling an energy dearth.

NADH$_2$ acts as an inhibitor of several enzymes, most of which can be regarded as coming early in sequences that involve its formation. It can therefore be regarded as a feedback inhibitor, although possibly its significance is rather to indicate the level of reducing power within the cell. Its inhibition of citrate synthase, the functioning of which produces NADH$_2$, can be a very important control. However, whereas Gram-negative organisms show the effect, Gram-positives do not. Furthermore, among Gram-negative species there is a further distinction between those in which the inhibition by NADH$_2$ is counteracted by AMP (obligately aerobic species such as *Pseudomonas*, *Azotobacter* and *Chromobacterium*) and those in which it is not (facultative anaerobes, the enterobacteria and related organisms). The reason for these striking differences may, perhaps, lie in a physiological peculiarity of the enterobacteria and probably of other facultative anaerobes. These organisms generate energy primarily by anaerobic glycolysis when growing on glucose, even under good aeration; the tricarboxylic acid cycle functions here primarily biosynthetically, as suggested by the fact that *Escherichia coli* mutants lacking succinic dehydrogenase or certain cofactors essential for oxidative phosphorylation grow aerobically on glucose much like the wild-type. However, organisms requiring the operation of the tricarboxylic acid cycle for energy production either cannot allow its functioning to be inhibited merely by accumulation of a by-product, as in the Gram-positives, or need to ensure that this inhibition can be reversed by a signal of low ATP level, as in the obligately aerobic Gram-negatives. In the facultatively anaerobic Gram-negatives, it is sufficient for end-product inhibition of citrate synthase to be counteracted (as in *Escherichia coli*) by its substrates, oxaloacetate and acetyl CoA antagonizing the effects of $NADH_2$ and α-oxoglutarate respectively.

Most of the work on the modulation of enzymes in the central routes of carbon dissimilation and energy production has been done with enterobacteria and we can expect that as other bacterial groups are studied new kinds of generalization will emerge. In this context, we may mention that a plausible explanation for the extreme complexity of the controls described for these pathways in bacteria in contrast to those in eukaryotes, reflects the isolation of the enzyme systems of the latter in mitochondria. This facilitates control of flow through different systems by keeping them physically apart. It has been suggested that the diminished extent of regulation of citrate synthase in Gram-positive in contrast with Gram-negative bacteria derives from the 'higher' type of organization of the tricarboxylic acid cycle enzymes into a membrane-bound system (possibly the mesosome, see Chapter 1, p. 119) in the former.

Another reason for complexity is the fact that many reactions are amphibolic. The enzymes concerned belong not only to Class I in making available biosynthetic intermediates and energy during growth on carbohydrate, but also to Class II in acting biosynthetically (in the production, for example, of cell wall components, ribose and deoxyribose, lipids and storage carbohydrates) when the bacteria are growing on a tricarboxylic acid cycle intermediate or related compound as sole carbon source.

Two generalizations that apply to the regulation of enzyme activity can be made about the reaction catalysed by the enzyme subject to modulation. It is thermodynamically irreversible and it comes immediately after a branch point of two (or more) metabolic routes. This is true of phosphofructokinase, pyruvate kinase, phosphoenolpyruvate carboxylase and citrate synthase, which have been mentioned as being subject to modulation. Fructose 6-phosphate can be phosphorylated via phosphofructokinase, forming fructose 1,6-diphosphate, or it can react with glutamine to yield glucosamine 6-phosphate in the first step of the biosynthetic pathway for the cell wall constituents *N*-acetylglucosamine and *N*-acetylmuramic acid; phosphoenolpyruvate can yield pyruvate *via* pyruvate kinase or oxaloacetate *via* phosphoenolpyruvate carboxylase; acetyl CoA can condense with oxaloacetate *via* citrate synthase, forming citrate, or with CO_2 *via* acetyl CoA carboxylase, forming malonyl CoA, in the first step of fatty acid biosynthesis.

These characteristics may be physiologically advantageous for the following reasons. Firstly, the extent of a reversible reaction will be to some extent dictated by the mass-action equilibrium between reactants and products; changes (caused by modulation) in the enzyme's affinity for the reactants will therefore be only partly effective in changing the rate of their conversion into products. However, where the reaction is irreversible, the flow through it is entirely determined by the catalytic activity of the enzyme; hence regulation is more efficient in the latter case. Secondly, suppose for illustration that the production of E in the sequence

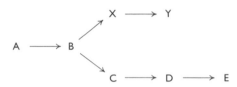

is to be regulated through inhibition of one of the enzymes by E. Clearly, it would be undesirable to inhibit the enzyme catalysing A → B, for then flow through the branch B → X → Y would be curtailed unnecessarily, and perhaps deleteriously. It would also be inefficient to modulate one of the enzymes after C, for then C or D would accumulate; not only would this waste B, which could otherwise pass into the B → Y branch, but also, since the inhibition can be partially overcome by increase in substrate concentration, when sufficient C had accumulated the reaction B → C would again proceed. The most effective point of control is, therefore, B → C, immediately after the branch point; this leaves the pathway from B to Y undisturbed, and any B that accumulates can be diverted into this branch.

Control of synthesis of the enzymes of glycolysis and the tricarboxylic acid cycle and the associated anaplerotic enzymes seems to be of less importance than control of enzyme activity. Most of the glycolytic pathway enzymes, in particular, seem to be constitutive in the enterobacteria although the AMP-activated pyruvate kinase is induced during growth on succinate, and the components of pyruvate dehydrogenase (see below, p. 470) are induced during growth on pyruvate. The tricarboxylic acid cycle enzymes, however, both in this group and in some *Bacillus* species, show some induction and repression under varying growth conditions; for instance, the enterobacterial enzymes are induced during growth on nutrient broth, when the cycle is essential for energy production, and also following a shift from anaerobic to aerobic conditions.

Regulation of the glycolytic pathway is better understood than are other pathways of glucose catabolism. In the pentose phosphate cycle in *Escherichia coli* (determining the distribution of glucose 6-phosphate as between the glycolytic and pentose phosphate sequences) the most significant control may be the inhibition of glucose 6-phosphate dehydrogenase by $NADH_2$; this is particularly interesting in view of a suggestion that the purpose of the considerable flow through this pathway in *Escherichia coli* is to generate $NADPH_2$ for biosynthetic purposes. In the organisms in which they are found, the enzymes of the Entner–Doudoroff and phosphoketolase pathways, on the other hand, are highly inducible; in addition, the glucose 6-phosphate dehydrogenase of *Pseudomonas aeruginosa* is (unlike the enzyme in *Escherichia coli*) inhibited by ATP.

Control of the peripheral pathways

Control of the enzymes of a peripheral sequence is mainly at the level of enzyme synthesis, although occasionally the first enzyme in the sequence may be inhibited by one of the compounds that indicate the energy state of the cell. In most cases these enzymes are inducible, most frequently by the carbon source itself but sometimes by the first intermediate in the sequence, and are often subject to catabolite repression (see p. 452). Although glucose is the commonest carbon source to produce the effect, catabolite repression of enzymes of glucose dissimilation by other carbon sources can occur; for instance, citrate represses the enzymes of the Entner–Doudoroff pathway in *Pseudomonas aeruginosa*.

Inducibility of synthesis of enzymes specific for the degradation of carbon sources that are only occasionally or rarely present in the cell's environment, allows the cell to produce the enzymes only when they can be useful. Catabolite repression also ensures that even if such a carbon source is present, it will not be metabolized if a more efficiently utilizable alternative is available. The advantages to the cell of these forms of control are therefore clear.

Sugar-degrading systems that have been studied in great detail in *Escherichia coli* include those for lactose, galactose, maltose and L-arabinose. These have in common that control is entirely at the level of enzyme synthesis; the inducer is the carbon source itself or a related compound rather than one of its metabolites; the enzymes are all induced or repressed together (*cf.* the *Pseudomonas fluorescens* systems for degrading certain aromatic compounds, described below); and catabolite repression is particularly strongly exerted by glucose, glucose 6-phosphate or gluconate.

Other well-studied systems showing the characteristics of control of enzyme synthesis rather than activity, induction by the carbon source itself, and marked susceptibility to catabolite repression, are the D-serine deaminase of *Escherichia coli*, which breaks down D-serine to NH_4^+ and pyruvate; the tryptophanase, also of *Escherichia coli*, which breaks down L-tryptophan to NH_4^+, indole and pyruvate; and the amidase of *Pseudomonas aeruginosa*, which splits amides, e.g. acetamide to acetate and NH_4^+, and is induced by them. The first two enzymes are repressed by glucose, and also, interestingly, by pyruvate, which can be regarded as the intermediate at which the degradation of both amino acids feeds into the central pathways; this strongly suggests end-product repression, but in view of the mechanism described for catabolite repression of the lactose proteins, this finding must be interpreted with caution. The amidase of *Pseudomonas aeruginosa* is repressed by succinate and propionate, among other compounds.

An interesting case which does not fit into our generalization, that the enzymes induced when an unfamiliar carbon source is utilized are those in a specific pathway for its catabolism leading to the common central sequences, is the utilization of acetate by *Escherichia coli*. Here, the enzymes that are induced are those of the glyoxylate cycle (see Chapter 3, p. 190), which serve an anaplerotic rather than only a catabolic function. They replenish the tricarboxylic acid cycle as intermediates are drawn off for biosyntheses. The apparent inducing effect of acetate probably depends on an antagonism of repression by phosphoenolpyruvate or some metabolite immediately derived from it.

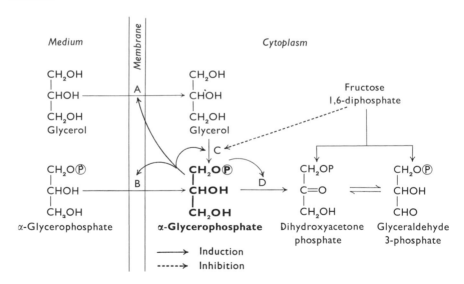

Figure 16. Pathway of glycerol utilization in *Escherichia coli* K-12.
  A 'Facilitator' (protein controlling the facilitated diffusion of glycerol into the cell, a process which is not energy-dependent).
  B L-α-glycerophosphate permease (controlling energy-dependent active transport of L-α-glycerophosphate into the cell).
  C Glycerol kinase.
  D L-α-glycerophosphate dehydrogenases (two enzymes for aerobic and anaerobic dissimilation of glycerol, respectively).

We come now to systems in which the inducer is not the carbon source itself but one of its metabolite derivatives, usually the product of the first reaction in the specific degradative pathway. *Product induction* of this kind can often be demonstrated by obtaining mutants which lack the enzyme catalysing the first reaction. In these mutants, the remaining enzymes controlled by the operon fail to be induced by the carbon source, but can be induced by the first metabolic product.

One such system in *Escherichia coli* is that for glycerol utilization. The pathway is shown in Figure 16. The inducer of the five proteins, the 'facilitator', the permease, kinase and dehydrogenases, is L-α-glycerophosphate. This is

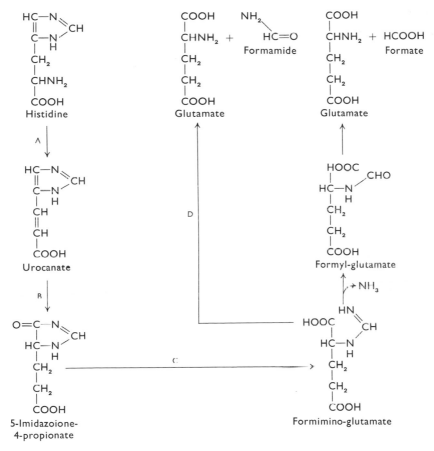

Figure 17. Pathways of histidine breakdown in *Aerobacter aerogenes* and *Pseudomonas aeruginosa*. The former yields glutamate+formamide, the latter glutamate+formate.

    A   Histidase
    B   Urocanase
    C   Imidazolone propionate hydrolase
    D   Formylglutamate transformylase

shown by the failure of glycerol itself to produce the permease and dehydro-genases in kinaseless mutants.

Product induction functions *in vivo* because the basal uninduced level of the first enzyme is appreciable; hence the presence of the carbon source in the medium results in sufficient intracellular accumulation of the first intermediate for induction of the pathway.

L-Histidine catabolism has been studied in the enterobacteria *Salmonella typhimurium* and *Aerobacter aerogenes*, in *Bacillus subtilis*, and in *Pseudomonas aeruginosa*. In all of these the pathway is inducible and subject to catabolite repression. The slightly differing pathways in *Aerobacter aerogenes* and *Pseudomonas aeruginosa* are shown in Figure 17. In these two organisms, urocanate rather than histidine is the inducer, histidine failing to induce the remaining enzymes in histidase-less mutants. In *Bacillus subtilis*, on the other hand, histidine is the inducer. In *Pseudomonas aeruginosa* the system is also interesting in showing control at the level of enzyme activity (these are commoner in amino acid-degrading pathways than in those for catabolism of sugars or polyhydric alcohols); the histidase is inhibited by pyrophosphate, accumulation of which indicates a high supply of energy within the cell, and this inhibition is counteracted by AMP and GDP.

Sequential induction

In the systems we have described so far, the enzymes of the specific catabolic pathways are all induced or repressed *en bloc* under conditions of induction or repression. This may be because the specific sequences have all been short, only three or four steps at most intervening between the carbon source in the medium and an intermediate in one of the central pathways within the cell. The pseudo-monads, with their well-known ability to degrade a wide variety of unlikely-looking organic materials, sometimes need many more steps to convert one of these into an intermediary metabolite. It is then often found that the enzymes of the degradative sequence are not induced together but in separate groups, the final product of one group inducing the enzymes of the following group, this effect being called *sequential induction*. The product may also repress the enzymes involved in its formation. An excellent example is provided by the system for L-mandelate breakdown in *Pseudomonas fluorescens* (Figure 18). The first seven enzymes (involved in degradation to acetate + succinate) are all inducible, but fall into three groups. Group 1 (enzymes A–C) is induced by mandelate; this is shown to be the true inducer by its ability to induce enzymes B and C in mutants lacking enzyme A. These enzymes are repressed by benzoate, the end-product of the group; by catechol, the end-product of group 2; and by succinate or acetate, which can be regarded as the end products of the entire sequence (β-oxoadipate and some of the other later intermediates could not be tested since they are not taken up by the cells). Group 2 consists only of enzyme

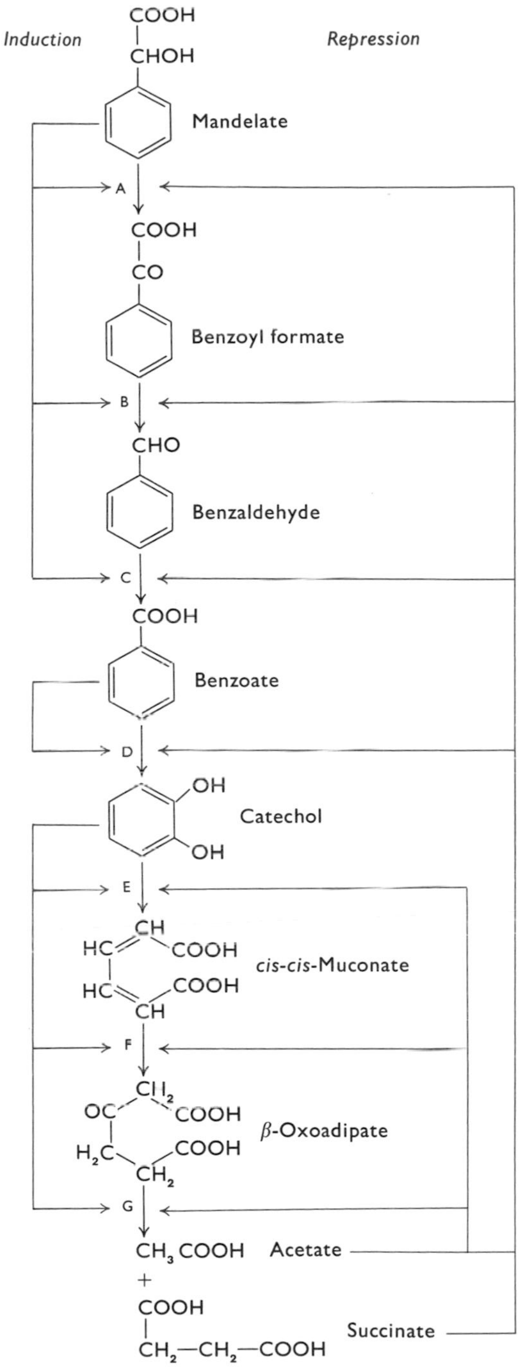

Figure 18. Pathway of mandelate breakdown in *Pseudomonas fluorescens*.

A   L-mandelate dehydrogenase
B   Benzoylformate carboxylase
C   Benzaldehyde dehydrogenase
D   Benzoate oxidase
E   Pyrocatechase
F   (4 enzymes)
G   (2 enzymes)

D; this is induced by benzoate (but not by mandelate in mutants lacking enzyme A), and is repressed by catechol and by succinate or acetate. Finally, group 3 consists of enzymes E, F and G; these are induced by catechol, and are repressed by succinate or acetate. The advantage of this system results from the fact that some of the degradation products of mandelate either occur in nature or are common to other catabolic pathways; for instance, benzoate and catechol may be encountered by the organism in its natural environment, and the latter is also an intermediate in its degradation of tryptophan. Hence if benzoate or catechol are in good supply the enzymes of mandelate catabolism will not be induced unnecessarily even if mandelate is also present.

Inorganic nitrogen metabolism
Pathways involved in assimilation of nitrogen from purely inorganic compounds into intermediary metabolites also show regulatory effects. In most bacteria, the important reactions for incorporation of NH_4^+ are its reaction with α-oxoglutarate to give glutamate, catalysed by the $NADPH_2$-dependent glutamate dehydrogenase, and its reaction with glutamate to give glutamine, requiring ATP and catalysed by glutamine synthase. In *Escherichia coli* the former enzyme is repressed when the nitrogen source is nutrient broth or glutamate rather than NH_4^+; the latter has been reported to be repressed in nutrient broth or NH_4^+-grown cultures and derepressed in glutamate-grown cultures. We may also mention that in two nitrogen-fixing organisms, *Azotobacter vinelandii* and *Clostridium pasteurianum*, the nitrogen fixation enzymes are repressed by NH_4^+. In the former system, nitrate (which the cells can also reduce to NH_4^+) inhibits, but does not repress, the nitrogenase. The latter system is inhibited by ADP, which is perhaps understandable in view of the large ATP requirement of the process.

Anaerobic respiration
If we take Class I reactions to include those which yield energy only as well as those that yield both energy and carbon skeletons for biosynthesis we should mention the control of the use of inorganic ions, such as nitrate or thiosulphate, as terminal electron acceptors for respiration, rather than O_2. (This is sometimes called *anaerobic respiration*.) As with the utilization of unusual carbon sources, control is primarily at the level of enzyme synthesis. For instance, the nitrate reductases of *Escherichia coli* and *Proteus mirabilis* are induced by nitrate and repressed by O_2, while the thiosulphate reductase of *Proteus mirabilis* is induced by thiosulphate and repressed by O_2 or nitrate. Repression of the latter enzyme is interesting because if an induced culture is exposed to O_2 or nitrate, not only is further synthesis of thiosulphate reductase repressed, but the enzyme present in the cells is rapidly inactivated (by a process presumably requiring protein

synthesis, since the inactivation is blocked by chloramphenicol). This *inactivation repression* effect is very common in eukaryotic microorganisms, but much less so among bacteria. It is not certain (in either the bacterial or eukaryotic cases) whether the inactivation involves inhibition of the enzyme or its selective degradation. If the latter is involved, then this effect exemplifies regulation by the control of breakdown of specific enzymes, which, as pointed out earlier, differs fundamentally from controls at the level of either enzyme activity or of enzyme synthesis.

THE CONTROL OF CLASS II REACTIONS

The control of synthesis of purine and pyrimidine nucleotides, and of those amino acids where this is regulated, is almost invariably by feedback inhibition and end-product repression, separately or together, as described in Chapter 4. The controls can be understood straightforwardly as designed to prevent wasteful flow through a pathway and the unnecessary synthesis of its enzymes when the end-product is in good supply. Feedback inhibition generally operates on enzymes coming immediately after metabolic branch points, and repression generally affects the enzymes leading from the branch point to the end-product. Reasons why inhibition should occur at this position have been mentioned above, but the examples among Class II reactions are particularly clear-cut. For instance, the first reactions in the unbranched pathways leading from common intermediary metabolites to pyrimidine nucleotides or to arginine (in *Escherichia coli* K-12) or histidine (in *Salmonella typhimurium* LT-2), are subject to end-product inhibition, and the enzymes catalysing this first step and all the other steps in the sequences are repressed by the end-product.

Branched pathways

In most Class II systems, however, complications are introduced through branching of the pathways. Consider a pathway with a single branch (Figure 19). One finds typically (at least in the enterobacteria) that end-products E and G each partially inhibit and/or repress enzyme 1; that each *may* also partially repress enzyme 2; and that in addition, E inhibits enzyme 3 and represses enzymes 3 and 4, while G may inhibit enzyme 5 and may repress enzymes 5 and 6. In other words, the end-products *in combination* control the shared part of the pathways. *Separately* they control their individual branches much as described above for unbranched pathways. Control of enzyme 1 can be *concerted, cooperative* or *cumulative*. Alternatively, the organism has two isozymes catalysing the A → B reaction, one of which is controlled by E, the other by G. Repression of enzyme 2, where it occurs, is usually concerted or cooperative. Examples of all these types have been given in Chapter 4. Another form of regulation, *sequential* feedback inhibition, was first reported for the aromatic amino acid system in *Bacillus subtilis* (Chapter 4, p. 209). In our scheme, it would involve E inhibiting enzyme 3, G inhibiting enzyme 5. Consequently C would accumulate thus inhibiting enzyme 1.

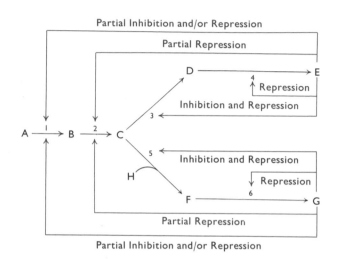

Figure 19. Branched pathways. Schematic diagram of branched pathways, and commonly found inhibition and repression patterns of control (see text).

Finally, *compensatory* antagonism of feedback inhibition can be exemplified by the control of carbamoyl phosphate synthase in *Escherichia coli* (Chapter 4, p. 241). Its product, carbamoyl phosphate, is utilized in both arginine and pyrimidine synthesis. UMP inhibits the enzyme, but ornithine, which combines with carbamoyl phosphate to yield citrulline, antagonizes this inhibition. In the schematic diagram (Figure 19) this represents the inhibition of enzyme 1 by E and its antagonism by H.

These controls in the shared portion of a branched pathway clearly avoid its complete closure where only one end-product is present (which would lead to starvation for the other end-product) and still retain some economy in its operation.

Biosynthesis of aromatic amino acids
Details of controls in the amino acid biosynthetic pathways of a number of bacterial species have been given in Chapter 4; some of the comparisons between different regulation patterns for a particular sequence justify Stadtman's dictum 'the regulatory function is more constant than the regulatory mech-

anism'. We would like to examine here one example a little further, namely the aromatic amino acid sequence the control of which is summed up in Chapter 4, p. 208. The findings for the pathway up to chorismate in *Bacillus subtilis* and the closely related *Bacillus licheniformis* are shown in Figure 20(a). Apart from the sequential feedback inhibition of phospho-2-keto-3-deoxyheptonate aldolase (A) previously mentioned, it is seen that *Bacillus subtilis* possesses two separate shikimate kinases (E) and two chorismate mutases (H), while *Bacillus licheniformis*, though it also possesses two H enzymes, has only one enzyme E. One of the E isozymes of *Bacillus subtilis* is inhibited by chorismate (7) or prephenate (8), like enzyme A; however, the single enzyme of *Bacillus licheniformis* is inhibited only by chorismate. In *Bacillus subtilis*, A, E and H activities are repressed by tyrosine; in *Bacillus licheniformis*, tyrosine represses only A and H activities, enzyme E being constitutive. Furthermore, aggregation patterns differ between the species; they will be described on p. 471.

It is worth noting that the pattern of control of A activity is highly distinctive for the various groups of bacteria. The sequential feedback inhibition of A activity is found in *Bacillus* species and also among staphylococci. Inhibition by a single amino acid, tryptophan, is observed in *Streptomyces* while in most *Pseudomonas* species the inhibitor is tyrosine. In these organisms there may be only one enzyme A with one inhibitor; it is, however, possible that isozymes exist but are so labile as to lose activity completely in extracts, or are totally repressed under the growth conditions employed. It would be hard otherwise to explain why, for instance, tyrosine is not a growth inhibitor for *Pseudomonas aeruginosa*, the A activity of which it reduces *in vitro* by 96%.

Regulation in the terminal pathway for tryptophan is shown for three non-enterobacterial species in Figure 20(b) (compare Chapter 4, Figure 3). In *Bacillus subtilis* the pattern is like that in enterobacteria, in that the first enzyme (anthranilate synthetase, enzyme N) is inhibited, and it and the other enzymes are repressed by tryptophan. In *Pseudomonas putida* and also in the Gram-negative rod *Chromobacterium violaceum*, enzyme N is inhibited by tryptophan, but controls of enzyme synthesis are very different (Figure 20). It is of interest that in *Pseudomonas* enzyme P is constitutive while enzyme R (the tryptophan synthase complex) is induced by its substrate, indoleglycerol phosphate. In *Chromobacterium violaceum* all the enzymes seem to be constitutive. Again, we shall discuss aggregation patterns later, on p. 469.

Finally, it may be mentioned that apparently considerable physiological differences in repression pattern may hide resemblances at the genetic level. An example is afforded by the arginine biosynthetic enzymes of *Escherichia coli* strains K-12 and B, which are repressed by arginine in the former and are constitutive or induced by arginine in the latter. The differences in behaviour are governed entirely by differences at a single genetic locus, *argR*, that determines the structure of the apo-repressor.

Control of vitamin and cofactor biosynthetic pathways

We know less about the control of these pathways than of amino acid pathways, partly because the vitamins and cofactors and the enzymes involved in their synthesis seem in most cases to be present in very low amount, making them harder to study. Like other Class II pathways, they are usually regulated by end-product inhibition and repression. For instance, five enzymes of biotin synthesis in *Escherichia coli* are repressed by biotin; while the first enzyme in haem synthesis in *Staphylococcus aureus*, δ-aminolaevulinate synthetase, is both inhibited and repressed by haemin.

Another interesting example is the control of NAD synthesis in *Bacillus subtilis* (Figure 21). The cofactor can be made from nicotinic acid if this is supplied exogenously, but otherwise is produced via quinolinate. In the former case, enzyme 2 is repressed; however, even in the latter case enzyme 7 is present, since it is required in the sequence (involving enzymes 5 and 6 as well) by which NAD is continually degraded and reconstituted.

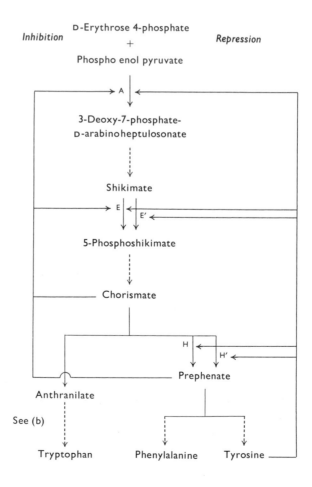

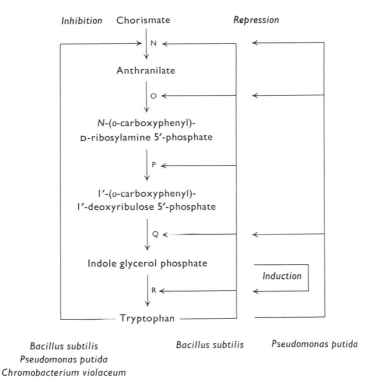

Figure 20. Control of the aromatic acid biosynthetic pathway.

(a) Tyrosine and phenylalanine pathway in *Bacillus subtilis*. Control in *Bacillus licheniformis* is similar but although it also possesses two H enzymes, it has only one E enzyme which is constitutive and is inhibited only by chorismate.

A	Phospho-2-keto-3-deoxyheptonate aldolase
E and E'	Shikimate kinase
H and H'	Chorismate mutase

(b) Tryptophan terminal pathway in *Bacillus subtilis*, *Pseudomonas putida* and *Chromobacterium violaceum*. In the latter tryptophan does not repress.

N	Anthranilate synthase
O	Carboxyphenylriboxylamine pyrophosphorylase
P	Carboxyphenylriboxylamine isomerase
Q	Indoleglycerol phosphate synthase
R	Tryptophan synthase

SPATIAL CONTROL OF CLASS I AND CLASS II
REACTIONS: ENZYME AGGREGATES

We have so far discussed enzymes as if these were randomly distributed with respect to one another within the 'soluble' components of the cell. However, even apart from the association of certain enzymes with the cell membrane (Chapter 2, p. 117) it is often found that enzymes of the same pathway *aggregate* to form complexes (held together by non-covalent bonds) that cohere even in cell extracts and through many stages of purification. Some examples are: among Class I reactions, the pyruvate dehydrogenase complex of *Escherichia coli*; and among Class II reactions, aggregates combining various activities in aromatic amino acid pathways, and the aspartokinase-homoserine dehydrogenase complex, both instances having been studied in several organisms. We shall discuss these examples, and then consider what their physiological significance might be.

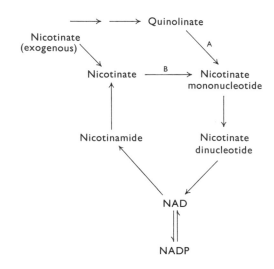

Figure 21. Pathways of synthesis and breakdown of nicotinic acid cofactors in *Bacillus subtilis*.

The pyruvate dehydrogenase complex of *Escherichia coli* can be isolated intact from cell extracts even following ultracentrifucation or protamine sulphate fractionation, and has a molecular weight of about four million. It catalyses the reactions shown in Figure 22. Dissociation into components with the individual activities E1 (decarboxylase), E2 (transacetylase) and E3 (flavoprotein), can be achieved partially by ammonium sulphate fractionation, and completely by chromatographic separation in the presence of 4 M urea or at pH 9·5—treatments that do not break peptide bonds; each of these components is in turn composed of subunits. The components can be made to reconstitute the active complex.

The complex and the component enzymes can be seen in electron micrographs as regular arrangements of subunits. This information and the biochemical data suggest that a single molecule of the complex contains 24 transacetylase polypeptide chains, 24 decarboxylase chains and 24 flavoprotein chains, the molecular weights of these chains being 40,000 90,000 and 55,000 respectively. The transacetylase forms eight aggregates, each of three chains, located at the vertices of a cube. Along the edges of this cube are arranged the decarboxylase and flavoprotein chains, one of each binding to one transacetylase chain.

(1)	Pyruvate + TPP-E$_1$	$\longrightarrow$	Hydroxyethyl-TPP-E$_1$ + CO$_2$
(2)	Hydroxyethyl-TPP-E$_1$ + Lipoyl-E$_2$	$\longrightarrow$	S-Acetyl dihydrolipoyl-E$_2$ + TPP-E$_1$
(3)	S-Acetyl dihydrolipoyl-E$_2$ + CoA-SH	$\longrightarrow$	Acetyl $\sim$ S-CoA + Dihydrolipoyl-E$_2$
(4)	Dihydrolipoyl-E$_2$ + FAD-E$_3$	$\longrightarrow$	Lipoyl-E$_2$ + FADH$_2$-E$_3$
(5)	FADH$_2$-E$_3$ + NAD	$\longrightarrow$	FAD-E$_3$ + NADH$_2$

Pyruvate + CoA-SH + NAD $\longrightarrow$ Acetyl $\sim$ SCoA + NADH$_2$ + CO$_2$

Figure 22. Reactions catalysed by the pyruvate dehydrogenase complex.

E1   Decarboxylase
E2   Lipoic reductase transacetylase
E3   Flavoprotein dihydrolipoic dehydrogenase
TPP   Thiamine pyrophosphate

We can guess that the speed and efficiency of reactions 2 and 4, which involve pairs of enzymes and enzyme-bound intermediates, will be much greater if the enzymes are juxtaposed. It is interesting that the decarboxylase and transacetylase are synthesized coordinately in equimolecular ratio, the proportions in which they occur in the complex. It would clearly be wasteful for one component to be made in excess and hence to remain uncomplexed.

The aromatic amino acid pathways that we have described display interesting differences in patterns of enzyme aggregation. The aggregation of chorismate mutase with prephenate dehydrogenase or prephenate dehydratase, of anthranilate phosphoribosyltransferase with the *trpE* gene product in anthranilate synthase, and of the A and B components of tryptophan synthase, in enterobacteria have already been described (Chapter 4, p. 210). In these examples, the aggregated enzymes catalyse successive steps in a pathway and the advantage, again, presumably lies in the ease with which a molecule produced in the first step becomes available as substrate for the second step, never leaving the enzyme aggregate (though proof of this is not often available). It may be noted that the catalysis of reactions P and Q (see Figure 1b, p. 207) by a single enzyme, although the exact opposite of enzyme aggregation, serves essentially the same purpose, namely the more efficient use of the product of a reaction by allowing it to transfer to the substrate-binding site for the ensuing reaction without diffusing away. We shall now consider some examples, harder to explain, where the aggregated enzymes do not catalyse successive steps in the pathway.

In *Bacillus subtilis*, DAHP synthase, one of the shikimate kinase isozymes, and one of the chorismate mutase isozymes form an aggregate; remarkably, there is no evidence for aggregation of any of these enzymes in *Bacillus licheniformis*. A similar association of DAHP synthetase and chorismate mutase has been described for *Staphylococcus epidermidis* and again both enzymes are inhibited by chorismate or prephenate.

In the tryptophan terminal sequence in *Bacillus subtilis*, *Pseudomonas putida* and *Chromobacterium violaceum*, there is no evidence for aggregation of anthranilate synthase and anthranilate phosphoribosyltransferase, and steps P and Q are catalysed by separate enzymes. In *Bacillus subtilis*, although there is a little evidence for an aggregate *in vivo* of enzymes O, P and Q, no aggregate can be demonstrated in cell extracts. In *Bacillus subtilis* and *Chromobacterium violaceum*, there are suggestions of aggregation of the two components of tryptophan synthetase as in the enterobacteria, but in *Pseudomonas putida* even this seems to be lacking.

The final examples that we shall give, the aspartokinase and homoserine dehydrogenase isozymes of *Escherichia coli*, resemble the enterobacterial enzyme that catalyses steps P and Q in the tryptophan terminal sequence in being multifunctional enzymes rather than aggregates of separate enzymes. In *Escherichia coli* K-12, the threonine-inhibited aspartokinase I and homoserine dehydrogenase I activities are associated in a single protein consisting of six identical polypeptide subunits, while the methionine-repressed aspartokinase II and homoserine dehydrogenase II activities are found in another protein consisting of four, probably identical, polypeptide subunits.

In such cases it is more difficult to point out the advantage to the cell of a single protein with more than one function, whether this takes the form of an enzyme aggregate or of a single multifunctional species. However, in the *Bacillus subtilis* case, an interesting suggestion may be put forward, namely that the aggregation of chorismate mutase with DAHP synthetase and shikimate kinase may serve to associate the latter two enzymes with a protein which has binding sites for chorismate (its substrate) and prephenate (its product). This might have facilitated the evolution of their apparently advantageous sensitivity to feedback inhibition by chorismate or prephenate. The unexpected susceptibility of the shikimate kinase to inhibition may be of selective value in that *Bacillus subtilis* is permeable to shikimate; shikimate kinase is the first enzyme in the further conversion of this compound and hence might be expected to need a mechanism for modulation. Interestingly, *Bacillus licheniformis* is impermeable to shikimate; this may explain why its shikimate kinase is not subject to control.

Many enzyme aggregates may exist *in vivo* of which we have no knowledge; the arguments for the advantages of association of enzymes catalysing successive steps in a pathway imply that it would be advantageous for all reactions in the pathway to be in juxtaposition. Just possibly this is actually what occurs in the living cell. An indication that this might be so is given by the observation that certain membrane-containing fractions from broken *Salmonella typhimurium* cells can catalyse the conversion of threonine into isoleucine, in other words, all the enzymes of the isoleucine-valine sequence are associated with the cell membrane.

Control of miscellaneous reactions involving small molecules

There are a number of reactions involving small molecules which cannot be said to belong to Class I or II but which are regulated by induction or re-pression. An example is the alkaline phosphatase, which hydrolyses phosphate esters to the corresponding hydroxyl compound and inorganic phosphate. In various species the latter represses enzyme synthesis. Two special classes of reactions are those involved in the neutralization of toxic materials in the medium, and those involved in the production of *secondary metabolites* (e.g. streptomycin by cultures of *Streptomyces griseus*) or of *bioderivatives* (e.g. the *bioconversion* of sorbitol to sorbose by *Acetobacter* species); they are often of industrial significance.

Probably the most important detoxification reaction is the splitting of the β-lactam ring of pencillin by the enzyme penicillinase (see Figure 23). This enzyme is produced by a variety of bacteria but has been most extensively studied in *Staphylococcus aureus*, *Bacillus licheniformis* and *Bacillus cereus*, and *Escherichia coli*. In the Gram-positive bacteria, the enzyme is induced by the

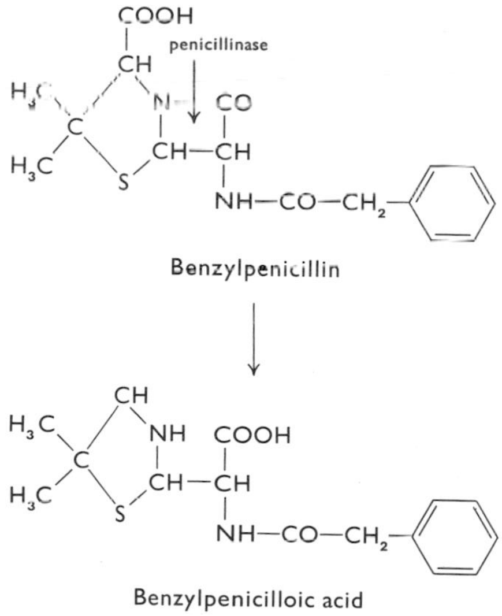

Figure 23. Reaction catalysed by penicillinase, illustrated with reference to benzylpenicillin.

presence of penicillin and is an exoenzyme, i.e. it is liberated (though to a varying extent) into the culture medium. In the Gram-negatives it is constitutive and is cell-bound, i.e. it is only found in the medium following damage to the cytoplasmic membrane.

Hardly anything is known about the control of production of secondary metabolites or bioderivatives by bacterial cultures, although probably conventional types of regulation occur. In most cases, synthesis begins only after the most rapid phase of growth ceases. It has been shown that the activity of some of the enzymes involved in streptomycin synthesis increases at this time, and this is consistent with a release from catabolite repression.

GENERAL COMMENTS OF THE CONTROL OF CLASS I AND CLASS II REACTIONS

We have given, in the last few sections, some idea of the variety of ways in which bacteria regulate the flow through the main catabolic and anabolic pathways, by ringing the changes on the simple possibilities of altering enzyme activity or enzyme synthesis. We have attempted to explain the existence of these elaborate control mechanisms in terms of their selective advantage in permitting the most economical exploitation of nutrients in the environment and the most rapid and efficient adaptation to changes in it. While there are often puzzling variations in patterns of control between closely related species (or even between different strains of the same species) some aspects of the system for regulation of a particular pathway are common to most or all members of a given group. Regulatory characteristics have thus taken their place among other bacterial properties used for taxonomic purposes.

The value of the control mechanisms to the organism is shown by the fact that a mutant of *Bacillus subtilis* in which the tryptophan enzymes are no longer repressible by tryptophan is at a selective disadvantage when grown in mixed culture with the wild-type. Similarly a mutant of *Escherichia coli* in which the proline pathway was no longer inhibited by proline was gradually overgrown by the wild-type.

APPLICATION OF THE OPERON MODEL TO ENTERO-BACTERIAL SYSTEMS OTHER THAN THE *ESCHERICHIA COLI* LACTOSE PROTEINS

Class I systems

No sequence, even in the enterobacteria, has been explored with the same thoroughness as the *Escherichia coli* lactose system. However, enough is known (a)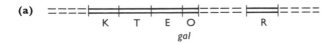

(b)

(c)

(d)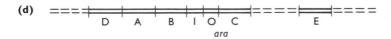

Figure 24. Arrangement in *Escherichia coli* of genes involved in some Class I pathways. Genes of unrelated function are shown by broken lines.

(a) The galactose system in *Escherichia coli* K-12.

- K   Galactokinase
- T   Galactose-1-phosphate uridyl transferase
- E   Uridine diphosphogalactose-4-epimerase
- O   Operator
- R   Regulator gene

(b) The glycerol system in *Escherichia coli* K-12. The order of *glpK* and *glpF* is not known.

- T   L-α-glycerophosphate permease
- A   L-α-glycerophosphate dehydrogenase (anaerobic)
- R   Regulator gene
- D   L-α-glycerophosphate dehydrogenase (aerobic)
- K   Glycerol kinase
- F   'Facilitator' of facilitated diffusion of glycerol into the cell

(c) The D-serine system in *Escherichia coli* K-12.

- C   Regulator gene
- A   D-serine deaminase

(d) The L-arabinose system in *Escherichia coli* B.

- D   L-ribulose 5-phosphate 4-epimerase
- A   L-arabinose isomerase
- B   L-ribulokinase
- I   Initiator (a region which may represent a binding site for the activator—see below, p. 477)
- O   Operator
- C   Regulator gene
- E   L-arabinose permease

to suggest many basic similarities to, and some differences in detail from, this particular case. We shall consider in particular four examples, namely the inducible systems for breakdown of galactose, glycerol, D-serine and L-arabinose in *Escherichia coli* K-12. Gene-enzyme relationships are shown in Figure 24. The enzymes are induced by the first compound, except for the glycerol system which is induced by the product glycerophosphate (see above). In all instances, constitutive mutants carrying regulator gene mutations can be obtained, in which the enzymes are synthesized at high rates irrespective of whether inducer is present or not.

The galactose system (comprising three enzymes) is very similar to the lactose system. The enzyme structural genes are clustered and are transcribed to a polycistronic messenger; the enzymes are synthesized coordinately; the regulator gene produces a repressor which subjects the structural genes to negative control; o^c-type mutations can be obtained that show *cis*-dominance; and DNA–RNA hybridization experiments show control to be at the level of transcription. The only difference from the lactose example is that the galactose regulator gene is not linked to the structural gene cluster; however, there is no necessity in the operon model for it to be so linked, since the effect of the regulator gene is mediated by the protein repressor which diffuses freely throughout the cell.

With the glycerol system we find a quite different genetic arrangement. The five structural genes are not completely linked; genes T and A are adjacent, as are K and F, but these pairs and D are scattered in widely separate regions of the chromosome. The regulator gene (which again produces a repressor) is, however, contiguous with the structural gene D. In such a situation, each separate structural gene or adjacent pair must have its own operator and promoter. It is quite possible for these operators to differ in affinity for repressor, and so coordinate control is not inevitable. In fact, in the glycerol system only those proteins whose structural genes are contiguous are synthesized coordinately. A collection of genes some or all of which are scattered, but which are regulated in parallel by the same compounds, is called a *regulon*. Each separate structural gene or pair of genes is presumably transcribed on to a separate messenger; this is indicated in the case of *glpK* and F by the existence of a mutant in which the increased levels of both proteins suggest that it carries a mutation in the common promoter.

The D-serine system consists solely of the inducible enzyme D-serine deaminase, with a conventional (repressor-producing, negatively-controlling) regulator gene. Its main interest lies in the fact that some regulator gene mutations leading to constitutivity also confer loss of sensitivity to catabolite repression. It may be, therefore, that the latter's mechanism is in this instance quite different from that of the *lac* system—although here again cyclic 3',5'-AMP antagonizes repression by glucose.

Finally, with the L-arabinose system we come to an instance where the function of the regulator gene is radically different. There are four proteins, three enzymes and a permease; the structural genes for the former are clustered, and the regulator gene, called the C gene, is adjacent to them. The permease gene, however, is unlinked to these. It is possible to isolate regulator gene mutants that differ from the wild-type (denoted C^+) either in having constitutive production of these proteins (C^c) or in having lost the inducibility of all of them (C^-). A study of the dominance relationships of these in partial diploids enables one to make some guess as to the function of the regulator gene. It is found that C^+ is dominant to both C^c and C^-, but C^c is dominant to C^-. The most plausible interpretation of these and other results is that the C gene product is a repressor in the absence of inducer, but that inducer combines with it to form an *activator* which is necessary for gene expression. This system therefore provides an example of *positive control*. The activator probably binds to a region of DNA next to the promoter and thereby facilitates attachment of the RNA polymerase. This mechanism has already been suggested for cyclic 3′,5′-AMP combined with its binding protein in the lactose system. In the arabinose system catabolite repression is similarly mediated *via* cyclic 3′,5′-AMP.

Class II systems

We have already described how repressible systems (which predominate among Class II sequences) can easily be accommodated within the operon model simply on the assumption that the effective repressor is a combination of regulator gene product with repressing metabolite rather than the regulator gene product alone. All other aspects of control enumerated previously could apply unchanged to this class. We shall examine the tryptophan systems of *Escherichia coli* and *Salmonella typhimurium*, the isoleucine-valine, pyrimidine, biotin and arginine systems of *Escherichia coli* and the histidine system of *Salmonella typhimurium*. Gene-enzyme relationships for the tryptophan, isoleucine-valine, and histidine systems are shown in Chapter 4, Figures 3, 10 and 12 (pp. 210, 228 and 233) respectively, and for the remainder in Figure 25 (p. 479).

Tryptophan

Gene-enzyme relationships in the tryptophan sequence have been referred to before (Chapter 4, p. 210). The control mechanism appears to be almost identical in the two species. The regulator gene ($trpR$) is defined by mutations ($trpR^-$) leading to constitutive (non-repressible) synthesis of the enzymes. The $trpR^-$ mutants are selected by a common procedure for isolating constitutive mutants in amino acid biosynthetic sequences, namely as being resistant to inhibitory analogues of the particular amino acid. Such analogues may inhibit by mimicking the amino acid in a variety of ways; in some cases, they may act as false feedback inhibitors of the first enzyme in the sequence, in which case the procedure is more likely to select mutants with altered forms of the enzyme that are no longer sensitive to feedback inhibition. Constitutive mutants are usually resistant because they overproduce the amino acid, which, being then in great excess over the analogue, will tend to exclude the latter from interactions that can involve either of them. The gene *trpR*, mutations in which lead to constitutivity, presumably codes for a repressor mediating negative control, since the wild-type *trpR*$^+$ is dominant to *trpR*$^-$ in partial diploids. The five structural genes form a cluster which is transcribed to a single polycistronic messenger in the direction *trpE→trpA*, as is shown by the occurrence of polar and o^c type mutations, the latter also demonstrating the existence of an operator at the *trpE* end. A thoroughly developed mRNA assay system by DNA–RNA hybridization has facilitated the demonstration that control is at the level of transcription (and has also provided much of our knowledge of the molecular explanation of polarity). However, one group using this system has suggested that some part of control operates at the level of translation.

A point of some interest is that control of these clustered genes is not coordinate, the repressibility of the enzymes encoded by *trpC*, *B* and *A* being less than that of the *trpE* and *D* enzymes (i.e. relatively more of the former are made under conditions of repression). This intriguing observation is explained by evidence suggesting the existence of a second promoter, between *trpD* and *C*, not followed by an operator.

Isoleucine and valine

In the isoleucine-valine system (see also Chapter 4, p. 228), we again have a gene cluster in *Escherichia coli*, but it seems likely that this is transcribed to more than one mRNA molecule. This is suggested by the fact that two kinds of *cis*-dominant o^c mutation have been isolated, one acting only on the *ilvB* gene and the other acting only on *ilvA*, *D* and *E* (the remaining gene, *ilvC*, presumably also has its own operator). This would mean that *ilvB* is transcribed to one monocistronic messenger, *ilvC* to another, and *ilvA*, *D* and *E* to a polycistronic messenger, the transcription of each messenger being under the control of a separate operator and promoter. As might be expected, *ilvA*, *D* and *E* are coordinately controlled, but this group, the *ilvB* enzyme, and the *ilvC* enzyme are not coordinate with one another.

Pyrimidine nucleotides

The pyrimidine pathway of *Escherichia coli* is an instance of mixed clustering and scattering of the structural genes (Figure 25(a)). The interesting point here is that not only are the clustered genes coordinately controlled, but also two of the three scattered genes (the exception being the first enzyme; see above, p. 435). This is of course readily explicable on the supposition that the coordination reflects similar affinity of different operators for the repressor;

however, the fact that coordination does not necessarily imply clustering is rather important.

Biotin

The biotin system in *Escherichia coli* (Figure 25(b)) exemplifies *divergent orientation* of transcription of its gene cluster *bioABFCD*, since *A* is transcribed leftwards and *BFCD* rightwards from a control region between *A* and *B*. It is

(a)

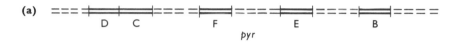

(b)

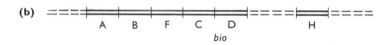

(c)

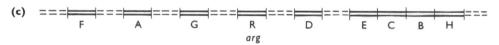

Figure 25. Arrangement in *Escherichia coli* of genes involved in some Class II pathways. Genes of unrelated function are shown by broken lines.

(a) The pyrimidine system.

    D Dihydroorotate dehydrogenase

    C Dihydroorotase

    F Orotidylate decarboxylase

    E Orotidylate pyrophosphorylase

    B Aspartate carbamoyltransferase

(b) The biotin system.

    A 7,8-Diaminopelargonate aminotransferase

    B Enzyme that converts dethiobiotin to biotin

    F 7-oxo-8-aminopelargonate synthase

    C Enzyme mediating unknown step prior to 7-oxo-8-aminopelargonate synthase

    D Dethiobiotin synthase

    H As C

(c) The arginine system.

    F Ornithine carbamoyltransferase

    A *N*-Acetylglutamate synthase

    G Argininosuccinate synthase

    R Regulator

    D Acetylornithine δ-aminotransferase

    E Acetylornithinase

    C *N*-Acetylglutamic γ-semialdehyde dehydrogenase

    B *N*-Acetylglutamate γ-phosphotransferase

    H Argininosuccinase

not yet known whether the promoter and operator sites for *A* and for *BFCD* are independent or whether there is some overlap.

Arginine

The arginine system (Figure 25(c)) also contains a mixture of clustered and scattered genes (it is interesting that some of the scattered, i.e. non-contiguous, genes in fact lie rather close together; there are other instances of this). Three of the clustered genes, *argC*, *B* and *H*, are coordinately controlled but this group is not coordinate with *argE* (with which it is contiguous) and neither the *CBH* group nor *argE* is coordinate with any of the scattered genes. It also seems that *argCBH* is transcribed to a polycistronic message and *argE* to its own mono-cistronic message. It is probable that, as with the biotin system, transcription of *E* and of *CBH* shows divergent orientation, but here with overlap of control sites. The pathway behaves conventionally with regard to the function of the *argR* regulator gene; *argR⁻* mutations, conferring non-repressibility by arginine, are recessive to the wild-type *argR⁺*, demonstrating that again negative control obtains. In this system also, some evidence suggests that regulation is partly at the level of translation. In the enterobacterial *Proteus mirabilis* (which is only distantly related to *Escherichia coli*) the cluster contains the *G* gene, the order being *ECBGH*.

Histidine

The histidine system in *Salmonella typhimurium* in many ways conforms to the simplest operon model (Figure 11, p. 444). The genes are contiguous and are transcribed to a single polycistronic messenger, polar and o^c type mutations showing the direction of transcription as *hisG→E* and also the presence of the operator. Coordinate control is found and this was, in fact, the first case where it was demonstrated. However, constitutive mutants obtained by selection for analogue resistance carry mutations which may map in any of five separate genetic sites. One of these may be a conventional regulator gene, specifying a protein repressor, but of the other four, one is the structural gene for the histidine-activating enzyme (histidyl-transfer RNA synthetase) while the remaining three may be genes coding for various species of tRNA^His. This suggests that histidyl-tRNA complex is somehow involved in repression by histidine of its biosynthetic enzymes. Possibly it rather than histidine acts as co-repressor; however, there are suggestions that regulation is more complicated. Genetic or physiological interference with the sensitivity to feedback inhibition of the first enzyme of the sequence may under certain circumstances affect repressibility of the whole cluster, and it has been found that this enzyme specifically attaches to histidyl-tRNA *in vitro*. We do not understand these results which may point to some form of regulation unlike any so far estab-lished; we may note that on current evidence, involvement of the appropriate amino acyl-tRNA's in control is probable also in the isoleucine-valine system, where the first enzyme of the sequence again is implicated.

Control systems in other prokaryotic organisms

Regulation systems in other bacterial species have not as yet been analysed in the same detail as some of those discussed above but some data are available.

One system in which dominance studies can be readily carried out is that of the pencillinase of *Staphylococcus aureus*, due to the fact that its structural and regulator genes are often carried on extrachromosomal genetic elements or plasmids (see Chapter 7, p. 420), more than one of which can, under certain circumstances, coexist within the same cell. Constitutive mutants, producing penicillinase at high level in the absence of penicillin, result from mutation in a locus i closely linked to the structural gene p. When two of the genetic elements are brought into the same cell, one carrying a mutant i^- and the other the wild-type i^+, an inducible phenotype results, showing that the i gene product exerts a negative control on penicillinase synthesis.

Another point of interest is that mutations are known, mapping between the constitutive i^- mutations and p^- mutations affecting the structural gene, which lead simultaneously to constitutivity and to lowering, to between 0.1% and 50%, of the maximum level of penicillinase produced under inducing conditions. These mutations can be explained by supposing that here the operator and promoter regions overlap, so that a single-step mutation can simultaneously affect binding to the DNA of both RNA polymerase and of the repressor; however, other explanations are possible.

Tryptophan

As regards Class II systems, there is a fair amount of information about the location of genes for enzymes of several amino acid pathways in *Pseudomonas*, *Bacillus* and *Micrococcus spp*. The linkage relationships of the structural genes for the six enzymes of the terminal tryptophan sequence have now been studied in several very different species. In *Bacillus subtilis* there are six clustered genes. It will be remembered that in the enterobacteria there are only five genes, one of the enzymes catalysing two successive steps (Chapter 4, p. 210). Let us denote the genes by the same letters as before, dividing up C into C_1 and C_2 as the structural genes for enzymes P and Q (Chapter 4, Figure 3, p. 210) respectively. The gene order in *Bacillus subtilis* can then be represented EDC_1C_2BA. This is the same as in the enterobacteria apart from the subdivision of C. A similar gene order is probable in *Staphylococcus aureus*. However, in *Micrococcus lysodeikticus*, there appear to be two unlinked clusters, EC_2BA and C_1D. In *Pseudomonas putida* genes E, D and C_2 are clustered, as are A and B, but these clusters and gene C_1 are unlinked to one another. It may be noted that these groupings correspond to the divisions in control of the enzymes.

Histidine

The histidine system has also been examined in several species. In *Staphylococcus aureus*, at least six genes have been found to map in a cluster with the same gene order as in *Salmonella*. In *Bacillus subtilis*, however, the genes are divided into at least two clusters, while in *Pseudomonas aeruginosa* they are scattered among at least five.

Isoleucine and valine

The isoleucine-valine genes have also been studied in these three species. In *Staphylococcus aureus*, they map in a cluster as in the enterobacteria, but in a different order which corresponds to the biosynthetic sequence. In *Bacillus subtilis*, they fall into at least three clusters (two of which are closely linked but apparently not contiguous); while in *Pseudomonas aeruginosa*, of the three genes investigated two are clustered but are not linked to the third.

Arginine

Finally, something is known about map positions of the arginine genes in *Bacillus subtilis* and *Pseudomonas aeruginosa*. In the former, they fall into either two or three groups (one gene maps very close to a cluster of several others, but it is not certain whether it is actually contiguous with them or not). In the latter, they are scattered among at least seven sites.

In few cases outside the enterobacteria is there definite evidence, in either Class I or Class II pathways, for negative, operator or promoter sites, a polycistronic messenger, or control at the level of transcription. However, this may be attributed to less intensive investigation of these species, and also to greater technical problems—there are good practical reasons why so much work in bacterial biochemistry has been done with *Escherichia coli*. One difficulty that is especially important is that of constructing partial diploids in most bacterial species; consequently dominance studies cannot be carried out. It may also be worth pointing out that the existence of polar mutations, which in enterobacteria are particularly useful in providing evidence for polycistronic messengers, may depend on the presence in these organisms of a rather unusual RNAse (see above, pp. 446–7); the absence of this enzyme in other species would presumably result in the non-appearance of these mutations.

Possible value of gene clusters

We shall consider one final point: the origin of the clustering of functionally related genes. As we have seen, although this is a marked bacterial characteristic, it seems to occur to different extents in different bacterial groups; in respect of amino acid biosynthetic sequences, at any rate, it would seem most marked for the enterobacteria and *Staphylococcus*, only slightly less so for *Bacillus*, and much reduced in *Pseudomonas*. We may ask whether this clustering is advantageous for reasons connected with the control of gene expression, is advantageous for some other reason, or is an evolutionary relic, and we should like to know why, whichever of these is correct, its manifestation varies between the different groups.

As usual with such arguments, we can only make informed guesses. Although clustering was originally regarded as an integral part of the operon system, facilitating coordination and economizing on promoter and operator regions (and hence on the amount of DNA needing to be replicated), this is less convincing in the light of the variety of systems now studied. In particular, even in *Escherichia coli* we seem to have clustering in some cases associated with a single messenger but without coordinate control, as in the tryptophan cluster, and with two or more messengers (this necessitating some duplication of promoter and operator regions) in the isoleucine-valine and arginine systems, while in the pyrimidine sequence there is apparently coordinate control of unclustered genes. Evidently the correlations between clustering and coordinate control and economy in use of DNA, though good, are not perfect.

What of other possible explanations? Two speculations can be mentioned. Firstly, it has been proposed that some degree of clustering is advantageous to organisms among which, as is true for the bacteria, genetic material is transferred only in small pieces. It may not be useful for an organism that has lost several catalytic functions of a biosynthetic pathway to acquire just one functional gene, whereas it may be greatly advantageous for it to pick up all the functional genes as part of a cluster. Secondly, the evolution of metabolic pathways has been suggested as proceeding via tandem duplication (giving rise to two contiguous copies) of the gene determining the last step, followed by independent evolution of one of these copies to become the gene determining the next-to-last step, and so on. This process clearly generates clusters of functionally related genes, which could then separate adventitiously.

THE CONTROL OF CLASS III REACTIONS

Here, in contrast to the detailed schemes described in previous pages, we are primarily concerned with the synthesis or breakdown of whole classes of macromolecules and these processes in most cases reflect the growth state of the cell.

We shall first describe some modes of regulation of transcription of DNA to give RNA which are to some degree specific but which may also reflect the physiological state of the cell. These have in common that they operate by changing the relative affinities of the RNA polymerase for different classes of promoter site. The RNA polymerase of *Escherichia coli*, and probably that of *Bacillus subtilis* too, is a multimer consisting of a *core enzyme* with a subunit composition that can usually be represented as $\alpha_2\beta\beta'$ (β and β' being similar in molecular weight but unrelated in function), to which there becomes attached another polypeptide chain called sigma (σ). The core enzyme initiates transcription *in vitro* unspecifically at any site on a DNA molecule; only when σ has bound to it (yielding the *holoenzyme*) is transcription initiated specifically at promoter sites, and in the correct direction. The σ factor is not required, however, for elongation of the growing RNA chain and may dissociate from the core once transcription has started.

Changes in RNA polymerase initiation specificity following infection by virulent phages

Under certain circumstances, one or more of these subunits may be altered with consequent changes in the frequency of transcription initiations by the RNA polymerase at some promoter sites. So far, the most thoroughly examined instances have involved the sequential transcription of phage genes following infection of *Escherichia coli* and *Bacillus subtilis*, by certain virulent DNA bacteriophages. The penetration into the host cell of the DNA from an infecting phage particle sets in motion a train of events that ends in lysis of the cell and release of phage progeny. At first, only those phage genes are expressed the products of which are needed for phage DNA production or for inhibiting host cell DNA, RNA or protein synthesis. Later there follows the expression of genes for products that are involved in the elaboration of the phage protein coat and, finally, in lysis. Groups of phage genes are, therefore, switched on in a definite sequence.

It is often found that host cell RNA polymerase holoenzyme only transcribes those phage genes that are expressed immediately after infection. Sometimes transcription of genes expressed later requires the synthesis of a new phage-coded RNA polymerase, as with the *Escherichia coli* phage T7. In other instances, e.g. phage T4 and *Bacillus subtilis* phage SPO1, successive modifications of the host enzyme are observed which seem to correlate with the appearance of new classes of phage-coded RNA molecules. For instance, soon after infection with T4 the α subunit of the core enzyme is 5'-adenylated and later the β' subunit is modified. It is thought that each change causes an alteration in the affinity of the holoenzyme for many different phage gene promoters; some which previously did not bind the RNA polymerase now do so, with consequent 'switching on' of the adjacent gene(s), whereas others lose their affinity for the enzyme, their genes being 'switched off'. It has also been suggested that a new, phage-coded σ factor is rapidly synthesized following infection, and that this is required for transcription of subsequent classes of RNA molecules while rendering the holoenzyme unable to transcribe host DNA (host cell σ factor activity seems to be lost during infection).

Changes in RNA polymerase initiation specificity during sporulation

In *Bacillus subtilis*, the sequence of events culminating in release of the mature spore (Chapter 9) requires transcription of genes for sporulation as well as the continued transcription of others needed for vegetative functions (e.g. production of the enzymes of the TCA cycle). This alteration in transcription is accompanied by an alteration in the RNA polymerase. One of the β subunits of the 'sporulation' enzyme is shorter than that of the 'vegetative' enzyme, and its σ factor may be modified or replaced. The 'sporulation' enzyme is also unable to transcribe the DNA of the virulent phage ϕe.

Such effects might provide a flexible way for achieving sequential patterns of gene expression. They have been intensively studied partly in the hope that these simple systems may prove to be models for sequential patterns in the differentiation of higher organisms.

Control of transcription of ribosomal and transfer RNA's

This is really a 'specific' rather than 'general' control, but is included here because of its similarity to the phenomena just discussed. Initiation of transcription of genes coding for ribosomal and transfer RNA molecules requires the association of the RNA polymerase holoenzyme with yet another polypeptide chain, ψ. The activity of ψ factor is inhibited by an unusual nucleotide called guanosine tetraphosphate or ppGpp, the cellular content of which seems to reflect its physiological state. This explanation is, however, only tentative.

Incidentally, it is interesting that the structural genes for the 16S and 23S ribosomal RNA species occur in contiguous pairs in all bacterial species so far studied; a pair is transcribed to the two separate RNA molecules, carrying excess RNA, and these are then broken down to give the two mature rRNA molecules, which are consequently produced in equimolecular amounts. The transcripts of tRNA genes also contain excess RNA, which is 'trimmed' to give the mature tRNA.

Termination of transcription

It appears that correct termination of transcription and release of the completed RNA chain at (presumably) some signal in the DNA base sequence require a protein ρ, the need for which in the complete *in vitro lac* mRNA-synthesizing system from *Escherichia coli* has already been noted. The factor is not inherently a regulatory element but may play such a role in the intracellular development of the temperate phage λ in *Escherichia coli* (see Chapter 7, p. 405). When the DNA molecule of the phage initiates the sequence of events that culminates in release of progeny λ particles from the lysed bacterium, two isolated genes *N* and *cro*—the meanings of these symbols do not concern us—are transcribed immediately, while others (a cluster, *cIII β exo int*, adjoining *N*, and another cluster *cII OPQ* adjoining *cro*) are transcribed slightly later and their transcription requires a functioning *N* gene product. It has been suggested that initially transcription of *N* stops at the *N-cIII* boundary, and that of *cro* at the *cro-cII* boundary, but that the *N* gene product in some way modifies the action of the *Escherichia coli* ρ factor so that in its presence termination fails to occur at the signals ending *N* and *cro*, and transcription continues as far as *int* and *Q*. There is evidence that a similar mechanism operates during T4 infections.

Controls linked to growth rate

The correlations between macromolecular synthesis and growth state of the cell have been described in Chapter 2. Here we shall summarize the results (obtained mostly with the enterobacteria).

DNA synthesis is geared not only to overall growth rate but also to the individual cell cycle. In *Escherichia coli* (except at generation times over 60 min when the rate of DNA replication decreases) the growing point moves at a constant rate along the replicating chromosome, taking about 40 mins to traverse it at 37°. At low growth rates, there may therefore be a gap between chromosome replications; at high growth rates, the cell may have to get more than one replication into a division cycle, and this is achieved by starting new replications before the previous one is complete (*dichotomous replication*; see Chapter 2, p. 158).

Synthesis of cell wall and membrane, septum formation, and cell division also follow the cell cycle. In *Escherichia coli* growing at 37° with generation times under 60 mins division takes place at a constant interval (about 20 mins) after the end of chromosome replication; the interval is longer in more slowly growing cells. In *Bacillus subtilis*, the relation between DNA replication and the division cycle is similar, except that the time elapsing between completion of chromosome replication and cell division seems to be constant even at very slow growth rates.

The faster the growth rate, the more the ratio of RNA to protein increases (primarily because of a relative increase in the number of ribosomes). There is also an increase in the number of chromosomes (replicating or resting) per cell, and the cell size, this last point implying a *lower* ratio of outer layer components, cell wall and membrane, to internal constituents, since the larger the cell, the smaller the surface/volume ratio.

Following shift-up (see Chapter 2, p. 147), synthesis of RNA, protein and DNA change to new rates in that order—in particular, there is a preferential synthesis of ribosomal components. Following shift-down, the order is reversed, there being a particularly long lag before protein and RNA begin to be synthesized.

Growth inhibitors acting on the synthesis of one macromolecular component may permit that of other macromolecules to continue for some time; in other words, synthesis of the various macromolecular species is not necessarily tightly coupled. An exception to this is the so-called *stringent* response shown by *Escherichia coli* and other bacteria following the exhaustion of a required amino acid by an auxotroph. Inevitably protein synthesis stops immediately. But synthesis of ribosomal and transfer RNA, formation of polyribosomes, and the uptake and phosphorylation of nucleosides and glucose also cease, and protein breakdown is stimulated. In *Escherichia coli*, *relaxed* mutants can be obtained mapping at a single locus *rel*, in which none of these changes occur. It appears that the stringent response involves the accumulation of guanosine tetraphosphate (p. 485) on amino acid starvation of the auxotroph; in similar circumstances in *rel* mutants levels of this nucleotide do not increase, although its formation in such strains is normal under all other conditions. It is not known exactly how *rel* mutations act.

Breakdown of protein and RNA

We shall enlarge here on a point that arises from experiments on shift-down; as described in Chapter 2, p. 149. During the lag before protein and RNA synthesis are resumed, considerable breakdown of protein and RNA may occur and we shall now consider the function of protein and RNA turnover.

Protein breakdown can be investigated by adding a radioactively labelled amino acid to the cells under study, and then transferring them to a medium containing a high concentration of the unlabelled amino acid as a 'trap'. It is assumed that when labelled amino acid incorporated into protein is liberated by proteolysis it will be so diluted in the 'trap' as to have negligible chance of being re-incorporated during protein synthesis. Hence the amount of labelled amino acid released in a given time is a measure of the rate of protein breakdown.

By such techniques it has been shown that some protein breakdown accompanies normal exponential growth although its accurate determination is difficult; for *Escherichia coli*, rates of 0·5% to 2·5% of the protein per hr have been suggested. However, under shift-down conditions, or when growth ceases due to the presence of an inhibitor or to the exhaustion of a nutrient, the rate of protein breakdown increases immediately, to about 5% per hr in *Escherichia coli* and to nearly 8% per hr in *Bacillus cereus*. It seems likely that during exponential growth, a limited group of proteins are broken down relatively rapidly, while in the more unfavourable conditions a wider range of proteins suffer degradation. On shift-down or during starvation, the amino acids freed are rapidly re-utilized, so that there is an efficient turnover process in which the cellular protein content remains more or less constant.

In parallel with the increased breakdown of protein attendant on shift-down or inhibition of growth, there is an enhanced rate of degradation of RNA (mostly ribosomal), which in *Escherichia coli* also reaches about 5% per hr. Again, most of the released nucleotides are rebuilt into RNA, but under certain conditions a fraction of them may be further broken down and the ribose used as an energy source.

These degradative processes, which accelerate under conditions of temporary or permanent cessation of growth, subside rapidly when growth resumes, but the control mechanisms responsible are as yet unknown.

It is easy to understand the usefulness of protein and RNA turnover as enabling adaptation to occur under non-growing conditions. Consider, for example, the diauxic growth of *Escherichia coli* on a mixture of glucose and lactose (p. 452). When the glucose is exhausted growth temporarily comes to a stop; the lactose cannot be utilized because induction of the lactose proteins has been prevented by catabolite repression on the part of the glucose. There ensues a lag during which breakdown of protein and RNA occurs. As conditions now favour induction of the lactose proteins, resynthesis utilizing amino acids and nucleotides liberated during this degradation results in the formation of *lac* mRNA and of the lactose proteins, so that gradually utilization of the lactose can begin and the second stage of growth can commence.

Similar events follow 'shift-down' from nutrient broth to minimal medium, but here the enzymes of many pathways, not merely one, need to be synthesized. Consequently, the lag may last several hours.

The mechanism of growth-linked controls of macromolecular synthesis

Returning to the way in which the generation time appears to set the rate of production of macromolecules, it is natural to enquire how this interrelation is achieved but lack of knowledge is virtually complete. However, there are three possibilities which are not necessarily mutually exclusive, and it is quite possible that they apply differently to the different macromolecular species.

Firstly, control mechanisms may exist of which we are as yet uncertain. This is particularly likely to be true for processes closely coupled to the cell cycle, such as synthesis of DNA, cell wall and membrane.

Secondly, no control mechanisms additional to those already described may be necessary. In other words, given the regulatory systems already described for Class I and II pathways which determine the availability of building blocks and other small molecules needed for Class III reactions, together with certain other factors such as the growth-controlled protein and RNA breakdown described above, the observed connections between growth rate and macromolecular synthesis follow automatically. The inference is that if we could programme a computer with details of the external environment of a cell, of its size and structural and biochemical make-up at a given time, of the kinetic parameters for all its enzymes and of the regulatory characteristics of these, then the actual rates of macromolecular synthesis could be calculated. As pointed out above, such a model constructed according to our present knowledge would lack the controls that determine the sequential events of the cell cycle but these could be supplied when eventually discovered.

The last type of explanation of the correlation of generation time with rate of macromolecular synthesis resembles the second in not requiring specifically regulatory processes, but differs from it in suggesting that the flow through the various pathways of a bacterial cell is necessarily geared to growth rate purely for reasons of chemical kinetics. This approach, put forward by Hinshelwood, is based on the idea that all cell reactions can be regarded as members of cycles of the type:

Rate of formation of B depends on the concentration of A,
,, ,, ,, ,, C ,, ,, ,, ,, ,, B,
. . . .
,, ,, ,, ,, Y ,, ,, ,, ,, ,, X,
,, ,, ,, ,, A ,, ,, ,, ,, ,, Y.

For instance, A and B may be intermediates in the biosynthesis of an amino acid, X may be the amino acid, and Y the enzyme catalysing the formation of A; or A and B may be intermediates in the formation of a pyrimidine base, X may be a species of transfer RNA, and Y again the enzyme catalysing the formation of A. On the assumption that the outer layers of the cell play a part in its permeability to nutrients, even the constituents of these can be brought into the scheme. It can be shown mathematically that under growth conditions in a system of this kind, the relative proportions of all its components eventually reach a steady state in which (a) the rate of increase of the components is maximal and (b) the increase is exponential. In addition, following a change in growth conditions or after addition of an inhibitor, these relative proportions alter automatically in such a way that (a) and (b) apply again.

Clearly, this hypothesis predicts adaptive changes such as enzyme induction or repression without invoking the elaborate mechanisms which we have seen to operate in bacteria. The existence of the latter has been taken to disprove the former, perhaps unjustifiably, since quite possibly the specific mechanisms allow a speed of response beyond the capacity of the self-adjusting mechanism, and would therefore be expected to have evolved in addition to it. It is unfortunately true that, because it does not predict specific control features that could be looked for, the hypothesis has proved very hard to verify experimentally.

So, in spite of our intimate understanding of many specific regulatory features of the bacterial cell, the origin of the larger-scale growth correlations so far escapes us; nor do we even have any very clear idea as to what kind of mechanisms might be involved in the latter. It is appropriate, however, to close by pointing out that the control systems operate in the context of the whole, metabolizing cell, and that their intricate details must be understood as having evolved to confer selective advantage on the living organism.

SUMMARY

The complex controls of bacterial metabolism are to be expected in view of the strong selective pressures on bacteria to utilize available nutrients as efficiently as possible, and to achieve optimal growth rates while doing so. These controls may be classed as *specific*, where flow through a particular pathway is regulated, or *general*, where formation or breakdown of a whole class of macromolecules is affected.

Specific controls act either on enzyme activity or on enzyme synthesis. In the former, an enzyme may be *activated* or *inhibited* (*modulated*); the activator or inhibitor (*effector*) molecule is usually structurally unrelated to its substrates or products, and binds to an *allosteric* site distinct from the active site. Enzymes subject to allosteric modulation have certain peculiar properties, notably a subunit structure and sigmoidal, rather than hyperbolic (Michaelis–Menten), kinetics, dependence of reaction velocity on substrate concentration in the presence (and often absence) of effector. This latter property creates a physiologically advantageous approximation to a threshold effect whereby the change from very low to near-maximal velocity takes place over a narrow range of substrate concentrations.

Specific controls of enzyme synthesis comprise *induction* or *repression*. A number of enzymes may be induced or repressed together; if their rates of synthesis are always proportionate, they are said to be *coordinately controlled*. These controls seem to act primarily at the level of transcription of the structural genes for the regulated enzymes into messenger RNA, rather than at the level of translation of the mRNA into protein. Their mechanism requires three genetic elements; a regulator gene, which may be located far from the structural gene(s), and promoter and operator DNA regions that lie adjacent to the structural gene(s), the operator lying between the promoter and the structural genes. The regulator gene, in the commonest situation (*negative control*), codes for a protein *repressor* that either alone (in inducible systems) or in combination with the repressing metabolite (in repressible systems) binds to the operator. RNA polymerase initiates transcription by binding to the promoter; however, its binding or its functioning is prevented when the repressor is attached to the operator, so that synthesis of the corresponding mRNA and protein(s) is reduced. An inducer combines with repressor to yield a product incapable of binding to the operator, so that mRNA and protein synthesis are possible. Catabolic systems often show *catabolite repression*, in which enzyme induction by a particular carbon source is antagonized by a more readily utilized carbon source. The effect depends on a diminution, during rapid metabolism of the latter, in the intracellular pool of cyclic $3',5'$-AMP; in such systems, initiation of transcription by RNA polymerase at the promoter is enhanced by cyclic $3',5'$-AMP attached to a specific binding protein.

Class I sequences can be divided into *central*, comprising the main routes of glucose breakdown and the TCA cycle and associated anaplerotic reactions, and *peripheral*, comprising the steps leading from other carbon sources to the central pathways. The central reactions are controlled primarily at the level of enzyme activity. Controls are of two types, those indicating the flux through regions of the sequence coming before or after the step at which control is exerted, and those indicating the availability of energy or reducing power within the cell. The former are most often of the *feedback inhibition* type, whereby a metabolite occurring late in a pathway inhibits an enzyme catalysing an earlier reaction; sometimes they are of the *precursor activation* type, whereby a metabolite occurring early activates an enzyme catalysing a later step. In the latter type, energy-yielding reactions are inhibited by nucleotide triphosphates or activated by nucleoside mono- and/or diphosphates. Again, $NADH_2$ inhibits certain reactions involved in sequences that generate the reduced coenzymes.

The peripheral sequences are principally controlled at the level of enzyme synthesis, by induction and catabolite repression (usually cyclic $3',5'$ AMP mediated). Although the initial compound of a peripheral sequence is usually the effective inducer, occasionally *product induction* is found, in which the product of the first reaction is the true inducer. In some cases, energy-linked controls also occur.

The most characteristic controls in Class II sequences are feedback inhibition, usually by the end-product of a pathway, of the activity of the first enzyme, and end-product repression of synthesis of all the enzymes. However, many varied patterns exist.

Multi-enzyme aggregates are often found in Class I and Class II sequences. These often, but not always, comprise enzymes mediating successive steps, in which case they probably promote efficient utilization of intermediates by eliminating the need for their diffusion throughout the cell.

Genes determining the enzymes of Class I and Class II pathways very often show some degree of clustering in bacteria; its extent, however, varies considerably for different sequences in any one organism, and for the same sequence in different organisms.

General controls involving Class III reactions are very often determined by growth conditions. An important means of regulation is through RNA polymerase, the subunit composition of which may alter or be modified. These changes affect the affinity of the enzyme for whole classes of promoter sites and thereby the rates of transcription of their structural genes. Such effects are responsible for some of the changes in transcription patterns which accompany sporulation or phage infection. They also modify rates of transcription of ribosomal and tRNA genes. Another important general finding is that the rates of degradation of cell components, particularly RNA and protein, increase under conditions unfavourable for growth, thus providing a supply of amino acids and nucleotides that may enable the cell to adapt to the changed environment. However, the mechanism governing many Class III reactions, especially those geared to the cellular division cycle, are unknown and it has even been suggested that the observed effects follow from the total interaction of all cell processes.

FURTHER READING

1 Atkinson D. E. (1970) Enzymes as control elements in metabolic regulation. In *The Enzymes*, Volume I (ed. Boyer, P.D.), 3rd edition, p. 461.

2 Beckwith J. R. and Zipser D. eds. (1970) *The Lactose Operon*. Cold Spring Harbor Laboratory of Quantitative Biology, Cold Spring Harbor, N.Y.

3 Clarke P. H. and Lilly M. D. (1969) The regulation of enzyme synthesis during growth. In *Microbiol Growth* (eds. Meadow P. and Pirt S. J.), Nineteenth Symposium of the Society for General Microbiology, p. 113. Cambridge University Press.

4 De Crombrugghe B., Chen B., Anderson W., Nissler P., Gottesman M., Pastan I. and Perlman R. (1971) *Lac* DNA, RNA polymerase and cyclic AMP receptor protein, cyclic AMP, *lac* repressor and inducer are the essential elements for controlled *lac* transcription. *Nature* **231**, 139 (see also the editorial comments, **231**, 129).

5 Dean A. C. R. and Hinshelwood Sir Cyril (1966) *Growth, Function and Regulation in Bacterial Cells*. Oxford University Press.

6 Epstein W. and Beckwith J. R. (1968) Regulation of gene expression. *Ann. Rev. Biochem.* **37**, 411–436.

7 Hartman P. E. and Suskind S. R. (1969) *Gene Action*, 2nd edition. Prentice-Hall, Englewood Cliffs, N.J. (chapters 5, 8 and 9 are particularly relevant).

8 Hershko A., Mamont P., Shields R. and Tomkins G. M. (1971) Pleiotypic response. *Nature* **232**, 206 (compares stringent control in bacteria with similar effects in mammalian cells).

9 Jacob F. and Monod J. (1961) Genetic regulatory mechanisms in the synthesis of proteins. *J. Mol. Biol.* **3**, 318.
   Jacob F. and Monod J. (1961) On the regulation of gene activity. *Cold Spring Harbor Symp. Quant. Biol.* **26**, 193 (two classical papers).

10 Koshland D. E. Jr. (1970) The molecular basis for enzyme regulation. In *The Enzymes*, Volume I (ed. Boyer P. D.), 3rd edition, p. 342. Academic Press, London.

11 Lavallé R. and De Hauwer G. (1970) Repression by tryptophan at the level of transcription and translation in *E. coli*. *Cold Spring Harbor Symp. Quant. Biol.* **35**, 491.

12 Martin R. G. (1969) Control of gene expression. *Ann. Rev. Genet.* **3**, 181.

13 Miller O. L. Jr., Beatty B. R., Hamkalo B. A. and Thomas C. A. Jr. (1970) Electron microscopic visualization of transcription. *Cold Spring Harbor Symp. Quant. Biol.* **35**, 505 (lends support to the picture of transcription and translation shown in Fig. 13).

14 Monod J., Changeux J.-P. and Jacob F. (1963) Allosteric proteins and cellular control systems. *J. Mol. Biol.* **6**, 306.
    Monod J., Wyman J. and Changeux J.-P. (1965) On the nature of allosteric transitions: a plausible model. *J. Mol. Biol.* **12**, 88 (two influential papers that clarify several concepts).

15 Morse D. E. and Guertin M. (1971) Regulation of mRNA utilization and degradation by amino-acid starvation. *Nature* **232**, 165.
    Imamoto F. and Kano Y. (1971) Inhibition of transcription of the tryptophan operon in *Escherichia coli* by a block in initiation of translation. *Nature* **232**, 169 (two papers on the mechanism of polarity; see also the editorial comments, **232**, 161).

16 Ornston L. N. (1971) Regulation of catabolic pathways in *Pseudomonas. Bact. Rev.* **35**, 87.

17 Pine M. J. (1970) Steady-state measurement of the turnover of amino acid in the cellular proteins of growing *Escherichia coli*: existence of two kinetically distinct reactions. *J. Bact.* **103**, 207 (refers also to earlier work on protein breakdown under shift-down or starvation conditions).

18 Ptashne M. and Gilbert W. (1970) Genetic repressors. *Scientific American* **272**, 36.

19 Reed L. J. and Cox D. J. (1970) Multienzyme complexes. In *The Enzymes*, Volume I (ed. Boyer P. D.), 3rd edition, p. 213. Academic Press, London.

20 Richmond M. H. (1968) Enzymic adaptation in bacteria: its biochemical and genetic basis. In *Essays in Biochemistry*, Volume 4 (eds. Campbell P. N. and Greville G. D.), p. 105. Academic Press, London.

21 Sanwal B. D. (1970) Allosteric controls of amphibolic pathways in bacteria. *Bact. Rev.* **34**, 20.

22 Stadtman E. R. (1970) Mechanisms of enzyme regulation in metabolism. In *The Enzymes*, Volume I (ed. Boyer P. D.), 3rd edition, p. 398. Academic Press, London.

23 Travers A. (1971). Control of transcription in bacteria. *Nature* **229**, 69 (refers to recent work on σ, ρ and ψ factors).

24 Umbarger H. E. (1969) Regulation of amino acid metabolism. *Ann. Rev. Biochem.* **38**, 323.

Chapter 9
Differentiation: Sporogenesis and Germination

Previous chapters have stressed two important physiological features of microbial cells: their metabolic versatility and their exceedingly efficient systems for controlling enzyme synthesis and enzyme activity. The present chapter will deal with a third characteristic of some bacteria and yeasts. This is the ability, under certain conditions, to undergo progressive physiological and morphological changes which end in the formation of dormant structures. These structures include (a) endospores, which are produced by the Gram-positive, spore-forming rods of the genera *Bacillus* and *Clostridium* and by the coccus *Sporosarcina*, (b) cysts which are formed by *Azotobacter*, (c) conidial spores formed by fragmentation of the ends of aerial mycelial strands of actinomycetes and (d) ascospores formed inside modified cells (asci) of yeast. Only endospores and ascospores will be dealt with here.

Bacterial endospores

Vegetative cells of spore-forming bacteria are, at the end of growth, converted to spores in a relatively synchronous fashion, through a time-ordered sequence of morphogenetic and biochemical steps. The transition from the vegetative to the dormant form constitutes an example of unicellular differentiation, unique in the biological world. The emergence of a vegetative cell from the spore under favourable physiological conditions takes place within a few minutes in response to specific chemical agents.

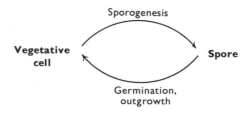

494

Bacterial sporogenesis thus represents a biological problem the regulation of which is intermediate in complexity between the regulation of bacterial enzymes on the one hand and the differentiation of specialized tissues in higher organisms on the other hand.

Morphological stages of spore development

At the end of the exponential phase of growth, cells of many bacterial species may become committed to sporulation provided that they have the genetic capacity and the appropriate environmental conditions. Instead of dividing,

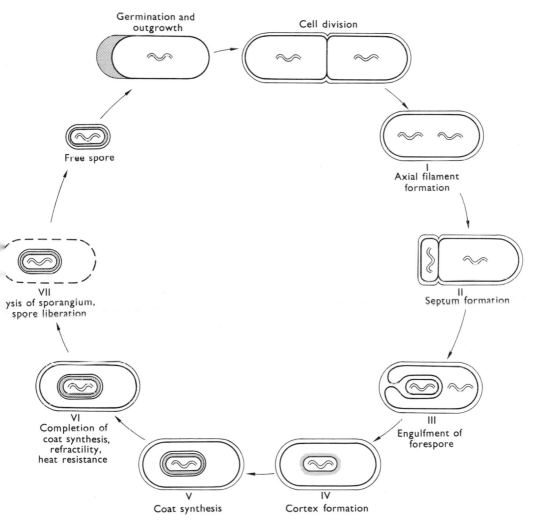

Figure I. Life cycle of a sporulating bacterium and morphological stages. (Modified drawing from A. Kornberg, J. A. Spudich, D. L. Nelson and Deutscher, 1968).

each cell will then produce a single intracellular spore, called an *endospore*, over a period of 8–10 hours.

The morphology of spore formation is now well established in many sporogenic aerobic and anaerobic bacteria. Figure 1 gives a schematic presentation of the life cycle of a sporogenic bacterium with the seven characteristic and distinctive structural steps recognisable in electron micrographs.

First an axial filament of chromatin material is observed, part of which is destined for the spore. This is followed (Stage II) by the inward folding of the cell membrane to form the forespore septum. This event coincides with the time when DNA is compartmentalized. Continued growth of the membrane and enclosure of the forespore leads in the next Stage (III) to the prespore. Stage IV marks the beginning of cortex formation and is also the period in which diaminopimelic acid, a constituent of the cortex, and cysteine, a constituent of the spore coats, are incorporated. This stage is followed by completion of the spore coats, accumulation of dipicolinic acid, uptake of calcium and the final development of the cortex and heat resistance of the spore (Stage V). The appearance of the mature spore within the sporangium (Figure 2) represents State VI. Finally lytic enzymes liberate the spore (Stage VII).

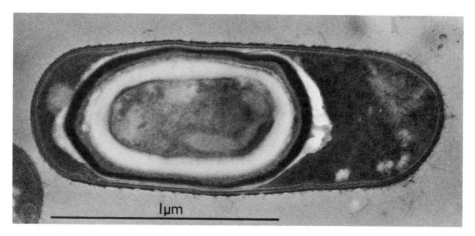

Figure 2. Structure of a sporulating cell. Electron micrograph of a mature spore of *Bacillus subtilis* surrounded by its sporangium.

All the events, genetic, morphological, biochemical and physiological, leading to the conversion of a vegetative cell into a spore are defined as sporogenesis. Those stages of sporogenesis which concern only the synthesis and assembly of spore components are defined here as sporulation although, in fact, the terms are often used synonymously. The final product, the dormant spore, differs cytologically, physiologically and biochemically from the vegetative cell. The biochemical changes occurring during sporogenesis are primarily concerned with the synthetic machinery of the cell, and are followed by changes in the synthesis of spore-specific components and structures.

Properties of bacterial spores

Structure

Even the light microscope showed that the bacterial spore was a complex structure. With the availability of the electron microscope it has been possible to study the finer details. The mature spore (Figure 3) contains a number of layers.

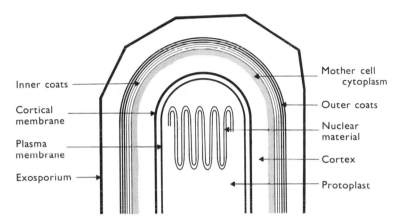

Figure 3. Diagrammatic cross-section showing ultrastructure of a typical bacterial spore.

The outermost envelope surrounding the spore is called the *exosporium*. This is a thin, delicate covering which lies outside the spore coat. In some spores, e.g. those of *Bacillus cereus*, it exists as a loose covering which is connected to the sides of the spore, whereas in others (e.g. *B. polymyxa*) it is stretched tightly over the surface of the spore. This material has an ordered structure, with hexagonal fragments resembling a crystalline configuration.

Beneath the exosporium lie the several layers of the spore coat, of which about 90% is made of protein containing the 20 common amino acids. From examination of carbon replicas, it is evident that the surfaces of different types of spores have different patterns. The spore coat is composed of several laminated layers of protein, each about 2·0 to 2·5 nm thick. Beneath these is a thin membrane which separates the spore coat from an area of low electron density which is composed of concentric layers of fine fibre-like bundles. This area, called the *cortex* occupies approximately half the volume of the spore. It is a specific spore component, disappears during germination, and is absent from vegetative cells. There exists some evidence that it is the location of dipicolinic acid, peptidoglycan polymers and calcium. The chemical composition of the cortical peptidoglycans seems to be very similar to that of the cell wall peptido-glycan of the vegetative cells. In *B. sphaericus* spores, however, diaminopimelic acid replaces the lysine present in the vegetative cell peptidoglycan. A thin membrane separates the cortex from the cytoplasm of the dormant spore the internal structures of which appear to be similar to those of the vegetative cell.

Chemical basis of spore resistance

The most characteristic property of bacterial spores is extreme resistance to heat, ultraviolet light, X-rays, organic solvents, chemicals and desiccation. The appearance of these types of resistance at different stages during sporulation and their loss during germination makes it clear that they are due to the par-ticular physicochemical composition of the dormant spore. The most con-spicuous chemical component is the chelating agent DPA (dipicolinic acid, or 2,6-pyridine dicarboxylic acid) which is always found in spores in about a 1:1 ratio with Ca^{2+}. The DPA–Ca content of different species of bacterial spores ranges from 5 to 15% of their dry weight but is zero in vegetative cells and in asporogenous mutants. It was long believed that DPA was essential for heat-resistance. However, mutants of *B. subtilis* have been isolated recently which are unable to synthesize DPA but still produce heat-resistant spores. Despite considerable investigation, the biochemical basis of heat-resistance re-mains obscure.

Another theory explaining the extreme heat resistance of bacterial spores supposes that a contractile pressure is exerted by the specialized multilayer spore envelope which squeezes the central core, and maintains it in a dry state which confers heat resistance (spores contain little, if any, water). The con-tractile pressure might be due to the effect of some positively-charged molecules on the acidic peptidoglycan polymer of the spore cortex.

Besides Ca^{2+}, spores also contain higher levels of other divalent metals such as Mg^{2+} and Mn^{2+} and S—S compounds. The latter occurs mainly as cystine in the coat fraction.

Metabolic changes associated with bacterial sporogenesis

The morphological development of the spore is accompanied by major bio-chemical changes. Some enzymes synthesized during sporulation are associated with the production of the spores but are absent from the dormant spores; others are components of the spore itself. An example of a metabolic change associated with the spore-forming machinery is the involvement of the tricar-boxylic acid cycle enzymes in this process. In vegetative cells of several *Bacillus* species grown in a complex medium containing glucose, these enzymes are re-pressed and therefore non-functional. The derepression of these enzymes is indispensable if sporulation is to take place. Mutants which are blocked in the synthesis of aconitase, one of the key enzymes of the tricarboxylic acid cycle, fail to sporulate although they are able to grow vegetatively in the presence of added glutamate. During sporulation there is also a several-fold increase of a particulate $NADH_2$ oxidase activity over that present during vegetative growth. In the asporogenous mutants the level of $NADH_2$ oxidase remains low and constant during the entire life cycle of the bacterium. Transformation of the mutants by wild-type DNA restores both the enzyme activities and the ability to sporulate. It appears therefore, that sporulation requires a rich supply of intermediates as well as energy from the tricarboxylic acid cycle for the synthesis of spore-specific components. That these pathways are part of the spore-forming mechanism, and not spore-specific components, is shown by experiments in which the spores are liberated during sporulation from the rest of the sporulating cell, the sporangium, by treatment with lysozyme. As can be seen (Table I) the enzymes of the tricarboxylic acid cycle remain in the sporangium and are not significantly incorporated into the spore.

Table I. Localization of the particulate $NADH_2$-oxidase and the TCA cycle enzymes

Cell fraction*	Particulate $NADH_2$ oxidase	Units of enzyme activity	
		Aconitase	Malic dehydrogenase
Sporangium	28,800	2250	17,500
Spores	0	0	500

*Samples prepared from sporulating cells of *B. subtilis* Marburg. Data from Szulmajster and Hanson (1965).

Synthesis of spore-specific components

Biochemical changes associated with sporulation

There is no doubt that an increase in the activity of the energy-yielding tricarboxylic acid cycle enzymes is required for the synthesis of specific proteins which are incorporated into the mature spore and indeed considerable protein synthesis takes place during sporulation. As a result, spores differ from vegetative cells not only in the pattern of enzyme activities, but sometimes also in the nature of the protein which catalyzes the same reaction in spores and vegetative cells. Differences in enzymes may thus be both qualitative and quantitative (Table 2). Some of these changes may be coincidental; others are associated with the sporulation process. As the medium becomes depleted of its carbon or nitrogen sources, changes in the metabolic patterns of the cells occur. The activities of some vegetative enzymes decrease (isocitric dehydrogenase, lysine

Table 2. Characteristics of sporulating cells and of spores.

(a) Enzyme activities which are increased in sporulating cells:
  $NADH_2$ oxidase (particulate)
  Citrate condensing enzyme
  Aconitase
  Fumarase
  Succinic acid dehydrogenase
  Adenosine deaminase
  Alanine racemase
  Alkaline phosphatase
  Glutamic-alanine-transaminase
  Glutamic-aspartic-transaminase
  Ornithine transcarbamylase
  Purine nucleoside phosphorylase
  *Glucose dehydrogenase
  *Acetoacetyl-CoA reductase
  *Ribosidase
  *DAP-adding enzyme in B. *sphaericus*
  *Dipicolinic acid synthetase
  *Extracellular proteases
  *Dihydrodipicolinic acid synthetase

(b) Specific components synthesized by sporulating cells:
  Toxins
  Antibiotics
  N-succinyl-L-glutamic acid

(c) Spore-specific components:
  Antigens
  DPA-Ca^{2+}
  Protein crystal (B. *cereus*, var. *alesti*, B. *thuringiensis*)
  Spore cortex peptidoglycan (B. *sphaericus*)
  Lactam of muramic acid
  Sulpholactic acid (B. *subtilis*)

(d) Components not detectable or only present in small amounts in spores:
  Isocitric dehydrogenase
  Fumarase
  Malic dehydrogenase
  $NADH_2$ oxidase (particulate)
  Aspartic transcarbamylase
  Cytochromes
  Succinic cytochrome c. reductase

---

*Not detectable in vegetative cells.

decarboxylase, aspartokinase, etc.) others (proteases, wall-lytic enzymes, TCA cycle enzymes) appear or increase.

From the extensive studies of the last ten years it can be concluded that dormant spores contain a great number of enzymes involved in the metabolism of carbohydrate, protein, nucleic acid and lipid. A remarkable difference is the low level in spores of enzymes associated with the electron transport system in the cell.

Some of the enzymes so far studied (inorganic pyrophosphatase, nucleoside phosphorylase, DNA polymerase) appear to be identical proteins when extracted from vegetative cells or from dormant spores. It is, therefore, quite possible that a majority if not all of the spore enzymes are similar in their primary structure to the homologous enzymes present in the vegetative cells, although they may have different physicochemical properties *in situ*. Were this so it would reflect the biological efficiency of the cell and would not require the destruction of all the vegetative enzymes during sporulation.

It is now well established, at least for several spore enzymes, that thermal resistance is due to stabilization inside the spore and not to intrinsically heat-stable molecules. The thermal resistance of some (alanine racemase, alanine dehydrogenase, ribosidase, aldolase, inorganic pyrophosphatase, etc.) can be explained in various ways. For example, the heat stability of alanine racemase of *B. cereus* spores has been shown to be due to the attachment of the enzyme to particles derived from the spore wall: detachment renders this protein as heat-labile as is the vegetative enzyme. Catalase is also an example of a spore enzyme which differs qualitatively from the analogous vegetative cell enzyme. In spores the enzyme is attached to particles and is heat-resistant, whereas catalase from vegetative cells is soluble and heat-sensitive. The two catalases differ in their kinetic properties and, what is more significant, are immunologically distinct. During sporulation, the appearance of spore catalase is coupled with a decrease in the vegetative enzyme.

Another reason for heat stability was suggested by the studies of the glucose dehydrogenase of *B. cereus*. The kinetic, immunological and chromatographic properties of this enzyme extracted from vegetative cells and from spores are identical. The heat resistance of this enzyme is due to two factors: reversible interconversion of monomer-dimer and high salt concentration. When the conditions are optimal heat resistance increases a million-fold. Finally, some enzymes (including aldolase and inorganic pyrophosphatase) are stabilized by the Ca^{2+} or Min^{2+} which are always present at high levels in dormant spores.

The extent to which unique species of macromolecules are produced during sporulation and incorporated into the spore is indicated by the presence of numerous soluble, spore-specific antigens. For instance, immunological examination of the proteins of *B. subtilis* separated by electrophoresis showed that of 12 recognizable antigens, four were common to the spore and to the vegetative cell, five were new heat-resistant proteins and three were new heat-sensitive proteins. A number of heat-sensitive antigens which were observed in extracts of vegetative cells disappeared during sporulation.

We do not as yet have a systematic study of the time at which many of the known physiological and biochemical properties appear in any one sporulating system. It is clear, however, that they occur in a definite sequence. For example, in *Bacillus cereus* T the spore-specific glucose dehydrogenase appears when growth stops and sporulation commences, whereas the spore-specific alanine racemase appears some 6 hr later. In another sporulating system, *Bacillus megaterium*, the synthesis of spore deaminase precedes the synthesis of spore ribosidase by several hours. Differences are also seen in the time at which various physiological properties of the spore appear. The spore develops resistance to X-rays when the sporulating cell synthesizes a structure rich in S—S bonds. Two hours later heat-resistance develops and is correlated with the synthesis of dipicolinic acid and the accumulation of calcium.

Induction of sporogenesis

The phenotypic expression of the spore genome depends upon a number of external factors. Spore formation normally begins after exponential growth has stopped because some nutrient in the medium has been expended. Thus the classical remark made by Knaysi 'spores are formed by healthy cells facing starvation' remains apt. Induction to sporulation may occur even in a primary cell, i.e. in one formed immediately after outgrowth of the spore, without any cell division. This process of spore induction bypassing cell division is called 'microcycle sporogenesis'.

There exist for each species optimum conditions for sporogenesis and these involve factors such as pH value, aeration, temperature, nutrients and the cations of the medium. Limitation of any one of a variety of nutrients can initiate sporogenesis. Since these factors can influence the metabolism of the vegetative cell, their exact role has never been clear but some generalizations can be made. Sporogenesis, which occurs either massively at the end of vegetative growth or at low frequencies in exponentially growing cultures, is strongly influenced by the carbon and nitrogen sources available. Metabolizable nitrogen compounds generally repress sporogenesis and in combination with glucose are even more effective. In general, carbon sources which are rapidly metabolized favour vegetative growth, whereas those which are more slowly metabolized stimulate spore formation. Many parallels exist between the repression of spore formation and the repression of inducible enzymes in bacteria.

Sequence of events during sporulation

At about the time when nutrients become exhausted in the medium, sporulation begins and the morphological changes already described take place in an ordered manner (Figure 4). These changes are the expression of the genetic and biochemical events associated directly or indirectly with this developmental process.

Little is known about the control and the nature of the sequence of events leading from a vegetative cell to a spore, but the events can be divided into two classes: early and late. The first class includes the appearance of enzymes of the TCA cycle and of those causing arginine catabolism, purine nucleoside phosphorylase, and some exoenzymes (protease, ribonuclease and amylase). These

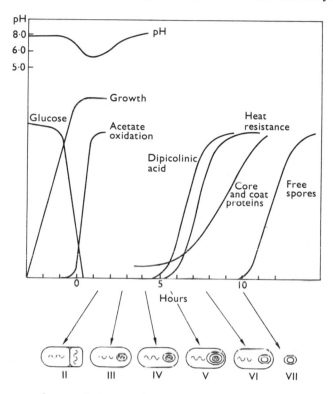

Figure 4. Sequence of events during sporulation of an aerobic bacillus.

enzymes are not specifically concerned with sporulation but are absent during vegetative growth under most conditions. The synthesis of some of these enzymes at the beginning of sporulation is probably a consequence of the derepression brought about by the exhaustion of carbon or nitrogen sources in the medium. Furthermore, at the end of exponential growth the concentration of intracellular metabolites increases and this may cause induction of enzymes absent during vegetative growth. Some of these non-specific enzymes are nevertheless indispensable for sporulation. Thus, mutants of *B. subtilis* deficient in some enzymes of the TCA cycle such as aconitase are able to grow vegetatively but cannot sporulate because sporulating cells require a functional TCA cycle for ATP production.

The late enzymes include those involved in the synthesis of spore components such as dipicolinic acid (DPA) and spore cortex peptidoglycan. There are certainly many other spore-specific enzymes involved in the later stages but these have not been identified.

Genetic control of sporogenesis

It is well established that the morphological and biochemical events leading to the formation of a mature spore from a vegetative are under the control of genes or, perhaps, operon-like groups of genes.

The problem of how many genes control sporogenesis and whether they are scattered over the chromosome or clustered in a few regions under the control of a few regulatory genes has been studied in *B. subtilis* (Marburg strain) although some genetic studies are also being made on other spore-formers. Asporogenous mutants either spontaneous or induced with mutagenic agents such as ultraviolet light, heat treatment, or *N*-methyl-*N'*-nitro-*N*-nitrosoguanidine have been isolated. Two types of mutants occur: asporogenous mutants (Sp^-) which are incapable of producing dormant spores, and oligosporogenous mutants (Osp) which produce spores at a low frequency (*e.g.* 10^{-5}) under normal conditions of sporulation. Either of these types may be blocked at any stage.

Restoration of sporulation capacity can be accomplished, in *B. subtilis* Marburg, in both groups of mutants by transformation or by transduction with phage PBS-1. The latter method has proved to be the more useful for mapping the spore genes since larger segments of the genome are transferred than in transformation.

Chromosomal location of sporulation mutants

The first classification of asporogenous mutants (Table 3) is based on their ability to synthesize 'early' products of sporulation, the extracellular protease

Table 3. Classification of asporogenous mutants

Mutant type	Phenotype			
	Protease	Antibiotic	Competence	Spore formation
Wild	+	+	+	+
Spo A	−	−	−	−
Spo B	±	+	±	−
Spo C	+	+	+	−
Spo D	−	+		±

(*Prot*) and antibiotics (*Ab*). The transformability or competence (*Co*) of the strains is also used as a trait in this classification. By correlating these traits with the morphological stages of spore development four classes of phenotypes were found. Mutants of class A have been found to be deficient in the production of a protease and an antibiotic and to have decreased competence for DNA uptake. Such mutants (*Spo A*) are generally blocked at stage O of sporulation. This finding suggests, but by no means proves, that these traits might be essential for early steps of the sporulation process. There are Sp⁻ mutants which are only deficient in one of these traits. Most Sp⁻ mutants capable of reaching stage II have been found to retain the ability to produce protease and antibiotic and are competent for transformation (*Prot*⁺, *Ab*⁺, *Co*⁺).

The class B mutants are much like *Spo A* mutants except that they produce a reduced protease activity as compared to the wild-type but are defective in antibiotic excretion. The *Spo A* and *Spo B* mutants, in common with other early blocked asporogenous mutants, have been found to be pleiotropic in the sense that they are deficient in enzymes involved in late events such as the biosynthesis of DPA. In these mutants, several apparently independent characters are lost by a single step mutation.

Spo C mutants are blocked in 'late' events, since they produce the 'early' products. *Spo D* mutants are oligosporogenous and defective in exocellular protease production.

There exists also a group of mutants in which the inability to sporulate is not due to an alteration of any of the genes governing sporulation-specific events. As mentioned earlier, the enzymes of the TCA cycle are part of the vegetative cell machinery and none of them, so far tested, is present in the dormant spore (see Table 1). However, certain mutants devoid of aconitase will grow if provided with glutamate, and these were found to be asporogenous even when grown in the presence of the amino acid. Furthermore, transfer of the defective gene to sporogenic strains by transformation caused these strains to become asporogenous.

Transduction analysis indicates that genes affecting sporulation are located in all four linkage groups into which the chromosome of *B. subtilis* is divided. The data clearly indicate that *Spo* genes are scattered throughout the genome, although clusters of some mutant types can be observed. A number of *Spo B*, *Spo C* and *Spo D* mutants are located in the area adjacent to the streptomycin-resistance locus and to genes coding for RNA. Another cluster of spore-defective mutant genes appears to be located in the terminal region of the chromosome (group IV) in a region which lacks auxotrophic markers adjacent to *lysine-1* gene.

In spite of the progress made in the study of the genetics of sporogenesis, the data available provide no information regarding either the primary locus for the initiation of the complex sporulation process or the number of regulatory genes involved. Mutations in more than 25 different genes of *B. subtilis* have been reported which prevent the development of spores without affecting vegetative growth but the precise function of any one of them in sporulation remains unknown.

The participation of many spore genes in the formation of the spore is also evident from examination of the products of gene expression. Sporogenesis is accompanied by extensive breakdown and resynthesis of proteins and RNA, and is dependent upon new protein and RNA synthesis. If inhibitors of protein synthesis (such as chloramphenicol) or of RNA synthesis (such as actinomycin D) are added early in sporogenesis, spore formation is inhibited. The proteins which are incorporated into the spore are synthesized *de novo* from amino acids.

We can, therefore, consider sporogenesis as a process in which a number of specific genes, which were repressed during vegetative growth, are activated. Studies of the RNA formed during sporogenesis support this concept. The types of RNA found in sporulating cells resemble those of exponentially growing cultures: all fractions of RNA, including stable (rRNA and tRNA) and labile (mRNA), are formed. The mRNA formed during sporogenesis differs from that produced during vegetative growth or during outgrowth from germinated spores. Firstly, it has a different composition. Secondly, RNA/DNA hybridization experiments give evidence of transcription of specific genes during sporulation since competition experiments indicate that the mRNA formed during sporulation differs from that found in the vegetative cells.

Transformation of dormant spores into vegetative cells

Three sequential processes are responsible for the transformation of a spore into a vegetative cell: activation, which conditions the spore to germinate in a suitable environment; germination, an irreversible process, which results in the loss of the typical characteristics of a dormant spore; and outgrowth, in which new classes of proteins and structures are synthesized leading to the conversion of spore into a new vegetative cell.

Activation

When a freshly prepared spore suspension is introduced into a medium supporting germination, it will germinate very slowly or not at all unless it has been previously activated by heat or other treatment. Activated spores are still dormant structures and retain the characteristics of spores. Activation is a reversible process which can be effected by treatment with reducing agents or exposure to low pH. What is known about it suggests that it may involve the reversible denaturation of a macromolecule, possibly protein. Whatever its nature it presumably alters specific sites in the spore so that they can respond to germinating agents.

Germination

Germination is an irreversible process in which a number of events take place shortly after exposure of activated spores to specific stimulants. It is accompanied by swelling of the spore, either rupture or absorption of the spore coat, and the loss of a number of typical physiological properties of the spore. Thus, there is loss of resistance to environmental stress, loss of refractility, increase in permeability, release of spore components (dipicolinic acid, Ca^{2+}, spore peptides) and increase in metabolic activity. Collectively the process is degradative and probably involves a number of enzymic reactions. Since spores will germinate normally in the presence of inhibitors of protein and nucleic acid synthesis, the enzymes responsible for germination are presumably already in the spore. Specific chemical stimulants trigger the rapid germination of activated spores. The choice of the germinating agent depends on the species of spore and to some extent on the age of the spores and the medium in which they were produced. Many chemicals e.g. L-alanine, adenosine, glucose and salts, can induce germination. Germinating agents are usually normal metabolites which probably interact with stereospecific sites on the spore and are consumed during germination. The nature of the triggering action suggests that an enzymic reaction is involved which eventually initiates the degradative series of reactions characteristic of germination. Although the trigger mechanism is unknown, a given spore preparation may be made to germinate by treating it in quite different ways. For example, germination may be induced by L-alanine; this requires heat-activation, or it may be induced by calcium dipicolinate, which does not. In addition to physiological germination, disruption of the outer structure of the spore by surface active agents or mechanical rupture may also lead to germination.

Outgrowth

Germination is followed by a period of active biosynthetic activity called outgrowth. During this phase, proteins and structures characteristic of vegetative cells are synthesized *de novo*. Outgrowth ends at the time of cell division and resumption of vegetative growth. The conditions for outgrowth are usually different from those for germination. Germination and outgrowth have different temperature optima and most spores need nutrients for outgrowth which are not required for germination.

Synthesis of macromolecules during outgrowth

RNA

Sporulation involves the expression of unique spore phenotypes: outgrowth involves the expression of vegetative phenotypes. The vegetative mRNA molecules required for protein synthesis during outgrowth are therefore either stored as stable molecules in the dormant spore or are early transcriptional products formed during outgrowth. One would expect that mRNA formed during sporulation would be spore-specific, whereas that formed during outgrowth would be specific for vegetative growth. There is no evidence that all mRNA species produced in sporulating cells are stable and retained in the spore. Examination of dormant spores has failed to reveal the presence of mRNA. Moreover, when spores are germinated in the presence of actinomycin D, an antibiotic which inhibits the synthesis but not the functioning of mRNA, the synthesis of proteins which normally commences at the onset of outgrowth is completely inhibited.

RNA is the first type of macromolecule to be synthesized in outgrowing spores. Under optimal conditions, spores of *Bacillus cereus* T commence germination 2 minutes after addition of the germinating agent, RNA synthesis is observed at 2·5–3·0 minutes and protein synthesis at 4 minutes.

All classes of RNA (transfer, ribosomal and messenger) are synthesized during outgrowth but the rate of synthesis depends on the germinating medium. Throughout the period of outgrowth, as well as during vegetative growth, addition of actinomycin D blocks further RNA synthesis immediately and, after a delay of several minutes, protein synthesis. The half-life and base composition of the mRNA synthesized during outgrowth are the same as in vegetative cells. It has been demonstrated that the mRNA molecules formed during outgrowth are different from those formed during sporulation since the two kinds did not compete in DNA-RNA hybridization experiments. It is now well established that spores which seem to be devoid of functional mRNA contain a DNA-dependent RNA polymerase which enables them to synthesize all the mRNA's required for the transcription of early functions during outgrowth.

It therefore appears that part of the genome characteristic of vegetative growth is transcribed both during outgrowth and vegetative growth, whereas that characteristic of spore formation is transcribed only during sporogenesis.

Proteins—enzymes

Since spores are devoid of functional mRNA, outgrowth provides another experimental system in which the order and function of transcription and translation of genetic information can be studied. Cytological studies show that the syntheses of structural elements do not occur at the same time but are ordered. In outgrowing spores different kinds of proteins appear in a definite sequence. This was first shown by following the appearance of vegetative cell antigens during outgrowth. More evidence was obtained by labelling the proteins of outgrowing spores at various time intervals and by analyzing the different protein species by acrylamide gel electrophoresis.

The sequence of enzyme synthesis has been examined most extensively during outgrowth of spores of *B. cereus* T. Activated spores of this organism germinate completely in about 5 minutes and produce a population in which the developmental changes occur synchronously. Figure 5 shows the time course of synthesis of some of the enzymes during outgrowth; the synthesis of each begins at a specific time and continues for only a fraction of the total period. The next burst of enzyme synthesis does not occur until after cell division. Ribosome synthesis and vegetative cell wall synthesis start early, while DNA synthesis does not occur until just before cell division. Since the mRNA formed during outgrowth probably has a half-life of only a few minutes, the differences in the time of synthesis of macromolecules indicate that outgrowth, like sporogenesis, is a period in which developmental changes are determined by the time of transcription of particular genes.

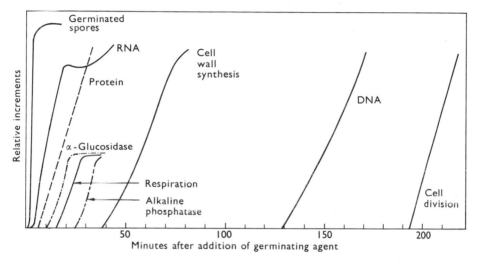

Figure 5. Synthesis of macromolecules during outgrowth of spores of *B. cereus* T in synthetic medium.

Cell wall formation

Cell wall synthesis in outgrowing spores can be detected 20 minutes after the induction of germination, at a time which coincides with the swelling of the germinated spore. The formation of cell wall material during outgrowth can be followed by the incorporation of radioactive diaminopimelic acid (DAP) into the hot acid-insoluble fraction. When chloramphenicol is added 10 minutes after the beginning of germination there is inhibition of the incorporation of ^{14}C-DAP, suggesting that synthesis of cell wall of the developing cell is dependent on the formation of certain enzymes. It has been demonstrated that some diaminopimelic acid incorporated in the new cell wall is derived from a DAP-containing peptide pre-existing in the spore and not completely hydrolyzed and excreted during germination.

Regulation of sporogenesis and outgrowth

During sporogenesis part of the genome is activated and part is repressed. During outgrowth the reverse occurs. Outgrowth and sporogenesis are similar in that both occur in the absence of DNA replication, are dependent upon new mRNA synthesis, and are accompanied by the ordered synthesis of new classes of proteins. Moreover, both have many similarities to the process of sequential enzyme synthesis in bacteria and it is possible that eventually induction and repression may explain the control of these processes.

It could be assumed that the spore genome is under either negative or positive control. The negative control mechanism implies that a regulatory protein acts as inhibitor to stop transcription of the spore genome unless an inducer is present. In the positive control mechanism it is assumed that a regulatory protein is produced which promotes the transcription of the spore structural genes.

The discovery that a protein factor (*sigma factor*), normally associated with *E. coli* DNA-dependent RNA polymerase confers template specificity on the enzyme, provides a specific positive control mechanism for regulation at the transcription level related to systems in which genes are activated in a sequential manner as is the case during sporulation or after virus infection.

It is also important to consider the role of the physiological control on the expression of the spore genes at the translation level. Such a control may be operating throughout all the stages of sporulation and outgrowth by different mechanisms: amount and stability of mRNA produced, repression of spore enzyme synthesis, feedback inhibition of the activity of these enzymes, and finally catabolic repression of some specific spore inducible enzymes.

Medical, veterinary and industrial importance of spore-forming bacteria

The importance of spore-forming bacteria in industrial technology and their medical and veterinary significance is now well established. Spore-formers as insect pathogens are now widely used as effective microbial insecticides (in particular *B. popilliae* and *B. thuringiensis*). On the other hand, certain species of the genera Bacillus and Clostridium are highly pathogenic to members of the animal kingdom. This pathogenicity is frequently associated with a high degree of mortality resulting from the production of extremely potent toxins in the infected animal or in its food. These toxins are the cause of death in cases of tetanus, botulism and anthrax and they account for the importance of spore-forming bacteria and spores in medicine and in veterinary science.

ROLE OF SPORES IN NATURE

The evolutionary development of controls for sporogenesis and outgrowth is best understood by considering the role of spores in nature. It is generally assumed that spores are important for survival and dissemination. Although bacterial spores are characterized by their resistance to adverse conditions, it is not really known whether these conditions are encountered naturally. For example, there is little evidence that resistance to high temperature or even to those antibiotics found in soil are significant factors in survival. Resistance to desiccation, however, may be of greater importance. In a study of the microbial flora of the Egyptian desert, it was found that at times of drought, three spore-formers (*B. subtilis*, *B. licheniformis* and *B. megaterium*) predominated, whereas few non-spore formers were recovered.

The formation of a spore may tide the organism over an unfavourable period when a reduction in the supply of certain nutrients occurs. This type of adaptation confers selective advantage on organisms which live in a fluctuating environment.

YEASTS

In yeast, sporulation leads to the formation of ascospores within modified yeast cells, called asci, through a process which involves reductive nuclear division (meiosis)*. The number of spores per ascus, their appearance, shape and the manner in which they are formed are characteristics of the *Saccharomycetaceae*. Yeast of the genera *Saccharomyces*, *Saccharomycodes*, *Endomycopsis*, *Pichia*, *Hansenula* and *Hanseniaspora* produce four spores per ascus. In this section we will discuss primarily sporulation in the well-studied *Saccharomyces cerevisiae*. This yeast is diploid in the vegetative phase (Figure 6). Vegetative cells divide by budding (mitosis). In appropriate conditions (see below) meiosis occurs and precedes sporulation. Four haploid spores are produced and formed within the ascus. Unless these spores are isolated they will fuse with their opposite mating-type neighbours in the ascus to restore the diploid state. In heterothallic strains†, clones of cells from a single spore will grow for many generations in the haploid state. Lindegren and Lindegren showed, by carrying out the appropriate experiments, that they could identify a single pair of alleles, *a* and *α*, which control the mating response (mating-type). In homothallic strains the haploid spores give rise to clones containing primarily diploid cells. These diploids contain *a* and *α* mating-type alleles.

---

* During meiosis a diploid cell, *i.e.* one containing two sets of chromosomes, divides twice in succession to give four grand-daughter cells. In the same period the chromosomes divide only once. The net result is that the number of chromosome sets per cell becomes reduced from two to one, *i.e.* the cells become haploid.

† In heterothallic strains mating can occur only between certain pairs of *different* individuals. This compatibility is genetically determined and allows the individuals to be divided into mating-types. In homothallic strains haploid derivatives from the *same* individual are able to mate.

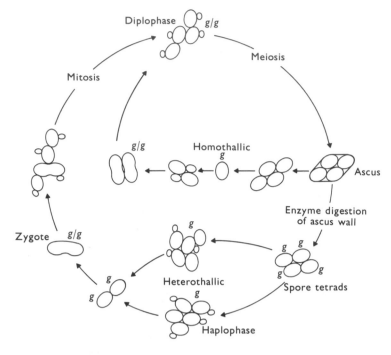

Figure 6. Life cycle of *Saccharomyces* species.

Properties of yeast spores

In contrast with the bacterial endospore, yeast spores are not heat-resistant. They do, however, differ in their chemical composition from vegetative cells of yeast. A superficial lipid layer is contained in the spore wall but is absent from the walls of vegetative cells. This layer can be stained with Sudan Black B. Further, spore walls stain more readily with Gallocyamin, are more resistant to digestion by snail enzymes, and contain higher concentrations of glucose, mannan, and trehalose than do vegetative cells. Spores contain less RNA and a lower free ribonucleoside content but higher contents of proline and glutamic acid than do vegetative cells.

Metabolic changes associated with yeast sporulation

Meiosis in yeast is a process of intracellular differentiation wherein cells stop growth by budding and mitosis and enter meiosis. A particular physiological state must be established prior to the commitment to sporulation. When cells are transferred to sporulation medium (potassium acetate) the ability to sporulate is dependent upon the physiological age of the culture. Wild-type diploid cells in sporulation medium arrive at a point of commitment to meiosis which may be observed by returning portions of sporulating cultures to glucose nutrient medium in which cells do not ordinarily sporulate. Cells returned to nutrient medium early in sporulation revert to mitotic division: later the cells become committed to meiosis and give rise to asci in nutrient medium. The development of structures during meiosis is shown in Figure 7. The fraction of committed cells increases with time and is highly correlated with the percentage of cells at or beyond the binucleate condition. From these findings, it is clear that the point of commitment to meiosis coincides closely with the onset of the reductional division.

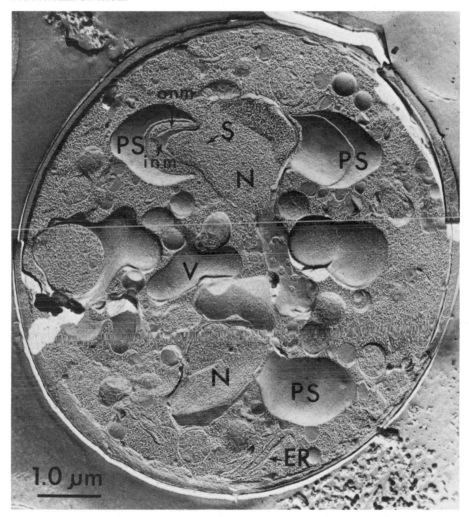

Figure 7. Freeze-etch preparation of a yeast cell in an early stage of ascospore formation: onm = outer nuclear membrane; inm = inner nuclear membrane; S = spindle fibres; N = nucleus; PS = pre-spore; ER = endoplasmic reticulum. Magnification = 18,700× (E. Guth, T. Hashimoto and S. F. Conti, unpublished micrograph).

Sporulation has long been recognized to depend upon the environmental conditions favouring respiratory activity. In particular the time at which acetate metabolism is developed correlates with the physiological age at which sporulation begins. This competency appears during later stages of growth when glucose is exhausted and the cells have adapted to metabolize the end-products of glucose utilization. In some strains a preliminary period of growth on acetate considerably improves the ability to sporulate.

Figure 8 shows the sequence of biochemical events occurring during sporulation. After cells are transferred to acetate medium, there is an immediate rise in dry weight followed four hours later (T_4) by a meiotic round of DNA replication. This DNA synthesis achieves a percentage increase equivalent to the total ascus production (see Figure 8). Accumulation of glycogen and lipid commences in the early stages. RNA and protein content per cell increase to a

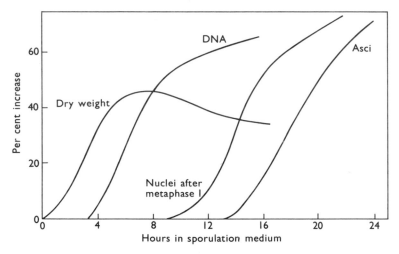

Figure 8. Changes in cellular characteristics during sporulation in *Saccharomyces cerevisiae*.

maximum level at about T_{10} and then decline. The measurements of the rate of protein synthesis during sporulation revealed two periods of maximum synthetic activity: an early phase coincidental with increases in the ratio of cellular DNA/RNA and a later phase during ascus spore formation. Protein synthesis is required throughout sporulation process, and the overall process is characterized by extensive protein and RNA turnover. Thus the proteins required for the completion of ascus production are continuously produced until the last stages of the process.

Genetic control of sporulation

A number of genes have now been recognized which regulate sporulation and meiosis in yeast. In *Schizosaccharomyces pombe* Bresch and his co-workers have demonstrated that there are distinct, multiple genes responsible for the meiosis associated with sporulation and for the dissolution of the separate cell walls during regenerative fusion. Esposito and Esposito have isolated a number of temperature-sensitive recessive mutants which are blocked in meiosis but undergo mitotic divisions. These results suggest that the two modes of nuclear division are under separate genetic control. Sporulation-deficient mutants can be recognized by their inability to form viable asci detectable by microscopic examination. Some mutants permit meiosis to proceed through the first nuclear division while others proceed through the second meiotic nuclear division but do not complete ascus formation.

The availability of a number of temperature-sensitive mutations of meiosis permits one to analyze the meiotic process by temperature shift experiments. Shifting cultures from a non-permissive to the permissive temperature during meiosis defines the time of expression of a temperature-sensitive function. The reverse experiment defines the end and thus the duration of the dependence on this gene product. Figure 9 shows a plot for such temperature-sensitive mutants of meiosis in *Saccharomyces cerevisiae*.

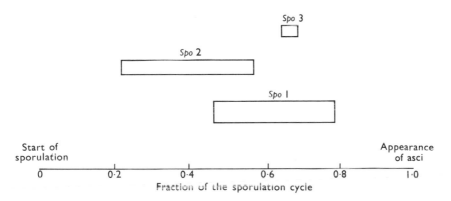

Figure 9. A physiological map of the temperature sensitive periods of three conditional meiotic mutants in *Saccharomyces cerevisiae*.

If non-viability is due to commitment, commitment occurs during the reductional division and the fraction of non-viable should be equal to the fraction of multinucleate cells in the culture. In some of the mutants, lethality precedes the temperature-sensitive period and therefore cannot be due to events occurring in the temperature-sensitive period. Commitment to meiosis in yeast may actually precede the reductional division as has been shown in the *Lilium* where commitment to meiosis occurs in the late G2 of the premiotic interphase.

The conditional mutants of meiosis in addition to indicating unique elements in the meiotic process, also provide a promising approach for studying development. Meiosis would appear to be a developmental process in which different elements are expressed at unique periods of the meiotic cycle. Little is known at the moment of the biochemical nature of these components.

ACKNOWLEDGEMENT

Figure 2, from Madame A. Ryter, Pasteur Institute, Paris and Figure 7 from Dr. S. A. M. Conti, University of Kentucky.

FURTHER READING

Bacteria

1   Halvorson H. O., Vary J. C. and Steinberg W. (1966) Developmental changes during the formation and breaking of the dormant state in bacteria. *Ann. Rev. Microbiol.* **20**, 169.
2   Murell W. G. (1967) The Biochemistry of the Bacterial Endospore. *Adv. Microbiol. Physiol.* **1**, 77.
3   Campbell L. L. (ed.) (1969) *Spores IV*. American Society for Microbiology.
4   Schaeffer P. (1969) Sporulation and Production of Antibiotics, Exoenzymes and Exotoxins. *Bacteriol. Res.* **33**, 48.
5   Gould G. W. and Hurst A. (1969) *The Bacterial Spore*, p. 724, Academic Press, London.
6   Mandelstam J. (1969) Regulation of Bacterial Spore Formation. *Symp. Soc. Gen. Microbiol.* **19**, 377.
7   Hanson R. S., Peterson J. A. and Yousten A. A. (1970) Unique Biochemical Events in Bacterial Sporulation. *Ann. Rev. Microbiol.* **24**, 53.
8   Szulmajster J., Kerjan P. and Maia J. C. C. (1971) Regulation of Sporogenesis in *Bacillus subtilis*. In *Genetics of Ind. Microorganisms* (eds. Vanek Z., Hostalek Z. and Cudlin J.) Czech. Acad. Sc. publ.
9   Kornberg A., Spudich J. A., Nelson D. L. and Deutscher M. P. (1968) *Ann. Rev. Biochem.* **37**, 51.

Yeasts

1   Fowell R. R. (1969) Sporulation and Hybridization of Yeast. In *The Yeast* (eds. A. H. Rose and J. S. Harrison), p. 303. Acad. Press, London.
2   Croes A. F. (1967) Induction of Meiosis in Yeast. I. Timing of Cytological and Biochemical Events. *Planta* **76**, 209.
3   Esposito M. S. and Esposito R. E. (1969) Genetic Control of Sporulation in Saccharomyces. *Genetics* **61**, 79.
4   Bresch C., Muller G. and Egel R. (1968) Genes involved in Meiosis and Sporulation of Yeast. *Mol. Gen. Genet.* **102**, 301.
5   Esposito M. S., Esposito R. E., Arnaud M. and Halvorson H. O. (1970) Conditional Mutants of Meiosis in Yeast. *J. Bacteriol.* **104**, 202.

Appendix A
Bacterial Classification

There are two questions to answer before bacterial classification can be considered in detail: what are bacteria and how are they related to the other major groups of micro-organisms? Traditionally micro-organisms have been divided into protozoa, algae, fungi and bacteria. This division is quite arbitrary and unrealistic, but although Haeckel suggested in 1866 that all micro-organisms should be grouped together in the 'Protista' it has taken until the present time for the idea to be accepted that this group should include all unicellular organisms and those multicellular organisms in which the cells are the same and there is no tissue differentiation.

Prokaryotic and eukaryotic cells

Micro-organisms are still divided into protozoa, fungi, algae and bacteria but recent cytological and biochemical studies have demonstrated a more fundamental division within this group. Two distinct cell types have been observed and these have been called *prokaryotic* and *eukaryotic* cells.

All living organisms have a basic pattern of metabolic processes which enables them to grow and divide. In the eukaryotic cell many of these complex processes are associated with definite intracellular structures. Two examples of these are the mitochondria which contain enzymes concerned with production of energy and the chloroplasts which are associated with the utilization of light energy for biosynthetic purposes. These organelles are absent from prokaryotic cells and the corresponding metabolic processes are associated with cell membranes. There are fundamental differences in both the structure and division of nuclear (genetic) material. In prokaryotic cells the nuclear material consists only of DNA which probably exists as a single circular double strand attached to the plasma membrane, but not separated from the remainder of the cell contents (cytoplasm) by a membrane. Eukaryotic cells are different in that the nuclear DNA is associated with proteins and occurs in the form of definite structures (chromosomes) surrounded by a membrane. During division there is a complex sequence of events to ensure that the genetic material is correctly distributed between the two daughter cells. In eukaryotic but not in prokaryotic

Table I. A comparison of Prokaryotic and Eukaryotic cells.

	Prokaryotic cells	Eukaryotic cells
Cellular Structures		
Cell wall composition	peptidoglycan present	peptidoglycan absent
Membrane composition	sterols absent from bacterial membranes	sterols abundant in mammalian membranes
Mitochondria	absent	present
Chloroplasts	absent	present in photosynthetic organisms
Nuclei		
Nuclear membrane	absent	present
Chromosome number	I	more than I
Mitotic apparatus	absent	present
Nucleolus	absent	present
Histones	absent	present
Golgi apparatus	absent	present
Mesosomes	present	absent
Microtubules	absent	present
Ribosomes (sedimentation coefficient)	70 S	cytoplasmic ribosomes 80 S chloroplast and mitochondrial ribosomes similar to 70 S
Movement		
Cytoplasmic streaming	does not occur	may occur
Amoeboid movement	does not occur	may occur
Flagella	if present are simple structures	if present are of the '9+2' type
Metabolism		
Oxidative phosphorylation associated with	membranes	mitochondria
Photosynthesis associated with	membranes	chloroplasts
Reduced inorganic compounds as energy source	may be used	cannot be used
Non-glycolytic mechanisms for anaerobic energy generation	may occur	do not occur
Poly β-hydroxybutyrate as a reserve storage material	may occur	does not occur
Nitrogen fixation	may occur	does not occur
Peptidoglycan synthesis	occurs	does not occur
Exo- and endo-cytosis	do not occur	may occur

cells the cytoplasmic contents often stream in an organized manner. Both types of cells may possess flagella: in prokaryotic cells these are relatively simple structures but in eukaryotic cells the flagellum is a complex arrangement of filaments surrounded by a membrane. These major structural differences between the two types of cells are summarized in Table 1 as are also a number of metabolic differences. It is very important to remember that the major differences between the two cell types are organizational and not functional.

Micro-organisms showing the typical prokaryotic cellular organization are further subdivided on the basis of mechanism of movement and general cellular metabolism into Blue-green algae and Myxobacteria which have no obvious organelle associated with movement, but move by 'gliding'; Spirochaetes, where undulations of the cells are caused by axial filaments lying between the membrane and the cell wall; and bacteria which, if motile, have characteristic flagella. The discussion in this appendix will be limited to the bacteria, but the general principles of biological classification are applicable to all organisms.

Some general properties of bacteria

When we examine the bacteria we are at once impressed by their ubiquity and diversity. The range of environmental conditions in which bacteria will grow is wide: growth temperatures can range from less than $0°$ to more than $60°$; salt concentrations from that of the Dead Sea to that of distilled water; and pH values from 0 to more than 10. The ability of bacteria to grow in a wide variety of environmental conditions is a reflection of their metabolic versatility. All bacteria have the same fundamental requirements for growth—a source of energy, of reducing power, of carbon, nitrogen and other elements—but there is considerable variation in the sources that can be utilized for these purposes. Energy for bacterial growth may be obtained from light, from oxidation reduction reactions involving inorganic compounds for example H_2S, Fe^{2+} and H_2, or from oxidation-reduction reactions involving organic compounds. The sources of carbon may be carbon dioxide or any one of a large number of organic compounds. The hydrogen donors can be either inorganic or organic compounds. However, these metabolic differences are in fact modifications and extensions of a basic metabolism common to all bacteria.

Structurally there is again great diversity and bacteria can vary in shape from a sphere to a cork-screw shaped rod, and in size from 0.1 μm^3 to 500 μm^3. Internally there are varying degrees of cellular organization often associated with membranous bodies (see Chapter 1). All these differences, although superficially large are variations on the basic 'prokaryotic theme'. Nevertheless the bacteria comprise a complex group of organisms and the necessity for bringing order into this rather confusing picture requires that we should attempt to classify them.

The function of classification

The object of a classification is to group together bacteria having similar properties and to separate those with dissimilar properties. This can obviously be an arbitrary process and the form it takes may be decided by the interests of the person making the classification. For example a medical bacteriologist might choose to start by dividing bacteria into saprophytic (free-living) and parasitic organisms and then to divide the latter into pathogens and non-pathogens. Similarly, the number and size of groups in the sub-divisions will depend on the number of common properties that bacteria must have before we decide to include them in any particular group. If a large number of properties is required the groups will consist of a small number of bacterial types, whereas if there are only a few properties in common, large numbers of bacterial types will be included in that group. This difference in group size is the basis for a hierarchical classification. A classification has the advantage over a catalogue in that it will show relationships between bacteria and between groups of bacteria. Once a bacterium has been assigned to a group it should also be possible to predict other characters that it might possess. The aim of most systematists is that the classification should not be arbitrary but 'natural', i.e. it should show the relationship between various groups of bacteria. It is implicit in this definition that relationship means relationship by ancestry, i.e. phylogenetic, and if we assume that all bacteria existing today are derived from a common ancestral type then the degree of similarity between various groups will depend on the time at which their evolutionary pathways diverged. We can only speculate on the ancestry of bacteria and it is preferable to define relationship as 'overall similarity' and to adopt Gilmour's definition that a natural classification is a general arrangement intended for general use by all scientists.

The basic unit for most biological classifications is the individual organism. Similar organisms are grouped together to form a species, related species are grouped into a genus, related genera are grouped into a family, related families into an order, and so on. This hierarchical system is satisfactory for plants and animals but are we justified in applying it directly to bacteria and, if we do, will the taxonomic groups (taxa) have the same significance for microbiologists as they have for other systematists? It is unrealistic to use the individual bacterium as the basic unit, so the isolated bacterial culture derived from a single cell is the fundamental unit in bacterial classifications. Isolates having similar properties are grouped together to form the next taxonomic unit, the species. In *The New Systematics* (1940) a species is defined as a self-perpetuating unit which has a definite geographical distribution area, which is morphologically distinguishable from other groups, and which does not normally interbreed with related groups. This definition cannot be applied to bacteria and there is no agreement among bacterial systematists as to a suitable definition of a 'bacterial species'.

The difficulty in defining a bacterial species arises from the failure of bacteriologists to use similar criteria when classifying different types of bacteria. Ravin (1963) has distinguished three types of 'bacterial species'; these are: *taxospecies* in which the bacteria are phenotypically similar (this is the normal sense in which the term species is used); *genospecies* in which the bacteria can exchange genetic information; and finally *nomenspecies* where we have a group of bacteria bearing a particular binomial name irrespective of any other claim it may have to be a species. Later in this appendix we will consider precise alternatives for 'species' and 'genus', but until then these terms will be considered as names for two convenient taxonomic groups.

Properties used in bacterial classification

Before we draw up a classification, we have to consider the various characteristics that are used to classify bacteria. A hundred years ago knowledge of bacteria was limited to their size and shape, and early classifications were based entirely on these properties. Later, the classifications were modified to include physiological properties. Today there is an extremely wide range of properties used in classifying bacteria.

Morphological characters
These include the size, shape, arrangement of the individual bacterial cells and the occurrence of extracellular appendages. For example, the occurrence of fimbriae and flagella and their arrangement on the bacterial cell are characters used extensively in classifications. Detailed knowledge of bacterial structure has increased very rapidly during the past few years as a result of improvements in the techniques of electron microscopy. This has extended and in some ways supplanted earlier methods using specific staining reactions for various cellular constituents and components.

Cultural characteristics
The majority of bacteria are able to grow under a wide variety of cultural conditions in the laboratory and in many cases the appearance of the culture under particular conditions of growth may be characteristic of the organism and hence of value in classification.

Resistance tests
Some bacteria are renowned for their ability to resist adverse conditions by spore formation (see Chapter 9). Other bacteria can withstand high temperatures or the presence of antimicrobial agents. Resistance tests reflect metabolic or structural characteristics of the different bacteria and are useful in bacterial classification.

Metabolism and nutrition

The range of compounds that can be used for growth is characteristic of any given organism as are the products of metabolism which are often excreted into the growth medium. Although many bacteria can utilize relatively simple compounds as sources of carbon or nitrogen they may also require minute amounts of specific growth factors which they cannot synthesize. As for oxygen requirements, bacteria show a complete range from the aerobic bacteria which will grow only in the presence of oxygen, through the facultatively anaerobic bacteria which will grow both in the presence and absence of oxygen, to the obligate anaerobes which are inhibited even by traces of oxygen.

Biochemical reactions

There are many specific and characteristic enzymic reactions which can be easily measured.

Molecular structure

In this category we can include DNA base composition and hybridization data, comparisons of ribosomal proteins and RNA composition.

Genetic relationship

Genetic information may be transferred from one strain of bacteria to another by conjugation, transformation or transduction (Chapter 7) and, in general, gene transfer is much more frequent between phenotypically similar strains than between strains showing considerable differences in their phenotype.

Conventional methods of classification

Bergey's system

These then are the types of character by which bacteria may be classified and at this point we must outline a scheme in which bacteria are grouped satisfactorily and the relationships between the different groups are demonstrated. The most widely-used system of classification is that described in Bergey's *Manual of Determinative Bacteriology* which attempts to show the phylogenetic relationships between different bacteria. The classification in the 7th Edition is based on the questionable assumption that the photolithotrophic bacteria that exist today are probably more closely related than other existing bacteria to the primordial types. Bacterial taxonomy is undergoing extensive revision at the present time and as an 8th Edition of 'Bergey' is in preparation the complex formal classification of the 7th Edition will not be described. The manual does however describe all genera and species of bacteria and for this reason is a very good reference book. For most purposes it is sufficient to have a simple scheme

Table 2a. Gram-positive bacteria.

Shape and arrangement	Spores	Sugar fermentation (end products)	Oxygen requirements	Misc.	Genera	Families
	−		Aerobic		Sarcina (*Sarcina lutea*)	
	+		Aerobic		Sporosarcina	
	−		Anaerobic		Zymosarcina	Micrococcaceae
	−				Micrococcus (*Micrococcus lysodiekticus*)	
	−				Staphylococcus (*Staphylococcus aureus*)	
	−	Lactic acid			Streptococcus (*Streptococcus faecalis*)	
	−	Lactic acid			Leuconostoc (*Leuconostoc mesenteroides*)	Lactobacteriaceae
	−	Lactic acid			Lactobacillus (*Lactobacillus bulgaricus*)	
	−	Propionic acid			Propionibacterium	Propionibacteriaceae
	−	Oxidative, only weakly fermentative			Corynebacterium (*Corynebacterium diphtheriae*)	Corynebacteriaceae
	+		Aerobic		Bacillus (*Bacillus subtilis*) *Bacillus megaterium*)	Bacillaceae
	+		Anaerobic		Clostridium (*Clostridium welchii*)	

Table 2b. Gram-negative bacteria.

Shape and arrangement	Spores	Sugar fermentation (end products)	Oxygen requirements	Misc.	Genera	Families
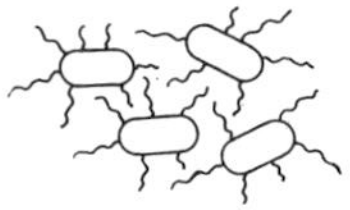	—		Aerobic	Nitrogen fixers (free living)	Azotobacter (*Azotobacter vinelandii*)	Azotobacteriaceae
	—		Aerobic	Nitrogen fixers (symbiotic)	Rhizobium	Rhizobiaceae
	—		Aerobic	Oxidize inorganic compounds	Nitrosomonas Nitrobacter Thiobacillus (*Thiobacillus thiooxidans*)	Nitrobacteriaceae Thiobacteriaceae
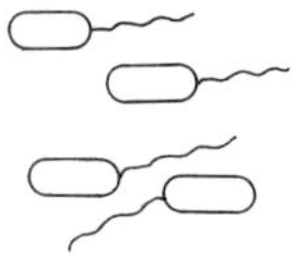	—		Aerobic	Oxidize organic compounds	Pseudomonas (*Pseudomonas aeruginosa*) Acetobacter	Pseudomonadaceae
	—		Facultatively anaerobic	Oxidize organic compounds	Photobacterium Aeromonas	
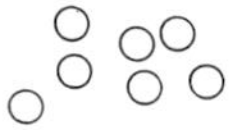	—		Aerobic		Neisseria	Neisseriaceae
	—		Anaerobic		Veillonella	
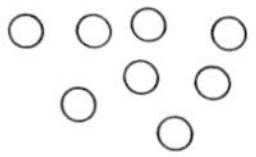					Brucella Pasteurella (*Pasteurella pestis*) Haemophilus (*Haemophilus influenzae*)	Parvobacteriaceae

		Genera (representative species)	Family
Mixed acids	Facultatively anaerobic	Escherichia (*Escherichia coli*) Salmonella (*Salmonella typhimurium*) Shigella Proteus (*Proteus vulgaris*)	Enterobacteriaceae
Butylene Glycol	Facultatively anaerobic	Aerobacter (*Aerobacter aerogenes*)	
Butylene Glycol	Facultatively anaerobic	Serratia (*Serratia marcescens*)	
	Aerobic	Vibrio (*Vibrio cholerae*) Desulphovibrio	Spirillaceae
	Anaerobic	Spirillum (*Spirillum serpens*)	

N.B. Motility and distribution of bacterial flagella are shown diagrammatically in the first column and representative species are shown in parenthesis beneath the genera.

The classification given in Tables 3a and 3b is based on that of Stanier et al (1963).

which will enable us to separate the major groups of bacteria and to outline their properties.

Stanier, Doudoroff and Adelberg's system

A number of schemes of this type have been suggested and the one presented in this appendix is based on that described by Stanier, Doudoroff and Adelberg (1971). Its function is to outline the major groups of bacteria and the properties used to distinguish them.

The first major division we can make is into the Gram-positive and the Gram-negative bacteria. The Gram-positive bacteria are those from which the dye, crystal violet, cannot be removed by washing with ethanol whereas from Gram-negative bacteria it can. Within the Gram-positive group we can distinguish a number of morphological types: the cocci, the non-flagellate bacilli and the flagellate bacilli. The morphological characters, together with pathways for sugar metabolism, enable us to divide Gram-positive bacteria into the major families. The families are then further subdivided into genera on the basis of arrangement of cells after cell division, ability to produce resistant spores, and oxygen requirements (Table 2a).

The classification of Gram-negative bacteria is more complicated. These organisms do not produce resistant spores but as with Gram-positive bacteria many distinguishing features are associated with the shape of cells and the number and position of flagella. Among the Gram-negative bacteria we find the Pseudomonadaceae and Spirillaceae, two large groups in which either a single flagellum or a tuft of flagella is found at either one or both ends of the cell and the flagella are not distributed over the whole bacterial surface. The ability to grow in the presence or absence of oxygen together with the mechanism of sugar metabolism is an important factor in distinguishing various groups. The Gram-negative bacteria show a great versatility in their ability to utilize various substrates and this is used to distinguish the Nitrobacteriaceae which oxidize inorganic nitrogenous compounds, and the Azotobacteriaceae and Rhizobiaceae which fix gaseous nitrogen. Members of the Azotobacteriaceae are free-living organisms whereas the Rhizobiaceae live in association with plants and this symbiotic association is of mutual benefit (Table 2b). This scheme is intended only as a guide to the genera and families of bacteria, not as a definitive 'natural classification'. The list of genera in Tables 2a and 2b is not comprehensive but includes only the commonly-occurring groups of bacteria. It is important to remember that although the majority of members of a particular genus or family will have the properties given in the table there will be exceptions. For example there are motile strains of Lactobacilli and non-motile strains of Salmonellae but as these strains have so many other properties in common with members of the Lactobacilli and Salmonellae they are included in these genera.

Adansonian principles: numerical taxonomy

In the scheme we have described, and in many others proposed in the past, undue importance has been given to certain features. It is this indiscriminate weighting of characters that has led to so much confusion and controversy among bacterial systematists. The alternative is to assume that all observable characters are of equal importance and to assign them equal weight in the classification scheme. This principle is not new and was first proposed by Adanson in 1757. Adanson's views on the construction of taxonomic groups remained in relative obscurity until their recent revival by Sneath and other systematists as the bases for a 'Numerical Taxonomy' (see Sokal and Sneath, 1963).

The Adansonian principles have been clearly defined by Sneath (1958) and Sokal and Sneath (1963). They are as follows:

(1) All characters are of equal importance in creating natural groups.

(2) These groups should be based on as many features as possible.

(3) The relationship between the groups is a function of the similarities of the characters which are being compared.

It follows from these principles that phylogenetic considerations are not taken into account in constructing taxonomic groups and the groups are constructed in an empirical manner. If the Adansonian principles are accepted, classification becomes a mathematical exercise in handling data from studies of bacterial properties. This approach to classification is now called 'Numerical Taxonomy' which is defined as 'Numerical evaluation of the affinity or similarity between taxonomic units and the ordering of these units into taxa on the basis of their affinities' (Sokal and Sneath, 1963). The affinity (or similarity) between strains is defined as the ratio of the number of characters in common/ total number of characters compared. This value is usually expressed as a percentage. Characters which are negative in both strains of bacteria may or may not be included. Before an attempt can be made to classify bacteria within a given group the affinities between all strains must be determined and then strains grouped together on the basis of these affinities. Presentation of the data in such a way that 'natural groupings' are immediately obvious is difficult, particularly when large numbers of bacterial types are involved and the procedure usually adopted is best understood by considering a simple example. If 8 bacterial strains are examined and affinities between then determined we can tabulate the results as in Table 3a. From an examination of the table it is difficult to find subdivisions within this group. The information can however be rearranged by computer so that strains with high affinities are grouped together and separated from strains which have lower affinities. The resulting similarity matrix (Table 3b) shows two distinct groups; strains 1, 7, 5 and 6 are in one group and strains 3, 2, 4 and 8 are in the other. The similarities of the bacteria within the two groups all exceed 70%, but with bacteria from the different groups the values do not exceed 40%.

Similarity matrices

Table 3a. The affinities (characters in common/characters tested × 100) of strains I to 8 are tabulated randomly and there are no obvious groupings among the strains. In Table 3b the similarity matrix has been re-arranged so that related strains are grouped together. There are two groups A and B; the affinities of the strains within the groups exceed 70% and between the groups does not exceed 40%.

Strain Number	1	2	3	4	5	6	7	8
1	100							
2	5	100						
3	10	95	100					
4	0	90	95	100				
5	80	15	35	15	100			
6	70	25	40	10	80	100		
7	95	10	20	10	90	75	100	
8	5	90	75	80	15	25	5	100

Table 3b.

		A				B		
Strain Number	1	7	5	6	3	2	4	8
A — 1	100							
A — 7	95	100						
A — 5	80	90	100					
A — 6	70	75	80	100				
B — 3	10	20	35	40	100			
B — 2	5	10	15	25	95	100		
B — 4	0	10	15	10	95	90	100	
B — 8	5	5	15	25	75	90	80	100

One of the first examples of a computer-based classification was that of the Chromobacterium carried out by Sneath (1957). He examined 45 strains for properties ranging from the size and shape of the bacteria to the presence or absence of certain enzymes. He then used the data to calculate the affinities between the various strains and, after arrangement by computer so that strains with highest affinities were grouped together, obtained the full similarity matrix shown in Figure 1. Here instead of tabulating percentage similarity as in Table 3, the values are represented by different degrees of shading and the 'natural groups' stand out very clearly. Similarity matrices of this type are **but** one of many ways in which these data may be presented, and readers are referred to *The Principles of Numerical Taxonomy* by Sokal and Sneath (1963) for description of other methods.

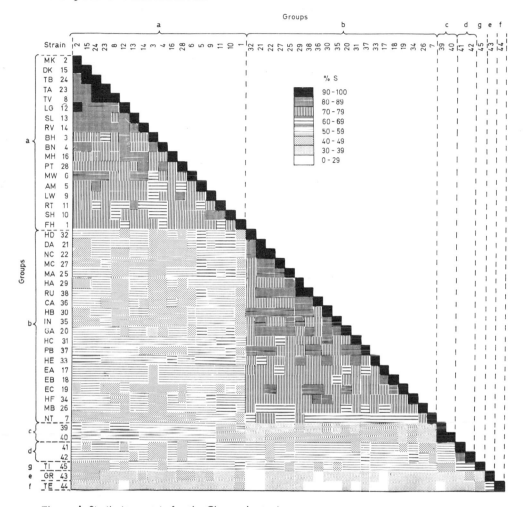

Figure I. Similarity matrix for the Chromobacteria.

The percent similarity for the 45 strains of Chromobacteria studied was based on 105 features and the similarity matrix is shaded to represent the degree of overall similarity.

Taxonomic groups can be constructed from similarity matrices by defining limits within which organisms must fall if they are to be included in any particular group. Organisms with the highest affinities will be in the smallest group and a hierarchical system can be built up depending on the limits imposed. The smallest groups of bacteria would have a very high degree of similarity and the larger groups would have less. It would be possible to continue to use the names 'species', 'genus', 'tribe', 'family' but to avoid confusion it is prefer-able to adopt the nomenclature suggested by Sneath and Sokal (1962) where groups based on cell phenotype are called 'phenons'. There is no need to coin other words in building up a hierarchical system of classification as the various taxonomic groups can be clearly defined by prefacing the word 'phenon' with a number giving the level of similarity. For example the 90-phenon would be a small group of organisms having affinity values of 90% or more, whereas the 70-phenon would include a larger number of bacterial types (Figure 2).

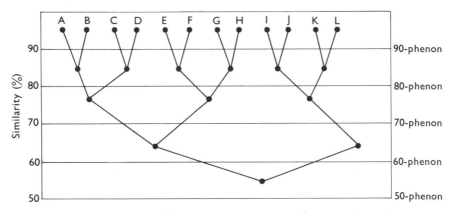

Figure 2. A hierarchical system of classification based on the percentage similarities.

A to H represent groups of organisms in which the affinities between individual bacterial strains exceeds 90%, i.e. these are '90-phenons'. The groups are related to each other in that the affinities between A and B, C and D etc. are between 80–90% and these 90-phenons form part of an 80-phenon. Similarly the affinities of all bacteria in groups A to H exceeds 60% and all the strains are included in the 60-phenon.

Other possible modes of classification

Will systematists decide eventually that cell phenotype is unsatisfactory and that genotype must be used for microbial classification? In a number of species of bacteria the basic information necessary for survival and growth is in one large replicon and additional genetic information which may be required to enable the organism to fill a specific ecological niche is found in smaller repli-cons or plasmids (see Chapter 7). These plasmids may contain 1–2% of the total DNA of the cell. Any attempt to classify bacteria on the basis of nucleotide sequences of the cellular DNA must take this genetic flexibility into account. It may be some time before it becomes possible to determine a complete nucleo-tide sequence for a replicon and even then the amount of work necessary to determine sequences for the replicons of all bacteria would be fantastic.

Analytical techniques are currently available which apparently give some measure of the genetic relatedness between different micro-organisms. The ratio of the bases guanine + cytosine/adenine + thymine is not constant for all bacteria and varies from 1:3 to 3:1. Examination of the data for a number of genera, e.g. *Micrococcus* and *Vibrio* shows a considerable variation in the DNA base ratios of organisms within these groups, whereas all the Enterobacteria have similar DNA compositions (Figure 3). Results of this type typify the difficulties associated with bacterial identification and classification, as it is to be

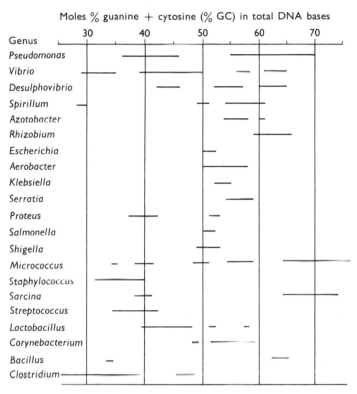

Figure 3. Variation of DNA base ratios within selected genera of bacteria. The horizontal lines represent the range of % GC values reported for each genus. Only genera for which at least two species or strains have been investigated for DNA base composition are included (Data from Hill 1966).

expected that closely related bacteria (i.e. those within the same genus) would have similar base compositions. Comparison of the structure of DNA molecules from different bacteria by hybridization methods (Chapter 5) gives some measure of the similarities of base sequences. Studies of this type have obvious potential uses. A system of bacterial classification could be devised where taxonomic groups are arbitrarily defined by the degree of hybridization between heterologous DNA's. The information might be used as a check on existing bacterial classifications since, where it has been possible to compare hybridization data with phenetic similarities, correlations have been found (Figure 4).

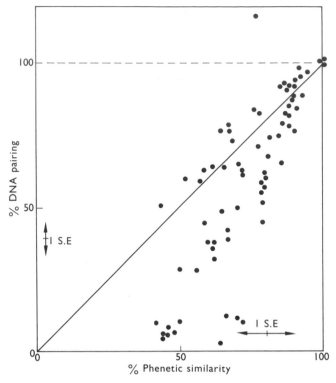

Figure 4. Congruence between DNA pairing and phenetic similarity for strains of the genera Pseudomonas, Xanthomonas Agrobacterium, Rhizobium, Vibrio and Chromobacterium. The arrows show standard errors and it can be seen that the scatter is of the same order of magnitude as twice the standard errors (approximating to 95% confidence limits). (Taken from Jones and Sneath, Ref. 10.)

Finally these hybridization data might provide an insight into possible phylogenetic relationships between organisms and the mechanisms responsible for evolution. Studies of this type are still in their infancy but it has already been demonstrated that a correlation exists between the degree of hybridization and the time that has elapsed since the evolutionary paths for various animals have diverged (Figure 5).

Finally some reference must be made to nomenclature and identification. The correct identification and naming of bacteria are necessary so that scientists can compare their observations, and as a result there are international regulations for the proposal and selection of names for bacteria. These are embodied in *The International Code of Nomenclature of Bacteria and Viruses* (1959). The object of this code is to see that bacterial types are precisely defined by Latin binomials which provide characteristic internationally recognizable labels. A binomial system has one advantage in that the relationship between units is indicated by the first or generic name but apart from this it gives little information of the properties of bacteria included in the group. Cowan (1965) suggested that a more elaborate descriptive code might be adopted for labelling bacteria (Table 4). In this system certain 'important' characters are given numbers which then form a sequential code for a particular bacterium. This system is somewhat analogous to the recommendations of the 'Enzyme Commission' for enzyme nomenclature. One important objection to this system is that al-

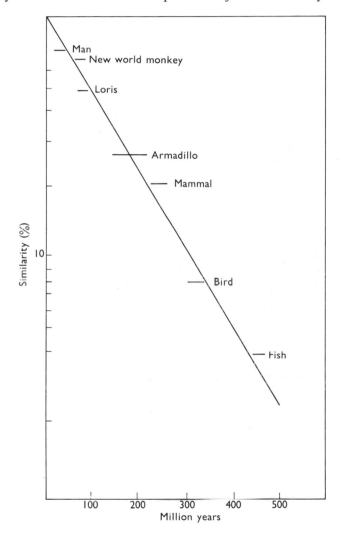

Figure 5. Correlation of hybridization data with time of divergence of evolutionary pathways.

The similarity of DNA's from various sources to DNA from the Rhesus monkey was determined from their ability to form hybrid double stranded DNA. The logarithm of this value was then plotted against the time that has elapsed since the evolutionary pathways diverged from that of the Rhesus monkey, and a linear relationship was found between these values. (Taken from R. J. Britten, Ref. 3.)

though it is often difficult to remember Latin binomials it is almost impossible to remember a sequence of ten numbers.

Table 4. A descriptive code for bacterial nomenclature. (Cowan, 1965.)

Number	Character	
	Before dash	After dash
1	Gram positive	Catalase positive
2	Gram negative	Catalase negative
3	Sphere	Oxidase positive
4	Rod	Oxidase negative
5	Acid fast	Glucose not attacked
6	Not acid fast	Glucose attacked by fermentation
7	Spore forming	Glucose attacked by oxidation
8	Not spore forming	Gas produced from glucose
9	Motile	Gas not produced from glucose
0	Non-motile	Does not grow in air

The number after the colon is an arbitrary number allotted to a species.
Examples: 13680–1469:1 *Staphylococcus aureus*
  13680–2469:1 *Streptococcus pyogenes*
  14670–1479:7 *Bacillus anthracis*
  24689–1469:1 *Salmonella typhimurium*
  14679–24590:1 *Clostridium tetani*

The identification of an unknown bacterium depends on comparison of its properties with those of known strains in an existing classification. Cowan (1965) describes three methods by which bacteria can be identified: the blunderbuss approach in which many tests are made before attempting to compare the unknown with known bacteria; the intuitive approach where we think we know the answer; and the progressive approach using dichotomous keys. Identification, unlike classification depends largely upon the constancy of a limited number of characters. For example, most clinically important bacteria can be correctly identified on the results of not more than ten tests. If more than a limited number of tests are to be used in identifying unknown bacteria then dichotomous keys and diagnostic tables become complicated and mechanical aids are required to handle the information. These can be simple systems in which the characters are recorded on punched cards which are then compared with cards containing the characters of known organisms, or better still an electronic computer can be programmed to do this and so bring the 'blunderbuss' approach up-to-date.

FURTHER READING

1   Ainsworth G. C. and Sneath P. H. A. (eds.) (1962) Microbial Classification. *Twelfth Symposium of the Society for General Microbiology.* Cambridge University Press.

2   Breed R. S., Murray E. G. D. and Smith N. R. (eds.) (1957) *Bergey's Manual of Determinative Bacteriology,* 7th edition. Williams & Wilkins, Baltimore.

3   Britten R. J. (1963) In *Carnegie Institution of Washington Year Book,* **62**, 303.

4   Britten R. J. (1964) In *Carnegie Institution of Washington Year Book,* **63**, 366.

5   Charles H. P. and Knight B. C. J. G. (eds.) (1970) Organisation and Control in Prokaryotic and Eukaryotic cells. *Twentieth Symposium of the Society for General Microbiology.* Cambridge University Press.

6   Cowan S. T. (1965) Principles and practice of bacterial taxonomy, a forward look. *J. Gen. Microbiol.* **39**, 143.

7   Hill L. R. (1966) An index to deoxyribonucleic acid base compositions of bacterial species. *J. Gen. Microbiol.* **44**, 419.

8   Huxley J. S. (ed.) (1940) *The New Systematics.* Clarendon Press, Oxford.

9   *International Code of Nomenclature of Bacteria and Viruses* (1958). Iowa State College Press, Ames, Iowa. (Reprinted in 1959 with a few corrections.)

10 Jones D. and Sneath P. H. A. (1970) Genetic transfer and Bacterial taxonomy. *Bacteriol. Revs.* **34**, 40.

11 McCarthy B. J. and Bolton E. T. (1963) An approach to the measurement of genetic relatedness among organisms. *Proc. Natl. Acad. Sci., U.S.* **50**, 156.

12 Mandel M. (1969) New approaches to Bacterial Taxonomy: Perspective and projects. *Ann. Rev. Microbiology* **23**, 239.

13 Marmur J., Falkow S. and Mandel M. (1963) New approaches to bacterial taxonomy. *Ann Rev. Microbiol.* **17**, 329.

14 Skerman V. B. D. (1959) *A Guide to the Identification of the Genera of Bacteria.* Williams & Wilkins, Baltimore.

15 Sneath P. H. A. (1957) The application of computers to taxonomy. *J. Gen. Microbiol.* **17**, 201.

16 Sneath P. H. A. (1958) Some aspects of Adansonian classification and of the taxonomic theory of correlated features. *Ann. Microbiol. Enzimol.* **8**, 261.

17 Sokal R. R. and Sneath P. H. A. (1963) *Principles of Numerical Taxonomy.* W. H. Freeman & Co., San Francisco.

18 Stanier R. Y. (1961) Towards a definition of the bacteria. In *The Bacteria,* Vol. 5: Heredity (eds. Gunsalus I. C. and Stanier R. Y.), p. 445. Academic Press, New York.

19 Stanier R. Y., Doudoroff M. and Adelberg E. A. (1971). *General Microbiology,* 3rd edition. Macmillan & Co., London.

20 Steel K. J. (1965) Microbial identification. *J. Gen. Microbiol.* **40**, 143.

Appendix B

Enzyme Mechanisms: Functions of Vitamins and Co-enzymes

CLASSES OF ENZYMES

A living organism depends on its catalytic proteins for its identity and continued existence. At first sight it appears enormously complicated but on analysis it is found that biological reactions are of six relatively simple kinds— such as oxidation by removal of a pair of hydrogen atoms. Complex changes and formation of intricate molecules are usually brought above in a step-wise manner. Many proteins (*apo*-enzymes) require the co-operation of a non-protein *co-enzyme* or *prosthetic group* and many of these are derivatives of the animal vitamins known as the B group. The same mechanisms involving the same coenzymes occur in all kinds of organisms including bacteria.

Although this cannot be a treatise on enzymology, we can perhaps illustrate the mechanisms of the six kinds of reaction and give examples of each together with the involvement of the coenzymes where appropriate. Many coenzymes and prosthetic groups are dinucleotides containing adenosine-5′-monophate (AMP) as one half. This will be represented thus AMⓅ in what follows:

The six classes of enzyme are:

(1) Oxido-reductases.   (4) Lyases.
(2) Transferases.   (5) Isomerases.
(3) Hydrolases.   (6) Ligases.

(1) Oxido-reductases

$$XH_2 + Y \rightleftharpoons X + YH_2$$

Dehydrogenases: the immediate acceptor, Y, is not directly oxygen except in oxidases. Usually the pair of H atoms (or electrons) passes through a series of carriers of increasing oxidation-reduction potential ultimately reaching O_2. Carriers include nicotinamide adenine dinucleotide (NAD) and the corresponding phosphorylated compound (NADP), flavine mononucleotide (FMN) and flavine adenine dinucleotide (FAD), and the cytochromes which are proteins with haem prosthetic groups in which the iron can be reversibly oxidized and reduced.

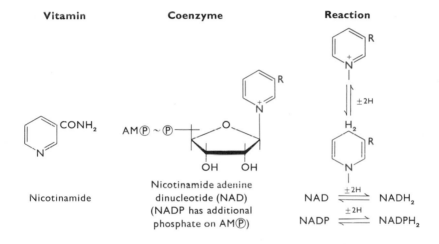

Vitamin	Coenzyme	Reaction
Nicotinamide	Nicotinamide adenine dinucleotide (NAD) (NADP has additional phosphate on AM℗)	NAD $\xrightleftharpoons{\pm 2H}$ NADH$_2$ NADP $\xrightleftharpoons{\pm 2H}$ NADPH$_2$

Vitamin	Prosthetic group	Reaction
Riboflavin	Flavin adenine dinucleotide (FAD) (Flavin mononucleotide (FMN) lacks the AM℗)	FAD $\xrightleftharpoons{\pm 2H}$ FADH$_2$ FMN $\xrightleftharpoons{\pm 2H}$ FMNH$_2$

Carrier   **Reaction**

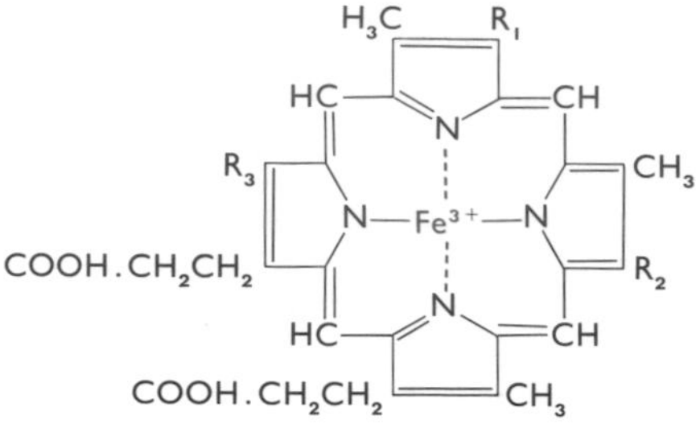

H$_3$C R$_1$

HC —— CH

R$_3$ — N — CH$_3$

N — Fe^{3+} — N

COOH.CH$_2$CH$_2$ — R$_2$

HC —— CH

COOH.CH$_2$CH$_2$ —— CH$_3$

$$Fe^{3+} \underset{}{\overset{\pm e^-}{\rightleftharpoons}} Fe^{2+}$$

Cytochrome(s)

A common sequence is for the pairs of H's to pass from substrate to NAD, to FAD, through several cytochromes (as electron transport) and finally to reduce oxygen to water:

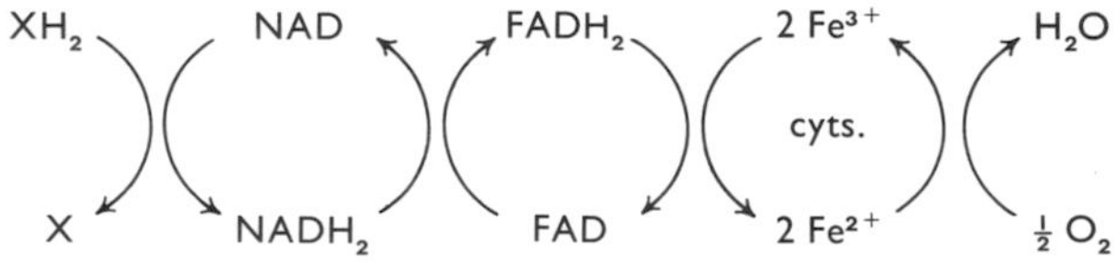

XH$_2$ ⟩⟨ NAD ⟩⟨ FADH$_2$ ⟩⟨ 2 Fe^{3+} ⟩⟨ H$_2$O

cyts.

X ⟩ NADH$_2$ ⟩ FAD ⟩ 2 Fe^{2+} ⟩ $\frac{1}{2}$ O$_2$

The overall energy of the oxidation $XH_2 + \frac{1}{2} O_2 \rightarrow X + H_2O$ is thus released in several stages and coupled reactions can make use of it to form ATP. In the step-wise oxidation of $NADH_2$ by oxygen 3 moles of ATP are made:

$$NADH_2 + \tfrac{1}{2}O_2 + 3\,ADP + 3\,\textcircled{P} \longrightarrow NAD + H_2O + 3\,ATP$$

This is referred to as *oxidative phosphorylation* and is the principal way in which free energy of chemical reactions is converted into a form utilizable for biological purposes such as chemical synthesis, mechanical work (movement), light production, transport against a concentration gradient, etc.

Examples of dehydrogenases:

Lactic

$$CH_3CHOH.COOH + NAD \rightleftharpoons CH_3CO.COOH + NADH_2$$

Lactic acid   Pyruvic acid

Alcohol

$$CH_3CH_2OH + NAD \rightleftharpoons CH_3CHO + NADH_2$$

Ethanol   Acetaldehyde

Succinic

$$COOH.CH_2CH_2COOH + FAD \rightleftharpoons COOH.CH{=}CH.COOH + FADH_2$$

Succinic acid   Fumaric acid

N.B. NAD was earlier called DPN or CoI and NADP was TPN or CoII.

Lipoic acid

Lipoic acid (thioctic acid) may also act as a H-carrier (and as an acyl carrier):

e.g. Pyruvic oxidase (and α-oxoglutaric oxidase):

(2) Transferases

$$XR + Y \rightleftharpoons YR + X$$

where R is acyl, amino, formyl, hydroxymethyl, methyl, glycosyl, or phosphate.

Transacetylases, transaminases, transmethylases, transglycosidases and kinases come under this heading. The group transferred is frequently carried by an appropriate co-enzyme.

Vitamin	Coenzyme	Reaction
$HO.CH_2$	$AM\textcircled{P}-\textcircled{P}-CH_2$	
$CH_3\overset{\mid}{C}.CH_3$	$CH_3\overset{\mid}{C}.CH_3$	
$\overset{\mid}{C}HOH$	$\overset{\mid}{C}HOH$	
$\overset{\mid}{C}=O$	$\overset{\mid}{C}=O$	
$\overset{\mid}{N}H$	$\overset{\mid}{N}H$	
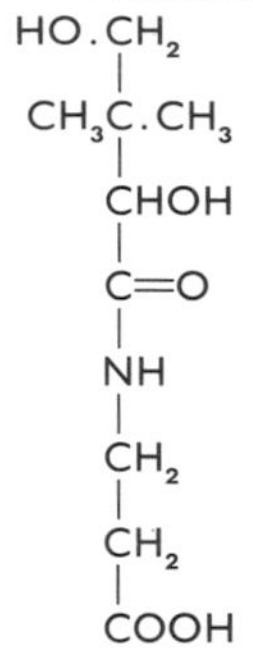 $\overset{\mid}{C}H_2$	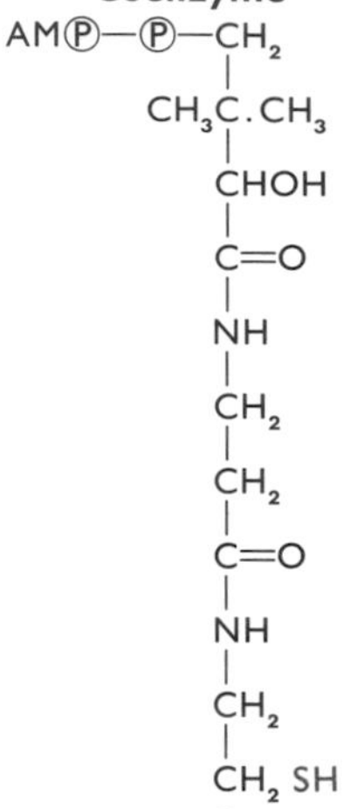 $\overset{\mid}{C}H_2$	$CoA.SH \rightleftharpoons CoA.S \sim CO.R$
$\overset{\mid}{C}H_2$	$\overset{\mid}{C}H_2$	
$\overset{\mid}{C}OOH$	$\overset{\mid}{C}=O$	
	$\overset{\mid}{N}H$	
	$\overset{\mid}{C}H_2$	
	$\overset{\mid}{C}H_2 SH$	
Pantothenic acid	Coenzyme A (CoA.SH)	

Examples of transferases:

Phosphotransacetylase

$$CH_3CO \sim S.CoA + \textcircled{P} \rightleftharpoons CH_3CO \sim \textcircled{P} + CoA.SH$$
$$\text{Acetyl CoA} \qquad\qquad \text{Acetyl phosphate}$$

Condensing enzyme

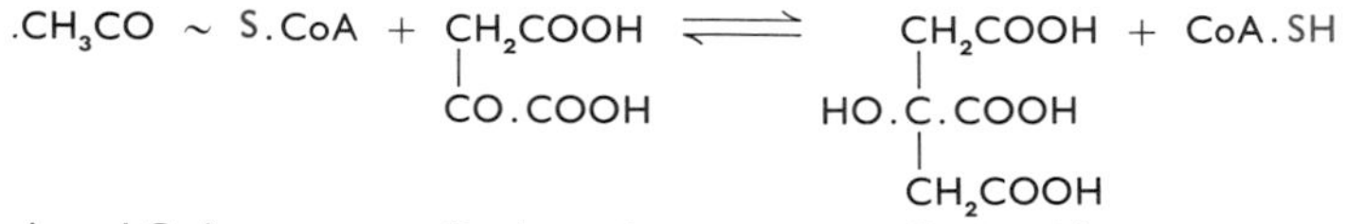

$$.CH_3CO \sim S.CoA + \underset{\overset{\mid}{CO.COOH}}{CH_2COOH} \rightleftharpoons \underset{\overset{\mid}{CH_2COOH}}{\underset{\overset{\mid}{HO.C.COOH}}{CH_2COOH}} + CoA.SH$$

Acetyl CoA   Oxaloacetic   Citric acid
                 acid

Vitamin	Coenzyme	Reaction	
CH_2OH	CHO	CHO	CH_2NH_2
(pyridine ring with HO, CH_3, N, CH_2OH)	(pyridine ring with $CH_2O\textcircled{P}$, N)		
Pyridoxin (P-in)	Pyridoxal phosphate (P-al $\textcircled{P}$)	P-al $\textcircled{P}$	P-amine $\textcircled{P}$

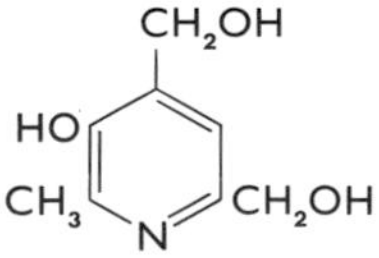

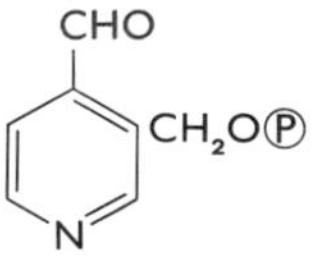

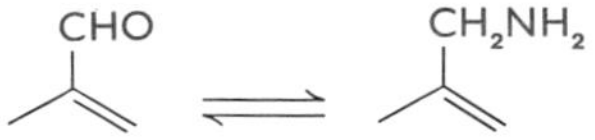

Pyridoxal phosphate is the coenzyme in a variety of other reactions of amino acids (see below) as well as those involving transamination.

Example of transamination:

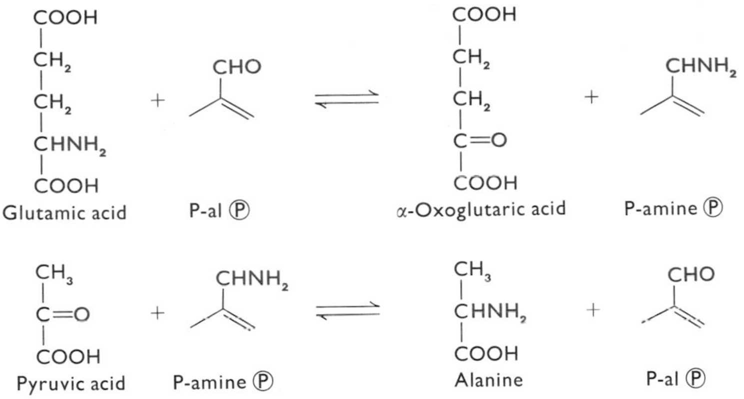

$$
\begin{array}{ccc}
\text{COOH} & & \\
| & & \\
\text{CH}_2 & \text{CHO} & \\
| & & \\
\text{CH}_2 & + \quad \diagup\!\!\diagdown & \rightleftharpoons \\
| & & \\
\text{CHNH}_2 & & \\
| & & \\
\text{COOH} & & \\
\text{Glutamic acid} & \text{P-al } \textcircled{P} &
\end{array}
\qquad
\begin{array}{ccc}
\text{COOH} & & \\
| & & \\
\text{CH}_2 & & \text{CHNH}_2 \\
| & & \\
\text{CH}_2 & + \quad \diagup\!\!\diagdown & \\
| & & \\
\text{C}=\text{O} & & \\
| & & \\
\text{COOH} & & \\
\alpha\text{-Oxoglutaric acid} & \text{P-amine } \textcircled{P} &
\end{array}
$$

$$
\begin{array}{ccc}
\text{CH}_3 & & \\
| & \text{CHNH}_2 & \\
\text{C}=\text{O} & + \quad \diagup\!\!\diagdown & \rightleftharpoons \\
| & & \\
\text{COOH} & & \\
\text{Pyruvic acid} & \text{P-amine } \textcircled{P} &
\end{array}
\qquad
\begin{array}{ccc}
\text{CH}_3 & & \\
| & & \text{CHO} \\
\text{CHNH}_2 & + \quad \diagup\!\!\diagdown & \\
| & & \\
\text{COOH} & & \\
\text{Alanine} & \text{P-al } \textcircled{P} &
\end{array}
$$

$$\text{Glutamic acid} + \text{Pyruvic acid} \rightleftharpoons \alpha\text{-Oxoglutaric acid} + \text{Alanine}$$

Vitamin   **Coenzyme**

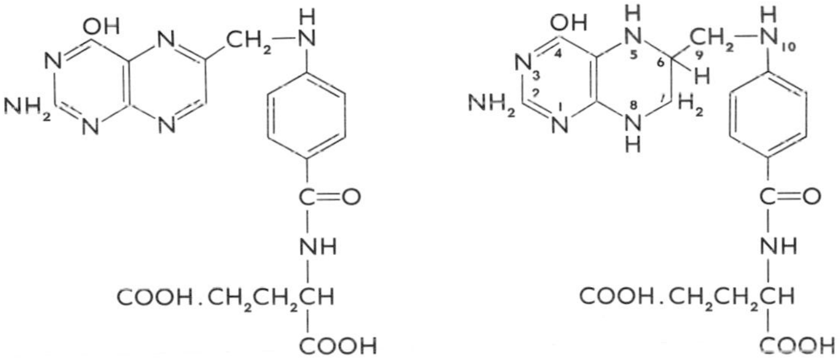

Folic acid   5,6,7,8 tetrahydro-folic acid (FH$_4$) Coenzyme F (CoF)

Reaction

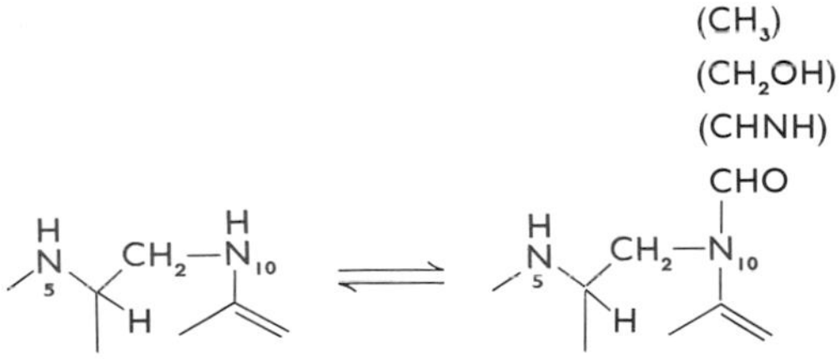

$$
(\text{CH}_3) \\
(\text{CH}_2\text{OH}) \\
(\text{CHNH}) \\
\text{CHO}
$$

Coenzyme F(FH$_4$) carries 1-C fragments—formyl, formamino, hydroxymethyl and methyl. Frequently this is on the N^{10} position but it can also be on N^5 or, after dehydration, as a ring between N^5 and N^{10}.

Examples of 1-C transfer:

$$CH_2NH_2COOH + N^5,N^{10}\text{-methylene-FH}_4 \rightleftharpoons FH_4 + CH_2OH.CHNH_2COOH$$
Glycine   Serine

$$\left\{ \begin{array}{l} \text{Vitamin B}_{12} + N^5\text{-methyl-FH}_4 \longrightarrow \text{Methyl-B}_{12} + FH_4 \\ \text{Methyl-B}_{12} + \text{Homocysteine} \longrightarrow \text{Methionine} + B_{12} \end{array} \right.$$

$$\text{Phosphoribosyl-glycineamide} + N^5,N^{10}\text{-methenyl-FH}_4 \longrightarrow$$
$$\text{5'-phosphoribosyl-}N\text{-}N'\text{-formyl-glycineamide} + FH_4$$

Carrier	Reaction
ATP	$ATP + Y \longrightarrow ADP + Y\text{\textcircled{P}}$

Kinases transfer phosphate from a nucleoside di- or tri-phosphate to an acceptor.
    Example:

Hexokinase

$$\text{Glucose} + ATP \longrightarrow \text{Glucose-6-phosphate} + ADP$$

(3) Hydrolases

$$X{-}Y + HOH \longrightarrow X{-}H + Y{-}OH$$

Peptidases, glycosidases, esterases, phosphatases, cause hydrolysis of the corresponding compounds. Coenzymes are *not* needed.
    Examples of hydrolases:

Dipeptidase

$$H_2N.CH.CONH.CH.COOH + HOH \longrightarrow H_2N.CH.COOH + HNH.CH.COOH$$
$$\quad | \qquad \quad | \qquad\qquad\qquad\qquad\qquad | \qquad\qquad\qquad |$$
$$\quad R_1 \qquad\; R_2 \qquad\qquad\qquad\qquad\qquad R_1 \qquad\qquad\qquad R_2$$

β-Galactosidase

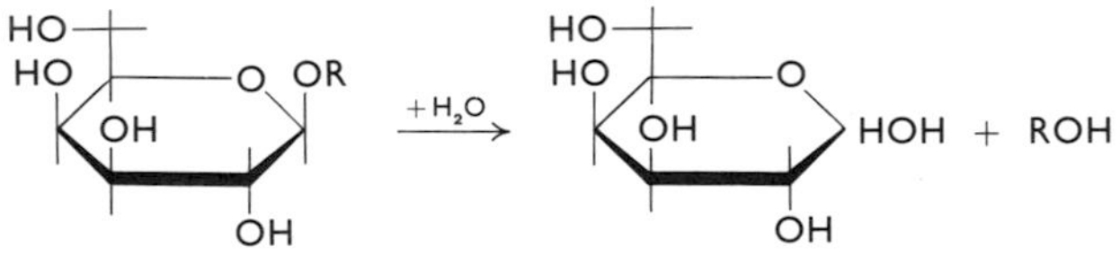

β-Galactoside   Galactose   Alcohol

Lipase

$$
\begin{array}{lll}
CH_2OCO.R_1 & & CH_2OH + R_1COOH \\
| & & | \\
CHOCO.R_2 & \xrightarrow{+3\,H_2O} & CHOH + R_2COOH \\
| & & | \\
CH_2OCO.R_3 & & CH_2OH + R_3COOH \\
\text{Glyceryl tri-ester} & & \text{Glycerol} \quad \text{Fatty acids}
\end{array}
$$

Glucose-6-phosphate phosphatase

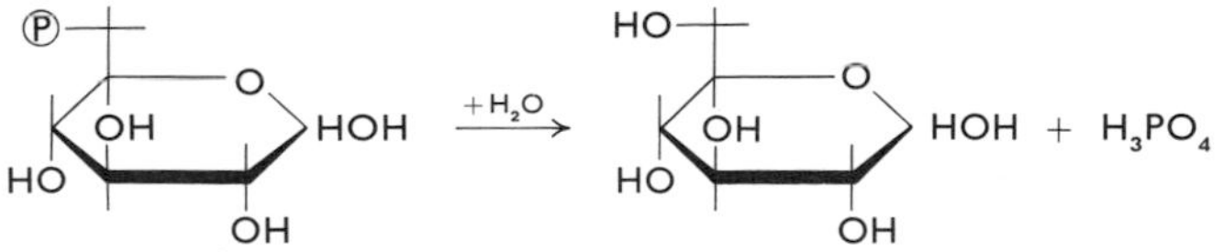

Glucose-6-phosphate $\xrightarrow{+H_2O}$ Glucose + Inorganic phosphate

(4) Lyases

$$X-Y \longrightarrow X + Y$$

Some decarboxylases, deaminases, aldolase and ketolase.

Vitamin

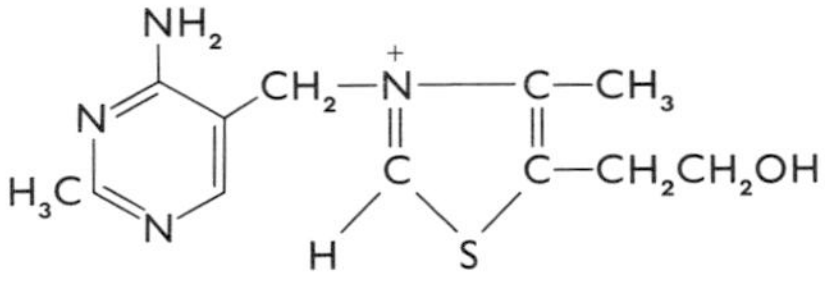

Thiamine

Coenzyme

$$-CH_2CH_2O\textcircled{P} \sim \textcircled{P}$$

Thiamine pyrophosphate (TPP)

Reaction

$$R.CO.COOH + TPP$$

$$\downarrow$$

$$(R.CHO.TPP) + CO_2$$

Some decarboxylases and transketolase use thiamine pyrophosphate as coenzyme

Examples:

Carboxylase

$$CH_3CO.COOH + TPP \longrightarrow (CH_3CHO.TPP) + CO_2$$

Pyruvic acid

$$\downarrow$$

$$CH_3CHO + TPP$$

Acetaldehyde

Transketolase

$$R.CHOH.CO.CH_2OH + R'.CHO \rightleftharpoons R.CHO + R'.CHOH.CO.CH_2OH$$

| Ketose | Aldose | Aldose | Ketose |

Some deaminases use pyridoxal phosphate as coenzyme (see p. 541)

Examples:

Serine dehydrase

$$CH_2OH.CHNH_2COOH \longrightarrow CH_2=C(NH_2)COOH + H_2O$$

Serine

$$CH_3CO.COOH + NH_3 \xleftarrow{+H_2O} CH_3C(=NH)COOH$$

Pyruvic acid

Cysteine desulphurase

$$CH_2SH.CHNH_2COOH \longrightarrow CH_2=C(NH_2)COOH + H_2S$$

Cysteine

$$CH_3CO.COOH + NH_3 \xleftarrow{+H_2O} CH_3C(=NH_2)COOH$$

Pyruvic acid

Aldolase does not need a coenzyme

Example:

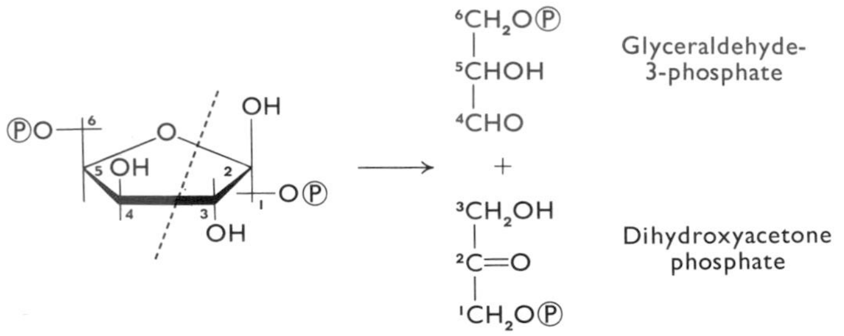

Glyceraldehyde-3-phosphate

$$^6CH_2O\text{(P)}$$
$$^5CHOH$$
$$^4CHO$$
+
$$^3CH_2OH$$
$$^2C=O$$
$$^1CH_2O\text{(P)}$$

Dihydroxyacetone phosphate

(5) Isomerases

$$X.R-Y.S \rightleftharpoons X.S-Y.R$$

Examples:

Triosephosphate isomerase

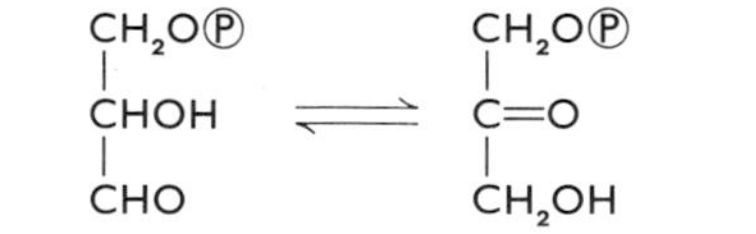

$$\begin{array}{ccc} CH_2O\text{(P)} & & CH_2O\text{(P)} \\ | & & | \\ CHOH & \rightleftharpoons & C=O \\ | & & | \\ CHO & & CH_2OH \end{array}$$

Glyceraldehyde 3-phosphate   Dihydroxyacetone phosphate

D, L-Alanine racemase

$$\text{D-Alanine} \underset{}{\overset{\text{Pyridoxal phosphate}}{\rightleftharpoons}} \text{L-Alanine}$$

Methylaspartate and methylmalonyl-CoA mutases require a vitamin B_{12} derivative as coenzyme.

Vitamin   **Coenzyme**

In vitamin B_{12} as isolated R = CN but this may be an artefact.

In coenzyme, R = adenosine but other derivatives are active.

Examples of isomerases using B_{12} coenzymes:

Glutamic acid β-Methyl aspartic acid

Methylmalonyl-CoA   Succinyl-CoA

(6) Ligases (synthases)

$$X + Y + ATP \longrightarrow X-Y + \begin{matrix} AMP + \text{Ⓟ} \sim \text{Ⓟ} \\ \text{or} \\ ADP + \text{Ⓟ} \end{matrix}$$

The mechanism may be as follows:

$$X + ATP \longrightarrow X.AMP + \text{Ⓟ} \sim \text{Ⓟ}$$

$$X.AMP + Y \longrightarrow X-Y + AMP$$

Frequently the inorganic pyrophosphate is rapidly hydrolysed thus establishing a reaction with the equilibrium position to the right.

Example:

Amino acid activating enzymes

$$\underset{\text{Amino acid}}{H_2N.\underset{R}{CH}.COOH} + ATP \longrightarrow \underset{\text{Amino acyl-AMP}}{H_2N.\underset{R}{CH}.CO} \sim AMP + \text{Ⓟ} \sim \text{Ⓟ}$$

$$\underset{\text{Amino acyl-AMP}}{H_2N.\underset{R}{CH}.CO} \sim AMP + tRNA \longrightarrow \underset{\text{Amino acyl-tRNA}}{H_2N.\underset{R}{CH}.CO-tRNA} + AMP$$

$$H_2N.\underset{R}{CH}.COOH + tRNA + ATP \longrightarrow H_2N.\underset{R}{CH}.CO-tRNA + AMP + \text{Ⓟ} \sim \text{Ⓟ}$$

Some carboxylations occur by means of a ligase with biotin acting as carrier.

Vitamin   **Prosthetic group**

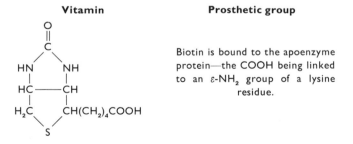

Biotin is bound to the apoenzyme protein—the COOH being linked to an ε-NH$_2$ group of a lysine residue.

Example:

$$CO_2 + ATP + Biotin-Enz \rightleftharpoons Biotin-CO_2-Enz + ADP + \text{Ⓟ}$$

$$Biotin-CO_2-Enz + \underset{\text{Pyruvic acid}}{CH_3CO.COOH} \rightleftharpoons \underset{\text{Oxaloacetic acid}}{COOH.CH_2.CO.COOH} + Biotin-Enz$$

$$CO_2 + Pyruvic\ acid + ATP \rightleftharpoons Oxaloacetic\ acid + ADP + \text{Ⓟ}$$

BACTERIAL GROWTH FACTORS

The following members of the B group of vitamins play roles in enzymic reactions (see above) and if a bacterium is unable to synthesize any of them, they must be provided in the growth medium unless all possible products of the enzymic reactions are supplied.

Thiamine
Nicotinamide
Riboflavin
Pyridoxin
Pathothenic acid
Folic acid
Biotin
Vitamin B_{12}
Lipoic acid

Some bacteria are unable to synthesize certain amino acids or purines, etc. and may require them. These also must be supplied as *growth factors*.

FURTHER READING

I   Dixon M. and Webb E. C. (1964) *Enzymes*, 2nd edition. Longmans, London.

Appendix C

Glossary

Abortive transduction. Transduction in which the donor DNA is not integrated with the recipient chromosome (as in *complete* transduction) but persists as a non-replicating fragment.

Allele. A form of a gene. Two genes which are derived from the same gene by mutation, and which are alternative occupants of the same chromosomal locus, are described as *allelic* with respect to each other.

Allosteric effect. A change produced in the properties of an enzyme by the specific action of a small molecule acting at a site other than the active site.

Antibiotic. Substance produced by one living organism which is toxic to one or more other types of organism.

Antibody. A protein produced in an animal when a substance normally foreign to its tissues gains access to them: the antibody combines chemically with the foreign substance (antigen).

Anti-codon. A sequence of three nucleotides (in an amino acid transfer-RNA) complementary to the codon triplet in a messenger-RNA.

Antigen. A substance capable of stimulating production of an antibody. Each antigen evokes the synthesis of a specific protein in an animal.

Auxotroph (auxotrophic mutant). A mutant strain differing from the normal or wild type of the organism in having an additional nutritional requirement.

Bacteriophage. A virus infecting a bacterium.

Capsule. A surface component of the bacterial cell lying outside the cell wall.

Chromatinic body. This structure is the equivalent of the plant or animal cell nucleus and contains the bacterial DNA. It is not surrounded by a nuclear membrane.

Chromatophore. A structural component of photosynthetic bacteria surrounded by a unit membrane and containing chlorophyll.

Chromosome. A structure carrying genes in a linear order. In bacteria probably a single large molecule of DNA, or a sequence of such molecules joined end-to-end without the joins being evident.

Cis. The arrangement in which two mutations are on the same chromosome or chromosome fragment, and two corresponding non-mutant sites are present together on another chromosome or fragment in the same cell (cf. *trans*).

Cistron. A chromosomal segment is called a cistron if any two mutatins within it give a mutant phenotype in the *trans* configuration but a normal phenotype in the *cis* configuration (cf. *cis* and *trans*). Not an infallible definition of the gene because of the phenomenon of complementation between alleles.

Clone. Genetically identical cells derived by successive divisions from a single cell.

Codon. A sequence of three nucleotides (in a nucleic acid) which codes for an amino acid or the end of a polypeptide chain.

Complementation. Complementary action of different mutant genomes or homologous fragments of genomes. This is the general rule for mutations in different genes, but it can occur, though usually not strongly, between alleles of the same gene.

Conjugation. Attachment of bacterial cells, permitting the transfer of genetic material from one to another.

Co-ordinate regulation. The induction or repression of a group of enzymes in a strictly proportional manner.

Cross. An encounter between organisms which permits the formation of genetic recombinants. In bacteria crosses can be performed by conjugation, transduction or transformation.

Deletion. Loss of a chromosome segment.

Enzyme. Biological catalyst of high specificity; protein with or without additional non-protein prosthetic group or coenzyme (see Appendix B).

Episomes. Dispensable pieces of genetic material (probably always DNA), endowed with the capacity of independent replication; able to multiply independently of the chromosome but capable of reversible integration into the chromosome.

Exchange (in the genetic sense). The exchange between chromosomes of homologous segments of DNA.

Exponential phase (*log phase*). State of growth in which population doubles regularly each mean generation time (q.v.) i.e. increases 1, 2, 4, 8, 16,... in equal time intervals.

Feedback inhibition. Inhibition of the activity of an enzyme specifically caused by a small molecule which is usually the end-product of a biosynthetic pathway.

Fimbriae. Surface appendage of certain Gram-negative bacteria, probably composed of protein sub-units and approximately 7 nm in diameter. They are shorter and thinner than flagella. Also called *pili*.

Flagellum. Bacterial flagella are long, thread-like structures made up of protein molecules of flagellin: the diameter varies between 12 nm and 30 nm.

Gene. The smallest genetic unit having an independent function. A length of DNA transcribable into a length of RNA of discrete function, i.e. a specific transfer-RNA or ribosomal-RNA or a messenger specifying a specific polypeptide chain.

Genetic code. The relationship between the sequence of nucleotides in DNA or RNA which specifies a sequence of amino acids. The code is 'triplet' in the sense that a run of three nucleotides codes for each amino acid.

Genome. A complete single set of genetic material—i.e. a complete set of genes.

Genotype. The genetic constitution, or ensemble of genes.

Haploid. Possessing a single genome.

Heterozygote. A cell carrying two different alleles of the same gene (cf. *homozygote*).

Homozygote. A cell carrying two identical copies of a gene (cf. *heterozygote*).

Induction. An increase in the rate of synthesis of an enzyme, specifically caused by a small molecule which is generally the substrate or a compound closely related to it.

L-form. Soft protoplasmic elements without defined morphology which no longer possess a rigid bacterial form.

Linkage. The tendency of genetic markers which are relatively close together on the chromosome to be transmitted together through crosses and not separated by recombination.

Log phase. See *exponential phase*.

Lysogeny. The harbouring by a (lysogenic) bacterial strain of a temperate bacteriophage, capable of uncontrolled multiplication and causing lysis in response to an inducing treatment or, occasionally, spontaneously.

Marker. A genetic mutation with a distinctive observable effect on the organism, serving to *mark* the chromosome locus at which it occurs, thus enabling the transmission of the locus through cell divisions and genetic crosses to be followed.

Mean generation time (M.G.T.) Time interval during which a population doubles in number.

Merozygote. A zygote in which there is one complete genome and a fragment of a second genome.

Mesosome. A membranous involution of the protoplast membrane.

Messenger-RNA (mRNA). The RNA which specifies the amino acid sequence for a particular polypeptide chain.

Micron (μ). 10^{-3} millimetre.

Minimal medium. Medium containing the minimal nutrients required by the wild-type of an organism—incapable of supporting growth of auxotrophs.

Mutagen. A substance capable of inducing mutation.

Mutant. A cell or clone of cells carrying a mutation.

Mutation. An abrupt and, usually, stably inherited change in the properties of the organism—usually a change in the DNA.

Operator. A region in the chromosome contiguous with and controlling the functioning of an operon.

Operon. A number of genes, adjacent in the chromosome, whose function is subject to a common control mechanism.

Phenotype. The observable characteristics of an organism. The phenotype does not necessarily reflect the complete genetic constitution or genotype, since the expression of some genes may be masked for one reason or another.

Phosphatide. A fat containing glycerol and one or two fatty acid residues linked to a nitrogenous base through a phosphoric acid residue. (A lipid containing phosphoric acid and a nitrogenous base.)

Polycistronic messenger. A length of messenger-RNA containing the information for the synthesis of more than one polypeptide chain.

Polynucleotide. Polymer consisting of nucleotides linked by phosphodiester links between positions 3′ and 5′ of successive sugar units:

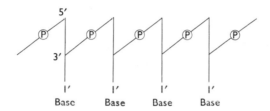

Polypeptide. Polymer consisting of amino acids linked by condensation of the carboxyl group of one with the amino group of the next:

$$NH_3CH.COOH + NH_2CH.COOH + etc. \quad \longrightarrow \quad NH_2CH.CONH.CH.CO...etc.$$
$$\quad\quad |\quad\quad\quad\quad\quad\quad |\quad\quad\quad\quad\quad\quad\quad\quad\quad\quad\quad |\quad\quad\quad\quad\quad |$$
$$\quad\quad R_a\quad\quad\quad\quad\quad\quad R_b\quad\quad\quad\quad\quad\quad\quad\quad\quad\quad\quad R_a\quad\quad\quad\quad R_b$$

Polyribosome (polysome, ergosome). Several ribosomes (q.v.) associated with the same strand of messenger-RNA.

Prophage. Temperate bacteriophage in its latent state.

Protein. Large molecule formed from one or more polypeptide chains and having 'native' secondary, tertiary and sometimes, quarternary structure. Many have non-protein prosthetic groups.

Protoplast. A structure derived from a vegetative cell by removal of the entire cell wall.

Prototroph. A strain of bacteria having the minimal nutritional requirements of the wild-type organism (cf. *auxotroph*).

Recombinant. A cell or clone of cells resulting from recombination.

Recombination. The formation of a new genotype by reassortment of genes following a genetic cross (see also *exchange*).

Replica plating. Replication of a pattern of colonies from one plate to another; a disc of sterile material (often velveteen) is pressed on the surface of the first plate, and the adhering bacteria are printed on the second.

Repression. A decrease in the rate of synthesis of an enzyme specifically caused by a small molecule which is usually the end-product of a biosynthetic pathway (e.g. an amino acid, nucleotide, etc.)

Ribosomal-RNA (rRNA). The RNA of the ribosome particles as distinct from mRNA and tRNA. 70 *S* ribosomes yield one molecule each of 16 *S* rRNA and 23 *S* rRNA (molecular weights 0.55×10^6 and 1.1×10^6) coming from the two sub-units, 30 *S* and 50 *S*.

Ribosome. Particle composed of protein and RNA to which messenger-RNA and amino acyl-tRNA are attached during synthesis of polypeptide chains. Usually designated by sedimentation coefficient, e.g. 70 *S*. These functional ribosomes can dissociate 70 $S \rightleftharpoons$ 30 $S + 50$ *S*.

Segregation. The separation of two distinct forms (or alleles) of the same gene, originally present in the same cells.

Shift-down. The transfer of a bacterial culture to a poorer medium, i.e. one supporting a lower growth rate.

Shift-up. The transfer of a bacterial culture to a richer medium, i.e. one supporting a higher growth rate.

Spheroplast. A form of a bacterium in which the cell wall structure has been modified but in which typical cell wall components are still present.

Suppressor mutation. A mutation which masks the effect of another mutation elsewhere in the genome.

Temperate bacteriophage. A bacteriophage capable of replicating in step with its host bacterium, being thus transmitted through cell divisions without necessarily causing lysis.

Trans. The arrangement of two mutations in the same cell on different chromosomes, or chromosome fragments, with each one linked to the non-mutant homologue of the other (cf. *cis*).

Transcription. The synthesis of RNA on a DNA template such that the bases in the product are complementary to those in the template.

Transduction. Transfer of a fragment of the genome from a donor to a recipient strain of bacteria by infecting the recipient with bacteriophage particles grown on the donor strain.

Transfer-RNA (tRNA). Molecule of RNA able to combine with a specific amino acid which becomes esterified to the terminal adenosine. Each species of tRNA has a specific trinucleotide sequence which interacts with a complementary sequence in mRNA. Also called soluble-RNA (sRNA) and acceptor-RNA.

Transformation. Transfer of a fragment of the genome from a donor to a recipient strain of bacteria by treating recipient cells with DNA isolated from the donor.

Transition. A mutation consisting of a change in one base pair of the DNA, a different purine being substituted for the purine and a different pyrimidine for the pyrimidine.

Translation. The synthesis of polypeptide whose amino acid sequence is specified by successive triplets of nucleotides in messenger-RNA.

Transversion. A mutation consisting of a change in one base pair of the DNA, a purine being substituted for the pyrimidine and a pyrimidine for the purine.

Zygote. The cell which is the immediate product of a cross—at least partly diploid.

Index

Italic figures refer to illustrations or tables; bold face figures refer to the more important page references of a series.

Abequose (3:6 Dideoxy D-galactose) *93, 96*, 99
Acetamide 460
Acetaldehyde *166*, 180, *181*, 538, 543
Acetic acid, in amino acid metabolism *221, 223*
 from breakdown of acetyl CoA 185
 from breakdown of amides 460
 as carbon source 141, 171, **190–3**, 195, 460, *503*
 as enzyme inducer 460, 462, 464
 in fatty acid metabolism 177
 fermentation product *166*, 179, 180, *181*
 sporulation medium 512, 514
Acetoacetyl coenzyme A *166*
Acetoacetyl coenzyme A reductase *500*
Acetoacetyl S.ACP 245, 246
Acetobacter 166, 473, *524*
Acetobacter xylinum 180
α-Acetohydroxyacid synthase *214–15*, 216, 225, *226, 228*
α-Acetohydroxyacid reductoisomerase (Dihydroxyacid synthase) *214–15*, 216, 225, *226, 228*
α-Aceto-α-hydroxybutyrate *215*, 216, *228*
Acetoin *166*
α-Acetolactate *166*, 225, *226, 228*
Acetone *166*
Acetyl coenzyme A, in amino acid metabolism *220, 223, 225, 226–7*
 condensation with oxaloacetate 28, **185–6**, *188, 191, 192, 455, 458, 540*
 deacetylation 185
 as enzyme activator 456, 457
 as enzyme inhibitor 189
 in lipid catabolism 182, 188
 in lipid synthesis 243–4, 246, 247, 458
 in pyruvate oxidation 28, *166, 169*, 189, 539
 from reduction of acetyl phosphate 180, *181*, 540
Acetyl coenzyme A carboxylase 244, 245, **246–7**, 458
N-Acetylgalactosamine 92, 94
N-Acetylglucosamine, in lipopolysaccharides 96, 97, 98
 in peptidoglycans 11, *12*, 37, 78, 79, *80–2*, 83, *84–6, 93*
 structure *79*
 synthesis *31*, 32, 457
 in teichoic acids 90, *93*, 94

N-Acetylglucosamine-1-phosphate 83
N-Acetylglucosamine-1-phosphate-glycerol phosphate 90
N-Acetylglutamate *220, 239*
N-Acetylglutamate γ-phosphotransferase (*N*-Acetyl γ-glutamokinase) *220–1, 239, 479*
N-Acetylglutamate synthase (Glutamate acetyltransferase) *218, 220–1*, 222, *239, 479*
N-Acetylglutamate γ-semialdehyde *221*
N-Acetylglutamate γ-semialdehyde dehydrogenase *221, 239, 479*
N-Acetyl γ-glutamokinase (*N*-Acetylglutamate γ-phosphotransferase) *220–1, 239, 479*
N-Acetyl-γ-glutamyl phosphate *221*
Acetyl lipoate 183–4
N-Acetylmuramic acid, in peptidoglycans 11, 37, 78, *79–82, 84–5*
 structure *78*
 synthesis *31*, 32, 458
N-Acetylornithinase *221, 239, 479*
N-Acetylornithine *221*
N-Acetylornithine δ-aminotransferase *221, 239, 479*
Acetyl phosphate 180, *181*, 540
Acetyl S.ACP *245*, 246
O-Acetylserine *223, 224*, 225
O-Acetylserine sulphhydrase *223–4*
Acetyl transacylase *244*
Aconitase (Aconitate hydratase) 186, 499, *500*, 503, 505
cis-Aconitate *185*, 186
ACP (Acyl carrier protein) *244–6*
Acridine half-mustard (ICR-191) *382–3, 412*
Acridine orange 336
Acridines, complexes with RNA 302, 311
 intercalation into DNA *302–3*
 mutagenic effects of *303*, 336, **382–3**, 411–412
Acrylate *166*
Actinomycetes 494
Actinomycin D, binding to DNA 302, 311
 effect on protein synthesis 149–50
 effect on RNA synthesis *150*, 302, 311, 324, 349, *350*
 effect on spores 506, 508
Activation, of amino acids 40–1, 42, **325–31**
 of enzymes 423, 424, **427–9**, 456, 490
 of spores 506

Activators, binding to enzymes 424, 427, *428*
  binding sites on DNA *475*, 477
Active transport *see* Permease systems
Acyl carrier protein (ACP) 244–6
Acyl lipoic acid 184, 539
Acyl ~ S.CoA 183, 184
Acyl ~ S.CoA glycerophosphate acyltransferases 247
Adenine, pairing with bromouracil 381, *385*
  pairing with cytosine 384, *385*
  pairing with thymine 18, 251, *252*, 255,
    259, 384, *385*, 531
  pairing with uracil 251, *252*, 276, *385*
  structure 16, *17*
  tautomers 384, *385*
Adenosine 507, 545
Adenosine deaminase *500*, 502
Adenosine 5′-diphosphate (ADP), balance
    with AMP and ATP **168–9**, 237, 456
  as enzyme activator 456
  as enzyme inhibitor 237, 464
  reduction of *243*
  synthesis 236
ADP-glucose *169*, 172
ADP-glucose pyrophosphorylase *169*, 172
Adenosine 5′-monophosphate (AMP),
    balance with ADP and ATP **168–9**, 237,
    456
  as co-enzyme 536, 537
  cyclic 451, **453–4**, 476, 477, 491
  as enzyme activator 189, 456, 458, 462
  as enzyme inhibitor 200, 219, 237
  synthesis of *232, 234–5*, 236
    control of 237–8, *239*
Adenosine 5′-triphosphate (ATP), in active
    transport 28, 117
  in amino acid synthesis 211, *212, 214, 218,
    220, 223*, 229, *230*
  as co-enzyme 190, 219, 288, 429–30, 546
  as energy store 27–8, 160, 164, **168–70**,
    189–90, 456, 457, 538
  as enzyme activator 241
  as enzyme inhibitor 189, 237, 459
  in lipid synthesis 244, 245
  in nitrogen fixation 464
  in nucleic acid synthesis 312
  in peptidoglycan synthesis 83, *84–5*, 86
  in photosynthesis 194, 196–200
  in protein synthesis 40–1, 327, 328, 331,
    333
  in purine synthesis *234–5*, 236, 237, 238,
    242
  in pyrimidine synthesis 240, 241, 242
  required for carbohydrate breakdown 162,
    163, 167, 177, *181*, 190
  yield from acetyl CoA breakdown 185
  yield from Entner-Doudoroff pathway
    179, 190
  yield from glycolysis 164, 165, 167, **189–
    190**, *191*

yield from oxidative phosphorylation 538
yield from pentose phosphate pathway 177
yield from phosphoketolase pathway 180,
  181, 190
yield from tricarboxylic acid cycle 187,
  189–90, *191*
ATP-ase 117
S-Adenosylmethionine 217
Adenylate deaminase 237
Adenylosuccinate *232, 235*, 236
Adenylosuccinate lyase *234–5*, 236, *239*
Adenylosuccinate synthase *234–5*, 236, 237,
  239
Adenylyl sulphate *223*
Adenylyl sulphate kinase *223, 224*, 225
ADP *see* Adenosine 5′-diphosphate
Aerobacter *166*, 525, *531*
Aerobacter aerogenes 237, *254*, 461, *462, 525*
Aeromonas 524
Agrobacterium 532
Alanine, as enzyme inhibitor 430
  in muramic acid synthesis *31*
  in peptidoglycans 80, *81–2*, *84–6*
  N-terminal amino acid 322, 343, 345
  structure *14*
  synthesis 29, *30*, 225, *226*, 541
    control of 225, 229
  tRNA, gene for 276–7, 338
    structure *264*, 338
  triplets for *339*
D-Alanine, in peptidoglycans 11, 78, *80–1*,
    83, *84, 86*
  in teichoic acids 91
L-Alanine aminotransferase 225, *226*
Alanine dehydrogenase 225, 501
D, L-Alanine racemase 88, 500, 501, 502,
    544
D-Alanyl-D-alanine ligase 83
Alcohol *see* Ethanol
Alcohol dehydrogenase 538
Alcohols, polyhydric 454
Aldolase (Fructose diphosphate aldolase)
    163, *169*, 177, 197, 501, **544**
Algae 517
  blue-green 78, 193, 325, 519
Alkaline phosphatase, mutants lacking 388,
    413, 414, 418–19
  repression of 473
  in spores *500, 509*
  structure 418–9
  test for 374
Alkyl thio-β-D-galactosides 431, *432, 434*
Alleles **49**, **418**, 436, 437, 440, 548
  complementation of 418–9
  dominance relations of 441, 446
  mating type 511
Allosteric effect *57*, 58, 246, 373, **424–9**, 548
  ligands 425, 427, 429
  sites *57*, 58, 424
Amber mutations *265*, *344–5*, 414

Amicetin 363, *364*
Amidase 460
Amidotransferase 229, *230–1, 233*
Amination **29**, **31**, 211, *212*, 217, *218*, 219, 464
See also Transamination
Amino acids, activation 40–1, 42, **325–31**
activating enzymes 41, 42, 327, 328, 329, 330, 375, 546
analogues 329, 375, 477–8, 480
aromatic family, synthesis 29, *30*, 202, *204–7*, 208
control of 209–11, 466–7, *469*, 471–2, 477–8, 481–2
aspartate family, synthesis 30, 31, 202, 211, *212–15*, 216
control of *216*, 217, *426*, 473
as carbon source 55
glutamate family, synthesis 29–31, 202, 217–19, *220–1*
control of 219, 221–3
incorporation into peptide chains 13, 15, 42–3, **331–4**, 351, **354–62**
optical activity 11, *14*
pyruvate family (branched), synthesis *30*, *31*, 202, 225, *226–7*
control of 225, *228*, 229, 478, 482
requirement for growth 141, 150, 369, 371, 488, 547
sequence, genetic control of 41–2, **409 14**
serine family, synthesis *30*, 202, *222–3*, 224
control of 224–5
structures of *14*
tRNA's for 42–3, *264*, 329–31, 338
triplets for *339*
Amino acyl-adenosine 329
Amino acyl-AMP 41, 327, 328, 329, 546
Amino acyl-phosphatidylglycerol *112, 113*
Amino acyl-tRNA, as enzyme repressor 480, 481
formation 42–3, **328–9, 331–3**, 546
interaction with ribosomes and mRNA 43, **356–64**
in peptidoglycan synthesis 84
Amino acyl-tRNA synthase (amino acid activating enzymes) 41, 42, **327**, 328, 329, 330, 375, **546**
Aminomannuronic acid *93, 94*
Aminopeptidase 345
5-Amino-1-(5′-phosphoribosyl)-imidazole 4-carboxamide 229, *230, 232, 234*
Aminopterin 381
2-Aminopurine 381–2, *385*, 388
Ammonia, in amino acid synthesis 29, *30*, 211, *212*, 217, *218*, 219, 464
oxidation 200
in purine and pyrimidine synthesis 33, *235*, 236, 240, *241*
as source of nitrogen 2, 3, 5, 29, *218*, 219, 464

Amoeboid movement *518*
AMP *see* Adenosine 5′-monophosphate
Amphibolic pathways **168**, **170**, 454, 457
Amylase 503
Anabolism *see* Enzymes, Class II; Reactions Class II
Analogues, amino acid 375, 477–8, 480
base 270, 295, *296*, 381–2, *385*, 388–9, 408–9
Anaplerotic enzyme systems 171, **190–3**, **454–5**, 456, 458, 460
Anthranilate phosphoribosyltransferase *207*, 210, 211, 471, 472
Anthranilate synthase *206–7*, 208, 209, 210, 211, 467, *469*, 471, 472
Anthranilic acid *207*, 208, 209, *468–9*
Antibiotics, cell wall synthesis *85*, 88–9, 145, 150, 151
definition 548
DNA synthesis 302, 311
protein synthesis 149–50, 329, *332*, 333, *351*, 356, 363–4, 465
RNA synthesis *150*, 302–3, 311, 324, 349, 350
spores *500, 504*, 505, 506, 508, 511
Antibody, definition 548
fluorescent 100–2, *103*
Anti-codon 263, *268*, 329, 330, 548
See also Ribonucleic acid, transfer
Antigens, capsule 104
cellular 508
definition 548
lipopolysaccharide *96*
peptidoglycan 100–3
spores *500*, 501
teichoic acid 92
Apo-enzyme 536, 546
Apo-repressor 221, *445*, 467
L-Arabinose, as carbon source 373, 454, 459
system *475*, 477
L-Arabinose isomerase *475*
L-Arabinose permease *475*, 477
Arginine, catabolism 503
decarboxylation 223
as enzyme inhibitor 222
as enzyme repressor 221, 242, 432, 467, 480
operon 221, *479*, 480, 482
structure *14*
synthesis *30*, 217, 219, *220–1*
control of 221–3, *241*, 432, 435, 466, 467, *479*, 480
triplets for *339*
Argininosuccinate *220*
Argininosuccinate lyase *220–1*, *239*, 479
Argininosuccinate synthase *220–1*, *239*, 479
Arthrobacter 190
Asci 494, 511
Ascospores
formation, genetic control of 514–16

Ascospores, formation—*cont.*
  metabolic changes during 512–14
    temperature and 515
  nature of 494, 511, 512
Asparagine, structure *14*
  synthesis 211, *212*
    control of 217
    triplets for *339*
Asparagine synthase 211, *212*, 217
Aspartate aminotransferase 211, *212*, *500*
Aspartate carbamoyltransferase 239, *240–1*,
    500
  control of 241, 242, 424, *426*, 435, *479*
Aspartate β-semialdehyde 211, *212*, *214*, *216*,
    217
Aspartate semialdehyde dehydrogenase *212*,
    214, *216*
Aspartic acid in arginine synthesis 217, *220*
  as enzyme inhibitor 455
  in nucleotide synthesis 32–3, 234–5, 236,
    237, 238–9, *240*
  structure *14*
  synthesis 30, 31, 211, *212*
    control of 217
    triplets for *339*
β-Aspartokinase *212*, *216*, 217
β-Aspartyl phosphate *212*, *216*, 217
Athiorhodaceae 194
ATP *see* Adenosine 5′-triphosphate
Aureomycin 363
Aurintricarboxylic acid 363, *364*
Autotrophs 193, 200
Auxotrophs 369, **371–3**, 487, 548
  selection for 372–3
  temperature-sensitive 373
  transduction mapping 403–4
2-Azatryptophan 329
7-Azatryptophan 329
Azotobacter 112, 188, 456, 494, *524*, *531*
Azotobacter vinelandii 464, *525*
Azotobacteriaceae *524*, 526

Bacitracin *85*, *88*
Bacillaceae *523*
Bacillus 71, 133, 481, 494, *523*, *531*
Bacillus anthracis 71, 106, *534*
Bacillus cereus 121, 473, 487
  spores 498, *500*, 501, 502, 508
Bacillus licheniformis 112, 217, *469*, 473
  amino acid metabolism 217, 467, *469*,
    472
Bacillus megaterium, capsule 68, 104, *105*
  cell wall 68, 70, 118, *119*
  protoplast formation 87, 111, 325
  membrane *112*
  phages 304
  spores 502
Bacillus polymyxa *69*, 217, 497
Bacillus popilliae 510

Bacillus sphaericus 498, *500*
Bacillus subtilis, amino acid metabolism
    208–9, 462, 466, 467, *469*, 472, 481
  co-enzyme synthesis 468, *470*
  DNA *130*, *271*
  generation time 486
  genetic mapping, transduction 304, 402–3
    transformation *368*, 391, 408–9
  membrane 112, *120*, *121*
  nuclease 286
  phages 304, 402–3, 408, 484
  ribosomes *131*
  spores 313, 485, 498, *500*, 501, 503, 504,
    505
  teichoic acids 91, 92, 94
Bacillus thuringiensis *500*, 510
Bacteriochlorophyll 194
Bacteriophages *see* Phages
Bactogen *see also* Chemostat 21, 142–3, 147
Base-pairing, in DNA synthesis 276–7, 295,
    315
  nucleic acid structure and 5–6, 18, 37–8,
    251–2, 255–6, 263
  mutations in 381–2, 383, *385*
  in protein synthesis 42, 263, **351–3**
  in RNA synthesis 305–6, 315, 415
Benzaldehyde *463*
Benzaldehyde dehydrogenase 462, *463*
Benzoate 462, *463*, 464
Benzoate oxidase *463*, 464
Benzoylformate *463*
Benzoylformate carboxylase 462, *463*
Benzylpenicillin *473*
Benzylpenicilloic acid *473*
Biosynthetic enzymes *see* Enzymes, Class II;
    Enzymes, Class III
Biotin, growth factor 547
  prosthetic group 244, 546
  synthesis 468, 479–80
Blue-green algae 78, 193, 325, 519
Bottromycin 363, *364*
5-Bromodeoxyuridine (BUDR) 270, 381,
    388–9, 408–9
5-Bromodeoxyuridine triphosphate 295, *296*
5-Bromouracil (BU) 270, 295, 381–2, *385*,
    388
Broth medium 141, 464, 488
Brucella *524*
Budding of yeast 511, 512
2,3-Butanediol *166*
Butanol *166*
Butyrate *166*
Butyryl coenzyme A *166*
Butyryl.S.ACP 245, *246*, 247

Caesium chloride 270, 278
Calcium, spores and 496, 497, 498, 501, 507
  dipicolinate *500*, 507
Calvin cycle 198–200

Camphor 454
Capsule, chemistry 7, 36, 86, 104–6
    structure 64, *68*, 103–4, *105*, 548
Carbamoyl aspartate (Ureidosuccinate) 239,
    240
Carbamoyl phosphate 220, 238, 239, *240*,
    435, 466
Carbamoyl phosphate synthase 239, *240–1*,
    242, 466
Carbon dioxide, as carbon source 193, 194,
    519
    fixation 194–200
    in lipid synthesis 244, 458
Carbon replicas 65
Carboxydismutase (Ribulose diphosphate
    carboxylase) 196, 200
Carboxylase (Pyruvate carboxylase) 190,
    456, 543
Carboxylation 546
1'-(*o*-Carboxyphenylamino)-1'-deoxyribulose
    5'-phosphate *207*, 210, *469*
Carboxyphenylriboxylamine isomerase 467,
    469, 471, 472
N-(*o*-Carboxyphenyl)-D-riboxylamine 5'-
    phosphate (*N*-(5'-Phosphoribosyl an-
    thranilate) *207*, 210, *469*
Carboxyphenylriboxylamine pyrophosphory-
    lase *469*, 472
Cardiolipin (Di-phosphatidylglycerol) *112*
Carotenoids 194
Catabolic enzymes *see* Enzymes, Class I
Catabolite repression 55–6, 451, **452–4**, 459,
    460, 476, 488, 491, 510
    cyclic AMP and 451, **453 4**, 476, 477,
    491
Catalase 501, *534*
Catechol 462, *463*, 464
CDP *see* Cytidine 5'-diphosphate
Cell
    composition 3, *4*, 11, 316–19
        effect of medium 104, 111, 112, 116,
        147–51, 487–9, 492
        growth rate and 144–6, 486
    dimensions 11, 519, 521
        age and *153*, 156
        growth rate and 144–6, 486
    fractionation of 316–9
    major structures 3, *4*, 11, 63–7
        see also Chromosome; Flagella; Cell wall
        etc.
    mass 3, 11, 145–6, *317–18*
    permeability 13, 117, 433, 489
Cell cycle 25, **154–9**
    cell division and 155, 486
    length of 156–7
    macromolecular synthesis and 25, 157–8,
        486, 488
    methods of study 155–6
Cell division, arrangement of cells after *523–*
    525, 526

cell cycle and 155, 486
cell wall and **101–3**, *104*, *105*, *118*, 119,
    121, 122–3
cytoplasmic membrane and 102, *105*, *118*,
    122–3, 155, 486, 488
DNA replication and 5, 20, *45*, 128, 284,
    386, 486
effect of medium on 145, 148, **153–4**
spore formation and outgrowth 495–6,
    502, 507, *509*
Cell membrane *see* Cytoplasmic membrane
Cell wall, cell division and **101–3**, *104*, *105*,
    118, 119, *121*, 122–3
    chemical components of 12–13, 36–7, **77–**
    100, *518*
        synthesis of **83–7**, **89–90**, **91–4**, 98–9,
        457
    effect of antibiotics *85*, 88–9, 145, 150,
        151
    effect of lysozome 71, *73*, *78*, *79*, 81–3,
        87–8
    function 11, 99–100
    isolation of 75–7
    lytic enzymes 500
    of *Halobacteria* 74–5, *76*
    phage receptors 377
    site of formation **100–3**, *104*, *105*, *118*, 119,
        121, 122, 123
    spore formation and outgrowth 495–6,
        502, 507, *509*
    structure 68–75
        Gram-negative bacteria 71–4, *75*
        Gram-positive bacteria 68–71
        sub-units of 74, *75*
Centrifugation, caesium chloride density gra-
    dient 270, *271*, 278, 280
    sucrose density gradient (zone centrifuga-
        tion) 317–20
Chemolithotrophs 193, 194, 196, 200
Chemo-organotrophs 193, 194
Chemostat *see also* bactogen 21, 142–3, 147
Chemosynthesis 193
Chloramphenicol, effect on protein synthesis
    150, 363, *364*, 465
    effect on spores 506, 509
Chlorobium 194
Chlorobium vesicles *126*
Chlorophyll 123, 124, 194
Chloroplasts 325, 517, *518*
Choline 249
Chondroids *see* Mesosomes
Chorismate mutase (Prephenate synthase)
    206–7, 208, 467, *469*
    aggregations of 209–10, 471, 472
Chorismate synthase *205–6*, *209*
Chorismic acid *205–6*, 208, 209, *210*, *468–9*
    as enzyme inhibitor 467, 472
Chromatinic body *see* Chromosome
Chromatium buderi 75
Chromatium okenii *134*

Chromatophores 64, 123–5, 548
Chromobacterium 456, 528–9, *532*
Chromobacterium violaceum 467, *469*, 472
Chromosome (Chromatinic body; Nuclear
  body)
  functional organization 409–11, 517–18
  integration of phage 406–7, 416–17
  integration of plasmids *391*, **399–400**,
    416–17, 419, 420
  mapping *see* Genetic mapping
  number/cell 128, 386, *518*
    growth rate and *144*, 145, 486
  pairing 391
  recombination 389–91
  replication 45, 131, **281–4**, 386
    rate of 157–8, 486
  size of 18, **127–8**, 370
  spore *495*, 496, 497
  state of DNA in 18, 53, **128–31**, 253–4,
    280, 369–70
  structure *4*, 44, 53, *68*, *122*, *283*, 306–8,
    370, 548
  transfer 395–9, 400, *401*
    factor for 394, 399–400, 420, 421
  viral *46*, 259–60, 273, 295–6, 369–70, 405
cis configuration 549
Cistron 415, 549
Citrate synthase (Condensing enzyme) *169*,
  185, *500*, **540**
  control of **188–9**, 247, 455–6, **456–7**, 458
Citric acid *27*, 28, *30*, *169*, **185–6**, 191, 456,
  458, 540
  as enzyme activator 246
  as enzyme repressor 459
Citric acid cycle *see* Tricarboxylic acid cycle
Citrulline *220*, *239*, 466
Classification
  Adansonian principles 528–9
    similarity matrices 527, 528–9
  Bergey 522, 526
  DNA composition and 530–2
  function 520
  hierarchical 520–1, 529–30
  identification 534
    dichotomous key 534
  nomenclature 532–4
  properties commonly used 521–2
  Stanier, Doudoroff and Adelberg *523–5*,
    526
  See also Enzymes, classes; Reactions,
    classes.
Clone 549
Clostridium 494, 510, *523*, *531*
Clostridium butyricum *166*, 246
Clostridium kluyveri 244
Clostridium pasteurianum 464
Clostridium perfringens 254
Clostridium propionicum 166
Clostridium tetani 534
Clostridium welchii *523*

Codons 411, 549
  amino acids *339*
  initiation 343–6
  recognition of anti-codon 270, 343, *344–5*,
    352–3
  termination 342–3, 362
Co-enzymes 536–46
Co-enzyme A 183, 184, 185, **540**
  function 183–4, 539–40
Co-enzyme F (5, 6, 7, 8 Tetrahydrofolic acid,
  FH$_4$) 224, 451–2
Co-enzyme I *see* Nicotinamide adenine dinucleotide
Co-enzyme II *see* Nicotinamide adenine dinucleotide phosphate
Colicins 421
Colony counts 140
Complementation tests 414–20, 549
  adjacent genes 417–18
  allelic 418–19
  in bacteria 415–16
  episomes and 416–17
  hybrid molecules from 418–19
  interallelic 419
  operons and 420
  in phages 415
Condensing enzyme *see* Citrate synthase
Conidial spores 494
Conjugation 45–6, 522, 549
  fertility factors and 394, 399, 416–17, **420**
  interrupted mating 395–8
  role of pili 107, 394, 420
  time sequence 398
  transfer of DNA during 8, 45, 395–9, 400,
    401, 485
Corepressor *445*, 480
Corynebacteriaceae 523
Corynebacterium 523, *531*
Corynebacterium diphtheriae 523
Corynebacterium poinsettiae *81*
Cross 549
Crossing-over *see* Recombination
Crotonyl.S.ACP 245, 246
Crotonyl.S.ACP reductase
Cryophilic bacteria 24
CTP *see* Cytidine 5'-triphosphate
Cyclase 229, *230–1*, *233*
D-Cycloserine *84*, 88
Cystathionase II *214*, *216*, 217
Cystathionine 211, *214*
Cystathionine γ-synthase *214*, *216*, 217
Cysteine, in methionine synthesis *214*
  in proteins 15
  in spores 496
  structure *14*
  synthesis *30*, *223*, 224
    control of 224–5
  tRNA 337
  triplets for *339*
Cysteine desulphurase 544

Cystine 15, 365, 498
Cysts 494
Cytidine 5'-diphosphate (CDP), synthesis *241*
CDP-abequose *93*, 98, *99*
CDP-diglyceride *248*, 249
CDP-ethanolamine 249
CDP-glycerol 91, 92
CDP-2-keto-3-deoxyoctonate *93*
CDP-ribitol 91, 92
CDP-tyvelose *93*, 98, *99*
Cytidine 5'-triphosphate (CTP), as enzyme inhibitor 219, 237, *239*, 241, 424, *426*, 430
  in lipid synthesis 247–9
  structure 247
  synthesis 239, *240*
Cytochromes 16
  system 164, 187, 200, **537–8**
  membranes and 117, *500*
Cytoplasm 3, 11, **63–4**, *66*, 111, 319
  streaming *518*, 519
Cytoplasmic membrane (Cell membrane; Protoplast membrane)
  composition of 111–14, *318*
    effect of medium 111, 112, 116, 145
    protein:lipid ratio 115, 116
  in division 102, *105*, 155, 486, 488
  DNA and 111, 117–18, 253, 298
  enzymes and 85, **117–18**, 374, 457, **470–2**, 517–18
  functions 117–18
  See also Permease systems
  spore formation 496
  structure *66*, *72–3*, *105*, *118*
    'unit-membrane' model, rigid interpretation 114–15
    liberal interpretation 115–17
  vesicles *see* Mesosomes
Cytosine, pairing with adenine 384, *385*
  pairing with aminopurine *381*
  pairing with guanine 18, 251, *252*, 255, 259, 276, *385*
  pairing with hypoxanthine 276, 383, *385*
  structure 16, *17*
  tautomers *385*

DAP *see* Diaminopimelic acid
Deacylase, hydrolytic 245
Deaminases 543, 544
Decanoate 246
Decarboxylases 543
Dehydrogenases 117, 537–8
5-Dehydroquinate dehydratase *204–5*, *209*
5-Dehydroquinate synthase *204–5*, 208, *209*
5-Dehydroquinic acid *205*, 208
5-Dehydroshikimate *205*
Deletions 273, **336**, 343, 362, 380, 386, 449

Deoxyadenosine 5'-diphosphate (dADP) 242, *243*
Deoxyadenosine 5'-triphosphate (dATP) 242, *243*
Deoxycytidine 5'-diphosphate (dCDP) 242, *243*
Deoxycytidine 5'-triphosphate (dCTP) 242, *243*
6-Deoxy-L-galactose (L-Fucose) *93*, 96
2-Deoxyglucose 453
Deoxyguanosine 5'-diphosphate (dGDP) 242, *243*
Deoxyguanosine 5'-triphosphate (dGTP) 242, *243*
6-Deoxyhexoses *93*, 96
6-Deoxy-L-mannose (L-Rhamnose) *93*
3-Deoxy-7-phospho-D-arabinoheptulosonate *204*, 208, *209*
Deoxyribonucleic acid (DNA), amount in cell 3, *4*, 127, 316, *317*
  associated protein 253
  base composition 18, *254*, 255, 522, 530–1
  base-pairing, effect on structure 5–6, 18, 37–8, **251–2**, 255–6
  in synthesis 276–7, 295, 315
  chemical nature of mutations 380–6
  denaturation 256–60, 369
  extrachromosomal, membrane-associated 111, 117–18, 253, 298
  in plasmids 399–400, 420–2
  genetic role of 5, 6, **44–53**, 277, 299, **306–308**, 334–5, **367–70**
  hybrids, with DNA 260–1, 269, **270–1**, **273–5**, 522, 531–3, *533*
    with RNA, cistron isolation and 271, 299–300, 376
    as mRNA assay **450–1**, 476, 478, 506, 508
    methods to prepare and estimate 269, 270–2
  in RNA synthesis 306, 309, 315
  lac-enriched 273–4, 450–1
  melting temperature 258, 369
  molecular weight 18, 253, *518*
  molecule, scale model *257*
    size of 18, 253–4, 281
  nucleotide sequence 16–8, 253, 254–5
    nearest-neighbour frequency analysis 290–1, 309
    notation for 338
  purification of 253
  radiation effect on 381, 384
  random coil formation 258, *259*
  renaturation 260–1, *271*
  repair mechanisms 287–8, **297–8**, 384, 391–2
    enzymes for 287–98, *390*, 392
  replication, cell division and 5, 20, 45, 128, 284, 386, 486
    conjugation and 400, *401*

Deoxyribonucleic acid, replication—*cont.*
  conservative mechanism   277, *279*
  direction of   284–6, 291–2
  discontinuous mechanism   284–7
  effect of antibiotics   302, 311
  energy for   280–1
  enzymes for   287–99
  growing points (replicating forks)   157–158, 281–7
  rate of   157–8, 284, **296, 298**
  semi-conservative mechanism   277–83
  strand separation   6, 38, 280–1, 293–4
  repressor binding   56, *442–5, 448–9*, 450, 451
  single-stranded   259–60, *273*
    replication of   260, 289, 293, 295–6
  spores   485, 492, 496, 509, 510, 514
  state of, in bacterial chromosome   18, 53, **128–31, 253–4**, 280, 369–70
    in viral chromosome   46, 259–60, 369–370, 405, 492
  structure   16–18, 253–5
    polarity of strands   *255*, 256, 284, 291–2
    secondary structure, Watson-Crick double helix   37–8, 254–6, *257*, 291–292
  synthesis, and cell cycle   157, 486
    in shift-up or shift-down   147–51, 487
    starvation   153
  visualization of   **128–31**, 254, 259, *260*, 281–2, *283*
DNA polymerase, catalytic sites   294–5
  direction of action   291–2, 293
  function *in vivo*   296–9
  need for primer   293, 296–7
  relation between template and product   276, 287–91
  RNA-dependent   315
Deoxyribonucleotides   5–6, 17–18, 38–9, 242, *243*, 288
2′-Deoxyribose   5, 16, *17*, 253, *261*, 457
Deoxythymidine 5′-diphosphate (dTDP)   242, *243*
Deoxythymidine 5′-monophosphate (dTMP)   242, *243*
Deoxythymidine 5′-triphosphate (dTDP)   242, *243*
Deoxyuridine 5′-diphosphate (dUDP)   242, *243*
Deoxyuridine 5′-monophosphate (dUMP)   242, *243*
Deoxyuridine 5′-triphosphate (dUTP)   242, *243*
Desulphovibrio   525, 531
Dethiobiotin synthase   *479*
2:8-Diaminoacridine (Proflavine)   302–3, 311, 382, 411
L-Diaminobutyric acid   79
7,8-Diaminopelargonate aminotransferase   *479*

Diaminopimelate-adding enzyme   83, *500*
meso-Diaminopimelate decarboxylase   *212–213, 216*, 217
Diaminopimelate epimerase   *212–13, 216*, 217
LL-α,ε-Diaminopimelic acid   *213*
meso-α,ε-Diaminopimelic acid (DAP), in cell walls   12, 78, 80, *85, 86*
  spheroplast formation   89, 151
  spores   496, 498, 509
  synthesis   30, 31, 211, *212–13*
Diauxie   24, 56, **252–3**, 488
Dichotomous keys   534
3:6-Dideoxy-D-galactose (Abequose)   *93*, 96, *99*
3:6-Dideoxyhexoses   *93*, 96, *99*
3:6-Dideoxymannose (Tyvelose)   *93*, 96
Diethylsulphate   383
Differentiation   59, 494, 509, 517
D-α,β-Diglyceride   247, *248*, 249
2,3-Dihydrodipicolinate reductase   *212–13, 216*, 217
2,3-Dihydrodipicolinate synthase   211, *212, 216*, 217, *500*
2,3-Dihydrodipicolinic acid   211, *213, 216*
Dihydrolipoyl dehydrogenase   184, 185
Dihydrolipoyl transacetylase   184, 185, 470–471
Dihydro-orotase   239, *240–1*, 242, *479*
Dihydro-orotate dehydrogenase   239, *240–1*, *479*
Dihydro-orotic acid   239, *240*
5,6-Dihydrouracil   *261*
Dihydrouridylic acid   263
Dihydroxyacetone   175, *198–9*
Dihydroxyacetone phosphate, as enzyme inhibitor   163
  in glycolysis   *161*, 163, *169, 460*, 544
  in lipid synthesis   243, *248*
  photosynthesis   197
Dihydroxyacid dehydratase   *214–15*, 216, 225, *226, 228*
Dihydroxyacid synthase (Acetohydroxyacid reductoisomerase)   *214–15*, 216, 225, *226, 228*
α,β-Dihydroxy-β-methylbutyrate   225, *226, 228*
α,β-Dihydroxy-β-methylvalerate   *215, 228*
cis-Dimethylcitraconitate   227
Dipeptidase   542
Di-phosphatidylglycerol (Cardiolipin)   *112*
1,3-Diphosphoglyceric acid   *161*, 163–4, 191, 196
2,3-Diphosphoglyceric acid   164–5
Dipicolinic acid (DPA, 2,6-Pyridine dicarboxylic acid)   496, 497, 503, 504, 505, 506
Dipicolinic acid synthase   *500*
Diplococcus   367, 391
  See also Pneumococcus

Diploids
  partial
    stable, construction of   53, 407, **416–17**,
      437–9, 482
      enzyme synthesis and   44, 418–19,
        438, 440–3, 476, 477
      transient   50, 53, 415–16
    total   511
Disaccharides   454
DNA *see* Deoxyribonucleic acid
Dominance, genetic   44, 53
  cis-dominance   441, *443*, 446, 476, 478
  enzyme synthesis and   44, 437, 477, 478,
    481
  of inducibility (*o*ᶜ mutations)   441–6
  testing   415, 482
DPA *see* Dipicolinic acid
DPN *see* Nicotinamide adenine dinucleotide
Drug resistance   378–80, 481, 505
  plasmids and   420, 421

Effectors   424, 455, 490
  binding to enzyme   424–5, 427–9
Electron acceptors   187, 194, 464, 537–8
Electron donors   194, 195
Electron transport   117, 187, 200, 501, **537–8**
Embden-Meyerhof pathway *see* Glycolysis
Endergonic process   5, 27, 28
Endoprotein   114
Endotoxin   249
Endomycopsis   511
Endonuclease   288, 293, 295, 297, 298
Endoribonuclease   447, 449
Endocytosis   *518*
Endospores, bacterial *see* Spores
Endoplasmic reticulum   *513*
Energy, as ATP   27–8, 160, 164, **168–70, 189–
    190**, 456, 457, 538
  charge   **168–70**, 189, 237, 456, 457
  from Entner-Doudoroff pathway   179, 190
  enzyme control and   **168–70**, 189, **237**, 455,
    456, 459, **491**
  for spores   499
  from glycolysis   168, 170, **189–90**, 191
  from light   193, 200, 517
  from oxidative phosphorylation   538
  from pentose phosphate pathway   177
  from phosphoketolase pathway   180
  from tricarboxylic acid cycle   **189–90**, *191*,
    456–7
Enolase   165
3-Enolpyruvyl-shikimate 5-phosphate   *205*
3-Enolpyruvyl-shikimate 5-phosphate aldolase   *205*, *209*
Enoyl-ACP dehydratase   245
Enterobacteriaceae   166, *238*, 456–7, 465–6,
  467, 472, *525*
Entner-Doudoroff pathway   177–9
  energy yield   179, 190

induction of   179, 459
Enzymes, active sites   16, 294–5, 365, 424,
    425, *429*
  activity
    activation   423, 424, **427–9**, 456, 490
    allosteric effect   57, 58, 246, 373, **424–9**,
      548
    concerted (multivalent) feedback inhibi-
      tion   203, 465
    control at branch points in pathway
      457–9, 465–6
    co-operative feedback inhibition   203,
      217, 465
    cumulative feedback inhibition   203,
      219, 237, 242, 465
    effector binding   424–5, 427–9
    false feedback inhibition   373, 478
    feedback inhibition   **202–3**, 206, 491,
      549
      aromatic amino acid metabolism
        208–211, 467, *469*
      aspartic acid family metabolism   *211–
        217*
      carbohydrate metabolism   189, 193,
        455–6, 459
      compensatory antagonism of   **203**,
        217, 241–2, 466
      glutamic acid family metabolism
        217–223
      histidine metabolism   232, *233*, 480
      purine metabolism   237–8
      pyrimidine metabolism   *239*, 241–2
      pyruvate amino acid family metabol-
        ism   225, *228*
      serine family metabolism   224–5
    inducer fit and   *429*
    modulation   424, 429, 455, 457–9, 490
    multimeric enzymes and   423–30
    precursor activation   170, 456, 491
    sequential feedback inhibition   **203**, 209,
      217, 466, 467
    sigmoid response   425, *426*, 427
    structural modification   219, 429–30
  affinity for substrate   41–2, 57, 424, 425–9
  aggregation of   183, 185, 210–11, 217, 247,
    468, **469–73**, 491
  autolytic   88, 90
  allosteric states   57–8, 246, 373, **424–9**, 548
  Class I, catabolic, degradative   4, 21, 26,
    27, 55–6
    mutations involving   373–4
    regulation of   55–6, 454–65, 469–71,
      473–7
  Class II, anabolic, biosynthetic   4, 21, 26,
    27, 56
    mutations involving   371–3
    regulation of   56–8, 465–9, 471–3, 477–
      483
  Class III, anabolic, biosynthetic   4–7, 21,
    58

Enzymes, Class III, anabolic—*cont.*
  mutations involving 374–7
  regulation of 58–9, 483–90
classification of bacteria 522
cofactors 536–47
constitutive 55, 242, 430, 458
controlled breakdown 424
energy-linked control **168–70**, 189, **237**,
    455, 456, 459, **491**
membrane-bound **85**, **117–18**, 374, 457,
    470–2, 517–18
multifunctional 472
multimeric 424–30
multiple (isofunctional) 203, 209, 217, 456,
    465, 467
in spores 499–502, 507, 508–9
sub-units **364–5**, 424, 425
synthesis, catabolite repression 55–6, 451,
    452–4, 459, 460, 476, 488, 491, 510
  cyclic AMP and 451, **453–4**, 476, 477,
    491
  co-ordinate repression 211, 225, 232,
    233, 237, 242, **433–5**, **440**, 466, 478–
    479, 480
  cumulative repression 205, 242, 465
  derepression 222, 225, 229, 236, 435
  differential rate of 431–2
  dispersion diagram 434–5
  end-product repression **57–8**, 204, 217,
    225, 237, 445, 446, 460, 465, 468
  inactivation repression 465
  induced phenotypic resistance 236
  induction, co-ordinate control of 433–5,
    440, 478, 549
    de novo protein synthesis in 436
    kinetics of 430, 433
  multivalent repression 205, 217, *228*,
    229
  negative control (repressor model) *56*,
    437, *438–9*, **440–6**, 481, 490
  positive control (endogenous inducer
    model) **437**, *438–9*, 477
  product induction 430, *460*, 461–2, 491
  repression, aromatic amino acid metabo-
    lism 208–11, 467, 477–8
    aspartic acid family metabolism 217,
      480
    false 373
    glutamic acid family metabolism
      219–22
    histidine metabolism 232–6
    kinetics of 430, 431, *432*, 433
    purine metabolism 237–9
    pyrimidine metabolism *239*, 478–9
    pyruvate acid family metabolism 225,
      228, 478
  sequential induction 462–4
  super-repression 440
  transcription-level control 441–2, 450–
    451, 476, 478, 490, 510

    translation-level control 442, 450, 451,
      478, 490, 510
  temperature-sensitive 144
Epiprotein 114
Episomes 420, 549
  integration into chromosome 301, **399–
    400**, 416–17, 419, 420, 421
Ergosomes *see* Polyribosomes
Erwinia amylovora 166
D-*Erythro*-imidazoleglycerol phosphate 229,
    231, *233*
Erythromycin 364, 375
Erythrose 4-phosphate 175, *176*, 180, 197,
    204, 208, *468*
Escherichia 525, *531*
Escherichia coli, L-arabinose system *475*, 477
  aromatic amino acid metabolism 209–11,
    477–8, 481
  aspartic acid family metabolism *216*, 217,
    472, 477, 480
  biotin system 468, 477, 479–80
  cell envelope *72*, *73*, 104, *112*, *122*
    components *80*, 81, *82*, 95
  chromosome 52–3, 127, *130*, 254, *307*, 370,
    386, 398
    replication of 157–8, 281–3
  conjugation 30, 394–8
    Hfr strains and 394–5, 399–400, *401*
  control of central pathways 455, 456, 459
  DNA 253, 254, 273–4, 370, 398
    synthesis 278–9, 289
  DNA polymerase 288–97
  exonuclease 285–6, 293
  galactose system 475–6
  generation time 486
  genetic map *52*, 53, *396–7*, 416–17
  glucose and glycogen synthesis 132, 170–1
  glutamic acid family metabolism 219, 221,
    472, *479*, 480
  glycerol metabolism 163, 460, *475*, 477
  glyoxylate cycle 193, 460
  growth 141–2, 144, *147*, *149*, *152*, 153, 371,
    487
  lactose system 273–5, 362, 431, 432, 433–4,
    435–54
  lipid synthesis 244–7, 249
  nitrogen metabolism 464
  nucleotide synthesis 237, *239*, 240–2, 477,
    479
  penicillinase 473–4
  pili 107, *108*
  polynucleotide ligase 287
  pyruvate dehydrogenase 183–5, 470–1
  pyruvic acid family metabolism 225, *228*,
    478
  RNA 263, *264*, *266*, 330
  RNA polymerase 308–9, 311–12, 313
  ribosomes *307*, *322–3*
  serine acid family metabolism 224
  D-serine system 459, *475*, 476

Esterases 542
Ethanol 162, *166, 172,* 179, 180, *181,* 189, 538
Ethanolamine *97,* 182, 249
Ethylethane sulphonate 383
7-Ethylguanine 383–4
Ethylmethane sulphonate 383
Eukaryotes 325, 457, 465, 517–18
Exchange, genetic 389–92, 549 *See also* Recombination
Exergonic process 5, 27, 28
Exoenzymes 474, 500, 503
Exonuclease 259, 285–6, 293, 294
Exosporium 497
Exponential phase 20–1, *22, 23,* 152–3

Facilitator *460,* 470
FAD *see* Flavin adenine dinucleotide
Family 520, *523–5,* 530
Fats *see* Lipids
Fatty acids, as carbon source 454
  in lipopolysaccharides 96, *98*
  in membranes 111–12
  β-oxidation of *182,* 188
  synthesis 243–7, 458, 542
Fermentation 165–7, *523, 525*
Fertility factors 394–5, 399–400, 416–17, 420, 450–1
Fimbriae *see* Pili
Flagella *4,* 16, 28, 64
  attachment 108–9, *110*
  distribution on cell 107, 108, 521, *523–5,* 526
  function 16, 28, 107
  structure 107–8, 109, *518*
Flagellin 16, 108, *109,* 117
Flavin adenine dinucleotide (FAD) 183, 184, 185, **537** 8
Flavin mononucleotide 537–8
Flavoprotein (FP) *182,* 187, 189, 200, 239
Flavoprotein dihydrolipoic dehydrogenase 470, *471*
Fluctuation test 378, *379*
Fluoride 165
p-Fluorophenylalanine 375
5-Fluorouracil (FU) 388–9
Folic acid 541–2, 547
Forespore *495,* 496
Formamide *461*
5-Formamido-1-(5′-phosphoribosyl)-imidazole 4-carboxamide *232, 235*
Formamino transfer 541
Formate *166, 461*
Formimino-glutamate *461*
N^{10}-Formyl-FH_4 *234,* 541
Formyl-glutamate *461*
Formylglutamate transformylase *461*
N-Formyl-methionine *339,* 346
  tRNAs for 343, 345

N-Formyl-methionine-tRNA (fMet-tRNA$_f$) 354–5, 356, 358–9
Freeze-etching 65–7
Fructose 28, *173*
Fructose 1,6-diphosphatase *169,* 171, 177, 197
Fructose 1,6-diphosphate, Calvin cycle 197, *198–9*
  as enzyme activator *169, 172,* 456
  glucose synthesis 171
  glycolysis *161,* 162–3, 167, *169,* 170, 191, 457
  pentose cycle 177
Fructose 1,6-diphosphate aldolase (Aldolase) 163, *169,* 177, 197, 501, **544**
Fructose 6-phosphate, Calvin cycle 197
  glucose synthesis 171
  glycolysis *161,* 162, 167, *169,* 170, 457
  pentose cycle *176*
  phosphoketolase pathway 180
L-Fucose (6-Deoxy-L-galactose) *93,* 96
Fumarate hydratase (Fumarase) 187, *500*
Fumaric acid, amino acid metabolism 31, *220*
  as enzyme inhibitor 455
  fermentation *166*
  glyoxylate cycle *192*
  purine synthesis *234–5, 236,* 237
  tricarboxylic acid cycle *185,* 187, 189, 538
Fungi 517
Furanose ring 329
Fusidic acid 363, *364*

Galactokinase *475*
Galactose 96, 98, 99, 100, 431, 542
  system 374, 404–5, 416–17, **454,** 459, *475,* **476**
Galactose 1-phosphate uridyl transferase *475*
β-Galactosidase *364, 365,* 431, *432,* 433–4, **542**
  See also Lactose system
β-Galactoside permease 117, 433–4
thio-*β*-Galactoside transacetylase 433–4
Gas vacuoles 125–7, *128*
Generation time (Mean generation time, MGT) **139,** 140, 141–2, 143, *147,* 550
Genes, alleles of **49,** 418–19, 436–7, 440–1, 491–2
  clustering of 210–11, 300, 436, **478–9, 480–1,** 491–2
  value of 482–3
  complementation of 414–20
  control of amino acid sequence by 409–14
  duplication of 263, 300, 306–8, 376, 483
  expression, control of 431–54
  See also Enzymes, synthesis
  number/cell *317*

Genes—*cont.*
　operator　229, *353*, 480, 482, 490, 551
　　constitutive mutations (*o*ᶜ mutations)
　　　441–6, 478
　　function　441–9, 478, 481
　　mapping　441–2
　　polar effect and　446–9
　　repressor binding　442, *443–5*, 446, *448–*
　　　449, 450, 481, 490–1
　　RNA polymerase and　*442*, 443, *444–5*,
　　　446, *448–9*, 450
　promoter, mutations in　450, 454
　　RNA polymerase and　307, 312–13, *353*,
　　　448–50, 454, 477, 481, 484, 491
　regulator　221, 301, *475*, *479*
　　negative control by products of 55, *56*,
　　　437–9, 445, 446, 476, 480, 490–1
　　on plasmids　481
　　positive control by products of　437, 477
　　size　6, *317*
　structural　6, 41, 44, 55, *56*, 301, 347, 550
　　linked functioning of, operons　225, 228,
　　　232–3, 272–5, 420, 478
　　　control, operon model　435–54
　　synthesis of　276–7, 338
　　visualization　306–8
Genetic code, deciphering of　338–42
　degeneracy of　269, 335, 413
　initiation codons　343–6, 355, 360
　table of base triplets　*339*
　termination codons　342–3, 346, 357, 359,
　　360
　　mutations in　*339*, *344–5*, 413–14, 447
　triplet nature of　7, 41, **335–7**, **411–13**, 550
　universality of　335, 337
Genetic mapping
　by complementation testing　414–20
　　of phages　415
　by conjugation　394–400
　general principles　370, 389–91, 392–3
　by hybridization of DNA　272–3
　linkage map of *E. coli*　52, *396–7*
　by overlapping deletions　393–4, 404
　by recombination　50–3, 390–2, 394–400
　by sequential replication　408–9
　three point crosses　404
　by transduction　399, 400–4, 407
　　special transduction　404–5, 407, 416–17
　by transformation　407–9
Genetic markers　370, 392–3, 550
　frequency of separation　52–3, 389–90,
　　392–3
　transfer of, cotransduction　402, 404
　　growth phase and　408
　　one way　394–5
　　starting point　398
　　timing of　398
　　transduction　400–4
　　transformation　407–9
　unselective　370, 394

Genetic relationship　522, 531–2, *533*
Genetics, population　367
Genome　44, 370, 550
　of spore　509, 510
Genospecies　520
Genotype　44, 367, 530–1, 550
Genus　520, *523–5*, 530
Gliding movement　519
Glucogenesis (Gluconeogenesis)　*169*, 170–2,
　188
Gluconic acid　179, 459
Glucosamine　32, 219
Glucosamine 6-phosphate　457
Glucose
　breakdown, control　168–70, 188–9, **454–**
　　459, 491
　　effect of aerobic/anaerobic conditions
　　　162, 165, 189, **456–7**
　　energy yield from　168–70, 177, 179, 180,
　　　189–90, *191*
　　fate of carbon atoms　163, **167–8**, 174,
　　　177–8, 179, 188
　　use in classification　165, *523–5*, *534*
　　　See also Glycolysis; Pentose phosphate
　　　cycle *etc.*
　as carbon source　3, **22–4**, 141, 373
　　diauxic growth　24, 56, **452–3**, 488
　　growth yield for　22, 23, 189–90
　　spores and　502, *503*, 507
　effect *see* Catabolite repression
　in lipopolysaccharides　96, 98, *100*
　in membranes　111
　synthesis　*169*, 170–1
　in walls of ascospores　512
Glucose dehydrogenase　*500*, 501, 502
Glucose 1,6-diphosphate　172
Glucose 1-phosphate　*169*, 172, *455*
Glucose 6-phosphatase　169, 543
Glucose 6-phosphate, Entner-Doudoroff
　pathway　*178*
　glucogenesis　171, 543
　glycolysis　*161*, 162, *169*, 459, 542
　oxidative decarboxylation (Pentose phos-
　　phate pathway)　173–7, 246, 459
　phosphoketolase pathway　*181*
Glucose 6-phosphate dehydrogenase　174,
　179–80, 459
α-Glucosidase　*509*
Glucosyl-glycerol phosphate　*90*
Glucuronic acid　92, *93*, 94
Glutamate acetyltransferase (*N*-Acetylgluta-
　mate synthase　*218*, *220–1*, 222, *239*,
　479
Glutamate dehydrogenase　219, 464
Glutamate γ-semialdehyde　*218*, 219
Glutamate semialdehyde dehydrogenase
　218, 221
D-Glutamic acid, in capsular polypeptide
　105–6
　in peptidoglycans　11, 31, 78, 79, 80, *81–6*

L-Glutamic acid, as amino donor *206*, 208, 211, *212, 213*, 215, 216, *221, 223*, 225, *226–7, 240*, 541
  effect on spores 499
  in glutamine deamination *32, 206*, 208, 219, *234–5*, 236, *240*
  from histidine breakdown *461*
  as nitrogen source 464
  structure *14*
  synthesis *30, 32, 169*, 219, *220*, 464
    control of 219–23
    triplets for *339*
Glutamic-alanine transaminase *500*
Glutamic-aspartic transaminase *500*
Glutamine, as amino donor 32, 33, *206*, 208, 219, 229, *230*, 232, *234–5*, 236, *238*, 239, *240*
  structure *14*
  synthesis *30*, 218, 219, 464
    control 219
    triplet for *339*
Glutamine synthase *218*, 219, 429–30, 464
γ-Glutamokinase *218*, 221
γ-Glutamyl phosphate *218*, 219
Glutathione 179
Glyceraldehyde 3-phosphate, in amino acid metabolism 208
  Entner-Doudoroff pathway *178*, 179
  as enzyme activator 172
  glycolysis *161*, 163–4, *169, 191, 460*, 544
  pentose phosphate cycle 174–7
  phosphoketolase pathway 180, *181*
  photosynthesis 197, *198–9*
Glycerate dehydrogenase *222–3*, 224
Glycerate 1,3-diphosphate *161*, 163–4, 191, 196
Glycerate 3-phosphate *see* 3-Phosphoglyceric acid
Glycerol 163, 171, 243, 542
  system *460, 461–2, 475*, 476
Glycerol kinase 163, *460, 461–2, 475*
Glycerol phosphate 90–1
  See also Teichoic acids
L α Glycerophosphate 163, 243, 247, *248, 460*, 461–2
L-α-Glycerophosphate dehydrogenase 163, 243, *460*, 461–2, *475*
L-α-Glycerophosphate permease *460*, 461–2, *475*
L-Glycero-D-mannoheptose (Heptose) *93*, 96, *97*, 98
Glyceryl triester 542
Glycine, as enzyme inhibitor 430
  nucleotide synthesis 32, 33, *235*, 236
  in peptidoglycans 78, 79, *80–2, 84–6*
  structure *14*
  synthesis *30*, 31, *222*, 224, 542
  triplets for *339*
Glycogenesis 7, 36, 172
Glycogen,

ascospores 514
  breakdown *169*, 172
  granules 64, 132
  structure 7, 36, 172
Glycogen synthetase *169*, 172
Glyceraldehyde, active 174–5
Glycolipids 11
Glycollate 193
Glycolysis (Embden-Meyerhof pathway) 27, 28, 161–70
  control 168–70, 454, *455*, 458
  energy yield from 168, 189–90, *191*, 457
  fate of carbon atoms 163, 167–8
Glycoproteins 114
Glycosaminopeptide *see* Peptidoglycan
Glycosidases 542
Glyoxylate cycle 190–3, 460
Glyoxylic acid 191, *192*
GMP *see* Guanosine 5'-monophosphate
Golgi apparatus *518*
Gougerotin 363, *364*
Gram stain 68
  classification *523–5*, 526, *534*
Granules, cytoplasmic 64, *124*, 132–4
Granulose (Glycogen granules) 64, 132
Green bacteria, gas vacuoles 125, *126*
  photosynthesis 123, 124–5, *126*, 194–5
Growth, balanced 137–9, 141–3, 154
  biphasic (diauxic) *24*, 56, **252–3**, 488
  in continuous cultivation 21, *23*, 141–3
  curves **21–5**, *140*, 151–4
  effect of environmental change 140–1, 143–5, 146–51
  effect of medium 25, 137–9, 140–1, **144–51**
  effect of temperature 24–5, 124, 144, 145
  exponential phase 20–1, *22, 23*, 152–3
  factors 150, 371, 487, 547
  false lag phase 151, *152*
  filamentous 145
  of individual cells 25, 137, 154–9
  generation time 21, 139, **156–7**, 284, 486
  inducer lag 430
  lag phase **21**, *22, 23*, 151, *152*, 488
  limitation of **23–4**, 59, 142–3, **148–51**, 153, 487
  mean generation time (doubling time, MGT) **139**, 140, 141–2, 143, *147*, 550
  measurement of 139–41, 151, *152*
  rate, cell composition and 144–6, 157
    chromosome number/cell and *144*, 145, 486
  rate constant 138–9
  rate, macromolecular synthesis and 147–151, 486–90
    ribosome content and 145–6
    regulation 54
  See also Enzymes, activity; Enzymes, synthesis
  starvation 153, 487–8, 502, *503*
  stationary phase *22*, **23–4**, 151, 153, 487–8

Growth—*cont.*
  synchronous 25, 155–6
  unbalanced 146–51
  yield for glucose 22, 23, 189–90
  yield, molar 189–90
Guanine, alkylation 383–4
  pairing with bromouracil 381, 382, *385*
  pairing with cytosine 18, 251, 252, *255*,
    259, 276, *385*
  pairing with thymine *385*
  pairing with uracil 263
  structure 16, *17*
  tautomers *385*
Guanosine 5′-diphosphate (GDP), as enzyme
    activator 456, 462
  synthesis 236
GDP-aminomannuronic acid *93*
GDP-fucose *93*
GDP-mannose 98, *99*
Guanosine 5′-monophosphate (GMP), as en-
    zyme inhibitor 237
  synthesis *234–5*, 236
    control of 237–8
GMP-PCP 355
GMP synthase (XMP: ammonia ligase)
    234–5, 236, 237, *238*
Guanosine tetraphosphate (ppGpp) 310,
    313, 485, 487
Guanosine 5′-triphosphate (GTP), as enzyme
    activator 241
  as enzyme inhibitor 237, *238*
  in nucleic acid synthesis 312
  in protein synthesis 331, 333, 354, **355**,
    356, *358–9*, 363
  in purine synthesis *235*, 236
  synthesis 236
GTP-ase 333, 355, 363
Guanylate reductase 237

Haem 16, 468, 537–8
Haemin 468
Haemoglobin 337
Haemophilus 367, *524*
Haemophilus influenzae *524*
Halobacteria, cell envelopes 74–5, *76*, 112
  gas vacuoles 125, *127*, *128*
Halobacterium halobium *76*, 116
Hanseniaspora 511
Hansenula 511
Haploids 50, 511, 550
Haplophase *512*
Heptose (L-Glycero-D-mannoheptose) *93*,
    96, *97*, 98
Heterolactic bacteria 167
Heteropolymers 34
Heterothallic 511, *512*
Heterotrophs 193
Heterotropic effects 425, *428*, 429
Heterozygote 550

Hexokinase 162, *169*, 170, **542**
Hexosamines *4, 5*, 27
  synthesis *31*, 32
Hexoses 96, 171, 177, 197, 200
Hinshelwood model for control of macro-
    molecular synthesis 488–90
Histidase *461*, 462
L-Histidinal *231*, 233
L-Histidine, activating enzyme 480
  breakdown *461*, 462
  as enzyme inducer 462
  as enzyme inhibitor 232, *233*, 425, 430
  as enzyme repressor 219, *233*, 480
  structure *14*
  synthesis *30*, 229, *230–1*
    control 232–3, 237, 480–1
    operon 232, *233*, 480–1, 482
      mutants in 232, 403–4, *412*, 417–18,
        419
    relation to purine synthesis 229, *230–1*,
      237
    triplets for *339*
L-Histidinol *231*, 233
Histidinol dehydrogenase *231*, 233
Histidinol phosphatase *231*, *233*, 412–13
L-Histidinol phosphate *231*, 233
Histidinol phosphate aminotransferase *231*,
    233
Histidyl-tRNA 480
Histidyl-tRNA synthetase 480
Histones *518*
Homocysteine 211, *214*, 542
Homocysteine methyltransferase 211, *214*,
    216, 217, 542
Homolactic bacteria 167
L-Homoserine 79, *81*, 211, *214*, *216*
Homoserine dehydrogenase 211, *214*, *216*,
    217, 429
  aggregation with aspartokinase 217, 470,
    472
Homoserine kinase 211, *214–15*, *216*, 217
Homoserine O-phosphate 211, *215*, *216*
Homoserine O-succinyltransferase *214*, *216*,
    217
Homopolymers 34
Homothallic 511, *512*
Homotropic effects 425, *428*, 429
Homozygote 550
Hydrocarbons 160
Hydrogen, acceptors 160, 537
  carriers 183, 537–8, *539*
  donors 519
  molecular 194, 195, 200, 519
Hydrogen sulphide 194, 195, 200, *223*, 224,
    519, 544
Hydrogenomonas 200
Hydrolases 537, 542
L-β-Hydroxyacyl.S.CoA *182*, 245
Hydroxyapatite columns 271–2
p-Hydroxybenzoate 454

D-β-Hydroxybutyryl.S.ACP 245, *246*
2-Hydroxyethyl-2-thiamine pyrophosphate
    183
Hydroxylamine 382, 383, 388–9
Hydroxymethyl.FH₄ 541
Hydroxymyristic acid 96, *98*
Hydroxyproline 335
Hydroxypyruvate *222*, 224
Hyperchromic effect 257, 262
Hypoxanthine *261*, 275, 383
    pairing with cytosine 276, 383, *385*

ICR-191 (Acridine half-mustard) 382–3, *412*
Identification 534
IDP (Inosine 5′-diphosphate) 275
Imidazoleacetol phosphate *231*, *233*
Imidazoleglycerol phosphate *232*
Imidazoleglycerol phosphate dehydratase
    231, *233*, 419
5′Imidazolone-4-propionate *461*
Imidazolone propionate hydrolase *461*
Immunofluorescence 100–1
IMP *see* Inosine 5′-monophosphate
Indole 460
Indoleglycerol phosphate *207*, *208*, 467
Indoleglycerol phosphate synthase *207*, 210,
    469, 471, 472
Inducers 430, 459
    catabolite repression and 453, 459
    endogenous 437, *438–9*
    gratuitous 431, 436
    interaction with repressor 437, *438–9*, *440*,
        443–4, 446, 490
    uptake 430, 433, 453
Induction, of enzymes *see* Enzymes, synthesis,
        induction *etc.*
    of phages 48–9, 405, 484
Inhibition *see* Enzymes, activity, feedback in-
        hibition *etc.*
Inhibitors 424, 427, *428*
Initiator 475, 477
Inosinate (IMP) dehydrogenase *234–5*, 236,
    237, *238*
Inosine *261*, *352*
Inosine 5′-diphosphate (IDP) 275
Inosine 5′-monophosphate (IMP; Inosinic
        acid), as enzyme inhibitor 237
    synthesis 232, *234–5*, 236
        control 237–8
Inositol 182, *248*
IPTG (Isopropylthio-β-D-galactoside) 431,
    432, 436, 451
Iron 3, 179, 200, 537–8
D-Isocitrate *184*, 186, 189, 191, *192*
Isocitrate dehydrogenase 186, 499, *500*
Isocitrate lyase 190–1, 192, 193
Isoenzymes (Multiple enzymes) 203, 209,
    217, 456, 465, 467, 472
L-Isoleucine, activating enzyme 329

structure *14*
synthesis *30*, 31, 211, *215*, 216, 225, 473
    control *216*, 217, *228*, 481
    isoleucine-valine operon *228*, 229, 478,
        481, 482
    triplets for *339*
Isomerases 536, 544–5
Isopropanol *166*, 195
2-Isopropylmalate 225, *227*, *228*
3-Isopropylmalate *227*, *228*
Isopropylmalate dehydrogenase *227*, *228*
Isopropylmalate isomerase *227*, *228*
Isopropylmalate synthase 225, *226–7*, *228*
Isopropylthio-β-D-galactoside (IPTG) 431,
    432, 436, 451
Isosteric interaction 58
Isozymes *see* Isoenzymes

Kanamycin 363, *364*
β-Ketoacyl-ACP-reductase 245
2-Keto-3-deoxyoctonate *93*, 96, *97*
α-Ketoglutarate *see* α-Oxoglutarate
Ketolase 543
Ketolisomerase *230–1*, *233*
Kinases 540, 542
Klebsiella 531
Krebs cycle *see* Tricarboxylic acid cycle

Lactate dehydrogenase 167, 538
Lactic acid, as carbon source 171
    classification and 167, *523*
    phosphoketolase pathway 179–81
    pyruvate fermentation 162, *166*, 167, 189,
        538
Lactobacillus *166*, *523*, 526, *531*
Lactobacillus arabinosus 91, 217
Lactobacillus bulgaricus *523*
Lactobacteriaceae *523*
Lactonase 174
Lactose, as carbon source 56, 373, *374*, 431,
    488
    system **431**, *432*, **433–4**, 435–6, 488
        catabolite repression and 452–4
        models of control 437–52
3-O-D-Lactyl-N-acetylglucosamine *See*
        N-Acetylmuramic acid
Lampropedia 156
Lecithin (Phosphatidyl choline) *248*
L-Leucine, structure *14*
    synthesis *30*, 225, *227*
        control 225, *228*
        triplets for *339*
Leucine aminotransferase *227*, *228*
Leuconostoc *523*
Leuconostoc mesenteroides 179, *523*
L-forms 89, 550
Ligases (Synthases) 536, 546
Light, energy from 193, 200, 517
    production 538

Lilium 515
Linkage, genetic 550
  See also Genetic mapping
Lipases 542
Lipid A *see* Lipopolysaccharide
Lipids, in ascospores 512, 514
  catabolism 163, 181–2
  in cell membranes 3, 12, **111–12**, 114–17,
    182, **242**
  in cell walls 95, 99
  granules 64, *124*, 132, **133**, 242
  synthesis 242–9
Lipoic acid **183–5**, 539, 547
Lipoic reductase transacetylase (Dihydro-
    lipoyl transacetylase) 184, 185, 470–1
Lipopolysaccharides, cell walls 12, 72, 74,
    95
  antigenicity *96*
  core polysaccharide 96, *97*, 98–9
  function 92, 99–100
  lipid A *95*, 96–7, *98*
  0-specific chains *95*, 96, 97, 98
  structure 95–8
  synthesis 98–9
Lipoproteins, cell membranes 114–17
  cell walls 12, 72, 74, 95
Lithotrophs 193
Locus, genetic *see* Genes
Lyases 536, 543
D-Lysine 80
L-Lysine, biotin binding 546
  cell walls 79, *80–6*, 498
  structure *14*
  synthesis 30, 31, 211, *213*
    control *216*, 217
  triplets for *339*
Lysine decarboxylase 499–500
Lysogeny **48–9**, **405–7**, 416–17, 484, 550
Lysozyme, effect on cell walls 12, 71, *73*, 78,
    79, 81–3
  to make protoplasts **87–8**, 111, 325
  to release spores 499
Lysyl-phosphatidylglycerol *112*

Macromolecules, in bacteria 3, *4*, 26, 314–19
  synthesis 5–6, 35–43
    control of, Hinshelwood model 488–90
    growth rate and 147–51, 486–90
  See also Deoxyribonucleic acid; Proteins
    etc.
Magnesium (Mg^{++}), cell walls and 92
  as enzyme cofactor 92, 162, 164, 165, 171,
    180, 183, 288
  mesosomes and 123
  in protein synthesis 333
  tRNA and 262, 266
  ribosomes and 318, 321, 323
  in spores 498

Magnesium fluorophosphate 165
Malate dehydrogenase *169*, 187, *499*, *500*
Malate enzyme (Malate dehydrogenase, de-
    carboxylating) *169*, 171, 187–8
Malate synthase 190–1
L-Malic acid *166*, *169*, *185*, 187, 189, *191–2*
  as carbon source 171
  as enzyme inhibitor 455
Maleate 187
Malonyl.S.ACP 245, 246
Malonyl.S.CoA 244, *246*, 458
Malonyl transacylase 244
Maltose 373, 454, 459
L-Mandelate 454, 462, *463*, 464
Mandelate dehydrogenase 462, *463*
Manganese (Mn^{++}), enzyme cofactor 83,
    84, 165, 244, 308
  enzyme stabilizer 501
Mannan 86, 512
Mannose 96, *99*, 111
Marker, genetic *see* Genetic marker
Mating, of bacteria *see* Conjugation
  of yeasts 511, *512*
Mean generation time (MGT) **139**, 140, 141–
    142, 143, 147, 550
  See also Growth
Medium, effect on cell composition 104, 111,
    112, 116, **147–51**, 487–9, 492
  effect on growth 25, 137–9, 140–1, **144–51**
  minimal 50, 371, 488, 550
  rich 24, 141
  selective 373–4
  simple 20, 24, 141
  spores and 499–500, 502, 507, 508, 512,
    514
Meiosis 511, 512–13
  genetic control 514–16
Melibiose *434*
Membranes, of bacterial cells *see* Cytoplasmic
    membrane
  of mammalian cells *518*
  nuclear 44, *513*, 517–18
  photosynthetic 123–5, *126*
Merodiploids 416, 419
  See also Diploids, partial
Merozygotes 415–16, 550
  See also Diploids, partial
Mesosomes 64, *518*, 550
  function *104*, *119*, 122–3, 155
  structure *118*, 119–22
  tricarboxylic acid cycle and 189, 457
Messenger RNA (mRNA) *see* Ribonucleic
    acid, messenger
Metabolism
  co-ordination 423–4
    biosynthetic pathways 466–9, 471–3,
      477–82, 483–90
    central pathways 454–9, 469–71
    importance of branch points 457–8,
      465–6

Metabolism, co-ordination—*cont.*
  mechanisms of 425–54, 469–73, 490–1
    See also Enzymes, activity; Enzymes,
      synthesis
    miscellaneous pathways 473–7
    peripheral pathways 459–65
    selective advantages 474, 482–3
  intermediary 26
  use in classification 521–2, *523–5*
Metachromatic granules (Polymetaphosphate
  granules; Volutin granules) 64, *132*,
  133–4
Methionine, as enzyme inhibitor *216*, 217
  as enzyme repressor *216*, 217, 472
  in protein synthesis 343–6, 357, *358–9*
  ribosomal protein 322
  structure *14*
  synthesis *30*, 31, 211, *214*, 542
    control *216*, 217
    triplets for *339*
β-Methylaspartate mutase 545
β-Methylaspartic acid 545
Methyl-B$_{12}$ 542
N^5,N^{10}-Methenyl-FH$_4$ *235*, 541
N^5,N^{10}-Methylene-FH$_4$ *222*, 224, 542
Methylene-tetrahydrofolate dehydrogenase
  238
N^5-Methyl-FH$_4$ 211, *214*, 442, 541
1-Methylguanine *261*
7-Methylguanine 383
ε-N-Methyl lysine 335
Methylmalonyl-CoA 545
Methylmalonyl-CoA mutase 545
N-Methyl-N'-nitro-N-nitrosoguanidine 383,
  504
Methyl thio-β-D-galactoside *434*
5-Methyl-tryptophan 329
Microcapsule 64
Micrococcaceae 523
Micrococcus 79, 481, *523*, 531
Micrococcus lysodeikticus, cell walls *80, 81*,
  94
  classification *523*
  DNA *129*, *254*
  genes 481
  mannan polymer 86
  membrane *112*
  metachromatic granules 133
Microelectrophoresis 25
Microscopy, electron 65–7, *69*
  fluorescent 100–1
  phase contrast 65
Microtubules *518*
Minimal medium 50, 371, 488, 550
Mitochondria 325, 457, 517, *518*
Mitosis 49–50, 511, 518
Modulators 424, 429, 455, 457–9, 490
  See also Enzymes, activity
Monosaccharides 454
Motility, classification and *523–5*, 526, *534*

energy for 28, 538
flagella and 16, 28, 107
modes of *518*, 519
cis-cis-Muconate *463*
Mucopeptide *see* Peptidoglycan
Muramic acid, cell walls 11–13, 94
  lactam of *500*
  structure 78
  synthesis *31*, 32
    See also N-Acetylmuramic acid
Muramic acid-6-phosphate 94
Murein *see* Peptidoglycan
Mutagen 550
Mutagenesis, acridines 49, 303, 382–3
  base analogues 49, 381–2
  non-replicating DNA and 342, 383–4
  radiation 49, 381, 384
Mutations
  alteration (point) 341–2, 343, 388, 449
    amber 339, 344–5, 414
    ochre 339, 414
    transitions 380, 381, 382, 383
    transversions 380, 381, 386
    umber 339
  auxotrophic *see* Auxotrophs
  catabolic (in Class I enzymes) 373–4
  in Class III enzymes 374–7
  close-linked 409–10, 420
  complementation of 414–19
  constitutive 476, 481
  definition 44, 551
  drug resistant 375–6
  expression of 386–9
    phenotypic lag 386
    phenotypic suppression 388–9
    segregation lag 386
    transcription lag 388
  frameshift 303, 362, 380, 382 3
    additions **336**, 343, 380
    deletions 273, **336**, 343, 362, 380, 386,
      449
    mutually compensating 411–14
  inversion of gene sequence 380
  operator constitutive (o^c) **441**, *442*, **446**,
    449–50, 478, 480
    dominance relations 441, *443–4*, 450,
      476
  phage resistant 377
  pleiotropic 515
  polarity of 232, **446–9**, 480, 482
  promoter 450, 454, 481
  recombination deficient (*rec*) 392
  regulator gene 476, 481
  relaxed 487
  spontaneous, apparent adaptation and
    378–80
    control 386
    mechanism 49, 384–6
    rate of 49, 370–1, 377–8
  sporulation (*spo*) 504–6, 515

Mutations—*cont.*
  super-repressed 440
  suppressor *264–5*, 343, *344–5*, 376–7, 414,
    522
  temperature-sensitive 373, 375
Mycobacteria 95
Mycobacterium phlei 309
Mycoplasma 116
Mycoplasma laidlawii 116
Myelin 115
Myxobacteria 519

NAD *see* Nicotinamide adenine dinucleotide
NADP *see* Nicotinamide adenine dinucleotide
    phosphate
Negative control *see* Enzymes, synthesis
Negative staining 65–7
Neisseriaceae 524
Neisseria 524
Neomycin 363, *364*
Nicotinamide 547
Nicotinamide adenine dinucleotide (NAD),
    amino acid synthesis *206*, 211, *227*, *231*
  breakdown 468, *470*
  glycolysis 164, *191*
  lipid synthesis 243
  mechanism of action 537–8
  phosphoketolase pathway 180, *181*
  polynucleotide ligase 288
  purine and pyrimidine synthesis *235*, 239,
    240
  reduced (NADPH₂), as enzyme activator
    200
    as enzyme cofactor 219, 225
    as enzyme inhibitor *169*, 188–9, 456–7,
      459, 491
    oxidase 499, *500*
    oxidation 164, 167, 191, 538
  synthesis 468, *470*
  tricarboxylic acid cycle *169*, 183, 184, 186,
    187, 189
Nicotinamide adenine dinucleotide phosphate
    (NADP), amino acid synthesis *205*, 211,
    212–15, 216, *218*, 219, *221*, *223*, 224, *226*,
    464
  breakdown *470*
  Entner-Doudoroff pathway *178*
  lipid synthesis 244, 245–6
  mechanism of action 537–8
  pentose phosphate pathway *173*, 174, *176*,
    177
  peptidoglycan synthesis *84*
  photosynthesis 196–7, *198–9*, 200
  reduced (NADPH₂), as enzyme activator
    172
  synthesis *470*
  tricarboxylic cycle 186–9
Nicotinate dinucleotide *470*
Nicotinate mononucleotide *470*

Nicotinic acid 468, *470*
Nitrate 200, 464
Nitrate reductase 464
Nitrite 200
Nitrobacter 524
Nitrobacteriaceae 524, 526
Nitrosomonas 524
Nitrocellulose membranes 272
Nitrogen, exhaustion 132, 172
  spores and 502, 503
  fixation 464, *518*, *524*, 526
  requirements 3, 5, 23, 519, 522
Nitrogenase 464
Nitrous acid 342, 383, *385*
Nomenclature 532–4
Nomenspecies 520
Nuclear body (Nucleoid) *see* Chromosome
Nuclease 265, 285–6, 297, 392
Nucleic acids, cellular content of 3, *4*, 18,
    127, 316, *317*
  hybridization 269–75
  notation for 338
  structure 5, 16–19, 251–69
  synthesis 5–6, 7, 277–315
  synthetic 275–7
  See also Deoxyribonucleic acid; Ribonuc-
    leic acid
Nucleolus *518*
Nucleosides 39, 487
Nucleoside phosphorylase 501
Nucleoside diphosphates 36, 456, 491, 542
Nucleoside diphosphate kinase 236, 239, *241*
Nucleoside monophosphates 456, 491
Nucleoside monophosphate kinase 236, 239,
    241
Nucleoside phosphorylase 501
Nucleoside tetraphosphates 310
Nucleoside triphosphatase *241*
Nucleoside triphosphates 329, *330*, 456, 491
Nucleotides 3, 16, 32–3, 39, 488

Octanoate 246
Δ¹¹-Oleic acid (*cis*-Vaccenic acid) 246
Oligodeoxynucleotides, synthetic 276–7
Oligonucleotides, notation for 338
OMP (Orotidine 5′-monophosphate) *241*
Operator *see* Genes, operator
Operon, L-arabinose (*ara*) 475, 477
  arginine (*arg*) 221, *479*, 480, 482
  biotin (*bio*) 479–80
  galactose (*gal*) 475, 476
  glycerol (*glp*) 475, 476
  histidine (*his*) 232, *233*, 362, 480–1, 482
  isoleucine-valine (*ilv*) *228*, 229, 478, 481,
    482
  lactose (*lac*) 362, 436
  leucine 228
  model *353*, 420, **435–54**, 551
    Class I reactions and 474–7

Operon, model—*cont.*
  Class II reactions and   477–82
  Class III reactions and   483–8
  pyrimidine (*pyr*)   478–9
  D-serine deaminase (*dsd*)   475–6
  tryptophan (*trp*)   210–11, 477–8, 481–2, 483
Order   520
Organelles   21, 64
  See also Granules; Mesosomes; Ribosomes etc.
Organotrophs   193
D-Ornithine   80
L-Ornithine, cell walls   79
  enzyme activator   466
  structure   *14*
  synthesis   219, *220–1*
  control   222–3, *239*
Ornithine carbamoyl transferase   *220–1, 239,* 431–2, 435, *479, 500*
Orotic acid   239, *240*
Orotidine 5'-monophosphate (OMP)   *241*
OMP decarboxylase   *239–41,* 242, *479*
OMP pyrophosphorylase   *239–41, 479*
Orthophosphate   *215,* 216, 219, *220,* 236, 239, *240*
Osmotic pressure, intracellular   12, 87
0-specific chains *see* Lipopolysaccharide
Oxaloacetate, amino acid synthesis   30, 211, *212,* 239, *240*
  decarboxylation   188
  glyoxylate cycle   171, 190, 191, *192,* 546
  pyruvate fermentation   *166*
  tricarboxylic acid cycle   27, 28, *169, 185,* 186, 187, 188, *455,* 456, 457, 540
Oxalosuccinate   *185,* 186
Oxidase   *534*
Oxidative phosphorylation   457, *518,* 537–8
Oxido-reductases   536, 537–9
α-Oxo acid aminotransferase   *214–15,* 216, *225, 226*
β-Oxoadipate   462, *463*
7-Oxo-8-aminopelargonate synthase   *479*
α-Oxobutyrate   *215,* 216, 225, *228*
2-Oxo-3-deoxy-6-phosphogluconate   178–9
2-Oxo-3-deoxy-6-phosphogluconate aldolase   178–9
α-Oxoglutarate, ammonia uptake and   29, *218,* 219, 464
  from deamination of glutamic acid   188, *206, 212–13, 215,* 221, 222, *226,* 231, *240,* 455, 541
  from deamination of glutamine   219
  as enzyme inhibitor   189, 247, 456, 457
  glutamic acid synthesis   30, *218,* 219
  tricarboxylic acid cycle   27, 28, *169, 185,* 186–7, *192, 455*
α-Oxoglutarate oxidase   186–7, 539
α-Oxo-γ-methylbutyrate   225, *226, 228*
α-Oxo-β-methylvalerate   *215,* 216, *228*

α-Oxo-γ-methylvalerate   *227, 228*
Oxygen, electron acceptor   160, 200, 464, 537–8
  requirements   160, 165, 456, 522, *523–5,* 526

Pactamycin   363, *364*
Palmitic acid   245
Pantetheine   244
Pantothenic acid   *288,* 540, 547
Parasites   520
Parvobacteriaceae   524
Pasteurella   524
Pasteurella pestis   524
Pelodictyon clathratiforme   126
Penicillin, effect on cell wall synthesis   85, **88–9,** *150,* 151
  enrichment media   371
  filamentous growth   145
  resistance   421, 481
Penicillinase   473–4, 481
Pentoses   96
Pentose phosphate cycle   27, 28, 173–7
  control of   459
  energy from   177
  fate of carbon atoms   174
Pentose phosphate isomerase   *173,* 174, 197
PEP *see* Phospho-enolpyruvic acid
Peptidases   40, 542
Peptides, as carbon source   454
  in cell walls *see* Peptidoglycans
  in spores   509
Peptidoglycans, amount in cell   75, 77, 78, 87
  capsule and   106
  digestion with lysozyme   12, 71, *73,* **78,** *79,* 81–3
  function   87, 89
  in Gram-negative bacteria   72, 74, 80, 81–3, 95
  in Gram-positive bacteria   75, 80–3, 94
  links to protein   95
  links to teichoic acids   94
  in spores   497–8, 504
  structure   7, 11–12, 35, **78–83**
    cross-linking in   80–3
  synthesis   *31,* 32, 36–7, **83–7,** *518*
    effect of antibiotics   85, 88–9, 151
    growth of polymer   89–90
Peptidyl-puromycin   *332,* 363
Peptidyl transferase   *331,* 333, 354, 356, *358–359*
  effect of antibiotics   363–4
Peptidyl-tRNA   *331, 332,* 358–9
Peripheral bodies *see* Mesosomes
Peripheral pathways, control   459–65
Permease systems   13, 28, 117, 225
  induction of   430, 433, 453, *475,* 477
  mutations in   374

Phages (Bacteriophages), complementation
    tests   415
  lytic, infection by   47–8
    DNA synthesis   284, 289, 295–6, 315,
      484
    protein synthesis   313–14, 345, 484
    RNA synthesis   304–5, 330, *348*, 486,
      492
    enzymes for   311, 313–15, 484–5
  mutations   336, 382, 386, 388, 415, 411–12
  receptors   47, 107, 377
  resistance to   377, 378
  structure   46–7
    nucleic acid of   258, *260*, 304, 338, 360,
      369–70
  temperate, induction of   48–9, 405, 484
    integration into chromosome   48–9,
      406–7, 416–17, 550
    transduction, general   49, **400–4**, 407, 451
    origin of particle responsible   405–7
    special   404–5, 407, 416–17
Phage α   304
Phage fd   311
Phage f$_2$ 345
Phage φe   313, 485
Phage φ80   405, 417
Phage φX174
  DNA synthesis   289, 295–6
  replicative form   *260*, 293, 295–6, 304
  mRNA   304, 309
  single-stranded chromosome   259–60, *273*
Phage *lambda*, complementation test   415
  chromosome of   *260*, 369–70, 405
  defective   405–6
  induction of   405, 486
  integration into bacterial chromosome
    406–7, 416–17
  transduction by   404–5, 407, 416–17, 451
Phage M-12   107
Phage MS-2   *108*
Phage PBS-1   402–3, 408, 409, 504
Phage P1   402, 405, 407
Phage P22 (PLT22)   401–2, 403, 405, 407, 416
Phage Qβ   313, 314, 338, 360
Phage R17   338, 360–2
Phage SP8   304
Phage SP01   484
Phage T1   378
Phage T2   *258*, 369
Phage T4, DNA   287, *289*, 305
  lysozyme   412–13
  mutations   336, 382, 386, 388, 411–13
  RNA synthesis   305, 311, 314, 484, 486
Phage T7   313–14, 484
Phenols   160
Phenon   530
Phenotype   44, 367, 386, 530, 551
L-Phenylalanine, activating enzyme (phenyl-
    alanyl tRNA synthase)   375
  structure   *14*

synthesis   30, *206*, 208
  control of   209–11, *469*
  triplets for   *339*
Phenylalanine aminotransferase   *206–7, 209*
Phenylethylthio-β-D-galactoside   *434*
Phenylpyruvate   *206*
Phenylthio-β-D-galactoside   *434*
Phosphatases   542
Phosphate
  inorganic   164, 200, 543
    as enzyme cofactor   180
    as enzyme repressor   473
  requirements   3, *4*
Phosphatides   95, 551
L-α-Phosphatidic acid   *112*, 247, *248*, 249
Phosphatidic acid phosphatase   247, 249
Phosphatidyl choline (Lecithin)   *248*
Phosphatidyl ethanolamine   111, *112–13,
    248*, 249
Phosphatidyl glycerol   *112–13, 248*, 249
Phosphatidyl inositol   *112–13, 248*
Phosphatidyl serine   *112–13, 248*, 249
Phosphatidyl serine decarboxylase   249
3′-Phosphoadenylyl sulphate   *223*
Phosphoadenylyl sulphate reductase   *223*,
    224
Phosphodiesterase   *271*
Phospho-enolpyruvate carboxykinase   *169*,
    171
Phospho-enolpyruvate carboxylase   *169*,
    171, 190, 193, 455–, 456–7, 458
Phospho-enolpyruvate synthase   171, 193
Phospho-enolpyruvic acid (PEP), amino acid
    synthesis   *204–5*, 208, *468*
  cell wall synthesis   83
  as enzyme inhibitor   170, 172, 193, 455
  as enzyme repressor   460
  glucogenesis   171
  glycolysis   *161*, 165, *169*, 170, *191*, 458
  tricarboxylic acid cycle   190, 458
Phosphofructokinase   162, *169*, 170, 455,
    456, 457
Phosphoglucokinase   172
Phosphoglucomutase   *169*, 172
6-Phosphogluconate dehydratase   178–9
6-Phosphogluconate dehydrogenase   180
6-Phosphogluconic acid   174, *176*, 178–9,
    181
Phosphoglycerate dehydrogenase   *222*, 224
Phosphoglycerate kinase   164
Phosphoglycerate phosphatase   *222*, 224
2-Phosphoglyceric acid   *161*, 164–5
3-Phosphoglyceric acid, amino acid synthesis
    222, 224
  glycolysis   *161*, 164–5, *191*
  photosynthesis   196, 197, *198–9*
Phosphoglyceromutase   164–5
Phosphohexose isomerase   162, *169*, 175
O-Phosphohomoserine   211, *215, 216*
Phosphohydroxypyruvate   *222, 224*

Phospho-2-keto-3-deoxyheptonate   *204*, 208, 209, 210, 467, *469*
Phosphoketolase   180
  pathway   179–81, 459
    energy from   190
    fate of carbon atoms   179
Phospholipids, in cell envelopes   95, 111–12, *113*
  synthesis   182, 247–9
C-55 Phospholipid   84–6, 92, 98–9
N-1-(5'-Phosphoribosyl)-AMP   *230*, *232*, *233*
Phosphoribosyl-AMP 1,6-cyclohydrase   *230–1*, *233*
N-1-(5'-Phosphoribosyl)-ATP   229, *230*, *232*, *233*
Phosphoribosyl-ATP pyrophosphohydrolase   *230–1*, *233*
Phosphoribosyl-ATP pyrophosphorylase   229, *230–1*, 232, *233*
5-Phospho-*β*-D-ribosylamine   229, *230*, *235*, 236, *238*
1-(5'-Phosphoribosyl)-5-aminoimidazole   *234*
Phosphoribosyl-aminoimidazole-carboxamide formyltransferase   *234–5*, 238
Phosphoribosyl-aminoimidazole carboxylase   *234*, *238*
1-(5'-Phosphoribosyl)-5-aminoimidazole-4-carboxylate   *234*
Phosphoribosyl-aminoimidazole synthase   *234–5*, *238*
N-(5'-Phosphoribosyl anthranilate (*N*-(*o*-Carboxyphenyl)-D-ribosylamine 5'-phosphate)   *207*, 210, *469*
Phosphoribosyl-anthranilate isomerase   *207*, 210
N-(5'-Phosphoribosyl-formimino)-5-amino-1-(5''-phosphoribosyl)-imidazole 4-carboxamide   *230*, *232*, *233*
5'-Phosphoribosyl-*N'*-formylglycinamide   *235*, *236*, *542*
5'-Phosphoribosyl-*N'*-formylglycineamidine   *235*
Phosphoribosyl-formylglycineamidine synthase   *234–5*, *238*
5'-Phosphoribosyl-glycineamide   *235*, 236, *542*
Phosphoribosyl-glycineamide formyltransferase   *234–5*, *238*
Phosphoribosyl-glycineamide synthase   *234–5*, 236, *238*
5-Phospho-*α*-D-ribosyl-pyrophosphate (PRPP)   229, *230*, *232–3*, *234*, 236, *238*
Phosphoribosyl-pyrophosphate amidotransferase   229, *230*, *234–5*, 236, 237, *238*
1-(5'-Phosphoribosyl)-4-(*N*-succinocarboxamido)-5-aminoimidazole   *234*
Phosphoribosyl-succinocarboxamido-aminoimidazole synthase   *234*, *238*
Phosphoribulokinase   196, 200

Phosphorylase   *169*, 172
Phosphorylation, oxidative   457, *518*, 537–8
Phosphoryl-acetylmuramic acid-pentapeptide   83–4
Phosphoserine   *222*
Phosphoserine aminotransferase   *222*, 224
Phosphoserine phosphatase   *222*, 224
5-Phosphoshikimate   *205*, *468*
Phosphotransacetylase   540
Photobacterium   *524*
Photolithotrophs   193, 196, 522
Photo-organotrophs   193
Photosynthesis
  carbon dioxide fixation   196–200
    Calvin cycle   198–200
    control   200
  electron donors   194–5
  energy from   193, 200, 517
  sites of   64, 123–5, *126*, 517, *518*
Phototrophs   194
Pichia   511
Pili (Fimbriae), in classification   521
  function   64, 107
  plasmids and   394, 420
  sex (F) pili   107, *108*, 394, 400, 420
  structure   106–7
Pilin, F   107
Piperideine-2,6-dicarboxylate   *213*
Plasma membrane *see* Cytoplasmic membrane
Plasmids, conjugation and   394, 399, **416–17**, **420**
  genes of   420–1, 481, 530
  integration into chromosome   301, **399–400**, 416–17, 419, 420, 421
  pili and   394, 420
  properties   420–1, 481
  structure   399, 421–2
  transduction of   422
Pleiotropism   505
Pneumococcus   46, 91, 106, 304, 391
Polarity of mutations   232, **446–9**, 400, 402
Polyamines   341
Polyarginine   341
Polycistronic mRNA   232, 343, *353*, **357–8**, **360–2**, 443, **446–7**, 476
Polydeoxyribonucleotides, poly dA   338
  poly dAT   289, *309*
  poly dC   289
  poly dG   289
Polydisperse   35, 346, 349
Poly-glutamic acid   341
Polyglycerophosphate *see* Teichoic acids
Poly-*β*-hydroxybutyrate granules   *124*, 133, *518*
  synthesis   195
C-55 Polyisoprenoid phosphate   84–6, 92, 98–9
Polylysine   341
Polymers, non-repetitive   34, 35–6, 37–43
  repetitive   34, 35, 36–7

Polymetaphosphate granules (Metachromatic granules) 64, *132*, 133–4
Polynucleotides, synthetic 275–7
  code-breaking 338–41
  strand separation 274, 304–5
  structure 551
Polynucleotide ligase 277, **287–8**, 298, *390*
Polynucleotide phosphorylase 275, **334**, 339
Polypeptide chains, in protein 13, 15, 16, **364–5**, 418–19, 424
  formation 43, 331–4, 342–6, 354–9
Polypeptides, capsular 104, 105–6
Poly-phenylalanine 334
Polyribitol phosphate *see* Teichoic acids
Polyribonucleotides, poly A 265, 275, 276, 338
  poly AAG 341
  poly AC 340
  poly AG 340, 341
  poly AU 339–40
  poly $AUGU_{0-4}A_n$ 345–6
  poly C 275, 276, 338
  poly G 274, 275, 276
  poly I 275–6
  poly U 265, 274, 275, 276, 334, 338
  poly UAUC 341
  poly UC 340
  poly UG 274, 305, 340, 341
  poly XU_n 340
Polyribosomes, attached to DNA 306–7, 353–4
  degradation of 318, *321*, 349, *353*, 354
  formation of 43, 353–4
  protein synthesis and 325, *326*, 353–4
  mRNA and 19, 131–2, 318, *320–1*, **323–5**, 334, 551
Polysaccharides, capsular 104–6
  as carbon source 454
  cell walls 94
  structure 34–5
  synthesis 7, **36**, 86
Potassium 323, 333, 512
Precursor activation 170, 456, 491
  See also Enzymes, activity
Prephenate dehydratase *206–7*, 209, 210, 471
Prephenate dehydrogenase *206–7*, 209, 471
Prephenate synthase *see* Chorismate mutase
Prephenic acid *206*, 208, *209*, *468*
  as enzyme inhibitor 467, 472
Prespore 496, *513*
Proflavine 302–3, 311, 382, 411
Proline, structure *14*
  synthesis 30, 217–19
    control 219, 221, 474
    triplets for *339*
Prokaryotes 325, 481, 517–19
Promoter *see* Genes, promoter
Prophage 405–7, 416, 551
Propionibacteriaceae *523*

Propionibacterium *523*
Propionic acid *166*, 460, *523*
Propionic acid bacteria *166*
Prosthetic group 536, 551
Proteases, extracellular 500, 503, 504, 505
Proteins, cellular content of 3, *4*, **13**, 316, *318*, 486
  cell walls 77, 94–5
  DNA-associated 253
  membranes 12, 112, 114, 115, 116, *318*
  multimeric 15, 16, **364–5**, 410–1, 424, 427–429
    hybrid molecules 418–19
  ribosomal 19, 308, *318*, 321, **322–3**, 343, 522
  spores 497, 499, 508–9, 510, 514
  structure 6, 13–16, 365, 551
    primary 15, 365
    quaternary 16, 365
    secondary 15, 365
    tertiary 15, 365, 419
  synthesis, amino acid activation 40–1, 42, **325–31**
    cell cycle and 25, 157–8
    effect of antibiotics 329, *332*, 333, 349, *351*, 356, **363–4**
    elongation (transfer) factors 333, **355–7**, *358–9*, 363
    growth rate and 145–6, 149–50, 486–7
    initiation *339*, 340, 343–6, 360, 363
      factors 354–5
    peptide chain formation 43, **331–4**, 342–6, **354–9**
    phage infection 313–4, 345, 484
    rate 332–3
      control *see* Enzymes, synthesis
    ribosomes and 322, 325, *326*, 333
      interaction with tRNA and mRNA 42–3, **351–4**, *358–9*, *364*
    role of base pairing 42, 263, 351–3
    role of mRNA 7, 41–3, 333–4, **346–51**, 360–3
    role of tRNA 40–2, **328–31**, 335, 337, *358–9*
    shift-up or -down 147–51, 487
    termination *339*, 342–3
      factors 357–60
    translocation 356–9
    transpeptidation 356–9
    turnover 148, 487–8, 506
Proteus mirabilis 464, 480
Proteus vulgaris 107, *525*
'Protista' 517
Protomers 427
Protoplasts, formation of 12, 87–8, 551
  properties of 11, **88**, *120*, 122, 123, 323, 369
Protoplast membrane *see* Cytoplasmic membrane
Prototrophs (Wild-type) 372, 404, 551

Protozoa 517
PRPP (5-Phospho-α-D-ribosyl-pyrophosphate) 229, *230*, *232–3*, *234*, 236, *238*
Pseudomonadaceae 524, 526, *532*
Pseudomonas 524, *531*
  aromatic amino acids 210, 467, 481
  carbon sources 160, 462
  Class I reactions 188, 190, 456
  gene clusters 483
Pseudomonas aeruginosa, amino acid metabolism 460, 462, 467, 482
  Class I reactions 459
  conjugation 399
  pyrimidine synthesis 242
Pseudomonas fluorescens 459, 462
Pseudomonas lindneri (*Zymomonas mobilis*) 179
Pseudomonas putida 210, 467, *469*, 472, 481
Pseudomonas saccharophila 177, 179
Pseudouridine 301
Pseudouridylic acid *261*
Psychrophils 112
Purines, growth and 141, 547
  structure 16–18
  synthesis 229, *234–5*, 236
    control of 237–8
    energy requirements 242
    origin of atoms in ring *32*, 33, 229
    relation to histidine synthesis 229, *230–231*, 237
    salvage pathway for 237
Purine nucleoside phosphorylase *500*, 503
Purple bacteria, photosynthesis *123–4*, 125, *126*, **194–5**
  sulphur granules 135, 195
Putrescine 222–3, 323
2,6-Pyridine dicarboxylic acid (Dipicolinic acid, DPA) 496, 497, 503, 504, 505, 506
Pyridoxal phosphate (P-al Ⓟ) 540, 541, 544
Pyridoxamine phosphate (P-amine Ⓟ) 540, 541
Pyridoxin (P-in) 540, 547
Pyrimidines, dimerization 384
  growth and 141, 547
  structure 16–18
  synthesis 238–41
    control of **240–2**, 435, 465, 466, **478–9**, 483
    energy requirements 242
    origin of atoms in ring *32*, 33, 238
    relation to arginine synthesis 238, *239*, 242, 435, 465, 466
Pyrocatechase *463*, 464
Δ'-Pyrroline 5-carboxylate *218*
Pyrroline 5-carboxylate reductase *218*, 221
Pyrophosphate, inorganic 178, 211, 236, 249, 327
  as enzyme inhibitor 462
  hydrolysis 546
Pyrophosphatase, inorganic 501

Pyruvate carboxylase 190, 454, 456, 546
Pyruvate decarboxylase 470–1, 542
Pyruvate dehydrogenase 169, 183–4, 189
  complex of 184–5, 470–1
Pyruvate kinase 165, *169*, 170, 456, 457, 458
Pyruvate oxidase 539
Pyruvic acid, amino acid synthesis 29, *30*, *207*, 208, 211, *212*, *214–15*, 216, 225, *226*, *228*, 541, 544
  Entner–Doudoroff pathway 178, *179*
  fermentation 164, 165–7, 180, *181*, 538
  glucose synthesis 170, 171, 188
  glycolysis *27*, 28, *161*, 164, **165**, *169*, 170, *191*, *455*, 458
  glyoxylate cycle 190, 546
  oxidative decarboxylation 183–5, 470–1
  tricarboxylic acid cycle 28, *169*, 183–5, 189

Q₁₀ 25
Quinolate 468, *470*

Reactions
  Class I 5, 26–8, 160, 200
    control of 170, 188–9, 200, 454–65, 474–7, 491
  Class II 5, 29–33, 204–49
    control of 202–7, 208–11, 216–17, 219, 221–3, 224–5, 228–9, 232–3, 237–8, *239*, 240–2, 465–74, 477–83, 491
  Class III 5–7, 34–43, 277–315, 325–34, 346–65
    control of 483–90, 492
Recessive genes 53
Recombinant 551
Recombination (Crossing-over), frequency of 50, **52–3**, 392–3
  genetic mapping and 50–3, 390–2, 394–400
  high-frequency 394–8, 399
  mechanism *51*, 384, **389–92**, 552
  overlapping deletions 393–4, 404
Reducing power 177, 189, 455, 456, 519
Regulator *see* Genes, regulator
Regulon 476
'Relaxed' strains 150–1, 487
Reovirus 265
Replica plating *372*, 373, **378–80**, 552
Replicon 530
Repression *see* Enzymes, synthesis, repression
Repressor, interaction with inducer 55, 56, 437, *438–9*, *440*, *443–4*, 446, 490
  interaction with operator 55, *56*, 442, *443–445*, 446, *448–9*, 450, 481, 490–1
  isolation of 272, 450–1
  production of *56*, 437–9, *475*, 476
Respiration *509*
Reticulocytes 337

Retro-inhibition *see* Enzymes, synthesis, feed-
back inhibition
L-Rhamnose (6 deoxy-L-mannose) *93*, 96,
99
Rhizobiaceae 524, 526
Rhizobium 524, 531, 532
Rhodopseudomonas 200
Rhodopseudomonas capsulatus 217
Rhodopseudomonas spheroides 217
Rhodomicrobium vannielii 126
Rhodospirillum rubrum 123–4, *125*, 133, 429
Ribitol phosphate 91
See also Teichoic acids
Riboflavin 537, 547
Ribonuclease (RNAase) 270–1, 272, 318,
321, 324, **365**, *449*, 482, 503
Ribonucleic acid (RNA)
    alkaline hydrolysis 310
    amount in cell 3, *4, 317*, 332
    base pairing in 251–2, 263, 265–6, *267*
        effect on synthesis 39, 305–6, 315
    genetic role 314–5, 338, 360–1, 369
    molecular weight 19, *300*
    reversible 'melting' 262
    structure, double helix in 265, 303, 360–1
        model of helical molecule *267*
        primary structure 6–7, 18, 261–3, 360–1
        secondary structure 264–9, 360–1
        unusual bases in *261*, 262, 263
    synthesis, DNA-dependent *see*
        Transcription
        growth rate and 147–51, 486–90
        in phage-infected cells 304–5, 313–14,
            330, *348*, 486, 492
        rate 311, 312
        RNA-dependent 299, **314–15**, 362
        shift-up or -down 147–51, 487
        spores 313, 506, 507–8, 514
    turnover 488, 506, 514
    types 19, *317*, 371
        sucrose gradient analysis of 317–19,
            347–9
RNA, messenger (mRNA), base composition
    346–9
    breakdown rate 301–2, *350, 353*, 354, 448,
        449
    lifetime *317, 347*, 348, 349, *350*
    polycistronic 232, 343, *353*, **357–8, 360–2**,
        443, **446–7**, 476
        intracistronic divide 362
    polyribosomes and 19, 131–2, 318, *320–1*,
        323–5, 334, 551
    properties *317*
    spores 506, 508, 509, 510
    synthesis *see* Transcription
    translation, control 442, 450, 451, 478,
        490, 510
        direction 340, 346, *353*, 413, *448–9*
        initiation sequences *339*, 340, **343–6**,
            355, 360, 363

recognition of tRNA 42, *268*, **269**, 342,
    351–3, 355, 356, 363, *364*
ribosomal binding 42–3, **351**, **353–4**,
    355, 356, 360, 363, *364*
summary diagrams *353, 358–9, 448–9*
termination sequences *339*, 342–3, 357,
    358, 360, 447
RNA polymerase, alterations in, control and
    151, 312–14, 483–5
    binding to operator *353*, **442–6**, 448–9,
        450, 481, 490–1
    binding to promoter *307*, 312–13, *353*,
        448–50, 454, 477, 481, 484, 491
    direction of synthesis by 310–11
    in phage-infected cells 311, 313–15, 484–5
    spores and 313, 485, 492, 508, 510
    structure 308, 484
        core polymerase 308, 311, 314, 484
        holoenzyme 308, 311–12, 314, 484
        molecular weight 308
        ψ_r factor, termination and 313, 314, 485
        ρ factor 312
        σ factor, initiation and 308, 309, 311–
            312, 484
    template for 308–9
        base composition of product and 309
        strand selectivity towards 304–6, 311–
            312
RNA replicase (RNA synthetase) 313, **314–
    315**, 360, 362
RNA, ribosomal (rRNA), base composition
    300
    genes for **299–300**, *317*, 376, 485
        tandem nature of 300, 306–8, 376
    properties of *300, 317*
    in ribosomes 42, 319, **321–2**, 323, 552
    spores 506, 508
    structure 262, 265, 552
        sequence 263, *266*
    synthesis 306–8, 312–13, *347*, 348–9
        control of 313, 314, 485, 487, 492
        post-transcriptional modification 301,
            334, 485
RNA, transfer (tRNA, soluble RNA, acceptor
    RNA)
    alanine *264*, 276–7, 338
    amino acid activation 42–3, 319, **330–3**,
        552
    base composition *300*
    formyl-methionine 343, 345
    genes for 300–1, 317, 376
    histidine 480
    molecular weight 19, *300, 317*, 329
    mutations *344–5*, 371, 375, 376–7, 414
    properties *317*, 328–31
    recognition of mRNA 42, **266–9**, 342,
        351–3, 355, 356, *358*, 363, *364*
        anticodon triplets 263, *268*, 329, 330
        'wobble' hypothesis *268*, 269, 352
    ribosomal binding **351**, **353**, 356, *358–9*

RNA, transfer (tRNA, soluble RNA, acceptor RNA)—*cont.*
  serine 330
  in spores 506, 508
  structure, clover leaf conformation **263–9**, 328, 329
    nucleotide sequence 262–3, 329
    terminal sequence 263, 328–9, *330*
    unusual bases in 328
  synthesis 313, *347*, 348–9
    control of 313, 314, 485, 487, 492
    post-transcriptional modification 301, 485
  tyrosine *264*
Ribonucleotide reductase 242
D-Ribose 6, 18, 33, 261–2, 457
Ribose 5-phosphate, histidine synthesis 229, *230*
  pentosephosphate pathway *173*, 174 5, *176*
  phosphoketolase pathway 180, *181*
  photosynthesis 197, *198–9*
  purine synthesis *234*, 236, *238*
Ribose phosphate pyrophosphorylase 229, *230–1*, *234*, 237, *238*
Ribosidase 500, 501, 502
Ribosomes
  cellular content 3, *4*, *64*, 111, 131–2, *318*
    growth rate and 145–6, 486
    shift-up or -down 148–51, 487
  effect of drugs 363, *364*, 375
  enzyme activity of 322, *331*, 333, 364–5
  interaction with mRNA and tRNA 42–3, **351–4**, *358–9*, *364*
  proteins of 19, 308, *318*, 321, **322–3**, 343, 522
  RNA of *see* RNA, ribosomal
  structure 19, 42, 132, 319–20, *323*
    sedimentation coefficients 42, *319*, 320–322, *324*, *518*
    sub-units 42, **320–2**, 325
Ribothymidylic acid *261*
L-Ribulokinase *475*
Ribulose 1,5-diphosphate 196, 197
Ribulose diphosphate carboxylase (Carboxy-dismutase) 196, 200
Ribulose 5-phosphate, pentose phosphate pathway *173*, 174, *176*, 245
  phosphoketolase pathway 180, *181*
  photosynthesis 196, 197, 200
L-Ribulose 5-phosphate 4-epimerase *475*
Rifamycin 312

Saccharomyces 511
Saccharomyces cerevisiae 511, *514*, *516*
Saccharomycetaceae 511
Saccharomycodes 511
Salmonella 96, *97*, 371, *525*, 526, *531*
Salmonella typhimurium *525*, 534

aromatic amino acids 209–11, 477
aspartic acid family 217, 473
Class I reactions 455, 456
flagella 108, *109*
genetic map 53, 399
growth of 144–6
histidine mutants 232, 403–4, *412*, 417–18, 419
histidine operon *233*, 362, 480–1
histidine synthesis 323, 425
phages of 401–2, 407, 416
plasmids 421
polysaccharides 95, 96, 98
pyrimidine synthesis *241*
pyruvate amino acids 225, *228*
serine family 224
Salmonella typhosa (*Salmonella typhi*) 101, 102
Saprophytes 520
Sarcina 156, *523*, *531*
Sarcina lutea 523
Sarcina ventriculi 166
Schizosaccharomyces pomba 514
Sedoheptulose 1,7-diphosphate 197
Sedoheptulose 7-phosphate 174–5, 180, 196, 197, 200
Segregation 50, 552
Segregation lag 386
Septum, cell division and 103, *104*, *105*, *118*, 486
  forespore *495*, 496
  synthesis of 119, *121*, 122, 123
D-Serine deaminase 459, *475*
D-Serine system *475*, 476
L-Serine, activating enzyme 330
  cell walls 78
  N-terminal amino acid 343, 345, 346
  phospholipids 182, *248*, 249
  purine synthesis 32, 33
  structure *14*
  synthesis *30*, *222*, 224
    control 224–5
  tRNAs 330, 352–3
  triplets for *339*
  tryptophan synthesis *207*, 208
Serine *O*-acetyltransferase *222–3*, 224, 225
Serine aminotransferase *222–3*, 224
Serine hydroxymethyltransferase *222*, 224
Serine phosphatidyltransferase 249
Serratia *525*, *531*
Serratia marcescens 525
Shift-down 25, 147, 552
  effect on cell composition 148–51, 487, 488
Shift-up 25, 147, 552
  effect on cell composition 147–51, 487
Shigella *525*, *531*
Shikimate *30*, *205*, 209, *468*, 472
Shikimate dehydrogenase *204–5*
Shikimate kinase *205*, 467, *469*, 472
Siomycin 363, *364*

Slime layer   64, 104, *106*
Sorbitol   473
Sorbose   473
Sparsomycin   363, *364*
Species   520–1, 530
Spectinomycin   363, *364*, 375
Spermidine   323
Spermine   323
Spheroplasts, formation   88–9, 552
   lysis   111, 323
   properties   *73*, 88–9, *125*, 552
Spindle   49, *513*
Spirillaceae   *525*, 526
Spirillum   *132*, *525*, 531
Spirillum serpens   *73*, *74*, *525*
Spirochaetes   519
Sporangium   *495*, 496, 499–502
Spores, activation of   506
   development, biochemistry of   498–502,
      503–4, 506
      cell division and   499, 502
      effect of medium on   8, 59, 64, 502
      genetic control   504–6, 510
         asporogenous mutants   499, 504–5
         mapping of genes   505–6
         oligosporogenous mutants   504, 505
      induction of   502
      morphological stages of   495–7
      RNA polymerase and   313, 485, 492, 510
      germination   59, 64, *495*, 506, 507
      organisms forming   *523–5*, 525, *534*
         medical and industrial importance   510
      outgrowth   59, 64, 506, 507–10
         cell wall synthesis   509
         macromolecular synthesis   507–9
      regulation of   508, 510
   properties   59, 64, 497, 498
      resistance to heat etc.   498, 501, 502, 511
   structure   *495*, *496*, 497
      antibiotics   *500*, *504*, 506, 508, 511
      antigens   *500*, 501
      coat   *495*, 496, 497
      cortex   *495*, 496, 497–8, *500*, 504
      crystal   *500*
      enzymes   499–502, 507
      exosporium   497
      nuclear material   *495*, 496, *497*, 509, 510,
         514
      toxins   *500*, 510
Sporogenesis *see* Spores, development
Sporosarcina   494, *523*
Staphylococcus   95, 422, 467, 483, *523*, *531*
Staphylococcus aureus   *523*, *534*
   amino acids   481, 482
   cell walls   80, *82*, 83, 91
   membranes   *112*
   penicillinase   473–4, 481
Staphylococcus epidermis   472
Staphylococcus lactis   76
Sterols   *518*

Streptococcus   95, 156, *166*, *523*, *531*
Streptococcus faecalis   *112*, 179, 242, *523*
Streptococcus pyogenes   101–2, *534*
Streptomyces   78, 467
Streptomyces griseus   473
Streptomycin, mode of action   363, *364*, 375
   production   473, 474
   resistance   378–80, 421, 422, 505
'Stringent' strains   150–1, 487
Succinic acid, amino acid metabolism   *213*,
      214
   as carbon source   141, 171, 458
   as enzyme inhibitor   193
   as enzyme repressor   460, 462, 464
   glyoxylate cycle   190–2
   mandelate breakdown   462–4
   tricarboxylic acid cycle   *166*, *184*, 188, 189
      455
Succinic dehydrogenase   187, 457, *500*, 538
Succinic cytochrome *c* reductase   *500*
Succinoxidase system   187
Succinyl CoA   187, *214*, 545
N-Succinyl-LL-α-ε-diaminopimelate   *213*
Succinyldiaminopimelate aminotransferase
   212–13, *216*, 217
Succinyldiaminopimelate desuccinylase   *213*,
   216, 217
N-Succinyl-L-glutamic acid   *500*
O-Succinylhomoserine   *214*, *216*
N-Succinyl-ε-oxo-L-α-aminopimelate   *213*
Succinyloxoaminopimelate synthase   *212–
   213*, *216*, 217
Sucrose gradient centrifugation   317–20
Sulphate, photosynthesis   195
   reduction   30, *223*, 224, 225
   requirements   3, *4*, 5, 23
Sulphate adenylyltransferase   *223*, *224*, 225
Sulphate permease   225
Sulphide   *224*
Sulphite   200, *223*
Sulphite reductase   *223*, *224*, 225
Sulpholactic acid   *500*
Sulphonamides   421
Sulphur bacteria   *132*, 134–5, 194, 195
   granules   64, *132*, 134–5
   photosynthesis   195, 200
Suppressor mutations   264–5, 343, *344–5*,
   376–7, 414, 552
Super-repression   440
Synthases (Ligases)   536, 546

Taxa   520, 527
Taxonomy, numerical   527–9
   See also Classification
Taxospecies   521
Teichoic acids, antigenicity of   92
   functions   92
   linkage to other wall components   92–4
   structure   90–1
   synthesis   86, 91–2

Teichuronic acid 92, *93*
Temperature, effect on growth 24–5, 124, 143–4, 519
 spores and 502, *503*, 506
Terramycin 363
Tetracyclines 150, 363, *364*
Tetrahydrofolic acid (CoF, FH$_4$) 224, 541–2
Thermophilic bacteria 24, 112
Thiamine 543, 547
Thiamine pyrophosphate (TPP) *175*, 180, 183, 185, *471*
 mechanism of action 543
2-Thiazolealanine 232, 236
Thin-sectioning 65–7
Thiobacteriaceae 524
Thiobacillus denitrificans 200
Thiobacillus thiooxidans 524
Thioctic acid (Lipoic acid) **183–5**, 539, 547
Thioesters 244
Thiorhodaceae 194
Thiostrepton *364*
Thiosulphate 464
Thiosulphate reductase 464–5
2-Thiouracil *261*
L-Threonine, activating enzyme 329
 as enzyme inhibitor 429, 472
 structure *14*
 synthesis *30*, 31, 211, *214–15*, 216, 473
  control of *216*, 217, *228*
 triplets for *339*
Threonine dehydratase *214–15*, 216, 217, *228, 426*
Threonine synthase *214–15*, 216
Thymidine diphosphate-rhamnose (TDP-rhamnose) *93*, 98, *99*
Thymidylic acid *261*
Thymine, dimerization 297–8
 pairing with adenine 18, **251**, *252*, 255, 259, 384, 385
 pairing with 2-aminopurine 381, *385*
 pairing with guanine *385*
 strains requiring 149, *150*, 281
 structure 16, *17*
 tautomers *385*
Tobacco mosaic virus (TMV) 369
Toxins, of spores *500*, 510
TPN *see* Nicotinamide adenine dinucleotide phosphate
TPP *see* Thiamine pyrophosphate
Trans configuration 415, 552
Transacetylases 540
Transaldolase *173*, 174, 175, 197
Transaminases 540, 541
Transamination 31, 32
Transcription, base-pairing 40, 305–6, 311, 315
 catabolite repression and 56, 453–4
 control 151, 312–14, 483–5
 direction **310–11**, *353*, 447, 480
 divergent orientation 479–80

effect of drugs 302–3, 311, 312
 initiation 305, 311–12, 313, 443, *444–5*
 operon model and 55–6, 435–51
 polar mutations 446–9
 pulse-labelling 323–5, 347–50
 template, base composition of product and 300, 309
  strand-selectivity towards 303–6, 308–310, 311–12
 termination 312, 485–6
 translation and *353, 448–9*
  See also RNA synthesis; RNA, ribosomal, synthesis; RNA, transfer, synthesis
Transduction, abortive 407, 416, 417
 chromosomal integration 406–7, 416–17
 genetic mapping 399, 400–5, 407, 416–17
 general 49, **400–4**, 407, 451
 origin of particle responsible 405–7
 special 404–5, 407, 416–17
Transferases 536, 540–2
Transfer-RNA *see* RNA, transfer
Transformation, amount of DNA required 367–8
 competence for 367
 efficiency 367–8
 genetic mapping 407–9
 mesosomes and 122
 role of DNA 8, 46, 258, 304, **369–70, 391–392**, 522, 553
Transformation of mammalian cells 315
Transglycosidases 540
Transition mutations *see* Mutations, alteration
Transketolase 173, 174–5, 197, 543
Translation *see* RNA, messenger, translation
Translocase *see* Permease
Transmethylases 540
Transversion mutations *see* Mutations, alteration
Trehalose 512
Tribe 530
Tricarboxylic acid cycle (Krebs cycle, Citric acid cycle) *27*, 28, 183–91
 control of 169, **188–9**, 454–8
 energy from **189–90**, *191*, 456–7
 fate of carbon atoms 188
 functions of 183, 188, 457
 intermediates for 171, 190–3, 460
 spores 485, 498, 500, 503, 505
Triglycerides 181–2, *248*, 542
Triose 28, *30,*
Triose phosphate 163, 167
Triose phosphate dehydrogenase 164, 167, 197
Triose phosphate isomerase **163**, *169*, 172, 177, 197, **544**
Tryptazan 329
L-Tryptophan, breakdown 459, 464
 as enzyme inhibitor *210, 237, 238*, 430, 467
 as enzyme repressor *210*, 219, *445*, 467

L-Tryptophan—*cont.*
  operon 210–11, 477–8, 481–2, 483
  structure *14*
  synthesis *30*, *206–7*, 208
    control 209–11, *445*, 467, *469*, 472
    triplets for *339*
Tryptophan synthase *207*, 208, *469*
  control of 210, 467
  sub-units of 210–11, 364–5, 410–11, 413,
    471, 472
Tryptophanase 459
Tyrosine, as enzyme inhibitor 467
  structure *14*
  synthesis *206*, 208
    control of 209–11, *468–9*
  tRNA 264
  triplets for *339*
Tyrosine aminotransferase *206–7*, 209
Tyvelose 96

UDP *see* Uridine diphosphate
Ultraviolet light (UV light), growth and 145
  mutations 384, 392
  spores 504
UMP *see* Uridine 5′-monophosphate
Uracil, pairing with adenine 251, *252*, 276,
    385
  pairing with guanine 263
  structure *18*
Ureidosuccinate (Carbamoyl aspartate) 239,
    240
Uridine 5′-diphosphate (UDP), as carrier
    36–7, 83–6, 92–3, 97–9, 172
  synthesis 238–9, *240–1*
    control *239*, 240–2
UDP-*N*-acetyl-2-*O*-enolpyruvylglucosamine
    83
UDP-*N*-acetylglucosamine 37, 83, 84, *85*,
    92, *93*, *100*
UDP-*N*-acetylmuramic acid 37, 83, *84*, *93*
UDP-Mur*N*Ac-peptides 37, 83–6, 88
UDP-galactose *97*, 98, *99*
UDP-galactose 4-epimerase *475*
UDP-glucose 36, 92, *97*, *100*, 172
UDP-glucuronic acid *93*
Uridine 5′-monophosphate (UMP)
  synthesis 238–9, *240–1*
    control *239*, 240–2, 466
Uridine 5′-triphosphate (UTP), cell wall syn-
    thesis 83, *84*
  synthesis 238–9, *240–1*
    control *239*, 240–2
Urocanase *461*
Urocanate *461*, 462
Uronic acid 104–5

cis-Vaccenic acid (Δ^{11}-Oleic acid) 246
Vacuoles, gas 125–7, *128*

L-Valine, activating enzyme 329
  as enzyme repressor *228*, 229
  isoleucine-valine operon *228*, 229, 478,
    481, 482
  structure *14*
  synthesis *30*, 225, *226*, 473
    control *228*, 229, 477, 478, 483
    triplets for *339*
Vancomycin *85*, 88
Vesicles, gas 125–7, *128*
  membranous *see* Mesosomes
  photosynthetic 123–5, *126*
Vibrio 525, *531*, *532*
Vibrio cholerae 525
Vibrio metchnikovii 107, *110*
Viruses, bacterial *see* Phages
  mammalian 315
Vitamin B_{12} 542, 545, 547
Vitamins 3, 5, 141, 547
  as enzyme cofactors 536, 537, 540–7
Volutin granules (Polymetaphosphate
    granules; Metachromatic granules) 64,
    132, 133–4

Wild-type (Prototrophs) 372, 404, 551

Xanthomonas 532
Xanthosine 5′-monophosphate (XMP *235*,
    236
XMP : ammonia ligase (GMP synthase) *234–
    235*, 236, 237, *238*
X-rays 384, 502
Xylose 96
Xylulosepimerase *173*, 174, 197
Xylulose 5-phosphate, pentose phosphate
    pathway *173*, 174–5, *176*
  phosphoketolase pathway 180, *181*
  photosynthesis 197, *198–9*

Yeasts, extract 141
  fatty acids 247
  fermentations 165, *166*
  ribosomes 325
  spores *see* Ascospores
  tRNA's *264*, 330, 338

Zinc 165
Zygote 392, 518, 553
  See also Merozygotes; Diploids, partial
Zymomonas 166
Zymomonas anaerobia 179
Zymomonas mobilis (*Pseudomonas lindneri*)
    179
Zymosarcina 523